Geophysical Monograph Series

Geophysical Monograph Series

199 **Dynamics of the Earth's Radiation Belts and Inner Magnetosphere** *Danny Summers, Ian R. Mann, Daniel N. Baker, and Michael Schulz (Eds.)*

200 **Lagrangian Modeling of the Atmosphere** *John Lin (Ed.)*

201 **Modeling the Ionosphere-Thermosphere** *Jospeh D. Huba, Robert W. Schunk, and George V. Khazanov (Eds.)*

202 **The Mediterranean Sea: Temporal Variability and Spatial Patterns** *Gian Luca Eusebi Borzelli, Miroslav Gacic, Piero Lionello, and Paola Malanotte-Rizzoli (Eds.)*

203 **Future Earth – Advancing Civic Understanding of the Anthropocene** *Diana Dalbotten, Gillian Roehrig, and Patrick Hamilton (Eds.)*

204 **The Galápagos: A Natural Laboratory for the Earth Sciences** *Karen S. Harpp, Eric Mittelstaedt, Noemi d'Ozouville, and David W. Graham (Eds.)*

205 **Modeling Atmospheric and Oceanic Flows: Insights from Laboratory Experiments and Numerical Simulations** *Thomas von Larcher and Paul D. Williams (Eds.)*

206 **Remote Sensing of the Terrestrial Water Cycle** *Venkat Lakshmi (Ed.)*

207 **Magnetotails in the Solar System** *Andreas Keiling, Caitriona Jackman, and Peter Delamere (Eds.)*

208 **Hawaiian Volcanoes: From Source to Surface** *Rebecca Carey, Valerie Cayol, Michael Poland, and Dominique Weis (Eds.)*

209 **Sea Ice: Physics, Mechanics, and Remote Sensing** *Mohammed Shokr and Nirmal Sinha (Eds.)*

210 **Fluid Dynamics in Complex Fractured-Porous Systems** *Boris Faybishenko, Sally M. Benson, and John E. Gale (Eds.)*

211 **Subduction Dynamics: From Mantle Flow to Mega Disasters** *Gabriele Morra, David A. Yuen, Scott King, Sang Mook Lee, and Seth Stein (Eds.)*

212 **The Early Earth: Accretion and Differentiation** *James Badro and Michael Walter (Eds.)*

213 **Global Vegetation Dynamics: Concepts and Applications in the MC1 Model** *Dominique Bachelet and David Turner (Eds.)*

214 **Extreme Events: Observations, Modeling and Economics** *Mario Chavez, Michael Ghil, and Jaime Urrutia-Fucugauchi (Eds.)*

215 **Auroral Dynamics and Space Weather** *Yongliang Zhang and Larry Paxton (Eds.)*

216 **Low-Frequency Waves in Space Plasmas** *Andreas Keiling, Dong-Hun Lee, and Valery Nakariakov (Eds.)*

217 **Deep Earth: Physics and Chemistry of the Lower Mantle and Core** *Hidenori Terasaki and Rebecca A. Fischer (Eds.)*

218 **Integrated Imaging of the Earth: Theory and Applications** *Max Moorkamp, Peter G. Lelievre, Niklas Linde, and Amir Khan (Eds.)*

219 **Plate Boundaries and Natural Hazards** *Joao Duarte and Wouter Schellart (Eds.)*

220 **Ionospheric Space Weather: Longitude and Hemispheric Dependences and Lower Atmosphere Forcing** *Timothy Fuller-Rowell, Endawoke Yizengaw, Patricia H. Doherty, and Sunanda Basu (Eds.)*

221 **Terrestrial Water Cycle and Climate Change Natural and Human-Induced Impacts** *Qiuhong Tang and Taikan Oki (Eds.)*

222 **Magnetosphere-Ionosphere Coupling in the Solar System** *Charles R. Chappell, Robert W. Schunk, Peter M. Banks, James L. Burch, and Richard M. Thorne (Eds.)*

223 **Natural Hazard Uncertainty Assessment: Modeling and Decision Support** *Karin Riley, Peter Webley, and Matthew Thompson (Eds.)*

224 **Hydrodynamics of Time-Periodic Groundwater Flow: Diffusion Waves in Porous Media** *Joe S. Depner and Todd C. Rasmussen (Auth.)*

225 **Active Global Seismology** *Ibrahim Cemen and Yucel Yilmaz (Eds.)*

226 **Climate Extremes** *Simon Wang (Ed.)*

227 **Fault Zone Dynamic Processes** *Marion Thomas (Ed.)*

228 **Flood Damage Survey and Assessment: New Insights from Research and Practice** *Daniela Molinari, Scira Menoni, and Francesco Ballio (Eds.)*

229 **Water-Energy-Food Nexus – Principles and Practices** *P. Abdul Salam, Sangam Shrestha, Vishnu Prasad Pandey, and Anil K Anal (Eds.)*

230 **Dawn–Dusk Asymmetries in Planetary Plasma Environments** *Stein Haaland, Andrei Rounov, and Colin Forsyth (Eds.)*

231 **Bioenergy and Land Use Change** *Zhangcai Qin, Umakant Mishra, and Astley Hastings (Eds.)*

232 **Microstructural Geochronology: Planetary Records Down to Atom Scale** *Desmond Moser, Fernando Corfu, James Darling, Steven Reddy, and Kimberly Tait (Eds.)*

233 **Global Flood Hazard: Applications in Modeling, Mapping and Forecasting** *Guy Schumann, Paul D. Bates, Giuseppe T. Aronica, and Heiko Apel (Eds.)*

234 **Pre-Earthquake Processes: A Multidisciplinary Approach to Earthquake Prediction Studies** *Dimitar Ouzounov, Sergey Pulinets, Katsumi Hattori, and Patrick Taylor (Eds.)*

235 **Electric Currents in Geospace and Beyond** *Andreas Keiling, Octav Marghitu, and Michael Wheatland (Eds.)*

236 **Quantifying Uncertainty in Subsurface Systems** *Celine Scheidt, Lewis Li, and Jef Caers (Eds.)*

237 **Petroleum Engineering** *Moshood Sanni (Ed.)*

238 **Geological Carbon Storage: Subsurface Seals and Caprock Integrity** *Stephanie Vialle, Jonathan Ajo-Franklin, and J. William Carey (Eds.)*

239 **Lithospheric Discontinuities** *Huaiyu Yuan and Barbara Romanowicz (Eds.)*

240 **Chemostratigraphy Across Major Chronological Eras** *Alcides N.Sial, Claudio Gaucher, Muthuvairavasamy Ramkumar, and Valderez Pinto Ferreira (Eds.)*

241 **Mathematical Geoenergy:Discovery, Depletion, and Renewal** *Paul Pukite, Dennis Coyne, and Daniel Challou (Eds.)*

242 **Ore Deposits: Origin, Exploration, and Exploitation** *Sophie Decree and Laurence Robb (Eds.)*

243 **Kuroshio Current: Physical, Biogeochemical and Ecosystem Dynamics** *Takeyoshi Nagai, Hiroaki Saito, Koji Suzuki, and Motomitsu Takahashi (Eds.)*

244 **Geomagnetically Induced Currents from the Sun to the Power Grid** *Jennifer L. Gannon, Andrei Swidinsky, and Zhonghua Xu (Eds.)*

245 **Shale: Subsurface Science and Engineering** *Thomas Dewers, Jason Heath, and Marcelo Sánchez (Eds.)*

246 **Submarine Landslides: Subaqueous Mass Transport Deposits From Outcrops to Seismic Profiles** *Kei Ogata, Andrea Festa, and Gian Andrea Pini (Eds.)*

247 **Iceland: Tectonics, Volcanics, and Glacial Features** *Tamie J. Jovanelly*

248 **Dayside Magnetosphere Interactions** *Quigang Zong, Philippe Escoubet, David Sibeck, Guan Le, and Hui Zhang (Eds.)*

249 **Carbon in Earth's Interior** *Craig E. Manning, Jung-Fu Lin, and Wendy L. Mao (Eds.)*

Geophysical Monograph 250

Nitrogen Overload
Environmental Degradation, Ramifications, and Economic Costs

Brian G. Katz

This Work is a co-publication of the American Geophysical Union and John Wiley and Sons, Inc.

WILEY

This Work is a co-publication between the American Geophysical Union and John Wiley & Sons, Inc.

This edition first published 2020 by John Wiley & Sons, Inc., 111 River Street, Hoboken, NJ 07030, USA and the American Geophysical Union, 2000 Florida Avenue, N.W., Washington, D.C. 20009

© 2020 American Geophysical Union

All rights reserved. No part of this publication may be reproduced, stored in a retrieval system, or transmitted, in any form or by any means, electronic, mechanical, photocopying, recording, or otherwise, except as permitted by law. Advice on how to obtain permission to reuse material from this title is available at http://www.wiley.com/go/permissions

Published under the aegis of the AGU Publications Committee

Brooks Hanson, Executive Vice President, Science
Carol Frost, Chair, Publications Committee
For details about the American Geophysical Union visit us at www.agu.org.

Wiley Global Headquarters
111 River Street, Hoboken, NJ 07030, USA

For details of our global editorial offices, customer services, and more information about Wiley products visit us at www.wiley.com.

Limit of Liability/Disclaimer of Warranty
While the publisher and authors have used their best efforts in preparing this work, they make no representations or warranties with respect to the accuracy or completeness of the contents of this work and specifically disclaim all warranties, including without limitation any implied warranties of merchantability or fitness for a particular purpose. No warranty may be created or extended by sales representatives, written sales materials, or promotional statements for this work. The fact that an organization, website, or product is referred to in this work as a citation and/or potential source of further information does not mean that the publisher and authors endorse the information or services the organization, website, or product may provide or recommendations it may make. This work is sold with the understanding that the publisher is not engaged in rendering professional services. The advice and strategies contained herein may not be suitable for your situation. You should consult with a specialist where appropriate. Neither the publisher nor authors shall be liable for any loss of profit or any other commercial damages, including but not limited to special, incidental, consequential, or other damages. Further, readers should be aware that websites listed in this work may have changed or disappeared between when this work was written and when it is read.

Library of Congress Cataloging-in-Publication Data is available.
Hardback: 9781119513964

Cover image: © Brian G. Katz
Cover design by Wiley

Set in 10/12pt Times New Roman by SPi Global, Pondicherry, India

Printed and bound by CPI Group (UK) Ltd, Croydon, CR0 4YY

10 9 8 7 6 5 4 3 2 1

This book is dedicated to Debra, Jeremy, and Lowell.

CONTENTS

Preface ix

Acknowledgments xi

1. Introduction 1
2. The Nitrogen Cycle 15
3. Sources of Reactive Nitrogen and Transport Processes 29
4. Methods to Identify Sources of Reactive Nitrogen Contamination 49
5. Adverse Human Health Effects of Reactive Nitrogen 71
6. Terrestrial Biodiversity and Surface Water Impacts from Reactive Nitrogen 91
7. Groundwater Contamination from Reactive Nitrogen 119
8. Nitrate Contamination in Springs 155
9. Co-occurrence of Nitrate with Other Contaminants in the Environment 175
10. Economic Costs and Consequences of Excess Reactive Nitrogen 197
11. Strategies for Reducing Excess Reactive Nitrogen to the Environment 221

Index 243

PREFACE

Nitrogen, an essential component of the building blocks of life (DNA, RNA, and proteins), can exist in many different forms. The various forms of nitrogen can move or cycle between the atmosphere, biosphere, and hydrosphere. The processes involved in the transfer of these different forms of nitrogen between these systems are collectively referred to as the nitrogen cycle. One form of nitrogen is di-nitrogen gas (N_2), which comprises 78% of all the gases in our atmosphere by volume. Ironically, this form of nitrogen is relatively nonreactive and not directly usable by higher life forms. Fortunately, certain specialized microorganisms can "fix" nitrogen in soils and convert nitrogen gas to inorganic nitrogen forms like ammonia that are available to plants, thereby providing sustenance to all animal life on this planet. This critical process is referred to as biological nitrogen fixation (BNF). Prior to the 19th century, the reactive nitrogen produced by microorganisms (BNF) and from nonbiological processes (e.g., lightning) was balanced by plant uptake and nitrogen losses (denitrification processes) and therefore did not accumulate in the environment. However, in the early 20th century, two scientists discovered a way to synthesize ammonia from nitrogen and hydrogen under high temperature and pressure (Haber–Bosch process). This has led to the production of artificial nitrogen fertilizers, which has grown exponentially since the 1950s and is projected to substantially grow into the future due to increasing demand and utilization. One certainly cannot discount the enormous benefits from nitrogen fertilizer use particularly in developing countries, which has led to an increase in food production and substantial reduction in malnutrition. However, the abundance of inexpensive fertilizer has led to its excessive use throughout the industrialized world, which created a surplus of nitrogen that was followed by substantial releases of reactive nitrogen to the environment.

While a considerable amount of attention has been devoted to the substantial increases in carbon dioxide in the atmosphere and its relation to climate change and alterations to the carbon cycle. Even though the carbon cycle has been impacted significantly by fossil fuel burning and other sources of carbon dioxide, we will see that the nitrogen cycle has been altered more than the carbon cycle or any other basic element cycle essential to life on earth. A better understanding of how the various forms of nitrogen are transported through the atmosphere, hydrosphere, and biosphere will ultimately lead to better management of reactive nitrogen and help to prevent further degradation of our important environmental systems.

This book discusses how human beings have substantially altered the natural nitrogen cycle and as a result how excess reactive nitrogen (e.g., ammonia, nitrate, nitrite, and nitrogen oxides) have caused widespread environmental degradation, adverse human health effects, and enormous economic costs associated with environmental and health impacts. A few brief examples of detrimental impacts to the environment include widespread harmful algal blooms, ecosystem impairment, loss of biodiversity, fish kills, contamination of drinking water aquifers and surface waters, acidification of soils and water bodies, and air pollution. These detrimental effects to the environment have resulted in various human health maladies including respiratory infections and heart disease linked to air pollution; links between nitrate in drinking water and thyroid disease, neural tube defects, several types of cancers in adults, and methemoglobinemia (blue-baby syndrome) in infants and respiratory illnesses associated with algal blooms. Economic consequences include staggering and rising costs (in the billions of dollars annually) for treatment of human health maladies, treatment of drinking water and wastewater, removing toxins from harmful algal blooms, restoration of impaired water bodies, improvements to agricultural best management practices, and loss of jobs and revenue related to declines in fishing, ecotourism, recreation, and real estate values.

During my long career as a research hydrologist and environmental scientist, I have investigated the sources, transport, and fate of reactive nitrogen and other contaminants in groundwater and surface water systems in in a variety of environmental settings. In many areas around the world, impacts to the environment from excess reactive nitrogen compounds have continually worsened. I wrote this book because I wanted to provide a better understanding of the environmental, human health, and economic consequences associated with the continual release of excessive amounts of reactive nitrogen from multiple sources, including fertilizers, animal wastes, disposal of human wastewaters, fossil fuel combustion, atmospheric deposition, and mining. Unfortunately, many people (including students, policy makers, teachers, environmental managers, and economic strategists) are not aware of the negative impacts from excess amounts of reactive nitrogen in the environment. This is surprising given that during the past 40 years, there have been numerous national and international scientific studies that have generated hundreds of published reports and articles on various aspects of environmental degradation, human health effects, and economic consequences related to the anthropogenic alteration of the natural nitrogen cycle with an overload of reactive nitrogen species. During the past decade, a wealth of information has been published from extensive studies in the United States, Europe, and China. The material presented in this book addresses this urgent and critical need for integrating and synthesizing previous and new sources of information on environmental, human health, and economic consequences of excess reactive nitrogen in our environment.

The first three chapters of the book present background information. Chapter 1 introduces the overall purpose and scope, and a concise overview of the various issues related to environmental, health, and economic consequences associated with excess reactive nitrogen. Chapter 2 presents an overview of the natural nitrogen cycle and how this natural cycle has been dramatically altered during the past 100 years by anthropogenic activities. Chapter 3 provides detailed information on the major point and nonpoint sources that contribute reactive nitrogen to the environment and the factors that affect nitrogen transport at regional and global scales. Chapter 4 builds on this introductory information and presents information on novel and innovative tools used to identify and quantify sources of nitrate contamination. The next several chapters are devoted to ramifications (and detrimental impacts) from the anthropogenic alteration of the nitrogen cycle including adverse human health impacts (Chapter 5); degradation and contamination of surface waters, groundwater, springs, and ecosystems, and how interactions between groundwater and surface water compound these problems in lakes, streams, and estuaries (Chapters 6–8). Chapter 9 discusses other contaminants that co-occur with nitrate contamination and originate from various anthropogenic sources. Chapter 10 presents detailed information on economic consequences and costs associated with environmental pollution, human health maladies, clean-up costs, drinking-water treatment costs, losses to commercial fishing, tourism, recreational activities, decreased real estate values, and cascading effects to other socioeconomic sectors. A major challenge for humanity is to reduce the significant losses of excess reactive nitrogen to the environment from agricultural activities in developed countries, while maintaining sustainable food production and food security in other parts of the world that have limited access to enough nitrogen to replenish crop uptake. The final book chapter (Chapter 11) discusses effective ways to reduce our nitrogen footprint (locally and globally) from human activities and presents practical strategies and recommendations for restoring water quality in impacted surface water bodies, groundwater and springs, and terrestrial and aquatic ecosystems. Three areas are highlighted where ongoing efforts are being focused in terms of increasing nitrogen use efficiency and sustainability in agricultural systems, reducing the per capita consumption of animal proteins, and decreasing fossil fuel combustion and replacing with alternative energy sources. Our society must be increasingly vigilant in our commitment to address the challenges associated with significantly reducing inputs of reactive nitrogen to prevent further degradation of our surface waters, aquifers, and ecosystems.

Brian G. Katz

ACKNOWLEDGMENTS

I am thankful for the opportunity to write this book, as I have been fortunate to have spent most of my career conducting research investigations on the sources, fate, and transport of reactive forms of nitrogen in a variety of environmental systems. I have received continuous editorial support from Rituparna Bose, Nithya Sechin, Bobby Kilshaw, Danielle Lacourciere, and Gunalan Lakshmipathy at Wiley. I am deeply grateful for the constructive review comments and suggestions from Jill Baron, Rick Copeland, Celeste Lyon, Kirstin Eller, J.M. Murillo, Mary Lynn Musgrove, Stephen Opsahl, Andrew O'Reilly, Carol Wicks, Ming Ye, and several anonymous reviewers. I am also thankful for the contributed figures and photos from Annette Long, Mark Long, Casey McKinlay, and John Moran.

1

Introduction

1.1. THE SIGNIFICANCE OF NITROGEN ON EARTH

Nitrogen is essential to sustain life on Earth, as it is a major component of certain essential amino acids and proteins, enzymes, vitamins, and DNA and RNA. Also, nitrogen is contained in chlorophyll, the green pigment in plants that is essential for photosynthesis. Nitrogen, as dinitrogen (N_2) gas is the most abundant element in the Earth's atmosphere making up 78% of all gases by volume. Galloway et al. (2003) estimated that the total amount of nitrogen in the atmosphere, soils, and waters of Earth is approximately 4×10^{21} g, which is more than the combined mass of all four other essential elements (carbon, phosphorus, oxygen, and sulfur) that are needed to sustain life. However, rather ironically, this bountiful form of nitrogen (N_2 gas) is unusable by most organisms. The strong triple bond between the nitrogen atoms in nitrogen gas molecules makes N_2 gas essentially unreactive. Fortunately, certain specialized nitrogen-fixing microorganisms in soils can transform or convert nitrogen gas to forms that are available to plants (e.g., ammonium, nitrate), thereby providing sustenance to all animal life on this planet. The process by which nitrogen gas is converted by microbes into inorganic nitrogen compounds, such as ammonia, is referred to as biological nitrogen fixation (BNF). BNF accounts for approximately 90% of this transformation, which is performed by certain bacteria (e.g., genus *Rhizobium*) and blue-green algae (cyanobacteria). Small amounts of nitrogen gas fixation can occur abiotically, through high temperature processes, such as lightning and ultraviolet radiation, both of which can break the strong triple bond in molecular nitrogen (N_2) in the atmosphere. Denitrification is another important process in which microorganisms consume nitrate and convert it to reduced forms of nitrogen and ultimately back to nitrogen gas. This and other key processes of the nitrogen cycle are discussed in more detail in Chapter 2.

1.2. CYCLING OF NITROGEN IN THE ENVIRONMENT

Nitrogen and its various forms can move or cycle between the atmosphere, biosphere, and hydrosphere (Fig. 1.1). The processes involved in the transfer of the different forms of nitrogen between these systems or reservoirs within these systems are collectively referred to as the nitrogen cycle. Prior to the 19th century, the reactive nitrogen produced by BNF and abiotic processes was balanced by plant uptake and denitrification processes and therefore did not accumulate in the environment. As human population increased substantially during the 19th century, there were two main needs for reactive nitrogen: fertilizers and explosives. To meet these needs, large amounts of nitrogen were mined from various sources including naturally occurring nitrate deposits, guano, and coal. The overall dependence on the use of these sources in Europe has been referred to as a "fossil nitrogen economy" (Sutton et al., 2011a). By the end of the 19th century, these sources could not support the growing needs for more reactive nitrogen. Not long after the end of the 19th century, a solution to this problem was discovered. A process was developed in a laboratory in the early 20th century by Fritz Haber (who received a Nobel Prize for Chemistry in 1918) that synthesized ammonia (NH_3) from nitrogen and hydrogen under high temperature and pressure. This process was industrialized by Carl Bosch (who received a Noble Prize in 1931) utilizing an iron catalyst along with high temperature (300–500°C) and high pressure (20 MPa) (Fowler et al., 2013). Thus, the combined discoveries of these two men became known as the Haber–Bosch industrial chemical process, which converts hydrogen and atmospheric nitrogen gases

---

Nitrogen Overload: Environmental Degradation, Ramifications, and Economic Costs, Geophysical Monograph 250, First Edition. Brian G. Katz.
© 2020 American Geophysical Union. Published 2020 by John Wiley & Sons, Inc.

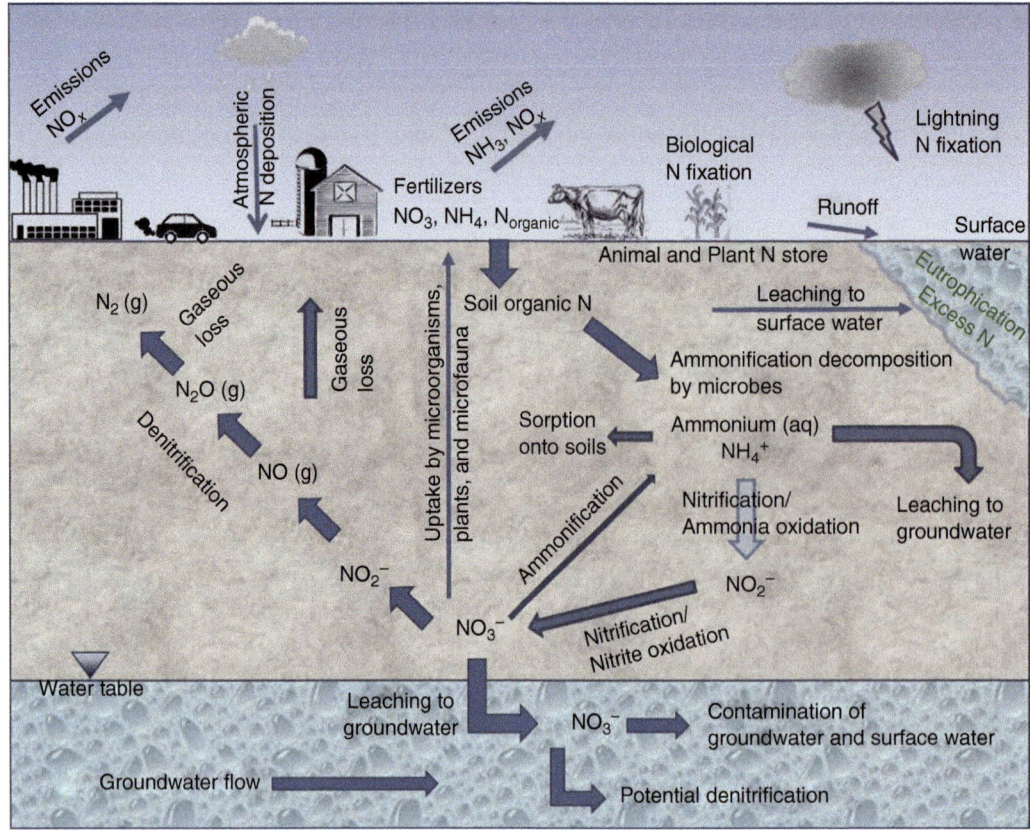

Figure 1.1 A conceptual diagram showing sources of anthropogenic and natural sources of reactive nitrogen, the various forms of nitrogen, and processes in the nitrogen cycle in the atmosphere, hydrosphere, and biosphere. *Source*: Modified from Rivett et al. (2008).

by chemical reactions into synthetic ammonia. This process is used mainly to create inorganic fertilizers, which started the extensive human alteration of the natural nitrogen cycle. An estimated 128 teragrams (Tg, 1×10^{12} g; 1 billion kg) of reactive nitrogen are synthetically produced annually from the Haber–Bosch process (Galloway et al., 2008) and used to make fertilizers for agriculture, lawns, and recreational turf grass. Sutton et al. (2011a) have referred to these large amounts of reactive nitrogen produced by the Haber–Bosch process as the "greatest single experiment in global geo-engineering that humans ever made." The production of artificial nitrogen fertilizers has grown exponentially since the 1950s and is projected to grow into the future due to increasing demand and utilization (e.g., Erisman et al., 2015; Nielsen, 2005). The abundance of inexpensive fertilizer led to its excessive use throughout the industrialized world, which created a surplus of nitrogen that was followed by substantial releases of reactive nitrogen to the environment. Global increases in reactive nitrogen have resulted from intensive cultivation of legumes, rice, and other crops that result in the conversion of nitrogen gas (N_2) to organic nitrogen compounds through human-induced BNF.

The production of reactive nitrogen from Haber–Bosch process was projected to increase from 120 Tg N/yr in 2010 to about 160 Tg N/yr in 2100 (Fowler et al., 2013).

The anthropogenic production of reactive nitrogen has increased steadily and significantly since the mid-20th century as world population increased substantially (Fig. 1.2). Accompanying this increase in population are steady increases in meat and grain production, BNF, and NO_x emissions. Galloway et al. (2003) estimated that reactive nitrogen inputs resulting from cultivation of legumes and other aforementioned crops (BNF) increased globally from approximately 15 Tg N/yr in 1860 to approximately 33 Tg N/yr in 2000. To increase productivity and crop yield per acre worldwide, increased amounts of fertilizer were being using along with fossil fuel burning machines that replace practices that involved manpower and the use of farm animals (Fig. 1.2). Throughout Europe, regional watersheds annually contribute approximately 3700 kg of reactive nitrogen per square kilometer, which is five times the background rate of natural N_2 fixation (Sutton et al., 2011a). On a global scale, as a result of intensive farming practices, nitrogen and other nutrients were and are being depleted in some areas and have been concentrated in

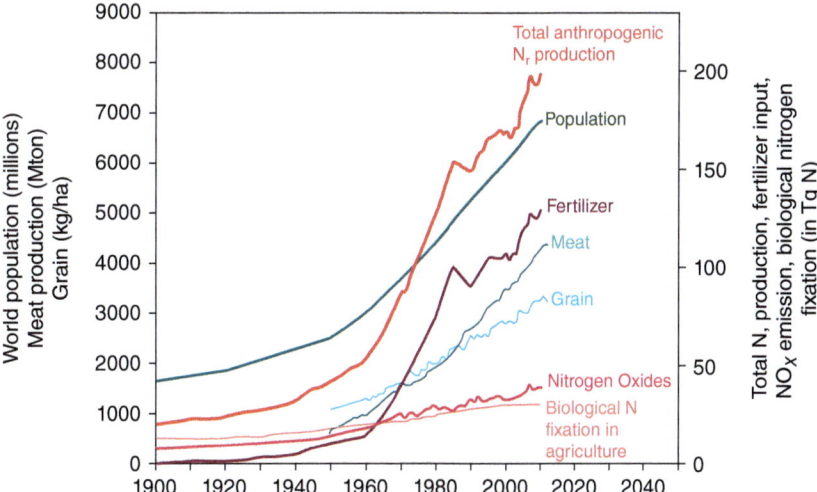

Figure 1.2 Plot showing increases in human population and increases in the creation of total anthropogenic reactive nitrogen between 1900 and 2012 (Mton, is millions of tons; kg/ha is kilograms per hectare; and Tg N is teragrams (10^{12} g) of reactive nitrogen). *Source*: With permission from Erisman et al. (2015). Reproduced with permission of WWF.

other areas, leading to a "cascade" of reactive nitrogen (Galloway et al., 2003) through the environment and creating a sequence of harmful environmental effects including ecosystem damages (loss of biodiversity, eutrophication of waters and soils, and soil acidification), increases in greenhouse gas emissions, contamination of drinking water, air pollution, human health maladies, and damages to the ozone layer (Galloway et al., 2008).

In addition to inputs of reactive nitrogen from the use of synthetic fertilizers, there are many other point and nonpoint anthropogenic sources that release reactive nitrogen to the environment (Fig. 1.1). Additional amounts of reactive nitrogen are contributed from fossil fuel combustion, which converts both atmospheric nitrogen and fossil nitrogen to reactive oxidized forms (NO_x). The amount of reactive nitrogen created from fossil fuel combustion increased from less than 1 Tg N/yr in 1860 to approximately 25 Tg N/yr in 2000 (Galloway et al., 2003; Nielsen, 2005). Anthropogenic inputs of nitrogen to the land surface from animal wastes, human wastewater (septic tanks, wastewater treatment facilities), industrial processes, and atmospheric deposition have also contributed to the substantial alteration of the cycling of nitrogen throughout the world.

The overall fate of anthropogenic reactive nitrogen in the terrestrial biosphere is not well understood. As Galloway et al. (2008) noted, the fate of only 35% of the reactive nitrogen inputs to the terrestrial biosphere were known, which included estimates of 18% exported to and denitrified in coastal ecosystems, 13% deposited to the ocean from the marine atmosphere, and 4% emitted as N_2O. Although tropical forests cover only 12% of the earth's surface, it is estimated that they emit about 50% of the N_2O to the atmosphere (Townsend et al., 2011). The approximately 65% of the remaining anthropogenic reactive nitrogen is potentially accumulated in soils, vegetation, and groundwater or possibly denitrified to nitrogen gas. However, these estimates have a considerable amount of uncertainty (Galloway et al., 2008).

Overall, freshwater nitrogen loadings worldwide have increased from 21 million tons in preindustrial times to 40 million tons per year in 2009 (Dodds et al., 2009), while riverine transport of dissolved inorganic nitrogen has increased from 2–3 million to 15 million tons. During the last part of the 20th century, increased nitrogen loading to the land surface has caused the natural rate of N fixation to double and the atmospheric deposition rates to increase more than tenfold. This so-called "greatest single experiment in global geo-engineering that human ever made" (Sutton et al., 2011a) has led to a legacy of detrimental consequences for human health, terrestrial and coastal ecosystems, and the economy. We now explore in more detail some of these adverse consequences from excess reactive nitrogen in our environment.

1.3. ENVIRONMENTAL CONSEQUENCES OF EXCESS REACTIVE NITROGEN

During the past 50 years, excess reactive nitrogen has had dire consequences for many varied types of environmental systems all over the world (Erisman et al., 2011, 2013, 2015; Payne et al., 2013; Townsend et al., 2003; B. Ward, 2012). As mentioned previously, only a small proportion of the Earth's biota can transform N_2 to reactive nitrogen forms. As a result, reactive nitrogen tends to be the limiting nutrient in most natural ecosystems and

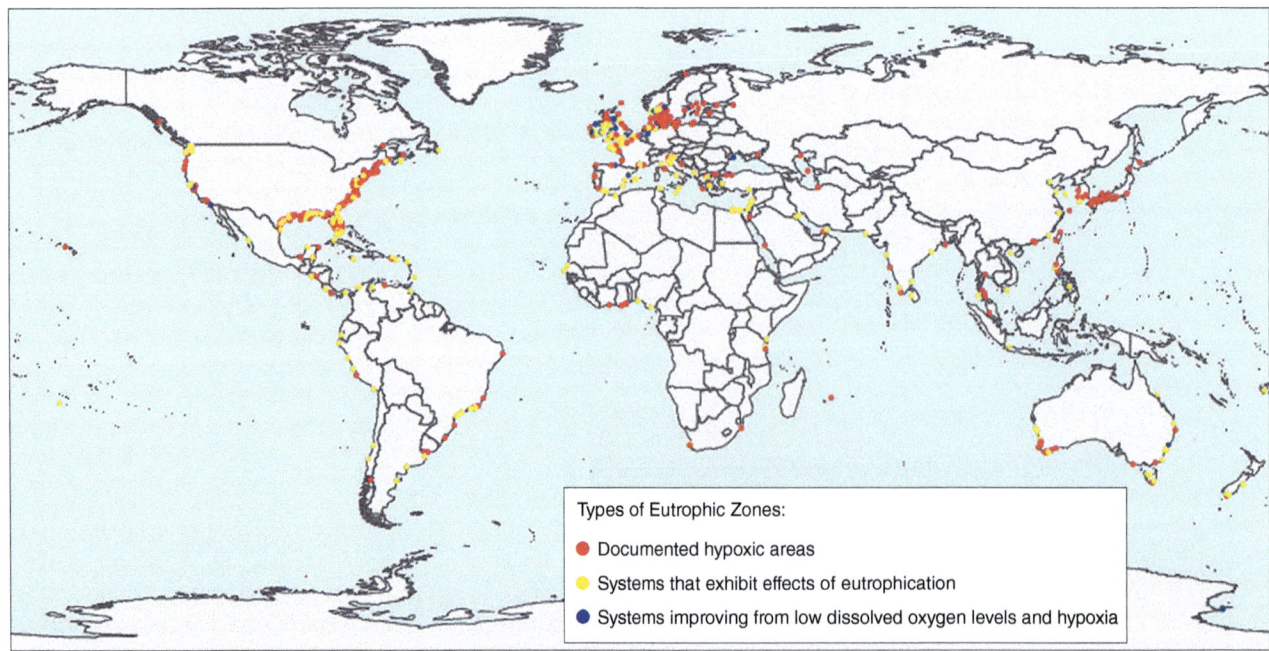

Figure 1.3 Map showing hypoxic (red circles) and eutrophic (yellow circles) coastal areas around the world. *Source*: Modified from the colored figure 4 in Erisman et al. (2015). Reproduced with permission of World Wildlife Fund.

typically for agricultural systems. Therefore, increased amounts of reactive nitrogen in the environment have led to shifts in plant species composition, decreases in species diversity, habitat degradation, eutrophication, decreases in water transparency, increased biomass of benthic and epiphytic algae, and changes in food-web processes in terrestrial ecosystems and in coastal waters (Howarth et al., 2000, 2011). These effects are more prevalent in certain areas, such as the highly affected areas in Central and Western Europe, Southern Asia, the eastern United States, and parts of Africa and South America (Erisman et al., 2015).

Excess nitrogen in water bodies have led to the proliferation of harmful (toxic) algal blooms, which have been reported in every state in the United States during the past 10 years (U.S. Environmental Protection Agency (U.S. EPA), 2009), as well in as many other parts of the world (Erisman et al., 2015). Nonaquatic animals also are affected when consuming waters-containing algae. Pets and livestock have died after drinking water containing algal blooms, including 32 cattle on an Oregon ranch in July 2017 (Flesher & Kastanis, 2017). Algal blooms have resulted in numerous beach closures in Florida and a state of emergency was declared in four coastal counties in southeast Florida in 2016. Toxic algal blooms have contaminated waterways from the Great Lakes to Chesapeake Bay, from the Snake River in Idaho to New York's Finger Lakes and reservoirs in California's Central Valley (Flesher & Kastanis, 2017). As of mid-August 2016, the U.S. EPA noted that states across the United States have reported more than 250 health advisories due to harmful algal blooms that year. The National Aquatic Resource Surveys conducted by the U.S. EPA and state and tribal partners in 2012 found that 34% of the lakes surveyed in the United States had high levels of nitrogen associated with harmful ecological impacts. A survey of rivers and streams in 2009–2010 found that 41% had high levels of nitrogen (https://www.epa.gov/national-aquatic-resource-surveys/data-national-aquatic-resource-surveys, accessed 19 October 2018). Large algal blooms occurred in the Black Sea in the 1970s and 1980s following the intensive use of fertilizers and livestock production in the Black Sea basin in the 1960s (Bodeanu, 1993; Mee et al., 2005). Most of Europe has a high potential risk of eutrophication of surface freshwaters (Grizzetti et al., 2011) as well as parts of the United States, South America, Australia, and Southeast Asia (Erisman et al., 2015) (Fig. 1.3).

Another severe impact of excessive amounts of reactive nitrogen is the depletion of oxygen (hypoxia), which has led to dead zones in estuaries and other large water bodies. In the United States more than 166 dead zones have been identified and have affected waterbodies such as the Chesapeake Bay (http://www.cbf.org/issues/dead-zones/, accessed 18 October 2018) and the Gulf of Mexico (Rabalais et al., 2002). The Gulf of Mexico dead zone grew to approximately 5840 mile^2 in 2013. This extremely large dead zone results from summertime nutrient pollution from the Mississippi River (excess reactive nitrogen most likely from fertilizers), which drains 31 upstream states (https://www.epa.gov/nutrientpollution/

sources-and-solutions). Recent studies have indicated that there could be more than 500 coastal dead zones worldwide (Fig. 1.3), with numbers possibly doubling each decade (Breitburg et al., 2018, Conley et al., 2009, Diaz & Rosenberg, 2008). Dead zones have increased in large coastal areas of the Adriatic Sea, the Black Sea, the Kattegat, and the Baltic Sea (Diaz & Rosenberg, 2008).

Nitrogen gases, aerosols, and dissolved compounds in air that contain NO, N_2O, nitrate, ammonia, and ammonium can have injurious and phytotoxic effects on the above ground parts of plants (e.g., trees, shrubs, and other vegetation). In the 1980s, foliar impacts to forests were prevalent in North America and Europe due to air pollution containing nitric and sulfuric acids; however, the situation has improved in these areas with air-pollution control legislation and pollution reduction strategies adopted by some industries. Increased amounts of nitrogen oxides in the atmosphere could also contribute to global climate change and stratospheric ozone depletion (Galloway et al., 2004). Ironically, one potential benefit from increased nitrous oxide emissions is stimulating the sequestration of global CO_2 by terrestrial and marine ecosystems. This effect could possibly lower the amounts of the greenhouse gas CO_2 released to the atmosphere (Fowler et al., 2015; Zaehle et al., 2011).

Atmospheric deposition of reactive nitrogen also leads to acidification of soils and surface waters. During 2000–2010, Peñuelas et al. (2013) reported that global land had received more than 50 kg/ha accumulated N deposition. They found that nitrogen additions to the land surface significantly reduced soil pH by 0.26 on average globally. Soils in many places around the world are particularly sensitive to nitrogen deposition, with buffering likely transitioning from base cations (Ca, Mg, and K) to nonbase cations (Al and Mn), which could have a toxic impact for terrestrial ecosystems (Bowman et al., 2008; Tian & Niu, 2015). Significant decreases in forest and grassland productivity have been observed when reactive nitrogen deposition increases above threshold levels. These adverse effects could promote biodiversity changes through the food chain, affecting insects, birds, and other animals that depend on these food sources (Erisman et al., 2013). Acidification of surface waters (lakes, streams, oceans) in many places around the world also has increased and has had detrimental effects on biota (Curtis et al., 2005; Doney et al., 2007).

Ammonia emissions to the atmosphere also have increased during the past several decades (Ackerman et al., 2018) and can result in the formation of fine particulate matter through reactions with nitric and sulfuric acids. These harmful compounds in the atmosphere can be dispersed over large areas of the world, resulting in eutrophication and acidification of terrestrial, freshwater, and marine habitats (Bobbink et al., 2010; Erisman et al., 2015). Also, deposition of ammonia can damage sensitive vegetation, particularly bryophytes and lichens that consume most of their nutrients from the atmosphere (Sheppard et al., 2011). Sala et al. (2000) noted that certain biomes (particularly northern temperate, boreal, arctic, alpine, grassland, savannah, and Mediterranean) are very sensitive to reactive nitrogen deposition because of the limited availability of nitrogen in these systems under natural conditions.

1.4. ADVERSE HUMAN HEALTH IMPACTS

The widespread alteration of the nitrogen cycle from the production and use of synthetic fertilizers has both positive and negative consequences for human health. One cannot discount the huge benefits from nitrogen fertilizer use in developing countries, which has led to an increase in food production and substantial reductions in malnutrition. With better nutrition, healthier diets can lead to more efficient immune response to parasitic and infectious diseases (Nesheim, 1993). Although, shortages of food and malnutrition still exist in many large parts of the world.

Increases in reactive nitrogen compounds in the atmosphere (air pollution) and drinking water have had direct negative effects on human health. Townsend et al. (2003) show schematically how the net public health benefit from human fixation and use of reactive nitrogen has peaked and declined as a result of an exponential increase in air and water pollution and ecological feedbacks to disease (Fig. 1.4). Increased emissions of nitrogen oxides in urban areas have resulted from fossil fuel combustion. In other areas, biomass burning and volatilization of fertilizers emit large quantities of reactive

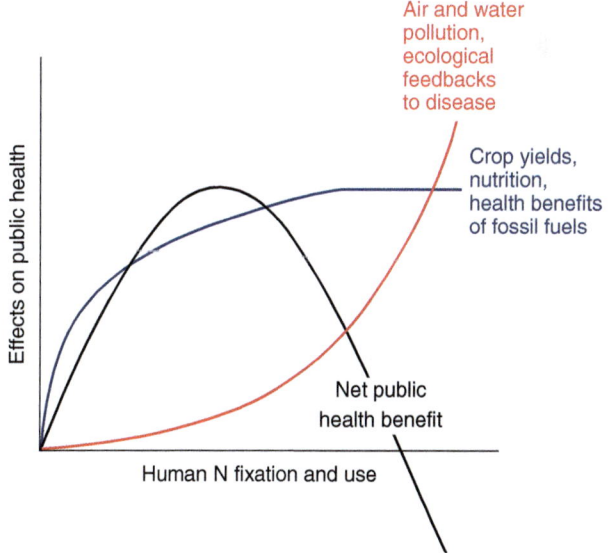

Figure 1.4 Conceptual diagram showing detrimental effects on human health from increasing amounts of excess reactive nitrogen from human fixation. *Source*: From Townsend et al. (2003). Reproduced with permission of John Wiley & Sons.

nitrogen, including ammonia and nitrogen oxides. Elevated nitrogen oxides enhance the production of tropospheric ozone (O_3) at low altitudes and aerosols, which can cause coughs, asthma, and other reactive airways disease, and mortality (Apte et al., 2015; Townsend et al., 2003). At high altitudes and in parts of the stratosphere, ozone can be destroyed due to the production of the catalyst NO as ultraviolet light breaks N_2O apart. The destruction of ozone in the stratosphere would allow more ultraviolet light to reach the Earth's surface, possibly causing more skin cancers. Kane (1998) noted that reductions in ozone may result in a 10–20% increase in ultraviolet-B radiation, and this could explain a 20–40% rise in skin cancer in the human population since the 1970s. Particulate air pollution has also been linked to cardiovascular diseases and overall mortality (Pope et al., 2002). N_2O is a potent greenhouse gas with a long residence time in the atmosphere of 120 years (Howarth et al., 2011) and therefore its harmful effects can persist for decades.

Humans are increasingly exposed to elevated levels of reactive nitrogen species (mainly nitrate) in drinking water. The U.S. EPA has set a maximum contaminant level of 10 mg/L (parts per million) for nitrate-N in drinking water. In the United States, it is estimated that 10–20% of groundwater sources may exceed 10 mg/L for nitrate-N (Dubrovsky et al., 2010). In 2015, the U.S. EPA reported that 183 community water systems exceeded allowable levels of nitrate in drinking water. Sutton et al. (2011a) estimated that about 3% of the population in 15 European Union countries (EU-15) that relies on groundwater as a drinking water source is exposed to nitrate concentrations exceeding the drinking water standard of 50 mg NO_3/L (equivalent to 11.2 mg N/L, expressed as nitrogen). In addition, it was estimated that 5% of the population using groundwater is chronically exposed to nitrate concentrations exceeding 5.6 mg N/L, which could double the risk of colon cancer for people that consume above median amounts of meat (Sutton et al., 2011a). In many other aquifers around the world, nitrate-N concentrations exceed the maximum contaminant level set by the World Health Organization. Many large aquifers in China have been contaminated with elevated nitrate-N concentrations (S. Wang et al., 2016b). About 90% of China's shallow groundwater is polluted, and nitrate is considered one of the pollutants of main concern (Qiu, 2011).

Numerous health effects have been associated with elevated nitrate levels in drinking water, including methemoglobinemia (elevated levels of methemoglobin in blood which leads to decreased oxygen availability to tissues, mostly affects babies), reproductive problems, and cancer. Recent studies during the past 20 years have reported that nitrate-N concentrations even below the maximum contaminant level of 10 mg/L can cause other health maladies including non-Hodgkins's lymphoma (M.H. Ward et al., 1996, 2005, 2018) and increased risk of bladder and ovarian cancers (Weyer et al., 2001).

A developing area of interdisciplinary research is investigating the links between ecosystem damages and human diseases. For example, Townsend et al. (2003) hypothesized that an increase in nutrient availability can favor opportunistic, disease-causing organisms and could lead to changes in the epidemiology of human diseases. Johnson et al. (2010) referred to studies that showed that ecological changes related to nutrient enrichment often aggravate infection and disease caused by "generalist" parasites with direct or simple life cycles. More recently, reports of attacks to the immune systems of sport fisherman in Lake Erie were linked to algal blooms (Flesher & Kastanis, 2017). Studies have reported associations between algal blooms with cholera outbreaks and correlations between inorganic nutrient concentrations and abundances of mosquitoes that can carry pathogenic microorganisms (Camargo & Alonso, 2006).

1.5. LEGACY REACTIVE NITROGEN STORAGE IN THE SUBSURFACE

Comparably disturbing, but not widely acknowledged, are the large amounts of reactive nitrogen, mostly in the form of nitrate, that leaches below the root zone in soils to the vadose (unsaturated) zone (the material between the base of the soil to the water table at the top of the saturated zone) and eventually to groundwater (aquifers) where nitrate can be stored (Walvoord et al., 2003; L. Wang et al., 2013, 2016a). This is especially prevalent in agricultural areas where there has been a surplus of nitrogen added to cropland. Ascott et al. (2017) assessed global patterns of nitrate storage in the vadose zone by using estimates of groundwater depth and nitrate leaching for the period 1900–2000. They estimated a peak global storage of 605–1814 Tg of nitrate in the vadose zone, with the highest storage per unit area in North America, China, and Europe (areas with thick vadose zones and extensive historical agricultural activities).

Several recent studies have shown that groundwater nitrogen applied decades ago can still be found in aquifers (Basu et al., 2012; McMahon et al., 2006; Meals et al., 2010; Puckett et al., 2011; Sanford & Pope, 2013; Sebillo et al., 2013). Legacy nitrogen stored in the subsurface over long periods of time can slowly be released to streams and large rivers (Van Meter & Basu, 2015, 2017; Van Meter et al., 2016, 2017). However, even though substantial decreases in fertilizer applications have been noted in the Netherlands, Denmark, and Germany (where the nitrogen surplus is back to the level of that of 1970), concentrations of nitrate-N in groundwater have not responded to this decrease of nitrogen surplus (Sutton et al., 2011a). Particularly noteworthy are cases in Eastern European countries where the nitrogen surplus has decreased by half (due to economic and political changes in the early 1990s), however, no improvements in water quality have been observed in streams (Sutton et al., 2011a). This is likely due to the large quantities of nitrate stored in aquifers and released very slowly over time, as a function of the groundwater residence time, which can range from weeks to several thousands of years (Alley et al., 2002; Tesoriero et al., 2013).

In arid environmental systems, an increased reliance on dryland ecosystems for agriculture may result in flushing of naturally occurring nitrate in the unsaturated zone into groundwater in regional basins due to irrigation and vegetation change (Robertson et al., 2017). Thus, the unsaturated zone, an aforementioned important reservoir for the storage and release of reactive nitrogen (mainly in the form of nitrate) over time, has significant implications for global warming, contamination of deeper aquifers used for drinking water supplies, and eutrophication of surface water bodies (Schlesinger, 2009).

1.6. ECONOMIC IMPACTS OF EXCESS REACTIVE NITROGEN

Environmental degradation from excess reactive nitrogen has led to severe economic consequences, mainly associated with treating drinking water and wastewaters, and human health issues associated with air pollution and surface water and groundwater contamination. Human-caused eutrophication of waterways from excess nutrients has resulted in large revenue losses to the commercial fishing industry (aquaculture, fisheries, and shellfish fisheries), tourism and recreational water usage, loss of real estate values for waterfront property, and threatened and endangered species. Dodds et al. (2009) provided a detailed economic analysis of the losses associated with environmental impacts related to eutrophication in the United States and reported a conservative estimate of potential losses totaling over $2.2 billion annually. This estimate did not include additional water treatment costs related to taste and odor problems. Sobota et al. (2015) estimated reactive nitrogen leakage for large watersheds in the United States and accounted for economic costs associated with mitigation, remediation, direct damage, human health, agriculture, ecosystems, and climate. They reported estimated annual damage costs of up to $2255/ha across watersheds, with a median value of $252/ha/yr. For the entire United States, Sobota et al. (2015) estimated environmental damages from reactive nitrogen in the early 2000s that ranged from $81 to $441 billion per year.

Based on information from multiple studies conducted during 2000 through 2012, the U.S. EPA compiled extensive information on costs associated with adverse impacts of nutrient pollution in the environment and reducing nutrient pollution from various point and nonpoint sources (U.S. EPA, 2015). The following categories were identified in the report and demonstrate how costs can have cascading economic impacts through many different socioeconomic sectors. For example, losses in tourism and recreation associated with algal blooms can include decreased restaurant sales and hotel stays, loss of tourism-related jobs, closure of lakeside businesses, and decreased tourism-related spending. Negative impacts to commercial fishing throughout coastal areas has included reduced harvests, fishery closures, loss of jobs, and increased processing costs related to elevated risks associated with shellfish poisoning. There also have been significant losses in property values of waterfront property and nearby homes associated with decreases in water clarity, algal mats, and increased pollutant concentrations. Algal blooms have caused a variety of injurious health effects in humans and animals through direct contact with skin during recreational activities, drinking water consumption, intake of contaminated shellfish (can cause neurotoxic shellfish poisoning and other serious health effects). A study in Florida, reported that increased emergency room visits cost Sarasota County more than $130,000 (in 2012 dollars) to treat respiratory illnesses associated with increased algal blooms (Hoagland et al., 2009). Waters containing algal toxins and with taste and odor problems have significant treatment costs. The U.S. EPA 2015 report also estimated the extensive mitigation and restoration costs for coastal and inland waterways contaminated with algal blooms. Also, various organic forms of nitrogen can be stored in bottom sediments of lakes, estuaries, and other waterbodies and then be released to the water column during storms and other weather-related events. In-lake treatment and mitigation costs may include aeration, alum treatments, biomanipulation, and dredging. There are enormous costs to Federal, State, and local agencies associated with restoring impaired waterbodies, including developing watershed management plans, total maximum daily loads (TMDLs), nutrient trading and other programs. High costs associated with nutrient pollution control from point and nonpoint sources include upgrades to municipal wastewater treatment plants (advanced wastewater treatment to meet TMDLs for impaired waters), conversion of septic tanks to centralized wastewater treatment systems, industrial and agricultural wastewater treatment, and the control of elevated levels of nitrogen in stormwater runoff.

Over the past two decades, the State of Florida has spent hundreds of millions of dollars on restoration of large springs and their spring runs with impaired water quality and adversely affected ecosystems. More than 1000 springs have been identified in Florida, and over the past several decades, their water quality has become degraded (Katz, 2004; Scott, 2002). Elevated nitrate-nitrogen concentrations have resulted in excessive algal growth that has depleted aquatic oxygen levels and negatively impacted the health of the spring ecosystem. About 30 of these springs have been given special status because they have been deemed by the Florida Legislature to be a unique part of the state's scenic beauty, provide critical habitat, and have immeasurable natural, recreational, economic, and inherent value. Studies have indicated that springs contribute significantly to Florida's economy, with more than 100 million dollars per year from visitor spending, jobs, and associated hotel stays and sales at restaurants, dive shops, and other expenses. In 2016, the Florida Legislature enacted the "Florida Springs and Aquifer Protection Act," which provides continued funding ($50 million per year) and special status and protection to historic first-magnitude springs (each discharging more than 2.8 m^3/s of water from the Floridan aquifer) and to other springs of special significance. The act requires the Florida Department of Environmental Protection to set restoration targets and adopt a basin management action plan for each water-quality impaired system to reduce nutrient loading and improve water quality over the next 20 years.

In England and Wales, Pretty et al. (2003) estimated the damage costs of freshwater eutrophication in to be $105–160 million per year (£75.0 million to 114.3 million). The damage costs were grouped into seven categories, each with costs of $15 million per year or more: (a) reduced value of waterfront dwellings, (b) drinking water treatment costs for nitrogen removal, (c) reduced recreational and aesthetic value of water bodies, (d) drinking water treatment costs for removing of algal toxins and decomposition products, (e) production of greenhouse and acidifying gases, (f) negative ecological effects on biota, and (g) net economic losses from the tourist industry. In China, Qiu (2011) estimates that billions of dollars will be spent cleaning up contaminated groundwater related to excess nutrients.

The European Nitrogen Assessment (ENA) program (Brink et al., 2011) estimated that the total annual environmental damage related to reactive nitrogen ranged between 70 and 320 billion euro (in terms of 2009 euro), which was equivalent to 150–175 euro per capita. About 75% of this estimate was attributed to health damage and air pollution. This constituted 1–4% of the average European income. The studies also reported the following six societal costs per kilogram of reactive nitrogen in pollutant emission (in euro/kg N): (a) health impacts (10–30), (b) health effects from secondary ammonium particles (2–20), (c) greenhouse gas balance effects of N_2O (5–15), (d) ecosystem impacts from nitrogen runoff (5–20), (e) nitrogen deposition (2–10), and (f) N_2O associated with stratospheric ozone depletion (1–3). Preliminary estimates of global damage costs associated with nitrogen pollution range from 200 to 2000 billion USD per year (Sutton et al., 2013).

Numerous studies have documented the enormous costs associated with the treatment of drinking water in the United States (to remove algal toxins and algal decomposition products) and advanced treatment of human wastewaters (municipal wastewater treatment facilities) and management practices to process animal wastes from confined animal feeding operations. Collectively, federal and state agencies have spent billions of dollars on trying to alleviate water-quality degradation from excess nitrogen; however, recent studies have shown little to no improvement in water quality. In 2004, the Drinking Water Infrastructure Needs Survey estimated that $150.9 billion will need to be spent on drinking water systems to provide safe treatment, storage, and distribution (U.S. EPA, 2004). A 2009 report by the State-U.S. EPA Nutrient Innovations Task Group (2009) noted that current nutrient control strategies are woefully inadequate and the rate of nutrient pollution will likely increase as the world population continues to increase. The population of the United States is projected to increase by more than 135 million over the next 40 years. As population grows, more nutrients (mainly reactive nitrogen compounds) will be released to the environment from municipal wastewater discharges and from other sources such as wastes from agricultural livestock and row-crop runoff are likely to grow.

1.7. PURPOSE AND SCOPE

Unfortunately, many people (including students of all ages, policymakers, teachers, environmental, and economic strategists) are not aware of the extensive ecological, human health, and economic consequences of excess amounts of reactive nitrogen in the environment. This is despite the fact that over the past 40 years, there have been numerous national and international scientific studies that have generated hundreds of published reports and articles on various aspects of environmental degradation, adverse human health effects, and the enormous economic costs related to the anthropogenic alteration of the nitrogen cycle with an overload of reactive nitrogen species. For more than 25 years, Galloway et al. (1995, 2003, 2004, 2008) have quantified global inputs of reactive nitrogen and documented the consequences of their cascading effects in the atmosphere, terrestrial ecosystems, freshwater and marine systems, and human health impacts. In Europe, there have been numerous studies that provide an historical foundation on the alteration of the nitrogen cycle. During the 1970s and 1980s, the sources and impacts of nitrogen were studied throughout Europe, including assessments by the Scientific Committee on Problems of the Environment (SCOPE) (Blackburn & Sorenson, 1988; Soderlund & Svensson, 1976) and by the Royal Society of United Kingdom (Stewart & Rosswall, 1982). Mosier et al. (2004) presented an evaluation of the alteration of the nitrogen cycle in agriculture.

Undeniably, one of the most comprehensive studies of the anthropogenic alterations to the nitrogen cycle and their consequences of excess reactive nitrogen was conducted by the ENA program (Sutton et al., 2011b). This 4-year program (2007–2011) presented the first detailed continental-scale (Europe) assessment of reactive nitrogen in the environment and provided a multidisciplinary overview of the key processes in the nitrogen cycle, issues associated with up-scaling from field, farm, and city to national and continental scales, and opportunities for more effective management at local and global levels. The ENA program also was designed to inform policymakers at local and global scales by providing an extensive review of the current scientific understanding of nitrogen sources in Europe, their effects and interactions, and analyses of current policies along with associated economic costs and benefits. In addition, detailed information was presented on current nitrogen challenges in Europe and for a global context, what is known about (a) nitrogen processing in the biosphere and associated uncertainties, (b) nitrogen flows and fate at various spatial scales, (c) managing nitrogen by identifying major societal threats, and (d) strategies for more effective nitrogen management. A more recent assessment and analysis of the global nitrogen cycle with a focus on terrestrial ecosystems and the atmosphere was published by the Royal Society of the United Kingdom (Fowler et al., 2013).

As one can clearly see, the magnitude and scope of the problems associated with excess reactive nitrogen in our environment are enormous. Even with this barrage of scientific information over the past two decades that effectively demonstrates the adverse environmental, human health, and economic impacts, the message about reactive nitrogen overuse and atmospheric emissions is not making any significant impact on a wide audience. Our society must become increasingly vigilant in our commitment to address the challenges associated with significantly reducing inputs of reactive nitrogen to prevent further degradation of our surface waters, aquifers, and ecosystems. More continued collaboration and partnerships and policies are needed to develop more effective means of reducing reactive nitrogen inputs to the land and our waterways.

There is an urgent need to continue to integrate information from various disciplines on the consequences of excess reactive nitrogen in our environment and effectively communicate this information in a more coherent way so that our society can better understand these problems and move forward to develop more effective strategies to reduce environmental, human health, and economic impacts from excess nitrogen in surface waters, groundwater, and the atmosphere. Recently, there has been increased attention to understanding the footprint of reactive nitrogen generated by the consumption of nitrogen in different areas around the world. Several nitrogen footprint tools have been used to educate people and institutions where people work to support policy changes and help people make better decisions regarding resource use that have contributed to the many problems involving reactive nitrogen in the environment (Galloway et al., 2014). The importance of emphasizing the value of clean water, our most precious resource, to our society cannot be overstated.

The material presented in this book addresses this urgent and critical need for integrating and synthesizing previous and new sources of information on environmental, human health, and economic consequences of excess reactive nitrogen in our environment. The book begins with the first three chapters that present background information. This first chapter (Chapter 1) presents a concise overview of the various issues and the environmental, health, and economic consequences associated with excess reactive nitrogen. The purpose and scope of the book are also described in this chapter. Chapter 2 presents an overview of the nitrogen cycle and how the cycle has been dramatically altered during the past 100 years by anthropogenic activities. Chapter 3 provides detailed information on the major point and nonpoint sources that contribute reactive nitrogen to the environment at large regional and global scales. Also discussed are factors that affect nitrogen transport to the environment at large regional and global scales. Chapter 4 builds on this introductory information and presents information on novel and innovative tools used to identify and quantify sources of nitrate contamination. The information presented in the first four chapters is provided to help the reader gain a better understanding of reactive nitrogen cycling. The next several chapters are devoted to ramifications (and detrimental impacts) from the substantial anthropogenic alteration of the nitrogen cycle including adverse human health impacts (Chapter 5); degradation and contamination of ecosystems, surface waters, groundwater, and springs, (Chapters 6–8). Chapter 9 discusses other contaminants that co-occur with nitrate-nitrogen contamination resulting from various anthropogenic sources. Chapter 10 presents detailed information on economic consequences and costs associated with environmental pollution (e.g., human health maladies, clean-up costs, drinking-water treatment costs, losses to commercial fishing, tourism, recreational activities, decreased real estate values, and cascading effects to other socioeconomic sectors). The final chapter

(Chapter 11) discusses effective ways to reduce our nitrogen footprint (locally and globally) from human activities and presents practical strategies, and recommendations for restoring water quality in impacted surface water bodies, groundwater and springs, and terrestrial and aquatic ecosystems. Three areas are highlighted where ongoing efforts are being focused that include increasing nitrogen use efficiency and sustainability in agricultural systems, reducing the per capita consumption of animal proteins, and decreasing fossil fuel combustion and replacing with alternative energy sources. In recent years, many countries have been making important strides in identifying and communicating societal threats, and in developing strategies for integrated approaches for tackling excess reactive nitrogen in our environment.

REFERENCES

Ackerman, D., Millet, D. B., & Chen, X. (2018). Global estimates of inorganic nitrogen deposition across four decades. *Global Geochemical Cycles, 33*, 100–107. https://doi.org/10.1029/2018GB005990

Alley, W. M., Healy, R. W., LaBaugh, J. W., & Reilly, T. E. (2002). Flow and storage in groundwater systems. *Science, 296*, 1985–1990.

Apte, J. S., Marshall, J. D., Cohen, A. J., & Brauer, M. (2015). Addressing global mortality from ambient $PM_{2.5}$. *Environmental Science and Technology, 49*, 8057–8066. http://dx.doi.org/10.1021/acs.est.5b01236

Ascott, M. J., Gooddy, D. C., Wang, L., Stuart, M. E., Lewis, M. A., Ward, R. S., & Binley, A. M. (2017). Global patterns of nitrate storage in the vadose zone. *Nature Communications*, 1–7. https://doi.org/10.1038/s41467-017-01321-w

Basu, N. B., Jindal, P., Schilling, K. E., Wolter, C. F., & Takle, E. S. (2012). Evaluation of analytical and numerical approaches for the estimation of groundwater travel time distribution. *Journal of Hydrology, 475*, 65–73.

Blackburn, T. H., & Sorenson, J. (Eds.) (1988). *Nitrogen cycling in coastal marine environments* (SCOPE Report No. 33). Wiley: Chichester.

Bobbink, R., Hicks, K., Galloway, J., Spranger, T., Alkemade, R., Ashmore, M., et al. (2010). Global assessment of nitrogen deposition effects on terrestrial plant diversity: A synthesis. *Ecological Applications, 20*, 30–59.

Bodeanu, N. (1993). Microalgal blooms in the Romanian area of the Black Sea and contemporary eutrophication conditions. In T. J. Smayda & Y. Shimizu (Eds.), *Toxic phytoplankton blooms in the sea* (pp. 203–209). Amsterdam: Elsevier.

Bowman, W. D., Cleveland, C. C., Halada, L., Hresko, J., & Baron, J. S. (2008). Negative impact of nitrogen deposition on soil buffering capacity. *Nature Geoscience, 1*. https://doi.org/10.1038/ngeo339

Breitburg, D., Levin, L. A., Oschlies, A., Grégoire, M., Chavez, F. P., Conley, D. J., et al. (2018). Declining oxygen in the global ocean and coastal waters. *Science, 359*, eaam7240.

Brink, C., Van Grinsven, H., Jacobsen, B. H., Rabl, A., Gren, I.-M., Holland, M., et al. (2011). Costs and benefits of nitrogen in the environment. Ch. 22. In M. A. Sutton, C. M. Howard, J. W. Erisman, G. Billen, A. Bleeker, P. Grennfelt, H. van Grinsven, & B. Grizzetti (Eds.), *The European nitrogen assessment: Sources, effects, and policy perspectives*. Cambridge, UK: Cambridge University Press, 664 p.

Camargo, J. A., & Alonso, A. (2006). Ecological and toxicological effects of inorganic nitrogen pollution in aquatic ecosystems: A global assessment. *Environment International, 32*, 831–849.

Conley, D. J., Paerl, H. W., Howarth, R. W., Boesch, D. F., Setzinger, S. P., Havens, K. E., et al. (2009). Controlling eutrophication: Nitrogen and phosphorus. *Science, 323*(5917), 1014–1015.

Curtis, C. J., Botev, I., Camarero, L., Catalan, J., Cogalniceanu, D., Hughes, M., et al. (2005). Acidification in European mountain lake districts: A regional assessment of critical load exceedance. *Aquatic Sciences, 67*, 237–251.

Diaz, R. J., & Rosenberg, R. (2008). Spreading dead zones and consequences for marine ecosystems. *Science, 321*, 926–929.

Dodds, W. K., Bouska, W. W., Eitzmann, J. L., Pilger, T. H., Pitts, K. L., Riley, A. J., et al. (2009). Eutrophication of U.S. freshwaters: Analysis of potential economic damages. *Environmental Science and Technology, 43*(1), 12–19.

Doney, S. C., Mahowald, N., Lima, I., Feely, R. A., Mackenzie, F. T., Lamarque, J.-F., & Rasch, P. J. (2007). Impact of anthropogenic atmospheric nitrogen and sulfur deposition on ocean acidification and the inorganic carbon system. *Proceedings of the National Academy of Sciences USA, 104*(37), 14580–14585. https://doi.org/10.1073/pnas.0702218104

Dubrovsky, N.M., Burow, K.R., Clark, G.M., Gronberg, J.M., Hamilton P.A., Hitt, K.J., et al. (2010). *The quality of our Nation's waters—Nutrients in the Nation's streams and groundwater, 1992–2004*. U.S. Geological Survey Circular 1350, 174 p. Retrieved from http://water.usgs.gov/nawqa/nutrients/pubs/circ1350, accessed 24 September 2019.

Erisman, J. W., Galloway, J. N., Dice, N. B., Sutton, M. A., Bleeker, A., Grizzetti, B., Leach, A. M., & de Vries, W. (2015). *Nitrogen, too much of a vital resource*. Science Brief. Zeist, The Netherlands: World Wildlife Fund.

Erisman, J. W., Galloway, J. N., Seitzinger, S., Bleeker, A., Dise, N. B., Petrescu, A. M. R., et al. (2013). Consequences of human modification of the global nitrogen cycle. *Philosophical Transactions of the Royal Society B, 368*, 20130116. http://dx.doi.org/10.1098/rstb.2013.0116

Flesher, J., & Kastanis, A. (2017). *Toxic algae: Once a nuisance now a severe nationwide threat*. Associated Press.

Fowler, D., Coyle, M., Skiba, U., Sutton, M. A., Capel, J. N., Reis, S., et al. (2013). The global nitrogen cycle in the twenty-first century. *Philosophical Transactions of the Royal Society B, 368*, 1621. https://doi.org/10.1098/rtsb.2013.0164

Fowler, D., Steadman, C. E., Stevenson, D., Coyle, M., Rees, R. M., Skiba, U. M., et al. (2015). Effects of global change during the 21st century on the nitrogen cycle. *Atmospheric Chemistry and Physics, 15*, 13849–13893. https://doi.org/10.5194/acp-15-13849-2015

Galloway, J. N., Aber, J. D., Erisman, J. W., Seitzinger, S. P., Howarth, R. W., Cowling, E. B., & Cosby, B. J. (2003). The nitrogen cascade. *BioScience, 53*, 341–356.

Galloway, J. N., Dentener, F. J., Capone, D. G., Boyer, E. W., Howarth, R. W., Seitzinger, S. P., et al. (2004). Nitrogen cycles: Past, present and future. *Biogeochemistry, 70*, 153–226.

Galloway, J. N., Schlesinger, W. H., Levy, H., II, Michaels, A., & Schnoor, J. L. (1995). Nitrogen fixation: Anthropogenic enhancement-environmental response. *Global Biogeochemical Cycles, 9*(2), 235–252.

Galloway, J. N., Townsend, A. R., Erisman, J. W., Bekunda, M., Cai, Z., Freney, J. R., et al. (2008). Transformation of the nitrogen cycle: Recent trends, questions, and potential solutions. *Science, 320*, 889–892.

Galloway, J. N., Winiwarter, W., Leip, A., Leach, A., Bleeker, A., & Erisman, J. W. (2014). Nitrogen footprints: Past, present, and future. *Environmental Research Letters, 9*. https://doi.org/10.1088/1748-9326/9/11/115003

Grizzetti, B., Bouraoui, F., Billen, G., van Grinsven, H., Cardoso, A.C., Thieu, V., et al. (2011). Nitrogen as a threat to European water quality. Ch. 17. In M. A. Sutton, C. M. Howard, J. W. Erisman, G. Billen, A. Bleeker, P. Grennfelt, H. van Grinsven, & B. Grizzetti (Eds.), *The European nitrogen assessment: Sources, effects, and policy perspectives* (pp. 379–404). Cambridge, UK: Cambridge University Press.

Hoagland, P., Jin, D., Polansky, L. Y., Kirkpatrick, B., Kirkpatrick, G., Fleming, L. E., et al. (2009). The costs of respiratory illnesses arising from Florida Gulf Coast Karenia brevis blooms. *Environmental Health Perspectives, 117*(8), 1239–1243.

Howarth, R., Anderson, D., Cloern, J., Elfring, C., Hopkinson, C., Lapointe, B., et al. (2000). *Nutrient pollution of coastal rivers, bays, and seas*. Issues in Ecology, no. 7.

Howarth, R., Swaney, D., Billen, G., Garnier, J., Hong, B., Humborg, C., et al. (2011). Nitrogen fluxes from the landscape are controlled by net anthropogenic nitrogen inputs and by climate. *Frontiers in Ecology and the Environment.* https://doi.org/10.1890/100178

Johnson, P. T. J., Townsend, A. R., Cleveland, C. C., Glibert, P. M., Howarth, R. W., & McKenzie, V. J. (2010). Linking environmental nutrient enrichment and disease emergence in humans and wildlife. *Ecological Applications, 20*(1), 16–29.

Kane, R. P. (1998). Ozone depletion, related UVB changes and increased skin cancer incidence. *International Journal of Climatology, 18*, 457–472.

Katz, B. G. (2004). Sources of nitrate contamination and age of water in large karstic springs of Florida. *Environmental Geology, 46*, 689–706.

McMahon, P. B., Dennehy, K. F., Bruce, B. W., Bohlke, J. K., Michel, R. L., Gurdak, J. J., & Hurlbut, D. B. (2006). Storage and transit time of chemicals in thick unsaturated zones under rangeland and irrigated cropland, High Plains, United States. *Water Resources Research, 42*(3), W03413. https://doi.org/10.1029/2005WR004417

Meals, D. W., Dressing, S. A., & Davenport, T. E. (2010). Lag time in water quality response to best management practices: A review. *Journal of Environmental Quality, 39*, 85–96.

Mee, L. D., Friedrich, J., & Gomoiu, M.-T. (2005). Restoring the Black Sea in times of uncertainty. *Oceanography, 18*, 32–43.

Mosier, A. R., Syers, J. K., & Freney, J. R. (Eds.). (2004). *Agriculture and the nitrogen cycle* (SCOPE Report No. 65). Washington, DC: Island Press.

Nesheim, M. C. (1993). Human health needs and parasitic infections. *Parasitology, 107*, S7–S18.

Nielsen, R. (2005). *Can we feed the world? Is there a nitrogen limit of food production?* Retrieved from http://home.iprimus.com.au/nielsens/nitrogen.html, accessed 23 September 2019.

Payne, R. J., Dise, N. B., Stevens, C. J., Gowing, D. J., Duprè, C., Dorland, E., et al. (2013). Impact of nitrogen deposition at the species level. *Proceedings of the National Academy of Sciences of the USA, 113*, 984–987.

Peñuelas, J., Poulter, B., Sardans, J., Ciais, P., van der Velde, M., Bopp, L., et al. (2013). Human-induced nitrogen–phosphorus imbalances alter natural and managed ecosystems across the globe. *Nature Communications, 4*, 2934. https://doi.org/10.1038/ncomms3934

Pope, C. A., Burnett, R. T., Thun, M. J., Calle, E. E., Krewski, D., Ito, K., & Thurston, G. D. (2002). Lung cancer, cardiopulmonary mortality, and long-term exposure to fine particulate air pollution. *Journal of the American Medical Association, 287*, 1132–1141.

Pretty, J. N., Mason, C. F., Nedwell, D. B., Hine, R. E., Leaf, S., & Dils, R. (2003). Environmental costs of freshwater eutrophication in England and Wales. *Environmental Science and Technology, 37*(2), 201–208.

Puckett, L. J., Tesoriero, A. J., & Dubrovsky, N. M. (2011). Nitrogen contamination of surficial aquifers- a growing legacy. *Environmental Science and Technology, 45*, 839–844.

Qiu, J. (2011). China to spend billions cleaning up groundwater. *Science, 334*, 745–745.

Rabalais, N. N., Turner, R. E., & Wiseman, W. J., Jr. (2002). Gulf of Mexico hypoxia, A.K.A. "The Dead Zone". *Annual Review of Ecology and Systematics, 33*, 235–263.

Rivett, M. O., Buss, S. R., Morgan, P., Smith, J. W. N., & Bemment, C. D. (2008). Nitrate attenuation in groundwater: A review of biogeochemical controlling processes. *Water Research, 42*, 4215–4232. https://doi.org/10.1016/j.watres.2008.07.020

Robertson, W. M., Bohlke, J. K., & Sharp, J. M., Jr. (2017). Response of deep groundwater to land use change in desert basins of the Trans-Pecos region, Texas, USA: Effects on infiltration, recharge, and nitrogen fluxes. *Hydrological Processes, 31*(3), 2349–2364.

Sala, O. E., Chapin, F. S., III, Armesto, J. J., Berlow, E., Bloomfield, J., Dirzo, R., et al. (2000). Global biodiversity scenarios for the year 2100. *Science, 287*, 1770–1774.

Sanford, W. E., & Pope, J. P. (2013). Quantifying groundwater's role in delaying improvements to Chesapeake Bay water quality. *Environmental Science and Technology, 47*, 13330–13338.

Schlesinger, W. H. (2009). On the fate of anthropogenic nitrogen. *Proceedings of the National Academy of Sciences of the USA, 104*, 203–208.

Scott, T. (2002). Florida's springs in jeopardy. *Geotimes.* Web feature. Retrieved from http://www.geotimes.org/may02/feature_springs.html, accessed 15 May 2019.

Sebillo, M., Mayer, B., Nicolardot, B., Pinay, G., & Mariotti, A. (2013). Long-term fate of nitrate fertilizer in agricultural soils. *Proceedings of the National Academy of Sciences of the USA, 110*(45), 18185–18189.

Sheppard, L. J., Leith, I. D., Mizunuma, T., Cape, J. N., Crossley, A., Leeson, S., et al. (2011). Dry deposition of ammonia gas drives species change faster than wet deposition of ammonium ions: Evidence from a long-term field manipulation. *Global Change Biology*, *17*(12), 3589–3607.

Sobota, D. J., Compton, J. E., McCrackin, M. L., & Singh, S. (2015). Cost of reactive nitrogen release from human activities to the environment in the United States. *Environmental Research Letters*, *10*(2015), 025006.

Soderlund, R., & Svensson, B. H. (1976). The global nitrogen cycle. In B. H. Svensson & R. Soderlund (Eds.), *Nitrogen, phosphorus and sulphur: Global cycles* (SCOPE Report No. 7) (Vol. *22*, pp. 23–73). Stockholm: Ecological Bulletin.

State-U.S. EPA Nutrient Innovations Task Group. (2009). *An urgent call to action*. State-U.S. EPA Nutrient Innovations Task Group Report to US EPA, August 2009, 170 p.

Stewart, W. D. P., & Rosswall, T. (Eds.). (1982). *The nitrogen cycle*. London: The Royal Society.

Sutton, M. A., Howard, C. M., Erisman, J. W., Billen, G., Bleeker, A., Grennfelt, P., van Grinsven, H., & Grizzetti, B. (2011a). Assessing our nitrogen inheritance. The need to integrate nitrogen science and policies. Ch. 1. In M. A. Sutton, C. M. Howard, J. W. Erisman, G. Billen, A. Bleeker, P. Grennfelt, H. van Grinsven, & B. Grizzetti (Eds.), *The European nitrogen assessment: Sources, effects, and policy perspectives*. Cambridge, UK: Cambridge University Press, 664 p.

Sutton, M. A., Howard, C. M., Erisman, J. W., Billen, G., Bleeker, A., Grennfelt, P., van Grinsven, H., & Grizzetti, B. (Eds.). (2011b). *The European nitrogen assessment: Sources, effects, and policy perspectives*. Cambridge, UK: Cambridge University Press, 664 p.

Sutton, M. A., Reis, S., Riddick, S. N., Dragosits, U., Nemitz, E., Theobald, M. R., et al. (2013). Towards a climate-dependent paradigm of ammonia emission and deposition. *Philosophical Transactions of the Royal Society B*, *368*. https://doi.org/10.1098/rtsb.2013.0166

Tesoriero, A. J., Duff, H. H., Saad, D. A., Spahr, N. E., & Wolock, D. M. (2013). Vulnerability of streams to legacy nitrate sources. *Environmental Science and Technology*, *47*(8), 3623–3629.

Tian, D., & Niu, S. (2015). A global analysis of soil acidification caused by nitrogen addition. *Environmental Research Letters*, *10*, 1–10. https://doi.org/10.1088/1748-9326/10/2/024019

Townsend, A. R., Cleveland, C. C., Houlton, B. Z., Alden, C. B., & White, J. W. C. (2011). Multi-element regulation of the tropical forest carbon cycle. *Frontiers in Ecology and the Environment*, *9*, 9–17.

Townsend, A. R., Howarth, R. W., Bazzaz, F. A., Booth, M. S., Cleveland, C., Collinge, S. K., et al. (2003). Human health effects of a changing global nitrogen cycle. *Frontiers in Ecology and the Environment*, *1*(5), 240–246.

U.S. Environmental Protection Agency. (2002). *Nitrogen—Multiple and regional impacts*. U.S. Environmental Protection Agency Clean Air Markets Division Report EPA-430-R-01-006, 38 p.

U.S. Environmental Protection Agency. (2004). *Drinking water costs and federal funding*; EPA 816-F-04-038. Washington, DC: U.S. EPA.

U.S. Environmental Protection Agency. (2009). *An urgent call to action*. Report of the State-EPA Nutrient Innovations Task Group, 41 p.

U.S. Environmental Protection Agency. (2013). Final aquatic life ambient water quality criteria for ammonia-Freshwater 2013. *Notice in Federal Register*, *78*(163), 52192–52194.

U.S. Environmental Protection Agency. (2015). *Case studies on implementing low-cost modifications to improve nutrient reduction at wastewater treatment plants*. EPA-841-R-15-004. Washington, DC. Retrieved from http://www2.epa.gov/nutrient-policy-data/reports-andresearch#reports, accessed 15 August 2018

Van Meter, K. J., & Basu, N. B. (2015). Catchment legacies and time lags: A parsimonious watershed model to predict the effects of legacy storage on nitrogen export. *PLoS One*, *10*(5), e0125971. https://doi.org/10.1371/journal.pone.o125971

Van Meter, K. J., & Basu, N. B. (2017). Time lags in watershed-scale nutrient transport: An exploration of dominant controls. *Environmental Research Letters*, *12*, 084017. https://doi.org/10.1088/1748-9326/aa7bf4

Van Meter, K. J., Basu, N. B., & Van Cappellen, P. (2017). Two centuries of nitrogen dynamics, legacy sources and sinks in the Mississippi and Susquehanna River basins. *Global Biogeochemical Cycles*, *31*(1), 2–23.

Van Meter, K. J., Basu, N. B., Veenstra, J. J., & Burras, C. L. (2016). The nitrogen legacy: Emerging evidence of nitrogen accumulation in anthropogenic landscapes. *Environmental Research Letters*, *11*. https://doi.org/10.1088/1748-9326/11/3/035014

Walvoord, M. A., Phillips, F. M., Stonestrom, D. A., Evans, R. D., Hartsough, P. C., Newman, B. D., & Striegl, R. G. (2003). A reservoir of nitrate beneath desert soils. *Science*, *302*(5647), 1021–1024.

Wang, L., Butcher, A., Stuart, M., Gooddy, D., & Bloomfield, J. (2013). The nitrate time bomb: A numerical way to investigate nitrate storage and lag time in the unsaturated zone. *Environmental Geochemistry and Health*, *35*, 667–681.

Wang, L., Stuart, M. E., Lewis, M. A., Ward, R. S., Skirvin, D., Naden, P. S., Collins, A. L., & Ascott, M. J. (2016a). The changing trend in nitrate concentrations in major aquifers due to historical nitrate loading from agricultural land across England and Wales from 1925 to 2150. *Science of the Total Environment*, *542*, 694–705.

Wang, S., Changyuan, T., Xianfang, S., Ruiqiang, Y., Zhiwei, H., & Yun, P. (2016b). Factors contributing to nitrate contamination in a groundwater recharge area of the North China Plain. *Hydrological Processes*, *30*(13), 2271–2285.

Ward, B. (2012). The global nitrogen cycle. In A. H. Knoll, D. E. Canfield, & K. O. Konhauser (Eds.), *Fundamentals of Geobiology* (pp. 36–48). Blackwell Publishing Ltd.

Ward, M. H., de Kok, T. M., Levallois, P., Brender, J., Gulis, G., Nolan, B. T., & VanDerslice, J. (2005). Drinking-water nitrate and health—Recent findings and research needs. *Environmental Health Perspectives*, *113*(11), 1607–1614.

Ward, M. H., Jones, R. R., Brender, J. D., de Kok, T. M., Weyer, P. J., Nolan, B. T., Villanueva, C. M., & van Breda, S. G. (2018). Drinking water nitrate and human health: an updated review. *International Journal of Environmental Research and Public Health*, *15*(7), pii: E1557. doi:10.3390/ijerph15071557.

Ward, M. H., Mark, S. D., Cantor, K. P., Weisenburger, D. D., Correa-Villasenor, A., & Zahm, S. H. (1996). Drinking water nitrate and the risk of Non-Hodgkin's Lymphoma. *Epidemiology*, *7*, 465–471.

Weyer, P. J., Cerhan, J., Kross, B. C., Hallberg, G. R., Kantamneni, J., Breuer, G., et al. (2001). Municipal drinking water nitrate level and cancer risk in older women: The Iowa Women's Health Study. *Epidemiology, 12*, 327–338.

Yang, S., & Gruber, N. (2016). The anthropogenic perturbation of the marine nitrogen cycle by atmospheric deposition; nitrogen cycle feedbacks and the ^{15}N Haber-Bosch effect. *Global Biogeochemical Cycles, 30*(10), 1418–1440.

Zaehle, S., Ciais, P., Friend, A. D., & Prieur, V. (2011). Carbon benefits of anthropogenic reactive nitrogen offset by nitrous oxide emissions. *Nature Geoscience, 4*, 601–605. https://doi.org/10.1038/ngeo1207

FURTHER READING

Billen, G., Silvestre, M., Grizzetti, B., Leip, A., Garnier, J., Voss, M., et al. (2011). Nitrogen flows from European watersheds to coastal marine waters. Ch. 13. In M. A. Sutton, C. M. Howard, J. W. Erisman, G. Billen, A. Bleeker, P. Grennfelt, H. van Grinsven, & B. Grizzetti (Eds.), *The European nitrogen assessment: Sources, effects, and policy perspectives*. Cambridge, UK: Cambridge University Press, 664 p.

Conant, R. T., Berdanier, A. A., & Grace, P. R. (2013). Patterns and trends in nitrogen use and nitrogen recovery efficiency in world agriculture. *Global Biogeochemical Cycles, 27*, 558–566.

Erisman, J. W., Sutton, M. A., Galloway, J. N., Klimont, Z., & Winiwarter, W. (2008). How a century of ammonia synthesis changed the world. *Nature Geoscience, 1*, 636–639.

Erisman, J. W., van Grinsven, H., Grizzetti, B. Bouraoui, F., Powlson, D., Sutton, M. A., Bleeker, A., & Reis, S. (2011). The European nitrogen problem in a global perspective. Ch. 2. In M.A. Sutton, C.M. Howard, J.W. Erisman Billen, G., Bleeker, A., Grennfelt, P., H. van Grinsven, & B. Grizzetti (Eds.), *The European nitrogen assessment: Sources, effects, and policy perspectives* (pp. 9–31). Cambridge, UK: Cambridge University Press.

Fields, S. (2004). Global nitrogen: Cycling out of control. *Environmental Health Perspectives, 112*(10), A556–A563.

Gu, B., Ju, X., Wu, Y., Erisman, J. W., Bleeker, A., Reis, S., et al. (2017). Cleaning up nitrogen pollution may reduce future carbon sinks. *Global Environmental Change, 48*, 55–66.

Joo, Y. J., Li, D. D., & Lerman, A. (2013). Global nitrogen cycle: Pre-Anthropocene mass and isotope fluxes and the effects of human perturbations. *Aquatic Geochemistry, 19*(5–6), 477–500.

Katz, B. G., Chelette, A. R., & Pratt, T. R. (2004). Use of chemical and isotopic tracers to assess sources of nitrate and age of ground water, Woodville Karst Plain, USA. *Journal of Hydrology, 289*, 36–61.

Katz, B. G., Sepulveda, A. A., & Verdi, R. J. (2009). Estimating nitrogen loading to ground water and assessing vulnerability to nitrate contamination in a large karstic spring basin. *Journal of the American Water Resources Association, 45*, 607–627.

Lassaletta, L., Billen, G., Garnier, J., Bouwman, L., & Valazquez, E. (2016). Nitrogen use in the global food system: Past trends and future trajectories of agronomic performance, pollution, trade, and dietary demand. *Environmental Research Letters, 11*(9), 095007. http://dx.doi.org/10.1088/1748-9326/11/9/095007

Lopez, C. B., Jewett, E. B., Dortch, Q., Walton, B. T., & Hudnell, H. K. (2008). *Scientific assessment of freshwater harmful algal blooms*. Washington, DC: Interagency Working Group on Harmful Algal Blooms, Hypoxia and Human Health of the Joint Subcommittee on Ocean Science and Technology, 65 p.

Lusk, M. G., Toor, G. S., Yang, Y., Mechtensimer, S., De, M., & Obreza, T. A. (2017). A review of the fate and transport of nitrogen, phosphorus, pathogens, and trace organic chemicals in septic systems. *Critical Reviews in Environmental Science and Technology, 47*(7), 455–541.

Mitsch, W. J., Day, J. W., Jr., Gilliam, J. W., Groffman, P. M., Hey, D. L., Randall, G. W., & Wang, N. (2001). Reducing nitrogen loading to the Gulf of Mexico from the Mississippi River Basin: Strategies to counter a persistent ecological problem. *BioScience, 51*(5), 373–388.

Munn, M. D., Frey, J., & Tesoriero, A. (2010). The influence of nutrients and physical habitat in regulating algal biomass in agricultural streams. *Environmental Management, 45*(3), 603–615.

Paerl, H. W., Gardner, W. S., Havens, K. E., Joyner, A. R., & McCarthy, M. J. (2016). Mitigating cyanobacterial harmful algal blooms in aquatic ecosystems impacted by climate change and anthropogenic nutrients. *Harmful Algae, 54*, 213–222. http://dx.doi.org/10.1016/j.hal.2015.09.009

Preston, S.D., Alexander, R.B., Woodside, M.D., & Hamilton, P.A. (2009). *SPARROW MODELING enhancing understanding of the Nation's water quality*. U.S. Geological Survey Fact Sheet 2009–3019, 6 p.

Puckett, L. J., Zamora, C. M., Essaid, H. I., Wilson, J. T., Johnson, H. M., Brayton, M. J., & Vogel, J. R. (2008). Transport and fate of nitrate at the ground-water/surface-water interface. *Journal of Environmental Quality, 37*, 1034–1050.

Reis, S., Bekunda, M., Howard, C. M., Karanja, N., Winiwarter, W., Xiaoyuan, Y., Bleeker, A., & Sutton, M. A. (2016). Synthesis and review: Tackling the nitrogen management challenge: From global to local scales. *Environmental Research Letters, 11* (2016), 120205. http://dx.doi.org/10.1088/1748-9326/11/12/120205.

Stueken, E. E., Kipp, M. A., Koehler, M. C., & Buick, R. (2016). The evolution of the Earth's biogeochemical nitrogen cycle. *Earth Science Reviews, 160*, 220–239.

Sutton, M. A., Erisman, J. W., Dentener, F., & Moeller, D. (2008). Ammonia in the environment: From ancient times to the present. *Environmental Pollution, 156*, 583–604.

Sutton, M. A., Nemitz, E., Erisman, J. W., Beier, C., Butterbach Bahl, K., Cellier, P., et al. (2007). Challenges in quantifying biosphere-atmosphere exchange of nitrogen species. *Environmental Pollution, 150*, 125–139.

Sutton, M. A., Reis, S., & Butterbach-Bahl, K. (2009). Reactive nitrogen in agro-ecosystems: Integration with greenhouse gas interactions. *Agriculture, Ecosystems and Environment, 133*, 135–138.

2
The Nitrogen Cycle

Nitrogen is the most abundant element in our atmosphere, comprising over 78% of all gases. It has unique chemical properties, as it occurs in both inorganic and organic forms with various oxidation states. Nitrogen and its various forms can move or cycle between the atmosphere, biosphere, and hydrosphere. The mostly microbially mediated processes involved in the transfer of the different forms of nitrogen between these three systems and among nitrogen reservoirs within these systems are collectively referred to as the nitrogen cycle (Fig. 2.1). This chapter presents an overview of the major processes that comprise the nitrogen cycle and a discussion of how anthropogenic activities have significantly altered the various processes in the natural nitrogen cycle.

During the past several decades, there has been a considerable focus of attention on substantial increases in carbon dioxide in the atmosphere and how this relates to the carbon cycle and climate change. Even though the carbon cycle has been impacted significantly by fossil fuel burning and other sources of carbon dioxide, we will see that the nitrogen cycle has been altered more than the carbon cycle or any other basic element cycle essential to life on earth (e.g., phosphorus, sulfur). The creation of reactive nitrogen from anthropogenic activities has doubled, whereas CO_2 concentrations in the atmosphere have increased by about 20–30% (Erisman et al., 2015). Anthropogenic sources are adding higher amounts of reactive nitrogen to the global nitrogen cycle than all other nitrogen sources combined. A better understanding of the nitrogen cycle and the transport of the various forms through the atmosphere, hydrosphere, and biosphere will help to encourage better management of reactive nitrogen and help to prevent further degradation of our environmental systems (discussed in greater detail in Chapter 11 of this book).

2.1. PREINDUSTRIAL "NATURAL" NITROGEN CYCLE

Prior to the industrial fixation of dinitrogen gas by the Haber–Bosch process, bacteria and other microorganisms mediated or controlled essentially all the processes involving the transformation of nitrogen (both organic and inorganic forms) in soils, the unsaturated zone, surface water bodies, and in aquifers. Stein and Klotz (2016) noted that recent advances in the postgenomic era have resulted in a better understanding of the nitrogen cycle. The main processes that cycle the various forms of nitrogen between the atmosphere, hydrosphere, and biosphere, include mineralization/ammonification, nitrogen fixation, nitrification, denitrification, dissimilatory nitrate reduction to ammonia (DNRA), and annamox reactions. The rate at which these transformations occur is dependent on many factors, including temperature, moisture content, types of microorganisms, oxygenated versus anoxic conditions, carbon sources, and the type of nitrogen source inputs. Although nitrogen cycling in tropical forests has been less studied, transformations of reactive nitrogen forms are affected by their unique soil conditions (including low pH, highly variable redox conditions, and large amounts of iron oxides, plant litter material and available nitrogen content). Furthermore, in these systems nitrogen availability is influenced by elevation, temperature, rainfall, and diversity of parent material and microorganisms in tropical forests (Pajares & Bohannan, 2016). However, a limited number of studies have investigated the immense diversity of novel microbial communities in tropical forests (Bruce et al., 2010; Ranjan et al., 2015).

2.1.1. Mineralization/Ammonification

During the process of mineralization or ammonification, microorganisms produce ammonia from the decomposition of organic nitrogen from manure, organic matter (leaf litter, decayed plants, and animals) and crop residues. The rates of mineralization or ammonification are dependent on soil temperature, litterfall nitrogen content, soil moisture content, and the amount of oxygen in soil pores.

Nitrogen Overload: Environmental Degradation, Ramifications, and Economic Costs, Geophysical Monograph 250, First Edition. Brian G. Katz.
© 2020 American Geophysical Union. Published 2020 by John Wiley & Sons, Inc.

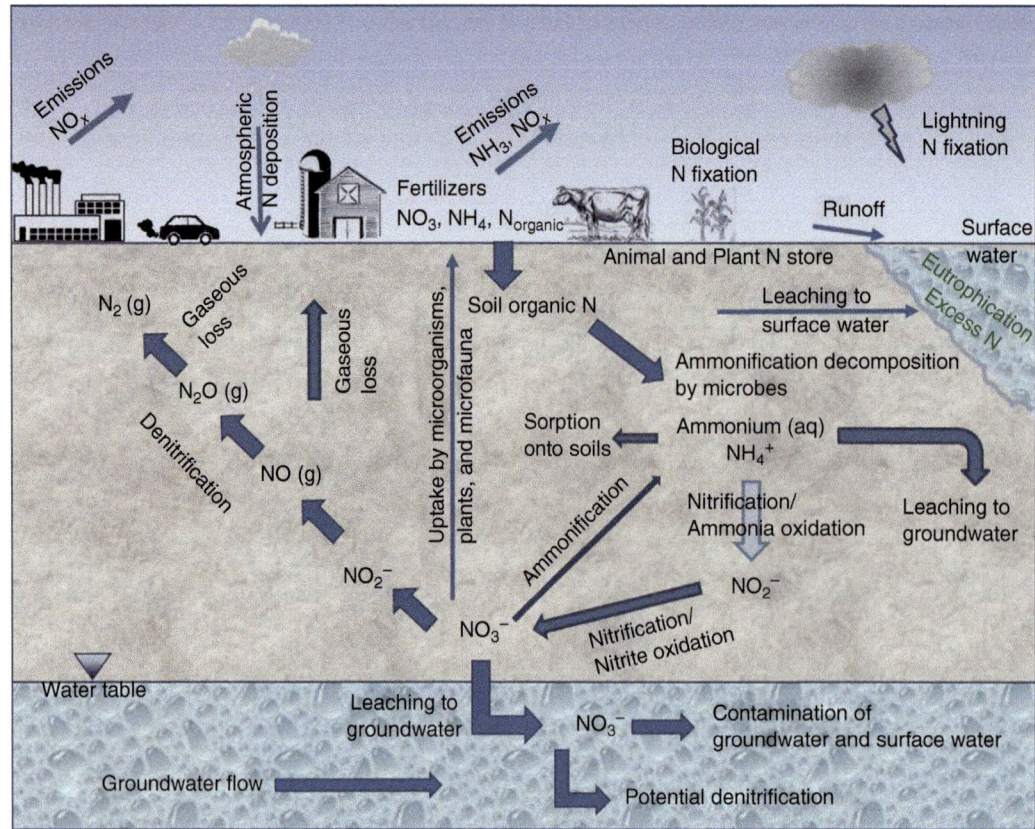

Figure 2.1 Main processes and reactions of the nitrogen cycle. *Source*: Modified from Rivett et al. (2008).

$$R\text{-}NH_2 \dashrightarrow NH_3 \dashrightarrow NH_4^+ \quad (2.1)$$

Organic $N\,(R = \text{organic monomer}) \to \text{ammonia} \dashrightarrow \text{ammonium}$

Recent research suggests that mineralization of nitrogen occurs in two steps (Butterbach-Bahl et al., 2011): the first involving the depolymerization of organic macromolecules to dissolved organic nitrogen, and the second step involving the mineralization/ammonification of dissolved organic nitrogen to ammonium (Equation 2.1). Nitrogen mineralization rates increase with increasing temperature (which promotes decomposition of soil organic matter) (Dessureault-Rompre et al., 2010; Fowler et al., 2015) and litterfall nitrogen flux (Pregitzer et al., 2008). The highest rates of mineralization occur in well-oxygenated and moist soils (as long as the soil is not saturated and becomes anaerobic). The optimal temperatures for mineralization of nitrogen are 20–35°C (Dessureault-Rompre et al., 2010).

The type of vegetation and litter quality are major controls on nitrogen mineralization. Other important vegetation parameters that affect mineralization include canopy structure, leaf geometry, and root distribution (Butterbach-Bahl et al., 2011). Mineralization rates can vary substantially from one ecosystem type to another and are dependent on the C:N ratios. For example, with similar soil organic matter concentrations, forest soils with relatively high C:N ratios had lower nitrogen mineralization rates than grassland soils that typically contain low C:N ratios. For example, average net N mineralization and nitrification rates in the Japanese forest soils were 0.62 ± 0.68 and 0.59 ± 0.65 mg nitrogen per kilogram per day (mg N/kg/d), respectively, and gross N mineralization and nitrification rates were 4.22 ± 3.59 and 0.98 ± 0.68 mg N/kg/d, respectively (Urakawa et al., 2016). Nitrogen mineralization rates in agricultural soils are quite variable depending on C:N ratios, soil moisture content, soil mineralogy (percent clays), tillage practices, and the abundance of heterotrophic microorganisms and other microbes. However, nitrogen mineralization rates in temperate agricultural systems are slower than rates in grasslands or forests (Booth et al., 2005). Johnson et al. (2005) estimated that on average 67–90 kg N/ha are mineralized each year from soil organic matter in New York State.

2.1.2. Nitrogen Fixation

Nitrogen fixation involves the conversion of N_2 gas into reactive nitrogen forms. Nitrogen fixation can occur in two ways, abiotically or through biological nitrogen fixation. Abiotic processes in nature include lighting, forest fires, and hot lava flows. The triple bonds of the N_2 molecules are broken into N atoms during these high-energy processes. The amounts of abiotic nitrogen fixed in nature are typically small but can be significant in certain areas. The largest source of abiotic nitrogen fixation is from the Haber–Bosch industrial process. Since the middle of the 20th century, this process has significantly increased the amount of reactive nitrogen in the environment and is described in more detail in the later section on the human alteration of the nitrogen cycle.

The process of biological nitrogen fixation involves the transformation of N_2 gas to ammonium (NH_4) (Equation 2.2) for making proteins and DNA. Only nitrogen-fixing organisms (diazotrophs) can use their metabolic processes and an enzyme complex called nitrogenase to convert nitrogen from the atmosphere into ammonium. Biological nitrogen fixation is the largest natural source of newly produced reactive nitrogen in most terrestrial ecosystems (Galloway et al., 2004). These nitrogen-fixing organisms (mostly different types of bacteria) include the genus Rhizobium, which can develop symbiotic relationships with the roots of certain plants (particularly legumes such as peas, beans, and clover) and *Anabaena* (a cyanobacterium). These symbiotic bacteria receive a carbon source and a favorable environment from the host plant root in exchange for some of the nitrogen that is fixed. Other forms of bacteria can fix nitrogen without plant hosts. These bacteria are referred to as free-living nitrogen fixers and blue-green algae (cyanobacteria, genus: *Anabaena*, Nostoc) live in aquatic environments. Other free-living types of bacteria and prokaryotes that are known to carry out nitrogen fixation include the following Genus and metabolic types (Bernhard, 2010): *Pseudomonas, Azotobacter, Methylomonas* (aerobic, chemoorganotrophic), *Alcaligenes, Thiobacillus* (anaerobic, chemolithotrophic), *Methanosarcina, Methanococcus* (Archaea, anaerobic, chemolithotrophic), *Chromatium, Chlorobium* (anaerobic, phototrophic), and *Desulfovibrio, Clostridium* (anaerobic, chemoorganotrophic).

$$2N_2 + 3H_2 \rightarrow 2NH_3 \qquad (2.2)$$

Nitrogenase (the enzyme catalyst that carries out nitrogen fixation) typically is deactivated in the presence of oxygen; however, nitrogenase genes have been found in many aerobic environments (including soils, lakes, and oceans) indicating that various nitrogen-fixing organisms have evolved over time and developed ways to protect their nitrogenase from oxygen over a broad range of environmental conditions (Bernhard, 2010). Butterbach-Bahl et al. (2011) presented ranges of nitrogen fixation rates for different types of ecosystems (Table 2.1).

Tropical forests contribute about 70% of terrestrial nitrogen fixation, although these ecosystems only occupy 12% of the Earth's surface (Townsend et al., 2011). Annual biological nitrogen fixation rates in tropical forests (15–36 kg N/ha/yr) are higher or close to estimates for temperate forest systems (7–27 kg/ha/yr) (Pajares & Bohannan, 2016). Recent research by Pajares and Bohannan (2016) indicates a much wider variety of microorganisms (including nitrogen fixing bacteria) than previously known, are present in tropical forest soils. The presence of biological nitrogen fixation "hotspots" (zones of high rates) contribute to the high degree of spatial and temporal variability in tropical systems (Pajares & Bohannan, 2016). Also, lichens, mosses, and other epiphytes associated with cyanobacteria in forest canopy communities can produce substantial inputs of reactive nitrogen to tropical forests (Benner et al., 2007).

Table 2.1 Ranges for biological nitrogen fixation in natural and managed ecosystems.

Ecosystem type	Crop	N fixation rate (kg N/ha/yr)	Source
Boreal forests and boreal woodland		1.5–2	Cleveland et al. (1999)
Temperate forests and forested floodplains		6.5–26.6	Cleveland et al. (1999)
Natural grasslands		2.3–3.1	Cleveland et al. (1999)
Mediterranean shrublands		1.5–3.5	Cleveland et al. (1999)
Agricultural ecosystems	Common bean	30–50	Smil (1999)
	Faba bean	80–120	Smil (1999)
	Soya bean	60–100	Smil (1999)
Managed grasslands	*Trifollium pratense* and *T. repens* mixture	31–171	Carlsson and Huss-Danell (2003)
Managed grasslands	Medicago sativa with different grasses	65–319	
Tropical forests		15–30	Pajares and Bohannan (2015)

Source: From Butterbach-Bahl et al. (2011).

Reiterating, in terrestrial systems biological nitrogen fixation is the dominant natural source of reactive nitrogen. This source produces ammonia in soils and makes up about 90% of the total amount of natural reactive nitrogen. Other natural terrestrial sources of reactive nitrogen include lightning and volcanic eruptions. All three sources are estimated to produce about 65 Tg (1 Tg = 10 E12 g) of reactive nitrogen per year (Tg/yr) (Fowler et al., 2013, 2015). Marine biological fixation produces about 140 Tg/yr of reactive nitrogen that is buried in sediments and the rest eventually being denitrified and released to the atmosphere as N_2 or N_2O (Erisman et al., 2015). Therefore, the current estimate of the total amount of reactive nitrogen created by terrestrial and marine systems is 205 Tg N/yr.

2.1.3. Nitrification

During nitrification, reduced forms of nitrogen (mainly ammonia) are oxidized to nitrite and nitrate. Biological nitrification occurs aerobically and is a two-step microbiological process, each step accomplished by two groups of Gram-negative, chemoautotrophic nitrifying bacteria. Both groups fix carbon dioxide for their major source of cell carbon and derive their energy and reducing power either from the oxidation of ammonia (ammonia-oxidizing bacteria) or nitrite (nitrite-oxidizing bacteria). Using energy obtained from metabolism of ammonia or nitrite, these groups of bacteria can build organic molecules. In the first step of the process, bacteria genus known as *nitrosomonas* is most commonly associated with this step that converts ammonia (NH_3) and ammonium (NH_4^+) into nitrite (NO_2^-) using two different enzymes, ammonia monooxygenase and hydroxylamine oxidoreductase. Other genera, including *Nitrosococcus* and *Nitrosospira*, can nitrify ammonia, as well as some subgenera, including *Nitrosolobus* and *Nitrosovibrio*, can also autotrophically oxidize ammonia (AWWA, 2002). In 2005, an archaeon was discovered that also oxidizes ammonia to nitrite (Koenneke et al., 2005). Ammonia-oxidizing Archaea have been found in oceans, soils, and salt marshes indicating that they are an important part of the biological nitrification process of the global nitrogen cycle.

The biochemical equation for this process is

$$NH_3 + O_2 \text{----}> NO_2^- + 3H^+ + 2e^- \quad (2.3)$$

The next step in the nitrification process involves the bacteria genus *Nitrobacter*; however, other genera, including *Nitrospina*, *Nitrococcus*, and *Nitrospira*, can also autotropically oxidize nitrite to nitrate (Watson et al., 1981). The enzyme nitrite oxidoreductase (NOR) is utilized to complete the nitrification process, converting nitrite (NO_2^-) to nitrate (NO_3^-) as follows:

$$NO_2^- + H_2O \text{----} \rightarrow NO_3^- + 2H^+ + 2e^- \quad (2.4)$$

The rate of nitrification occurs more rapidly in moist, well-aerated soils, and in warm soils (20–35°C). Nitrification essentially does not occur in temperatures below 5°C or above 50°C (Johnson et al., 2005). The optimal soil water content and soil pH for nitrification are typically in the range of 30–60% water-filled pore space and pH values of 5.5–6.5, respectively (Butterbach-Bahl et al., 2011). Other groups of heterotrophic bacteria and fungi are also capable of nitrification, but their rate is much slower than that of autotrophic organisms (Watson et al., 1981). Ammonia-oxidizing organisms have not been studied in much detail in tropical forest soils, and these microorganisms would be affected by wet conditions, small annual temperature fluctuations, low pH, high available nitrogen, and varying oxygen availability (Pajares & Bohannan, 2016).

Nitrification is an important process in water and wastewater treatment systems. In drinking water treatment systems, autotrophic nitrifiers considerably outnumber heterotrophic organisms. Conversely, in wastewater treatment systems, heterotrophic microorganisms predominantly carry out nitrification (Carey & Miglaccio, 2009). As can be seen from both Equations (2.3 and 2.4), hydrogen ions are products of the nitrification process. This can result in lower pH values and decrease in alkalinity in soils where wastewater or biosolids are applied to the land surface. Lower soil pH values could promote the mobilization of potentially harmful trace elements (e.g., copper, arsenic, and lead) into underlying aquifers that provide drinking water.

The production of nitrate from nitrification of excess ammonium has resulted in significant degradation to surface waters and groundwater systems around the world. Nitrate is a negatively charged ion and therefore not adsorbed to negatively charged clay mineral surfaces in soils. Consequently, due to its high solubility and mobility, nitrate is the primary form of nitrogen that is leached into groundwater systems and aquifers used for drinking water supplies. The rate of nitrate leaching depends on soil drainage properties, soil pH, dissolved oxygen, rainfall, recharge rate, the amount of nitrate in the soil, and plant uptake. Many of the potential health and environmental impacts associated with leaching of nitrate from point and nonpoint sources are discussed in more detail in subsequent chapters.

2.1.4. Denitrification

Denitrification is an important process in the nitrogen cycle that converts reactive oxidized forms of nitrogen (nitrate and nitrite) ultimately to inert nitrogen gas (N_2), which eventually is released to the atmosphere. During denitrification, nitrate is used as a terminal electron acceptor in anaerobic respiration by certain bacteria, such as *Pseudomonas*, and the overall reaction can be represented by the following equation.

$$2NO_3^- + 12H^+ + 10e^- \rightarrow N_2 + 6H_2O \qquad (2.5)$$

Actually, there are several intermediate steps during denitrification that occur with different enzymes used to catalyze each reaction (Kuypers et al., 2018):

$$NO_3^- + 2H^+ + 2e^- \rightarrow NO_2^- + H_2O\,(\text{Nitrate reductase}) \qquad (2.6)$$

$$NO_2^- + 2H^+ + e^- \rightarrow NO + H_2O\,(\text{Nitrite reductase}) \qquad (2.7)$$

$$NO + 2H^+ + 2e^- \rightarrow N_2O + H_2O\,(\text{Nitric oxide reductase}) \qquad (2.8)$$

$$N_2O + 2H^+ + 2e^- \rightarrow N_2 + H_2O\,(\text{Nitrous oxide reductase}) \qquad (2.9)$$

Nitrogen compounds produced during denitrification include nitric oxide (NO) (Equation 2.7) and nitrous oxide (N_2O) (Equation 2.8). Natural levels of these compounds are relatively low but would likely be substantially enhanced by the addition of fertilizer (Skiba et al., 1993). Nitrous oxide is considered to be a very potent greenhouse gas that has a 200- to 300-fold-stronger effect than carbon dioxide (CO_2) (Robertson et al., 2000). N_2O also contributes to air pollution and has the potential to harm the ozone layer. Denitrifiers are chemoorganotrophs and require a carbon source for denitrification to occur, and the optimal pH range for denitrification to occur is between 7.0 and 8.5. Combining equations for the intermediate chemical reactions, the overall denitrification reaction can be represented by the following equation (represented by methanol (CH_3OH) as the carbon source in this example):

$$6NO_3^- + 5CH_3OH \rightarrow 3N_2 + 5CO_2 + 7H_2O + 6OH^- \qquad (2.10)$$

Pseudomonas aeruginosa can reduce the amount of fixed nitrogen as fertilizer by up to 50% (Kuypers et al., 2018). In addition to *Pseudomonas*, other denitrifying bacteria include *Thiobacillus denitrificans*, *Micrococcus denitrificans*, some species of *Serratia*, *Achromobacter*. Some research suggests that denitrification can occur under aerobic conditions. Takaya et al. (2003) present a method for screening and characterizing natural aerobic denitrifiers that produce N_2 gas by reducing NO_3^- under oxic conditions. They studied bacterial strains that produced less N_2O under aerobic conditions than other aerobic denitrifiers produce, which could be used to construct an aerobic denitrifying system with low-level emissions of N_2O.

Denitrification has been studied extensively in a variety of hydrogeologic and ecologic settings. Some examples include the hyporrheic zone beneath streams (Harvey et al., 2013), riparian zones (Groffman & Crawford, 2003; King et al., 2016; Lowrance, 1992), saturated zone (Korom, 1992), wetlands (Hansen et al., 2016), and in anoxic zones in lakes and oceans (Ulloa et al., 2012). Gaseous losses of nitrogen (N_2O and N_2) from denitrification typically are higher in fine-grained soils (clayey or organic rich soils) that typically have high moisture content and tend to be anaerobic compared to well-drained aerated sandy soils that tend to have minimal denitrification.

Recent studies also have shown that nitrate reduction can involve multiple electron donors and end products in addition to simple organic compounds (Burgin & Hamilton, 2007). In terrestrial systems, denitrification not only occurs in soils, the vadose zone, riparian zones, and anoxic sediments but also occurs in aquifers (Rivett et al., 2008). Denitrification in most aquifers is dependent on solid-phase electron donors in the aquifer matrix (e.g., reduced iron and sulfur, and organic carbon), instead of solutes from surface sources (Green et al., 2008; Torrento et al., 2010, 2011). Denitrification in aquifers can potentially be a major component of regional and global nitrogen budgets, although estimates are poorly constrained and vary widely from 0 to 138 Tg N/yr (Seitzinger et al., 2006). Heffernan et al. (2012) found that denitrification in the karstic Floridan aquifer accounted for 32% of estimated nitrogen inputs across 61 sampled spring systems in Florida.

Denitrification is an important process in human-managed systems because it prevents excess nitrate and nitrite from being released into waterways and aquifers. Some examples of these managed systems are wastewater treatment plants, reactive barriers, bioreactors, and constructed or reclaimed wetlands (e.g., Bachand & Home, 1999; Seitzinger et al., 2006).

2.1.5. Dissimilatory Nitrate Reduction to Ammonia

DNRA is another anaerobic process that is another microbial pathway for nitrate attenuation in ecosystems. This process is catalyzed by fermentative bacteria that reduce nitrate to nitrite and to ammonium. DRNA as with denitrification tends to occur in similar soil conditions with low redox potential, high nitrate concentrations, and labile carbon availability (Butterbach-Bahl et al., 2011). This process commonly occurs in wetland ecosystems and in moist tropical soils with high clay content (Silver et al., 2001). Nitrogen fixation and DNRA also has been noted to affect nitrogen dynamics in estuaries (e.g., Gardner et al., 2006).

2.1.6. Plant Uptake, Assimilation, and N Immobilization

Plants assimilate nitrate and ammonium from soils by absorption through their root hairs. Nitrate taken up by the plant is reduced to nitrite and ammonium. Ammonium ions are then catalyzed in reactions into organic forms (e.g., nucleic acids, amino acids, and chlorophyll) by the assimilatory enzymes glutamine synthetase and glutamate synthase (Temple et al., 1998). The assimilation process involves the competition between microorganisms in the soil with plants or crops that need nitrogen for growth. Nitrate and ammonium can be converted by microorganisms to organic nitrogen compounds that are not available for crop uptake, as indicated by the following reaction:

$$NH_4^+ \text{ and/or } NO_3^- \rightarrow R\text{-}NH_2 (\text{organic N}) \quad (2.11)$$

However, when these microorganisms die, the organic nitrogen content in their cells is converted (remineralized) to plant-available nitrate by the processes mineralization and nitrification. Animals, fungi, and other heterotrophic organisms assimilate nitrogen from the conversion of organic nitrogen (proteins) into amino acids.

Nitrogen uptake by plants is highly species dependent as well as other factors, such as temperature, soil pH, and competition by different microbes (Jackson et al., 2008). Plant uptake efficiencies on production farms typically range from 18 to 49% (Cassman et al., 2002; Dubrovsky et al., 2010). In certain extreme environments, such as nitrogen-limited and cold ecosystems, plants can uptake amino acids and other organic monomers thus bypassing microbial nitrogen mineralization (Butterbach-Bahl et al., 2011). Where nitrogen is not limited, such as in agricultural soils or forests receiving nitrogen deposition, inorganic nitrogen remains the main nutrient for plant uptake (Harrison et al., 2007).

2.1.7. Decay of Plants and Animals

During the decomposition of dead plants and animals, putrefying bacteria and fungi produce ammonia from the organic nitrogen compounds contained in these organisms. Other breakdown products from the decay of animals include urea, allantoin, uric acid, and other forms of organic nitrogen. Bacteria will also convert these compounds to ammonium in soils. Ammonium can then be reassimilated by plants and microbes or converted into nitrate via nitrification.

2.1.8. Volatilization

There are several ways that nitrogen can volatilize and be released to the atmosphere. Ammonium (NH_4^+) can be converted to ammonia gas, which typically occurs at higher soil pH values and hot and windy conditions that promote evaporation. Volatilization losses can be significant for manures and urea fertilizers that are deposited on the land surface and not mixed into the soil by tillage or rainfall. Manure contains predominantly ammonium and organic nitrogen. Johnson et al. (2005) postulate that 65% of the ammonium nitrogen is retained if the manure is incorporated within 1 day, but if manure is incorporated after 5 days, most of the ammonium will be lost through volatilization. The organic nitrogen in manure does not volatilize. After time, this organic nitrogen will be mineralized by microorganisms and eventually become available for plant uptake.

2.1.9. Anammox Reactions

Ammonia oxidation was thought to occur under aerobic conditions; however, a new type of ammonia oxidation was discovered by Strous et al. (1999). In oceans, hot springs, hydrothermal vents, and several freshwater systems, ammonium can be converted by anaerobic ammonium oxidation (anammox) into nitrogen gas (N_2) using nitrite as the electron acceptor (Equation 2.12) by a bacterium named *Brocadia anammoxidans* (Butterbach-Bahl et al., 2011). Anammox was first found to occur in anoxic bioreactors of wastewater treatment plants, but now have been found in a variety of ecosystems containing anoxic sediments and anoxic water columns (coastal and estuarine sediments, mangroves, freshwater lakes, oceans) (Bernhard, 2010). Recent research has shown that anammox bacteria are important players in the global nitrogen cycle (for more detailed information on this process, refer to Huub et al., 2007).

$$NH_4^+ + NO_2^- \rightarrow N_2 + 2H_2O \quad (2.12)$$

2.1.10. Leaching to Groundwater

Leaching of nitrogen (mainly in the form of highly soluble and mobile nitrate) to groundwater occurs when nitrate moves below the root zone in soils and plants can no longer assimilate this excess nitrogen. The rate of leaching of nitrate is dependent on soil properties. High leaching rates typically occur in sandy well-drained soils; whereas low leaching rates are common in clayey, poorly drained soils. The rate of leaching in agricultural areas is influenced by other factors including rainfall, cultivation, irrigation, and soil management. The movement of nitrate into groundwater and its associated water quality degradation is discussed in Chapters 7 and 8.

2.2. HUMAN ALTERATION OF THE NITROGEN CYCLE

As mentioned in Chapter 1, there have been two noteworthy and substantial benefits from the human modification of the nitrogen cycle by significantly increasing the amounts of reactive nitrogen. These benefits include the ability to sustain the global human population of 7 billion through agricultural use of synthetic fertilizers from the Haber–Bosch process, and a stimulation of the sequestration of global CO_2 by terrestrial and marine ecosystems resulting from increased nitrous oxide emissions (Fowler et al., 2015; Zaehle et al., 2011).

The amount of reactive nitrogen continues to increase every year mainly from anthropogenic activities including the agricultural use of fertilizers, fossil fuel combustion, and the increased planting and cultivation of legumes and other crops that biologically fix nitrogen. Figure 2.2 shows increases in the production of reactive nitrogen from fossil fuel combustion, along with biological nitrogen fixation, and fertilizer production (Haber–Bosch process) from 1860 through the present with future projections to 2050 (Nielsen, 2005). Galloway et al. (2008) noted that reactive nitrogen increased from approximately 15 Tg N in 1860 to 187 Tg N in 2005 due mainly from increases in energy and food production. They attributed this dramatic increase to an increase in cereal production from 1897 to 2270 million tons (20%) (these plants convert nitrogen gas to ammonia, which is incorporated in the soil organic matter or can be nitrified to nitrate) and an increase in meat production from 207 to 260 million tons (26%) (Fig. 2.3). To meet these huge agricultural demands, reactive nitrogen produced from Haber–Bosch process increased from 100 Tg N/yr in 1995 to 121 Tg N/yr in 2005 (Galloway et al., 2008). Erisman et al. (2011) estimated that the number of humans supported by Haber–Bosch produced fertilizers applied to arable land has increased from 1.9 to 4.3 persons per hectare between 1908 and 2008. Nonagricultural uses of ammonia have been more difficult to quantify, but potentially significant amounts are used in the production of plastics, resins, glues, nylon, melamine, feed supplements for various animals, and explosives. In 2005, Galloway et al. (2008) estimate that approximately 23 Tg N was

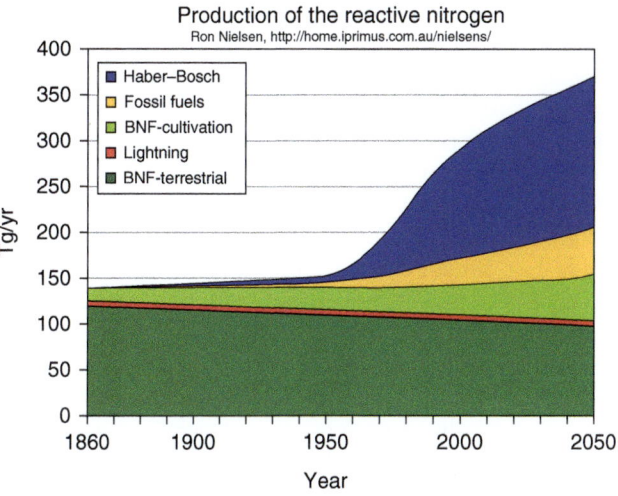

Figure 2.2 Plot showing increases in the creation of reactive nitrogen from natural (lightning) and anthropogenic sources from 1860 with projections to 2050. *Source*: From Nielsen (2005), with permission, constructed using data from Galloway et al. (2004), and Cowling et al. (2002).

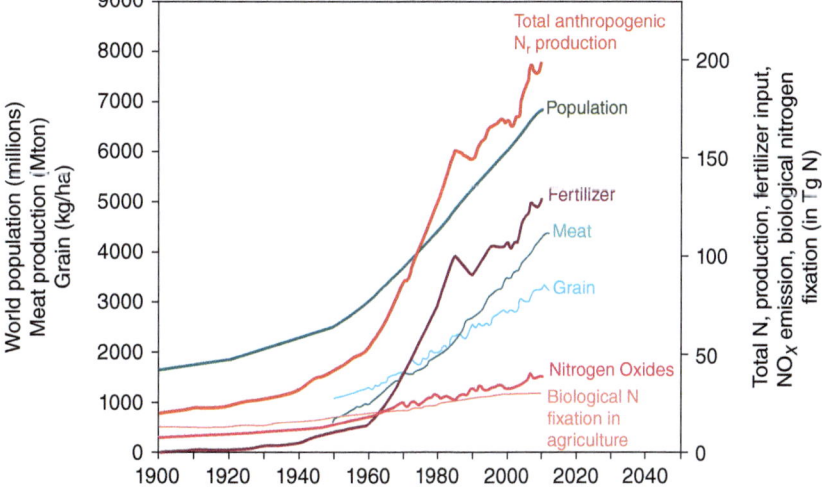

Figure 2.3 Plot showing increases in human population and increases in the creation of total anthropogenic reactive nitrogen between 1900 and 2012. *Source*: With permission from Erisman et al. (2015). Reproduced with permission of World Wildlife Fund.

produced from various industrial activities, which constitutes about 20% of the reactive nitrogen from the Haber–Bosch process.

Major changes in the global distribution and commodities transport of reactive nitrogen need to be considered when assessing anthropogenic alteration of the nitrogen cycle from one part of the world to another. Agricultural production of grains and meat may not be located in or near the areas that consume these products. Livestock practices have largely shifted to more confined animal feeding operations. These facilities are considered to be the world's largest user of land, but there has been a steady shift from grazing to the consumption of feed crops (Naylor et al., 2005). Therefore, the production areas bear the burden of environmental degradation (groundwater contamination, ammonia volatilization and particle emissions, and atmospheric deposition) and economic costs associated with the export of food to remote areas where the food is consumed.

Atmospheric deposition (dryfall and precipitation) of reactive nitrogen occurs over large parts of the world at rates much greater than natural background levels (Erisman et al., 2015). Inputs of reactive nitrogen to the atmosphere have increased due to energy production from coal, natural gas, and petroleum combustion. Although nitrogen emissions in the developed world have decreased in recent years due to legislative mandates and technological improvements, Galloway et al. (2008) estimated that approximately 25 Tg reactive N in nitrogen emissions were produced per year from 1995 to 2005. They also reported that annual average atmospheric deposition rates of reactive nitrogen in large regions of the world exceed 10 kg N/ha, and may increase to 50 kg/ha in some regions. By 2050, Galloway et al. (2008) estimated that atmospheric transport and deposition of reactive nitrogen globally will increase to 200 Tg N/yr. In addition to climate change and habitat conversion, atmospheric deposition of reactive nitrogen has contributed significantly to the loss of terrestrial biodiversity (Alkemade et al., 2009; Sala et al., 2000).

An increase in these estimates of reactive nitrogen was reported by Sutton et al. (2011). They estimated a total of 191 Tg of reactive nitrogen from the following anthropogenic annual nitrogen fixation amounts in teragrams per year: fossil fuel combustion and industry, 25; fertilizer production, 121; and biological crop fixation, 45. Fowler et al. (2015) estimated that the total anthropogenic reactive nitrogen fixed annually is 220 Tg N/yr (for 2010), which is slightly higher than the estimate of 205 Tg of reactive nitrogen per year (Erisman et al., 2015) produced from natural nitrogen fixation processes (as presented in the earlier section on nitrogen fixation in this chapter). About one third of the reactive nitrogen produced by humans originates from fossil fuel combustion and biomass burning. The remaining two thirds comes from nitrogen fixation processes for agriculture (Erisman et al., 2015) (Fig. 2.4).

2.2.1. Projected Changes in the Nitrogen Cycle During the 21st Century

Researchers are using various models to predict global changes in the nitrogen cycle during the 21st century. Fowler et al. (2015) presents a detailed review that includes current knowledge of the nitrogen cycle, modeling and sensitivity analysis of biological nitrogen fixation, emissions, atmospheric processing, and removal of reactive nitrogen compounds, along with uncertainty estimates. The total fixation of nitrogen through natural biological fixation, combustion of fossil fuels, and the Haber–Bosch processes will increase during the remainder of the 21st century. This would include an increase of 50% over values at the beginning of the century with an estimated total nitrogen fixation of 600 Tg N/yr (Table 2.2). Most of this fixed reactive nitrogen would be in the reduced form as ammonia or ammonium but could be transformed in the environment to other reactive nitrogen forms.

Overall, many of the biologically mediated processes that control the transfers of nitrogen between the atmosphere, terrestrial, and marine reservoirs are sensitive to changing climate conditions at local and global scales (such as temperature; amount, frequency, and seasonality of precipitation and humidity). Fowler et al. (2015) noted that some of these effects would include increases in fluxes of NO from soils and ammonia from vegetation with increasing temperatures. They further point out that to accurately assess the global interactions with changes in climate (along with associated land use changes) and the nitrogen cycle a coupled global climate and nitrogen-cycle model would need to be developed. However, some of these interactions have been modeled at regional scales and provide useful information about various biogeochemical interactions (Sutton et al., 2011; Zaehle, 2013). A detailed discussion of the various coupled models is included in Chapter 6.

2.2.1.1. Denitrification in Terrestrial Systems

Largely unknown are the processes that control the removal and attenuation of excess amounts of reactive nitrogen in the environment. Only about 35% of the reactive nitrogen inputs to the terrestrial biosphere were known in the mid-1990s (Galloway et al., 2008). Of this amount, they estimated that approximately 18% was exported to and denitrified in coastal ecosystems, 13% was deposited to the ocean from the marine atmosphere, and 4% was input to the atmosphere as emissions from fossil fuel combustion. The large unknown quantity of

Creation of reactive nitrogen

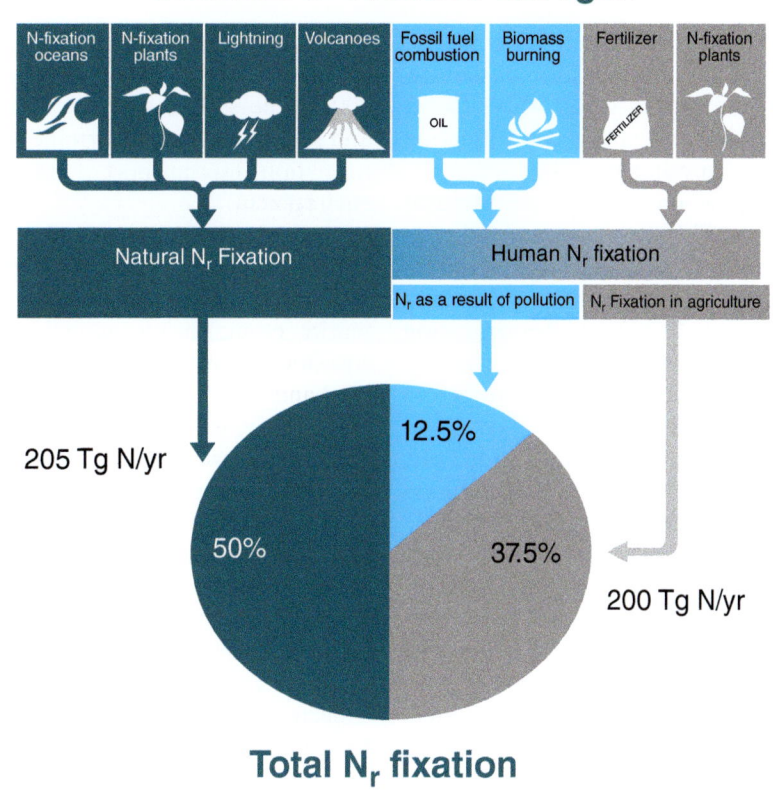

Figure 2.4 Various processes that create reactive nitrogen and the percentages of natural and human fixation of reaction nitrogen (N_r). *Source*: From Erisman et al. (2015) with permission. Reproduced with permission of World Wildlife Fund.

Table 2.2 Global terrestrial and ocean nitrogen fixation from various processes for 2010 and 2100 with uncertainty estimates along with changes in land and ocean fluxes.

Process category	2010 nitrogen fixation (Tg N/yr)	2100 nitrogen fixation (Tg N/yr)	Main driving factor(s) for change between 2010 and 2100
Biological nitrogen fixation terrestrial	128 ± 50%	170 ± 50%	Climate changes
Lightning	5 ± 50%	7 ± 50%	Climate
Fossil fuel combustion	40 ± 10%	20 ± 10%	Control technology on vehicles and industrial emissions
Fertilizer production	120 ± 10%	160 ± 10%	Increased agriculture demand
Agricultural biological N fixation	60 ± 30%	79 ± 30%	Increases in agricultural production
Ocean biological N fixation	120 ± 50%	166 ± 50%	Climate changes
Total	**473**	**602**	
Changes in land and ocean fluxes:			
Terrestrial NH_3 emissions	60	135	Climate changes, anthropogenic fixation due to increased demand for food, population increases, and diet changes
Soil NO emissions	9	11.5	
Atmospheric deposition (dry and wet)	100	120	
Terrestrial N cycling	240	328	
Marine N cycling	230	290	
Marine NH_3 emissions	5.7	1.7	Ocean acidification
Marine N_2O emissions	5.5	8	
Terrestrial N_2O emissions	1.2	14.4	

Source: Data from Fowler et al. (2015).

excess reactive nitrogen (approximately 65%) could be stored in soils, vegetation, and groundwater, or denitrified to N_2O or ultimately to nitrogen gas. More recent studies have attempted to better quantify denitrification rates in terrestrial and marine systems.

Quantifying denitrification on a global or regional scale has been problematic. There is a high variability in denitrification activity both temporally and spatially. Also, small areas (hotspots) and brief periods (referred to as hot moments) (McClain et al., 2003) of denitrification frequently account for a high percentage of N gas flux activity (Groffman, 2012). Furthermore, the many factors that control denitrification (microorganism types and abundance, oxygen, nitrate concentrations, carbon, pH, salinity, and temperature) are site-specific, which makes regional or global upscaling challenging.

Bouwman et al. (2013) developed a global, distributed flow-path model for terrestrial systems using soil nitrogen budgets to investigate denitrification and N_2O emissions from soils, groundwater, and riparian zones for the period 1900–2000 and for predictive scenarios for the period 2000–2050 based on the Millennium Ecosystem Assessment (https://www.millenniumassessment.org/en/index.html, accessed 30 March 2018). Model results indicated that total agricultural and natural N inputs from N fertilizers, animal manure, biological N_2 fixation and atmospheric N deposition increased from 155 to 345 Tg N/yr between 1900 and 2000, and inputs were estimated to increase to 408–510 Tg N/yr by 2050. N_2 production from denitrification increased from 52 to 96 Tg/yr between 1900 and 2000, and N_2O-N emissions from 10 to 12 Tg N/yr. Multiple scenarios predicted a future increase to 142 Tg N_2-N and 16 Tg N_2O-N/yr by 2050. Model results indicated that riparian buffer zones are an important source of N_2O from denitrification and contributed an estimated 0.9 Tg N_2O-N/yr in 2000. Groffman (2012) presents some new specific opportunities for making progress on quantifying denitrification on regional and global scales including novel measurement methods, new conceptual approaches for addressing hotspot and hot moment dynamics, and new remote sensing and geographic information system-based scaling methods.

Human management of soils and other terrestrial ecosystems has also affected ecosystem nitrogen cycling. For example, important factors to consider are drainage modification, fertilization, irrigation, tillage practices, amendment of soils with manure or lime, crop rotations, soil compaction or subsidence, animal grazing, climate changes, and local and regional atmospheric deposition of nitrogen species (Butterbach-Bahl et al., 2011). Land use change is another critical factor that affects the cycling of reactive nitrogen at site-specific and regional scales. For example, conversion of forest land to agriculture and associated increases in fertilization and manure spreading have resulted in substantial changes at regional and global scales (Allen et al., 2015; Galloway et al., 2003). These factors need to be evaluated on temporal and spatial scales to determine their effects on the transformation and transport of reactive nitrogen forms in the environment.

2.2.1.2. Denitrification in Marine Systems

DeVries et al. (2012, 2013) estimated denitrification in marine systems using a global ocean circulation model to simulate the distribution of N_2 gas produced by denitrifying bacteria in the three main suboxic zones in the open ocean. They inferred a globally integrated rate of water-column denitrification of 66 ± 6 Tg N/yr by fitting the model to measured N_2 gas concentrations. Furthermore, DeVries et al. (2012) estimated that the global rate of combined nitrogen loss from marine sediments and the oceanic water column amounted to approximately 230 ± 60 Tg N/yr. However, their findings of a net loss of approximately 20 ± 70 Tg of nitrogen from the global ocean each year would essentially match what would be expected from a balanced nitrogen budget (accounting for inputs from rivers and atmospheric deposition). This finding led to their conclusion that the marine N cycle is governed by strong regulatory feedback mechanisms.

Eugster and Gruber (2012) also estimated the global rates of marine N-fixation and denitrification and their associated uncertainties by combining marine geochemical and physical data with a new two-dimensional box model that separated the Atlantic from the IndoPacific basins. Their findings indicate a median water-column denitrification rate of 52 Tg N/yr (39–66 Tg N/yr, corresponding to the 5th–95th percentile) and a median benthic denitrification rate of 93 Tg N/yr (68–122 Tg N/yr). Using two different approaches, Eugster and Gruber (2012) estimated a global N-fixation rate of between 94 and 175 Tg N/yr, with a best estimate of 131 and 134 Tg N/yr, respectively. Their N-fixation and denitrification estimates along with recent estimates of atmospheric deposition and riverine input resulted in a preindustrial marine nitrogen cycle that is balanced to within 3 Tg N/yr (−38 to 40 Tg N/yr). Their marine nitrogen budget suggests a median residence time for fixed N of 4200 year (3500–5000 year).

2.2.1.3. Changes in Land Use and Management Practices

Other models are being used to investigate how land use changes and human management practices have significantly affected the cycling of reactive nitrogen. Some examples include changes in land use from the conversion of forest and grassland to row crop agriculture, which has resulted in significant loading of reactive nitrogen above baseline levels in the Mississippi and Susquehanna watersheds in the United States (Van Meter et al., 2016). Large nitrogen loads are transported from the Mississippi River into the Gulf of Mexico and from the Susquehanna River into Chesapeake Bay. Their process-based model estimated that soil legacy nitrogen loads (older than 10 years) contributed approximately 55% of the current annual nitrogen load in the Mississippi River Basin, and groundwater legacy nitrogen loads (older than 10 years) contributed 18% of the current annual nitrogen load in the Susquehanna River basin (Van Meter et al., 2016). The conversion of forest to agricultural soils in a humid subtropical region in China has resulted in a shift from fungi-dominated heterotrophic nitrification to bacteria-dominated autotrophic nitrification in upland fields planted with corn and fungi-dominated heterotrophic nitrification in rice paddy areas (Wang et al., 2015).

Joo et al. (2013) present a steady-state global nitrogen cycle model constructed for the pre-Anthropocene period with estimates for nitrogen fluxes from 1700 to 2050. The model estimates nitrogen masses using the C:N ratio from Terrestrial Ocean aTmosphere Ecosystem Model (TOTEM) I and II for 18 reservoirs for nitrogen, including biotic, dissolved organic and inorganic nitrogen, atmosphere, and particulate organic and inorganic nitrogen reservoirs on land and in oceans. The model is constrained using a nitrogen isotopic mass balance. Land use changes have resulted in major changes in the sizes of reactive nitrogen reservoirs and mass fluxes on land mainly due to inorganic fertilizer inputs. The model also indicated that leaching of reactive nitrogen into the ocean is affecting global ocean nitrogen productivity and the associated increase in denitrification and sediment burial. The studies cautioned that changes in ocean chemistry may have serious biological and environmental issues related to eutrophication.

REFERENCES

Alkemade, R., Van Oorschot, M., Miles, L., Nellemann, C., Bakkenes, M., & Ten Brink, B. (2009). GLOBIO$_3$: A framework to investigate options for reducing global terrestrial biodiversity loss. *Ecosystems, 12*, 374–390.

Allen, K., Corre, M., Tjoa, A., & Veldkamp, E. (2015). Soil nitrogen-cycling responses to conversion of lowland forests to oil palm and rubber plantations in Sumatra, Indonesia. *PLoS One, 10*(7), e0133325. https://doi.org/10.1371/journal.pone.0133325

American Water Works Association (AWWA). (2002). Nitrification. Prepared for the U.S. Environmental Protection Agency, Office of Ground Water and Drinking Water, 17 p.

Bachand, P. A. M., & Horne, A. G. (1999). Denitrification in free-water surface wetlands. I. Very high nitrate removal rates in a macrocosm study. *Ecological Engineering, 14*, 9–15.

Benner, J. W., Conroy, S., Lunch, C. K., Toyoda, N., & Vitousek, P. M. (2007). Phosphorus fertilization increases the abundance and nitrogenase activity of the cyanolichen *Pseudocyphellaria crocata* in Hawaiian montane forests. *Biotropica, 39*, 400–405. https://doi.org/10.1111/j.1744-7429.2007.00267.x

Bernhard, A. (2010). The nitrogen cycle: Processes, players, and human impact. *Nature Education Knowledge, 3*(10), 25.

Booth, M. S., Stark, J. M., & Rastetter, E. (2005). Controls on nitrogen cycling in terrestrial ecosystems: A synthetic analysis of literature data. *Ecological Monographs, 75*, 139–157.

Bouwman, A. F., Beusen, A. H. W., Griffioen, J., Van Groenigen, J. W., Hefting, M. M., Oenema, O., et al. (2013). Global trends and uncertainties in terrestrial denitrification and N_2O emissions. *Philosophical Transactions Royal Society of London Biological Sciences, 368*(1621), 20130112. https://doi.org/10.1098/rstb.2013.0112

Bruce, T., Martinez, I. B., Maia Neto, O., Vicente, A. C. P., Kruger, R. H., & Thompson, F. L. (2010). Bacterial community diversity in the Brazilian Atlantic forest soils. *Microbial Ecology, 60*, 840–849. https://doi.org/10.1007/s00248-010-9750-2

Burgin, A. J., & Hamilton, S. K. (2007). Have we overemphasized the role of denitrification in aquatic ecosystems? A review of nitrate removal pathways. *Frontiers in Ecology and the Environment, 5*, 89–96.

Butterbach-Bahl, K., Gundersen, P. Ambus, P., Augustin, J., Beier, C., Boeckx, P. et al. (2011). Nitrogen processes in terrestrial ecosystems. In M. A. Sutton, C. M. Howard, J. W. Erisman, G. Billen, A. Bleeker, P. Grennfelt, H. van Grinsven, & B. Grizzetti (Eds.), *The European nitrogen assessment: Sources, effects, and policy perspectives*. Cambridge, UK: Cambridge University Press, 664 p.

Carey, R. O., & Migliaccio, K. W. (2009). Contribution of wastewater treatment plant effluents to nutrient dynamics in aquatic systems: A review. *Environmental Management, 44*, 205–217. https://doi.org/10.1007/s00267-009-9309-5

Carlsson, G., & Huss-Danell, K. (2003). Nitrogen fixation in perennial forage legumes in the field. *Plant and Soil, 253*, 353–372.

Cassman, K.G., Doberman, A.R., and Walters, D.T., 2002, Agroecosystems, nitrogen-use efficiency, and nitrogen management. Ambio, v. *31*, no. 2, p. 132–140. Retrieved from http://www.jstor.org/stable/4315226, accessed 19 September 2019.

Cleveland, C. C., Townsend, A. R., Schimel, D. S., Fisher, H., Howarth, R. W., Hedin, L. O., et al. (1999). Global patterns of terrestrial biological nitrogen (N_2) fixation in natural ecosystems. *Global Biogeochemical Cycles, 13*, 623–645.

Cowling, E., Galloway, J., Furiness, C., & Erisman, J. W. (Eds). (2002). Optimizing nitrogen management in food and energy production and environmental protection. In *Report from the Second International Nitrogen Conference*, Potomac, MA, 14–18 October 2001. The Ecological Society of America. Washington, DC: A.A. Balkema and The ScientificWorld. ISBN: 9026519273, 1033 p.

Dessureault-Rompre, J., Zebarth, B. J., Georgallas, A., Burton, D. L., Grant, C. A., & Drury, C. F. (2010). Temperature dependence of soil nitrogen mineralization rate: Comparison of mathematical models, reference temperatures and origin of the soils. *Geoderma, 157*(3–4), 97–108. https://doi.org/10.1016/j.geoderma.2010.04.001

DeVries, T., Deutsch, C., Primeau, R., Chang, B., & Devol, A. (2012). Global rates of water-column denitrification derived from nitrogen gas measurements. *Nature Geoscience*, 5, 547–550. https://doi.org/10.1038/ngeo1515

DeVries, T., Deutsch, C., Rafter, P. A., & Primeau, F. (2013). Marine denitrification rates determined from a global 3-D inverse model. *Biogeosciences*, 10, 2481–2496. https://doi.org/10.5194/bg-10-2481-2013

Dubrovsky, N.M., Burow, K.R., Clark, G.M., Gronberg, J.M., Hamilton P.A., Hitt, K.J., et al. (2010). *The quality of our Nation's waters—Nutrients in the Nation's streams and groundwater, 1992–2004*. U.S. Geological Survey Circular 1350, 174 p. Additional information about this study is available at http://water.usgs.gov/nawqa/nutrients/pubs/circ1350, accessed 19 September 2019.

Erisman, J. W., Galloway, J. N., Dice, N. B., Sutton, M. A., Bleeker, A., Grizzetti, B., Leach, A. M., & de Vries, W. (2015). *Nitrogen, too much of a vital resource*. Science Brief. Zeist, The Netherlands: World Wildlife Fund.

Erisman, J. W., van Grinsven, H., Grizzetti, B. Bouraoui, F., Powlson, D., Sutton, M. A., Bleeker, A., & Reis, S. (2011). The European nitrogen problem in a global perspective. Ch. 2. In M.A. Sutton, C.M. Howard, J.W. Erisman Billen, G., Bleeker, A., Grennfelt, P., H. van Grinsven, & B. Grizzetti (Eds.), *The European nitrogen assessment: Sources, effects, and policy perspectives* (pp. 9–31). Cambridge, UK: Cambridge University Press.

Eugster, O., & Gruber, N. (2012). A probabilistic estimate of global marine N-fixation and denitrification. *Global Biogeochemical Cycles*, 26(4), 4013.

Fowler, D., Coyle, M., Skiba, U., Sutton, M. A., Cape, J. N., Reis, S., et al. (2013). The global nitrogen cycle in the twenty-first century. *Philophical Transactions of the Royal Society Biological Sciences*, 368, 1621. https://doi.org/10.1098/rstb.2013.0164

Fowler, D., Steadman, C. E., Stevenson, D., Coyle, M., Rees, R. M., Skiba, U. M., et al. (2015). Effects of global change during the 21st century on the nitrogen cycle. *Atmospheric Chemistry and Physics*, 15, 13849–13893. https://doi.org/10.5194/acp-15-13849-2015

Galloway, J. N., Aber, J. D., Erisman, J. W., Seitzinger, S. P., Howarth, R. W., Cowling, E. B., & Cosby, B. J. (2003). The nitrogen cascade. *BioScience*, 53, 341–356.

Galloway, J. N., Dentener, F. J., Capone, D. G., Boyer, E. W., Howarth, R. W., Seitzinger, S. P., et al. (2004). Nitrogen cycles: Past, present and future. *Biogeochemistry*, 70, 153–226.

Galloway, J. N., Townsend, A. R., Erisman, J. W., Bekunda, M., Cai, Z., Freney, J. R., et al. (2008). Transformation of the nitrogen cycle: Recent trends, questions, and potential solutions. *Science*, 320, 889–892.

Gardner, W. S., McCarthy, M. J., An, S., Sobolev, D., Sell, K. S., & Brock, D. (2006). Nitrogen fixation and dissimilatory nitrate reduction to ammonium (DNRA) support nitrogen dynamics in Texas estuaries. *Limnology and Oceanography*, 51(1, part 2), 558–568.

Green, C. T., Puckett, L. J., Bohlke, J. K., Bekins, B. A., Phillips, S. P., Kauffman, L. J., Denver, J. M., & Johnson, H. M. (2008). Limited occurrence of denitrification in four shallow aquifers in agricultural areas of the United States. *Journal of Environmental Quality*, 37, 994–1009. doi:10.2134/jeq2006.0419.

Groffman, P. M. (2012). Terrestrial denitrification: Challenges and opportunities. *Ecological Processes*, 1, 11. https://doi.org/10.1186/2192-1709-1-11

Groffman, P. M., & Crawford, M. K. (2003). Denitrification potential in urban riparian zones. *Journal of Environmental Quality*, 32(3), 1144–1149.

Hansen, A. T., Dolph, C. L., & Finlay, J. C. (2016). Do wetlands enhance downstream denitrification in agricultural landscapes? *Ecosphere*, 7(10), e01516. https://doi.org/10.1002/ecs2.1516

Harrison, K. A., Bol, R., & Bardgett, R. D. (2007). Preferences for different nitrogen forms by coexisting plant species and soil microbes. *Ecology*, 88, 989–999.

Harvey, J. W., Böhlke, J. K., Voytek, M. A., Scott, D., & Tobias, C. R. (2013). Hyporheic zone denitrification: Controls on effective reaction depth and contribution to whole-stream mass balance. *Water Resources Research*, 49, 6298–6316. https://doi.org/10.1002/wrcr.20492

Heffernan, J. B., Albertin, A. R., Fork, M. L., Katz, B. G., & Cohen, M. J. (2012). Denitrification and inference of ntirgen sources in the karstic Floridan aquifer. *Biogeosciences*, 9, 1671–1690.

Huub, J. M., den Camp, O., Mike, S. M., & Jetten, M. S. (2007). Anammox, Ch. 16. In H. Bothe, S. J. Ferguson, & W. E. Newton (Eds.), *Biology of the nitrogen cycle* (pp. 245–262). Amsterdam: Elsevier B.V.

Jackson, L. E., Burger, M., & Cavagnaro, T. R. (2008). Roots, nitrogen transformations and ecosystem services. *Plant Biology*, 59, 341–363.

Johnson, C., Albrecht, G., Ketterings, Q., Beckman, J., & Stockin, K. (2005). *Nitrogen basics: The nitrogen cycle*. Cornell University Cooperative Extension, 2 p.

Joo, Y. J., Li, D. D., & Lerman, A. (2013). Global nitrogen cycle: Pre-Anthropocene mass and isotope fluxes and the effects of human perturbations. *Aquatic Geochemistry*, 19, 477–500. https://doi.org/10.1007/s10498-013-9211-x

King, S. E., Osmond, D. L., Smith, J., Burchell, M. R., Dukes, M., Evans, R. O., Knies, S., & Kunickis, S. (2016). Effects of riparian buffer vegetation and width: A 12-year longitudinal study. *Journal of Environmental Quality*, 45(4), 1243–1251. doi:10.2134/jeq2015.06.0321.

Koenneke, M., Bernhard, A. E., de la Torre, J. R., Walker, C. B., Waterbury, J. B., & Stahl, D. A. (2005). Isolation of an autotrophic ammonia-oxidizing marine archaeon. *Nature*, 437, 543–546.

Korom, S. F. (1992). Natural denitrification in the saturated zone: A review. *Water Resources Research*, 28(6), 1657–1668. https://doi.org/10.1029/92WR00252

Kuypers, M. M. M., Marchant, H. K., & Kartal, B. (2018). The microbial nitrogen-cycling network. *Nature Reviews Microbiology*, 16, 263–276. https://doi.org/10.1038/nrmicro.2018.9

Lowrance, R. (1992). Ground water nitrate and denitrification in a Coastal Plain riparian forest. *Journal of Environmental Quality*, 21(3). https://doi.org/10.2134/jeq1992.0047242500210003017x

McClain, M. E., Boyer, E. W., Dent, C. L., Gergel, S. E., Grimm, N. B., Groffman, P. M., et al. (2003). Biogeochemical hot spots and hot moments at the interface of terrestrial and aquatic ecosystems. *Ecosystems*, 6(4), 301–312. https://doi.org/10.1007/s10021-003-0161-9

Naylor, R., Steinfeld, H., Falcon, W., Galloway, J., Smil, V., Bradford, E., Alder, J., & Mooney, H. (2005). Losing the links between livestock and land. *Science, 310*, 1621–1622.

Nielsen, R. (2005). *Can we feed the world? Is there a nitrogen limit of food production?* Retrieved from http://home.iprimus.com.au/nielsens/nitrogen.html, accessed 21 March 2018.

Pajares, S., & Bohannan, B. J. M. (2016). Ecology of nitrogen fixing, nitrifying, and denitrifying microorganisms in tropical forest soils. *Frontiers in Microbiology, 7*, 1045. https://doi.org/10.3389/fmicb.2016.01045

Pregitzer, K. S., Burton, A. J., Zak, D. R., & Talhelm, A. F. (2008). Simulated chronic nitrogen deposition increases carbon storage in northern temperate forests. *Global Change Biology, 14*, 142–157.

Ranjan, K., Paula, F. S., Mueller, R. C., Jesus, E. D. C., Cenciani, K., Bohannan, B. J. M., et al. (2015). Forest-to-pasture conversion increases the diversity of the phylum Verrucomicrobia in Amazon rainforest soils. *Frontiers in Microbiology, 6*, 779. https://doi.org/10.3389/fmicb.2015.00779

Rivett, M. O., Buss, S. R., Morgan, P., Smith, J. W., & Bemment, C. D. (2008). Nitrate attenuation in groundwater: A review of biogeochemical controlling processes. *Water Research, 42*(16), 4215–4232. https://doi.org/10.1016/j.watres.2008.07.020. Epub 23 Jul 2008

Robertson, G. P., Paul, E. A., & Harwood, R. R. (2000). Greenhouse gases in intensive agriculture: Contributions of individual gases to the radiative forcing of the atmosphere. *Science, 289*, 1922–1925. https://doi.org/10.1126/science.289.5486.1922

Sala, O. E., Chapin, F. S., III, Armesto, J. J., Berlow, E., Bloomfield, J., Dirzo, R., et al. (2000). Global biodiversity scenarios for the year 2100. *Science, 287*, 1770–1774.

Seitzinger, S., Harrison, J. A., Bohlke, J. K., Bouwman, A. F., Lowrance, R., Peterson, B., Tobias, C., & Van Drecht, G. (2006). Denitrification across landscapes and waterscapes: A synthesis. *Ecological Applications, 16*, 2064–2090.

Silver, W. L., Herman, D. J., & Firestone, M. K. (2001). Dissimilatory nitrate reduction to ammonium in upland tropical forest soils. *Ecology, 82*, 2410–2416.

Skiba, U., Smith, K. A., & Fowler, D. (1993). Nitrification and denitrification as sources of nitric oxide and nitrous oxide in a sandy loam soil. *Soil Biology and Biochemistry, 25*(11), 1527–1536.

Smil, V. (1999). Nitrogen in crop production: An account of global flows. *Global Biogeochemical Cycles, 13*, 647–662.

Stein, L. Y., & Klotz, M. G. (2016). The nitrogen cycle. *Current Biology, 26*(3), R93–R98. https://doi.org/10.1016/j.cub.2015.12.021

Strous, M., Fuerst, J. A., Kramer, E. H. M., Logemann, S., Muyzer, G., van de Pas-Schoonen, K. T., et al. (1999). Missing lithotroph identified as new planctomycete. *Nature, 400*, 446–449.

Sutton, M. A., Howard, C. M., Erisman, J. W., Billen, G., Bleeker, A., Grennfelt, P., van Grinsven, H., & Grizzetti, B. (2011). *The European nitrogen assessment: Sources, effects, and policy perspectives.* Cambridge, UK: Cambridge University Press, 664 p.

Takaya, N., Catalan-Sakairi, M. A. B., Sakaguchi, Y., Kato, I., Zhou, Z., & Soun, H. (2003). Aerobic denitrifying bacteria that produce low levels of nitrous oxide. *Applied Environmental Microbiology, 69*(6), 3152–3157.

Temple, S. J., Vance, C. P., & Gantt, J. S. (1998). Glutamate synthase and nitrogen assimilation. *Trends in Plant Science, 3*(2), 51–56.

Torrento, C., Cama, J., Urmeneta, J., Otero, N., & Soler, A. (2010). Denitrification of groundwater with pyrite and *Thiobacillus denitrificans*. *Chemical Geology, 278*, 80–91. https://doi.org/10.1016/j.chemgeo.2010.09.003

Torrento, C., Urmeneta, J., Otero, N., Soler, A., Vinas, M., & Cama, J. (2011). Enhanced denitrification in groundwater and sediments from a nitrate-contaminated aquifer after addition of pyrite. *Chemical Geology, 287*, 90–101. https://doi.org/10.1016/j.chemgeo.2011.06.002

Townsend, A. R., Cleveland, C. C., Houlton, B. Z., Alden, C. B., & White, J. W. C. (2011). Multi-element regulation of the tropical forest carbon cycle. *Frontiers in Ecology and the Environment, 9*, 9–17.

Ulloa, O., Canfield, D. E., DeLong, E. F., Letelier, R. M., & Stewart, F. J. (2012). Microbial oceanography of anoxic oxygen minimum zones. *Proceedings of the National Academy of Sciences, 109*(40), 15996–16003. Retrieved from www.pnas.org/cgi/doi/10.1073/pnas.1205009109

Urakawa, R., Ohte, N., Shibata, H., Isobe, K., Tateno, R., Oda, T., et al. (2016). Factors contributing to soil nitrogen mineralization and nitrification rates of forest soils in the Japanese archipelago. *Forest Ecology and Management, 361*, 382–396.

Van Meter, K. J., Basu, N. B., & Van Cappellen, P. (2016). Two centuries of nitrogen dynamics: Legacy sources and sinks in the Mississippi and Susquehanna River basins. *Global Biogeochemical Cycles, 31*, 2–23.

Wang, J., Liu, Q., Zhang, J., & Cai, Z. (2015). Conversion of forest to agricultural land affects the relative contribution of bacteria and fungi to nitrification in humid subtropical soils. *Acta Agriculturae Scandinavica, Section B-Soil and Plant Science, 65*(1). https://doi.org/10.1080/09064710.2014.967714

Watson, S. W., Valois, F. W., & Waterbury, J. B. (1981). The family nitrobacteraceae. In M. P. Starr, H. Stolp, H. G. Trüper, A. Balows, & H. G. Schlegel (Eds.), *The prokaryotes* (pp. 1005–1022). Berlin: Springer-Verlag.

Zaehle, S. (2013). Terrestrial nitrogen-carbon cycle interactions at the global scale. *Philosophical Transactions of the Royal Society Biological Sciences, 368*, 1621. https://doi.org/10.1098/rstb2013.0125

Zaehle, S., Ciais, P., Friend, A. D., & Prieur, V. (2011). Carbon benefits of anthropogenic reactive nitrogen offset by nitrous oxide emissions. *Nature Geoscience, 4*, 601–605. https://doi.org/10.1038/ngeo1207

FURTHER READING

Barton, L., McLay, C. D. A., Schipper, L. A., & Smith, C. T. (1999). Annual denitrification rates in agricultural and forest soils: A review. *Australian Journal of Soil Research, 37*, 1073–1093.

Billen, G., Garnier, J., & Lassaletta, L. (2013). The nitrogen cascade from agricultural soils to the sea: Modelling nitrogen transfers at regional watershed and global scales. *Philosophical Transactions of the Royal Society Biology Sciences, 368*, 20130123. https://doi.org/10.1098/rstb.2013.0123

Bothe, H., Ferguson, S. J., & Newton, W. E. (Eds.). (2007). *Biology of the nitrogen cycle*. Amsterdam: Elsevier B.V., 427 p.

Canfield, D. E., Glazer, A. N., & Falkowski, P. G. (2010). The evolution and future of Earth's nitrogen cycle. *Science, 330*, 192–196.

Chen, B., Liu, E., Tian, Q., Yan, C., Zhang, Y. 2014. Soil nitrogen dynamics and crop residues. A review. Agronomy for Sustainable Development, *34* (2): 429–442. Retrieved from https://hal.archives-ouvertes.fr/hal-01234828/document, accessed 19 September 2019.

Dun, Y., Tang, C., & Shen, Y. (2013). Identifying interactions between river water and groundwater in the North China Plain using multiple tracers. *Environmental Earth Sciences*. https://doi.org/10.1007/s12665-013-2989-4

Gruber, N. (2008). The marine nitrogen cycle: Overview and challenges. *Nitrogen Marine Environment, 2*, 1–50.

Hofstra, N., & Bouwman, A. F. (2005). Denitrification in agricultural soils: Summarizing published data and estimating global annual rates. *Nutrient Cycling in Agroecosystems, 72*, 267–278. https://doi.org/10.1007/s10705-005-3109-y

Munn, M.D., Frey, J.W., Tesoriero, A.J., Black, R.W., Duff, J.H., Lee, K., et al. (2018). *Understanding the influence of nutrients on stream ecosystems in agricultural landscapes*. U.S. Geological Survey Circular 1437, 80 p. https://doi.org/10.3133/cir1437.

Ward, B. (2012). The global nitrogen cycle. In A. H. Knoll, D. E. Canfield, & K. O. Konhauser (Eds.), *Fundamentals of geobiology* (pp. 36–48). Blackwell Publishing Ltd.

3
Sources of Reactive Nitrogen and Transport Processes

In Chapter 2, we saw how the global nitrogen cycle has been substantially altered by human inputs of reactive nitrogen from synthetic fertilizers, fossil fuel combustion, and biomass burning, and the increased cultivation of biological nitrogen fixing crops. However, in localized and regional areas, other sources of reactive nitrogen can have adverse impacts on ecosystems, human health, and economies. In this chapter, we will look more closely at the many sources that contribute reactive nitrogen in the environment. The many sources of reactive nitrogen can be grouped into three main categories: naturally occurring sources, anthropogenic point sources, and anthropogenic nonpoint sources.

3.1. NATURAL SOURCES

Natural sources of reactive nitrogen include lightning (which fixes atmospheric nitrogen into reactive nitrogen forms) and ammonia from volcanic eruptions. Another naturally occurring source of nitrogen may be derived from the weathering certain rock types that contain nitrogen. Allègre et al. (2001) estimated that the Earth's bulk rock nitrogen concentration is 1.27 (±1) mg N/kg, with roughly 20% of the global nitrogen pool tied up in rock (Schlesinger, 1997). This stored nitrogen can be released due to rain, wind, and subsequent erosion (Fig. 3.1). Some studies suggest that geologic nitrogen could be an important nitrate source to aquifers and surface waters (Holloway et al., 1999; Holloway & Dahlgren, 2002; Montross et al., 2013; Morford et al., 2011). Nitrogen is incorporated in the lithosphere by the retention of organic matter (such as kerogen) in sediment and volatilization from the mantle through volcanism. As a result nitrogen occurs in sedimentary, metamorphic, and igneous rocks in concentrations ranging from less than 200 mg N/kg to more than 1000 mg N/kg in some sedimentary and meta-sedimentary rocks. Holloway and Dahlgren (2002) discussed the role of nitrogen in different rock types and how this source could be a large and important reactive pool of nitrogen for augmenting human impacts on nitrogen cycling. An additional pool of reactive and leachable nitrogen could originate from the accumulation of wet and dry atmospheric deposition in arid to semiarid environments. Release of reactive nitrogen in some areas from rock weathering corresponds to nitrogen fluxes of 4–37 kg N/ha/yr. Holloway and Dahlgren (2002) also reported that a paired watershed study found that nitrate-N fluxes in streamwater were greater than 10 kg N/ha/yr (maximum fluxes approximately 20 kg N/ha/yr) in watersheds containing geologic nitrogen compared to values of less than 2 kg N/ha/yr for watersheds containing little or no geologic nitrogen.

Biological nitrogen fixation (BNF) by plants is an important source of naturally occurring reactive nitrogen. Ward (2012) lists the following most important fixers in the environment: endosymbionts with leguminous plants; cyanobacteria including filamentous types in the ocean, microbial mats, and lakes; anaerobes (methanogens, sulfate reducers, and fermenters; and free-living soil bacteria (e.g., Azotobacter). Prior to industrialization, BNF was estimated at approximately 260 Tg N/yr, which would have been balanced by denitrification processes that returned dinitrogen gas to the atmosphere (Fowler et al., 2015). In 2010, the global inputs of reactive nitrogen from BNF are 258 Tg N/yr, which would be the largest single global input of reactive nitrogen (Fowler et al., 2013). Galloway et al. (2004) estimated a global BNF value of 128 Tg N/yr for natural terrestrial ecosystems. Fowler et al. (2015) estimated that this value for 2010 would increase to 170 ± 50 Tg N/yr in 2100. They also predicted that BNF in oceans would increase from 120 ± 50 to 166 Tg N/yr in 2100. Cultivation of naturally N-fixing crops has replaced mixed vegetation in natural systems, which has greatly increased the total biological N fixation by 32–53 Tg N/yr from agriculture (Galloway et al., 2004; Ward, 2012). The total biological N fixation has increased, even though the natural rate has declined.

Nitrogen Overload: Environmental Degradation, Ramifications, and Economic Costs, Geophysical Monograph 250,
First Edition. Brian G. Katz.
© 2020 American Geophysical Union. Published 2020 by John Wiley & Sons, Inc.

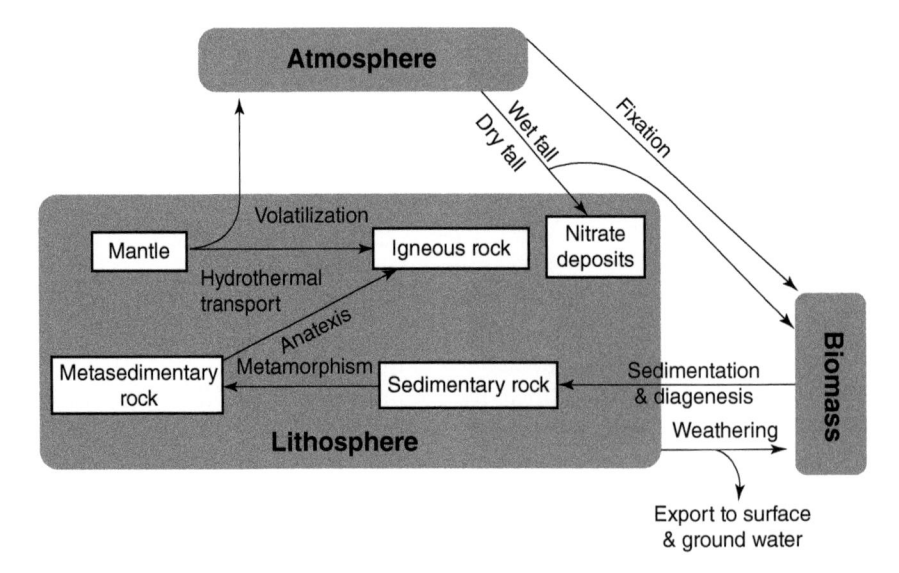

Figure 3.1 Nitrogen cycling in a geologic context including a summary of the processes involved in the transfer of nitrogen between major global pools. *Source*: From Holloway and Dahlgren (2002). Reproduced with permission of John Wiley & Sons.

3.2. ANTHROPOGENIC SOURCES

By far, anthropogenic point and nonpoint sources of reactive nitrogen in developed areas around the globe substantially overwhelm reactive nitrogen inputs from weathering of rocks. It is estimated that about 75% of anthropogenic production of reactive nitrogen results from industrial nitrogen fixation from the Haber–Bosch and other processes, and 25% from fossil fuel and biomass burning (in the form of nitrogen oxides, NO_x) (Erisman et al., 2015; Fowler et al., 2013; Galloway et al., 2008). The total amount of reactive nitrogen production in agriculture is more than double the preindustrial natural amount produced in terrestrial ecosystems.

3.2.1. Point Sources

Point sources of reactive nitrogen include discharges from wastewater treatment plants, land application of biosolids, landfill leakage, mining operations, industrial processes, and wastes from concentrated animal feeding operations.

3.2.1.1. Wastewater Treatment Facilities

Wastewater treatment facilities (WWTFs) discharge large amounts of treated effluent containing different forms of nitrogen to waterways and to groundwater through the use of sprayfields, rapid infiltration basins, reclaimed water irrigation, and biosolids deposited on the land surface. Wastewater treatment plants vary in their capability to remove nitrogen from their discharge effluent depending on the wastewater treatment method. Enhanced treatment systems produce discharge effluent that contains less nitrogen than plants using conventional treatment methods.

In the United States, people produce about 18 million tons of solid waste (feces) annually and municipal wastewater treatment plants in the United States treat about 34 billion gallons of wastewater per day (U.S. EPA, 2008). The American Society of Civil Engineers (ASCE, 2017) rates infrastructure in the United States and recently gave a D+ grade for wastewater treatment plants and public health. They reported that 76% of the U.S. population relies on the current number of 14,748 wastewater treatment plants for wastewater sanitation, and it is expected that more than 56 million new users will be connected to centralized treatment systems over the next two decades. This would require an estimated $271 billion to meet current and future demands. In the United States, there are more than 800,000 miles of public sewer lines and 500,000 miles of private lateral sewer lines. Each of these conveyance systems is susceptible to structural failure, blockages, and overflows. In fact, the U.S. Environmental Protection Agency (EPA) estimates that at least 23,000–75,000 sanitary sewer overflow events occur in the United States each year. It is estimated 532 new systems will need to be constructed by 2032 to meet future treatment needs. Wastewater treatment plants typically are located at the outlet of watersheds or near coastal and riverine areas. These areas are highly vulnerable to extreme flooding events and sea level rise. For example, during Hurricane Sandy in 2012, several wastewater treatment plants in New York and New Jersey were inundated with storm surge, and hundreds of millions of gallons of untreated sewage spilled into neighboring waterways. Many of these plants and others across the United States located in vulnerable areas have increased infrastructure fortification to protect against floods and storm surge.

In approximately 772 communities in the United States, wastewater and stormwater drain into the same treatment system (referred to as combined sewer overflows, CSOs). CSOs can experience capacity issues following heavy rain events, resulting in overflows containing stormwater as well as untreated human and industrial waste and other pollutants. After nonpoint source pollution (e.g., agricultural runoff and stormwater), CSOs are a leading source of water pollution in the United States (ASCE, 2017). Unfortunately, only limited data are available on CSOs and stormwater infrastructure. However, in 2016, EPA released a report to Congress on CSOs in the Great Lakes region. For the 184 CSO communities that discharge CSOs in the Great Lakes Basin, there were 1482 CSO events in 2014, that discharged an estimated 22 billion gallons of untreated wastewater into the Great Lakes Basin.

Maupin and Ivahnenko (2011) assembled wastewater nutrient data from about 118,250 municipal and industrial facilities (both active and inactive) for 45 states and the District of Columbia in seven major river basins in the United States (Fig. 3.2). They calculated annual point-source nitrogen loads for more than 26,600 facilities out of the total population of 118,250 for at least one of the 3 years of interest (1992, 1997, and 2002). Annual point-source nutrient loads for 2002 were calculated for almost 66% of major and 14% of minor facilities (Table 3.1). Densely populated regions in the United States or areas with large concentrations of industrial and municipal facilities produced the largest point-source nitrogen loads (total mass and on a mass per area basis), such as the New England and Mid-Atlantic, and the Upper Mississippi basin. Sewage systems contributed the greatest proportion of point-source nutrient loads overall compared to industrial sources, ranging from 51% for TN in the Missouri (MRB4) to 86% in the New England and Mid-Atlantic (MRB1).

A new approach has been developed by the U.S. Geological Survey for developing typical pollutant concentrations (TPCs) in the absence of measured total nitrogen (TN) or total phosphorus (TP) concentration data. The new tool, PSLoadEsT (Point-Source Load Estimation Tool), provides a user-friendly interface for point-source dischargers that generate reproducible calculations of nutrient loads to streams. This method is critical for estimating TPCs because only 20% of TN values were measured concentrations for the 16,967 point-source facilities analyzed for the 2012 estimates of point-source loads to streams. The TPC, which is estimated based on the type of treatment process, effluent, and industrial category, can be used to calculate nutrient loads from facilities that lack effluent nutrient concentration data. Skinner and Maupin (2019) used

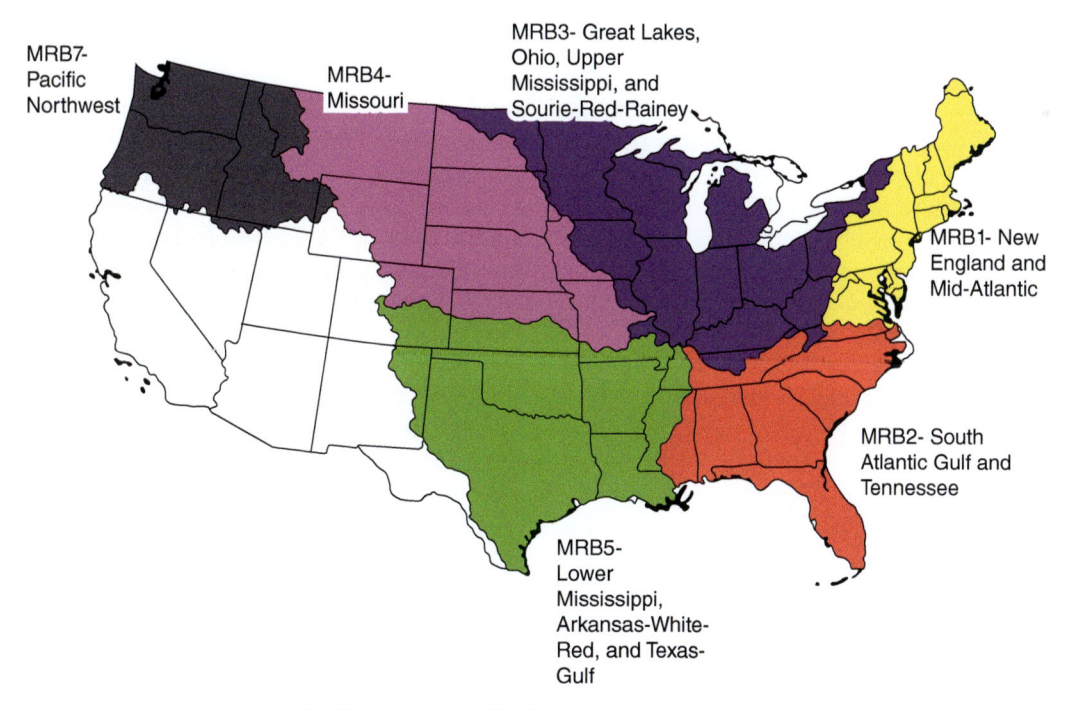

Figure 3.2 Major river basins studied by Maupin and Ivahnenko (2011) for their nutrient models. *Source*: Modified from colored figure 1 in Maupin and Ivahnenko (2011). Reproduced with permission of John Wiley & Sons.

Table 3.1 Point-source total nitrogen loads (2002) for all facilities, and distribution of total nitrogen loads among river basins in the United States (see Fig. 3.2).

MRB	Total TN load (kg/yr), all dischargers	Total TN areal load (kg/km²/yr), all dischargers	Median facility TN load (kg/yr)	Percentage of total TN load from minor facilities	Percentage of facilities that are sewage systems (SIC 4952)	Percentage of TN loads that were from sewage systems
1	12,07,99415	272.2	4985	6.5	70.7	86
2	5,62,11267	69.6	1450	8	61.2	77.6
3	25,20,55066	183.7	668	13.3	50.5	74.6
4	3,68,88359	27.9	906	47.4	63.6	50.7
5	8,71,53571	62.9	1470	11.5	64.6	62.6
7	2,30,74312	32.1	9992	5.7	59.1	82.4
Total	57,61,81990	95.3	1291	13	57.7	74.2

Source: Maupin and Ivahnenko (2011). Reproduced with permission of John Wiley & Sons.

PSLoadEsT to estimate point-source loads to streams in the conterminous United States for 2012. They estimated TN and TP loads for all major point-source facilities (which comprise major WWTFs and some industrial facilities) and for minor WWTFs that discharged to streams in the conterminous United States during 2012. Skinner and Maupin (2019) reported that the TN load contributed by major WWTFs to streams is about 15 times larger than that contributed by minor WWTFs even though there are almost three times as many minor WWTFs as major WWTF.

It is worth noting that wastewater treatment systems do not always operate properly (particularly during large storm events) or remove enough nitrogen and before discharging into surface waters. The U.S. EPA regulates point source discharges through programs such as the National Pollutant Discharge Elimination System (NPDES). However, in the United States, there are different types of wastewater treatment plants that discharge varying loads of nitrogen based on the treatment process. WWTFs receive influent total nitrogen concentrations of 20–70 mg/L (mostly as ammonia and organic nitrogen). Primary and secondary treatment (activated sludge including a nitrification step) systems can reduce the total nitrogen concentrations to 15–35 mg/L, and tertiary systems (with activated sludge and biological nutrient removal with advanced treatment) can reduce total nitrogen concentrations to <1 mg/L (Carey & Miglaccio, 2009).

Figure 3.3 shows the location of point sources of urban wastewater in Europe. Nitrogen input loads as high as 130,000 kg/km²/yr are in northern Germany, the Netherlands, and the southern part of the United Kingdom (Billen et al., 2011). In China, there are about 1944 municipal wastewater treatment plants across city/urban regions and 1599 municipal wastewater treatment plants across China's counties, accounting for processing capacities of 140 and 29 million cubic meter per day, respectively (https://www.waterworld.com/international/wastewater/article/16201297/chinas-13th-five-year-plan-what-role-will-wastewater-play, accessed 18 May 2019). Higher concentrations of facilities are located along China's coastal provinces such as Guangdong, Shandong, and Jiangsu, as well as in provinces that are located along the Yangtze River. Q.H. Zhang et al. (2016) noted that there were large differences in the efficiency of NH_3-N removal in WWTPs across cites and provinces. They indicated that removal efficiency for ammonia was good (average of 80%) in 656 WWTPs in 70 cities.

3.2.1.2. Biosolids

In the sewage treatment process, dissolved biological matter is converted into a solid mass using indigenous, water-borne bacteria (Q. Lu et al., 2012). This residual material is also referred to as sewage sludge or biosolids. Bradley (2008) estimated that countries with sewage treatment plants produce 27 kg of dry biosolids per person per year, with a global output that exceeded 10 million tons per year. Biosolids typically contain about 4–5% nitrogen (mostly organic nitrogen compounds and ammonia) and can be applied at agronomic rates to minimize leaching of nitrogen and to supply all the nitrogen required by a crop, without needing any additional fertilizer applications.

In the United States, the land application of biosolids is regulated under the Clean Water Act and state regulatory standards to protect surface water and groundwater from contamination. The U.S. EPA (2008) estimated that only 4% of the more than 15,500 municipal public operated treatment works in the United States have numeric limits for nitrogen. High application rates of biosolids have resulted in nitrate-N concentrations that have exceeded the maximum contaminant level for drinking water (nitrate-N; [NO_3-N] of 10 mg/L, U.S. EPA) in the United States with coarse-textured soils and mine reclamation areas (Q. Lu et al., 2012). The

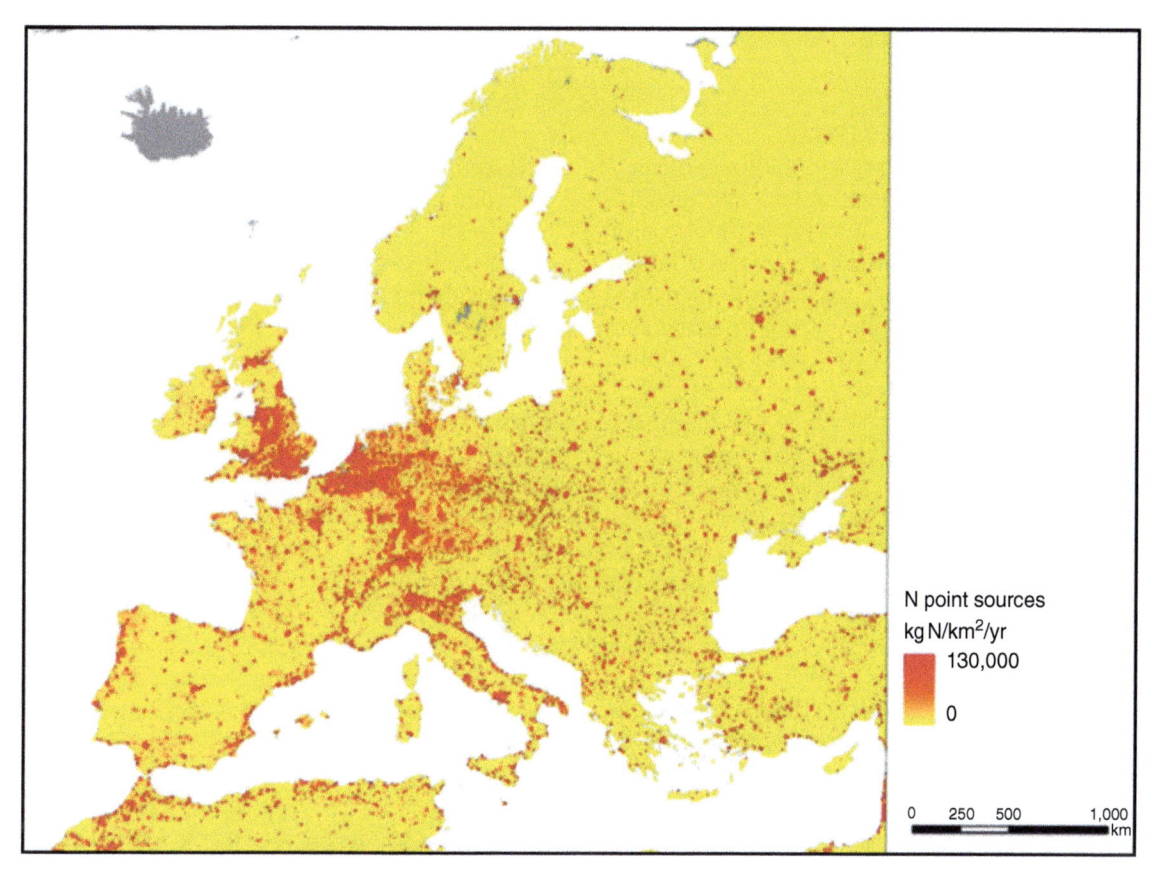

Figure 3.3 Urban point sources from wastewater in Europe. *Source*: From Bouraoui et al. (2009), in Billen et al. (2011). Reproduced with permission of Cambridge University Press.

application of biosolids has been shown to improve soil quality and fertility (Knowles et al., 2011), provide slow release of nutrients to crops, and result in similar or higher crop yields than inorganic fertilizers (Sullivan et al., 1997; Jaber et al., 2005).

3.2.1.3. Landfills

Several studies have shown that high concentrations of nitrogen in landfill leachate can contaminate groundwater in urban and developed areas (Kjeldsen et al., 2002). Ammoniacal nitrogen is the main form of nitrogen in leachate due to the anaerobic conditions in landfills. Wakida and Lerner (2005) report the following typical nitrogen concentrations in leachate: ammonium (0–1250 mg/L), nitrate (0–9.8 mg/L), and nitrite (1.5 mg/L). Several factors affect the quantity and quality of landfill leachate including liners, degree of waste compaction, climate variability, landfill age, and nitrogen content of waste products (Carey et al., 2013; Renou et al., 2008). In developing countries, open dumps can contribute elevated nitrogen concentrations to groundwater. For example, high nitrogen concentrations (84–124 mg/L) were found in groundwater near dumpsites and a municipal landfill in Lagos, Nigeria (Adelana et al., 2003; Longe & Balogun, 2010).

3.2.1.4. Mining

Elevated levels of nitrate were found in surface water and groundwater near an open pit mine in the Limpopo Province of South Africa (Bosman, 2009; Ihlenfeld et al., 2009). Bedrock from the mine is excavated and overburden rock and waste rock containing nitrogen is disposed in large dumps. Exposure to the atmosphere, erosion, and weathering enhance nitrification of naturally occurring nitrogen in the rock to nitrate. Also, blasting (with use of ammonia-based explosives) and blasting residues that remain behind contribute to elevated nitrate levels in groundwater and surface water. The use of explosives in road construction has also contributed to nitrate contamination in groundwater (Gascoyne & Thomas, 1997; Degnan et al., 2016).

3.2.1.5. Industry

Wakida and Lerner (2005) noted that various nitrogen compounds are commonly utilized in many industrial processes including plastic and metal treatments, explosives, raw materials for the textile industry, particleboard and plywood, household cleaning, and the pharmaceutical industry. The most common nitrogen compounds used are ammonia, nitric acid, urea, and ammonium nitrate. Contamination of groundwater and surface waters may occur from inadequate handling and disposal at these sites; however, documented cases are not widely reported.

3.2.1.6. Agricultural Livestock Wastes

Livestock wastes are one of the largest sources of reactive nitrogen pollution to surface waters and groundwater. In the United States, the 2012 agricultural census (USDA, 2012) estimated the following numbers of major livestock production: 90 million head of cattle and calves (17.5 million dairy cows), 66 million head of swine, and 1.5 billion broilers and 350 million laying hens. Manure generated from livestock production results in more than 3.6 million kilograms per day. Much of the manure is applied as organic fertilizer for crops; however, substantial amounts are transported to surface waters and groundwater (State-EPA Nutrient Innovations Task Group, 2009). Concentrated animal feeding operations (CAFOs) are considered point sources under the Clean Water Act in the United States and are regulated by the NPDES program if wastes are discharged to surface waters. Threshold numbers for CAFOs in the United States are greater than (or at least) 700 dairy cows, 1000 beef cattle; 2500 swine (each 55 kg or more); 10,000 swine (each under 25 kg); 82,000 laying hens (other than liquid manure handling systems); 55,000 turkeys; 125,000 chickens (30,000 chickens if the facility has a liquid manure handling system), 5000 ducks (liquid manure handling systems); 30,000 ducks (other than liquid manure handling systems); 500 horses, or 10,000 sheep or lambs. However, many farms do not meet the requirements for a CAFO and remain largely unregulated. The U.S. EPA (2001) estimated that CAFOs directly contributed 16% to the total agricultural impact of 59% to impairment for rivers in the United States. The report pointed out that the impact of agriculture to groundwater is relatively unknown, and there are no estimates for impairment due to CAFOs. Nutrients and other contaminants can reach groundwater from various CAFO sources, such as leakage from lagoons, breaches in piping or barn infrastructure, and land application of liquid and solid wastes.

3.2.2. Nonpoint Sources

There are many anthropogenic nonpoint sources of reactive nitrogen released to the environment. Nonpoint sources from agricultural practices include excess fertilizer applied to cropland, pastures, and fields, manure generated from grazing animals, and manure applied to fields. Nonpoint sources of reactive nitrogen from residential areas include septic tanks, lawn fertilizers, fertilizers applied to turfgrass (e.g., golf courses, parks, and sports fields), yard and pet wastes, and certain household cleaning products and soaps that contain nitrogen. In urban areas, leaky sewer lines can contribute nitrogen to soils and groundwater. Atmospheric deposition is another important nonpoint source of reactive nitrogen, which originates from the combustion of fossil fuels (coal-fired power plants, automobiles, and industrial emissions), and volatilization of ammonia from fertilizer and manure (U.S. EPA, 2002). In the United States, studies have estimated that nonpoint sources of nitrogen are the most widespread cause of nutrient pollution contributing more than 90% of the nitrogen released to the environment (Puckett, 1995).

3.2.2.1. On-Site Sewage Disposal and Treatment Systems

Septic tanks, cesspools, and pit latrines are commonly used for on-site sewage disposal in different parts of the world. Septic tanks are common in the United States, Canada, Australia, and in developing countries (Wakida & Lerner, 2005). About 20–25% of households in the United States rely on a septic tank for waste disposal. Most septic systems in the United States are conventional gravity flow septic systems that use gravity to move household wastes through the treatment system. The main components of a gravity flow system are a septic tank and a drainfield (Fig. 3.4). It is estimated that one person produces approximately 4.1 kg N/yr in feces and urine (urine accounts for 91% of the total nitrogen in raw wastewater) (Lusk et al., 2017). However, the nitrogen concentration in a septic tank and drainfield is dependent on the per capita daily water use in the household, diet, and garbage disposal. The per capita nitrogen generated is slightly higher in the United States, Denmark, and Germany compared with other countries, which is possibly related to the consumption of a nitrogen-rich diet (Lusk et al., 2017). Based on a literature review of nitrogen concentrations in septic tank effluent, Lowe et al. (2007) reported a large total nitrogen range of 26–124 mg/L. The organic nitrogen in the raw wastewater in the septic tank is mineralized to ammonium through ammonification. Due to the anaerobic conditions in a septic tank, little conversion of ammonium occurs as the wastewater effluent flows to the drainfield. Immediately below the drainfield under aerobic conditions, transformation of ammonium to oxidized forms of nitrogen occurs referred to as nitrification (mainly nitrate and nitrite). The amount of transformation of ammonium depends on the soil conditions beneath the drainfield, such as soil pH, temperature, soil drainage, oxygen content, water content, microbial community, and availability of organic carbon. Sorption of ammonium can also occur on soil surfaces, particularly on clay minerals, but ammonium sorption has also been observed on coarse-grained soils with little clay (De & Toor, 2015). Most properly functioning septic tank systems rapidly convert ammonium to nitrate, which can be transported

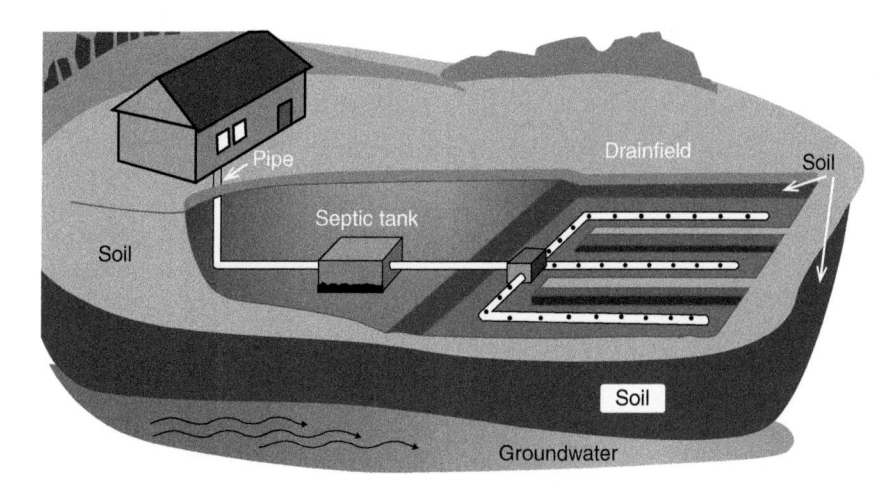

Figure 3.4 A conventional gravity on-site sewage treatment and disposal system (septic tank). *Source*: Modified from Toor et al. (2017).

to groundwater and or surface waters. Septic tanks are an important source of nitrogen pollution of surface waters and groundwater in many areas in the United States (Katz et al., 2011; Kaushal et al., 2011; Panno et al., 2007). For example, the Chesapeake Bay (the largest estuary in the United States) receives about 5.7 kg of nitrogen annually (4% of the total nitrogen loading) from septic systems within its watershed (Chesapeake Bay Program, 2011). Many other places around the world have reported contamination of waters from septic tanks (Beal et al., 2005; LaPointe et al., 2017; Lusk et al., 2017).

Another issue associated with septic systems is tank or drainfield failure rates, which can range from 5 to 40% (Carey et al., 2013). System failures from clogged systems or subsurface failures can result in effluent entering receiving waters, or percolate into groundwater. In areas with high water tables (such as coastal communities) there can be a higher risk of water-quality degradation from malfunctioning septic systems.

In the Pearl River Delta region, a rapidly urbanized region of south China, raw sewage was the main source of groundwater nitrate contamination (Q. Zhang et al., 2015). Rapid population and economic growth has led to deterioration of surface water quality and declining river discharge. As a result, the exploitation of groundwater in this area has increased substantially in recent years. In this area, domestic sewage is directly discharged into nearby ground or ditches without treatment and the water table is shallow in this region (average depth is about 2 m). About 26% of the 899 groundwater samples collected in this study exceeded World Health Organization (WHO) standards for nitrate in drinking water. In areas near landfills, about 42% of groundwater samples exceeded WHO standards. In another large region of China, the north China Plain, groundwater is the major source of water for anthropogenic activities. This area contains 65% of China's agricultural land but only 24% of its water (S. Wang et al., 2016). More than 50% of all groundwater in investigated areas of the North China Plain exceed the WHO limit of 44 mg/L for nitrate in drinking water (W. Zhang et al., 1996). Likely sources include a combination of wastewater irrigation and lateral flow of groundwater in the alluvial fans of the North China Plain (S. Wang et al., 2013).

3.2.2.2. Leaky Sewer Systems

In urban and suburban areas, sewer systems transport large quantities of waste to wastewater treatment plants. In many areas that were urbanized during the late 1800s and 1900s, older sewer systems have degraded over time leading to leakage (exfiltration into the surrounding ground) that has contributed to elevated nitrogen concentrations in groundwater (R.G. Taylor et al., 2006; Wakida & Lerner, 2005). Leaky sewer lines in urban areas can contribute inorganic and organic forms of nitrogen to the subsurface and to groundwater. In the United Kingdom, over 94% of the population is served by a sewerage system and monitoring has found widespread contamination of shallow subsurface and shallow groundwater from leaking sewers (Reynolds & Barrett, 2003). Using depth-specific hydrochemistry profiles from multilevel piezometers, Rueedi et al. (2009) calculated that 5–10% of total sewer volumes are lost to groundwater in a suburban area of Doncaster, U.K. It was estimated that about 30–40% of total recharge in this study area could be originating from losses from mains and sewer pipe networks. Also, other studies have shown that sewer leakage to urban streams has been significant (Divers et al., 2014; Hopkins & Bain, 2018; Kaushal et al., 2011). In a study of leaking sewer infrastructure in an urban watershed in Pittsburgh, Pennsylvania, Divers et al. (2013) found dissolved inorganic nitrogen (sum of ammonium, nitrite and nitrate concentrations) contributions from sewage ranging from

6 to 14 kg N/ha. It was noted that the importance of including sewage inputs to nitrogen budgets in urban streams, particularly as sewer systems age in urban areas. Risch et al. (2015) used a value of 10.9 kg DIN/d to estimate leakage from sewer lines in town in the South of France. This value was based on a geometric mean for various modeling scenarios investigated by Divers et al. (2013). Hopkins and Bain (2018) inferred sewer leakage hotspots in an urban area using a geographical information system that accounted for spatial patterns in sewer age, material, and proximity to surface waters.

3.2.2.3. Urban Fertilizer Use

Fertilizer use on lawns, home gardens, community gardens, and sports and recreational facilities (e.g., golf courses, sports fields, and parks) can result in a significant source of nitrate leaching to groundwater in urban and residential areas. Milesi et al. (2005) estimate that turfgrass covers about 16 million hectares or 35% of the total urban land in the United States. Lawn care practices are highly variable and are related to neighborhood characteristics, community expectations, and property values (Carey et al., 2012; Cook et al., 2012; Kinzig et al., 2005; Zhou et al., 2008). Based on a survey of homeowners in two watersheds in the Baltimore, Maryland, area, Law et al. (2004) estimated a mean fertilizer application rate of 97.6 kg N/ha/yr with a standard deviation of 88.3 kg N/ha/yr. The amount and rate of nitrate leaching from urban fertilizer applications depends on soil properties, nitrogen source, application rate and timing, and irrigation/rainfall. For example, C. Lu et al. (2015) have shown that turfgrass systems lose more nitrogen as they age. For recently planted turgrass or sod, Lusk et al. (2018) found that less than 3% of the nitrogen in leachate originated from newly applied nitrogen fertilizer, although dissolved organic nitrogen in leachate was derived from soil or sod biomass pools. Elevated nitrogen concentrations have been found in drainage below the root zone and in groundwater (Guillard & Kopp, 2004; Pionke et al., 1990; Telenko et al., 2015; Tucker et al., 2014). Other factors that affect nitrogen export from turfgrass include the fertilizer type (e.g., soluble vs. controlled release), application rates, timing of fertilization, and irrigation practices (Carey et al., 2013).

3.2.2.4. Home Construction

Wakida and Lerner (2005) noted that nitrate leaching resulting from house building is potentially equivalent to that due to plowing of a pasture, which can be a major source of nitrate in groundwater beneath agricultural land. As soil is disturbed during the building process, aeration accompanied by soil microbial activity can promote mineralization and nitrification processes. This in turn could lead to nitrate leaching if nitrate is not assimilated by plants. Other factors that influence nitrate leaching from home construction include nitrogen content in soil, previous land use, and the time of year of construction. The average potential nitrate loss from home construction sites was estimated to be 59 kg N/ha/yr (Wakida & Lerner, 2005). Carey et al. (2013) reported an average total nitrogen export of 36.3 kg N/ha/yr to stormwater runoff from a construction site in North Carolina.

3.2.2.5. Urban Stormwater Runoff

The population in the United States is mainly consolidated in urban areas with 80% of the population living on less than 10% of the land (State-EPA Nutrient Innovations Task Group, 2009). During storms, precipitation that falls on impervious surfaces (roofs, parking lots, and streets) collects fertilizers and other sources of nitrogen, and these contaminants are transported in stormwater runoff to retention areas (e.g., wetlands, infiltration basins, recharge basins, ponds, and swales) and receiving waters. The National Water Quality Inventory: 2000 Report to Congress (U.S. EPA, 2002b) ranked 11 pollution source categories and urban runoff was the fourth leading source of impairment in rivers, third in lakes, and second in estuaries. They reported that typical concentrations of organic nitrogen plus ammonia in urban runoff ranged from 1.2 to 1.9 mg/L and nitrate-N concentrations ranged from 0.57 to 0.74 mg/L in commercial and residential areas, respectively. Schueler (2003) reported typical nitrogen concentrations of 2.0 mg/L in urban stormwater runoff. In urban and suburban areas, there are large numbers of dogs, cats, and other animals that can contribute nitrogen in urine and feces, that can be transported via runoff. Groffman et al. (2004) estimated nitrogen inputs of 17 kg/ha/yr from pet waste in a suburban watershed in Baltimore, Maryland. In some large urban cities around the world, there are large numbers of dairy cows, oxen, and buffaloes. Feces and urine produced by these animals have contributed to high nitrate concentrations in shallow groundwater in India (Wakida & Lerner, 2005). Several studies have addressed ways of attenuating various forms of nitrogen in urban runoff through various management scenarios (e.g., Li & Davis, 2014; O'Reilly et al., 2012; G.D. Taylor et al., 2005).

3.2.2.6. Fertilizers and Reactive Nitrogen Compounds

There are numerous N-containing compounds that can be used as N-fertilizers (Follett & Walker, 1989). However, when availability, economics, convenience, and effectiveness are considered, the compounds usually considered are those included in Table 3.2. Anhydrous ammonia, or gaseous NH3, is a very important direct-application N-fertilizer. When in contact with moist soil, ammonia dissolves and reacts with soil water to form ammonium (NH_4^+) and hydroxide ions (OH^-). As a result, the pH can increase substantially and immediately around the application zone of anhydrous ammonia. Depending on the buffering capacity of the soil and the impacted soil pH, equilibrium concentrations of ammonia and ammonium will be attained. Gaseous ammonia can be lost by volatilization to the atmosphere.

All of the compounds shown in Table 3.2 are highly water soluble. For those with NH_4 as part in their chemical formula, the NH_4^+ will adsorb to the soil based on the cation exchange capacity (CEC, the total amount of exchangeable positively charged ions (cations) that a soil can adsorb. Negatively charged surfaces on clay particles and organic matter in the soil bind cations that can be replaced by or exchanged with other cations. Therefore, the primary transport mechanism for NH_4^+ ions is in association with eroding sediments since they are adsorbed by the soil. Urea and calcium cyanamide (Table 3.2) are organic forms of N that, when applied to soil, are acted upon by enzymes to mineralize the N in them to NH_4^+ ions. Once in NH_4^+ form, the N in these two fertilizers is also adsorbed to soil particles and is subject to the soil-erosion transport process described above. Also, the N in other organic materials such as manures and crop residues is also mineralized to NH_4^+, again being subject to transport with eroding sediments.

For compounds in Table 3.2 that have nitrate (NO_3) as part of their chemical formula, the nitrate ion (NO_3^-) would not sorb onto the negatively charged surfaces of soil particles. Therefore, the primary transport mechanism for NO_3^- ions is with percolating water by leaching or surface runoff (including return flow, which is surface or subsurface water that moves off an agricultural field following the application of irrigation water).

Table 3.2. Nitrogen fertilizer materials, their formulas, and percent nitrogen.

Material	Chemical formula	Chemical analysis (%N)
Anhydrous ammonia	NH_3	82
Ammonium nitrate	NH_4NO_3	33.5
Ammonium sulfate	NH_4SO_4	21
Diammonium phosphate	$(NH_4)_2H_2PO_4$	18–21
Monoammonium phosphate	$NH_4^+H_2PO_4$	11
Calcium nitrate	$Ca(NO_3)_2$	15
Calcium cyanamide	$CaCN_2$	20–22
Potassium nitrate	KNO_3	13
Sodium nitrate	$NaNO_3$	16
Urea	$CO(NH_2)_2$	45
Urea-ammonium nitrate	$CO(NH_2)_2 + NH_4NO_3$	28–32

Source: From Follett and Walker (1989). Reproduced with permission of Elsevier.

Nitrate that is leached below the crop root zone often ends up as a pollutant in groundwater supplies. Nitrate can also be dissolved in surface runoff water or in return-flow water that returns to the surface to become part of the runoff.

Fertilizers typically are applied at agronomic rates; however, agricultural production of crops generally has an efficiency of less than 30% (Galloway et al., 2003). Any excess nitrogen not taken up by the crop can volatilize into the air, leach, and infiltrate into groundwater, or be transported off the land in stormwater runoff. Böhlke (2002) noted that studies in different areas indicate that the magnitudes of groundwater nitrate recharge rates typically are about 10–50% of fertilizer nitrogen application rates beneath heavily fertilized well-drained fields.

3.2.3. Atmospheric Inputs and Deposition of Reactive Nitrogen

Atmospheric inputs of reactive nitrogen in gaseous and particulate forms of nitrogen oxides (NO_x) are emitted into the air from burning of fossil fuels during electric power generation (coal, petroleum), other combustion processes, and industrial processes (Fig. 3.5). Nitrogen oxides tend to have a small impact close to their sources because they are emitted as nitrogen monoxide and nitrogen dioxide at low dry deposition rates. After these nitrogen oxides are converted into nitric acid, deposition of reactive nitrogen is more efficient (Hertel et al., 2011). Globally in 2010 about 40 ± 10 Tg N/yr originates from combustion (Fowler et al., 2015). In the United States, vehicles account for approximately 55% of NO_x emissions to the atmosphere and stationary sources account for the remainder (State-EPA Nutrient Innovations Task Group, 2009).

Agricultural activities are the dominant source of global ammonia emissions (Fig. 3.5). Livestock and fertilizer contribute about 80–93% of total ammonia emissions based on a review of ammonia inventories reported for China, European Union, and the United States (Reis et al., 2009). Hertel et al. (2011) notes that ammonia deposition has a high dry deposition rate near its sources; therefore, ammonia may have a significant impact on ecosystems in areas with intense agricultural activities. These land uses tend to be in rural areas with sensitive ecosystems. Other nonagricultural sources of ammonia emissions to the atmosphere are vehicles with three-way catalytic converters, electrical generating units with certain catalytic and noncatalytic reduction technologies (Felix et al., 2014). Both ammonia and gaseous nitrogen oxides tend to form aerosol phase compounds (ammonium and nitrate), and these compounds can be transported over large distances, possibly up to more

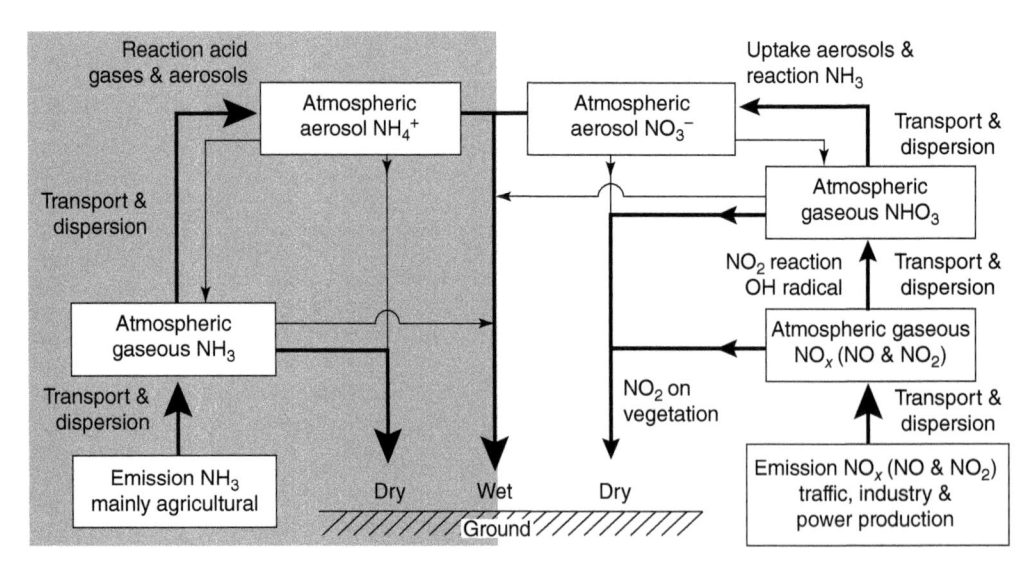

Figure 3.5 Conceptual diagram showing pathways of reactive nitrogen in the atmosphere. *Source*: From Hertel et al. (2011). Reproduced with permission of Cambridge University Press.

than 1000 km (Hertel et al., 2011). Ammonium concentrations in precipitation have increased at 90% of monitoring stations in the United States during 1985–2002 and ammonia is expected to constitute 69% of nitrogen deposition by 2020 (Davidson et al., 2012).

The amount of deposition of reactive nitrogen varies considerably around the world. A large number of models are being used for air pollution studies and transport of ammonia and NO_x. Some of these models incorporate near-source dispersion or process studies, and others generate information on a global scale (Hertel et al., 2011). For example, the model documentation system for the European Topic Centre on Air and Climate Change Mitigation lists 123 different models, developed around the world (http://acm.eionet.europa.eu/, accessed 29 March 2018). The European Monitoring and Evaluation Program (EMEP) and the European Environment Agency compile inventories of the annual mean emissions on a grid with a spatial resolution of 50×50 km. The Emissions Database for Global Atmospheric Research (EDGAR; http://edgar.jrc.ec.europa.eu/, accessed 29 March 2018) and Global Emissions Inventory Activity (GEIA; http://www.geiacenter.org/, accessed 29 March 2018) databases are available on $1 \times 1°$ resolutions. The US EPA maintains a similar list (https://www.epa.gov/air-quality-management-process/managing-air-quality-air-quality-modeling, accessed 29 March 2018). For a more complete discussion of the various models used to assess emissions of reactive nitrogen and wet and dry deposition (see Simpson et al., 2011).

Schwede and Lear (2014) developed a total deposition (TDEP) model for nitrogen and sulfur deposition across the United States for the U.S. EPA. Their hybrid model combines data from 2000 to 2012 for wet deposition from the National Atmospheric Deposition Program (NADP) and the National Trends Network (NTN) with modeled dry deposition data from several monitoring networks across the United States (including the Clean Air Status and Trends Network (CASTNET), the NADP Ammonia Monitoring Network (AMoN), and the Southeastern Aerosol Research and Characterization (SEARCH) network along with modeled data from the Community Multiscale Air Quality (CMAQ) model for the years 2002–2009). TDEP data are summarized for 12×12 km grid cells and are available for a range of modeled years. The highest rates of atmospheric deposition of total nitrogen are in the upper Midwest and eastern United States (10 to >20 kg N/ha/yr) for 2010 (Fig. 3.6). In some parts of the United States, decreases in nitrogen oxide emissions from 1993 to 2001 were associated with decreasing or stable inputs from atmospheric deposition during 1993 to 2003, with some of the largest decreases were observed in the Northeastern and Western States (Dubrovsky et al., 2010).

3.3. REACTIVE NITROGEN TRANSPORT PROCESSES

The various chemical, physical, and biological processes affecting nitrogen transport throughout different parts of the environment are shown in Figure 3.7. Reactive nitrogen can be transported regionally through the atmosphere by burning of fossil fuels and volatilization from fertilizer and manure. Nitrogen can then return to the land surface in precipitation and dry deposition. Runoff from agricultural and urban areas can flow directly into streams. Recharge from precipitation and or irrigation can transport reactive nitrogen from various sources on the land surface into groundwater systems.

SOURCES OF REACTIVE NITROGEN AND TRANSPORT PROCESSES   39

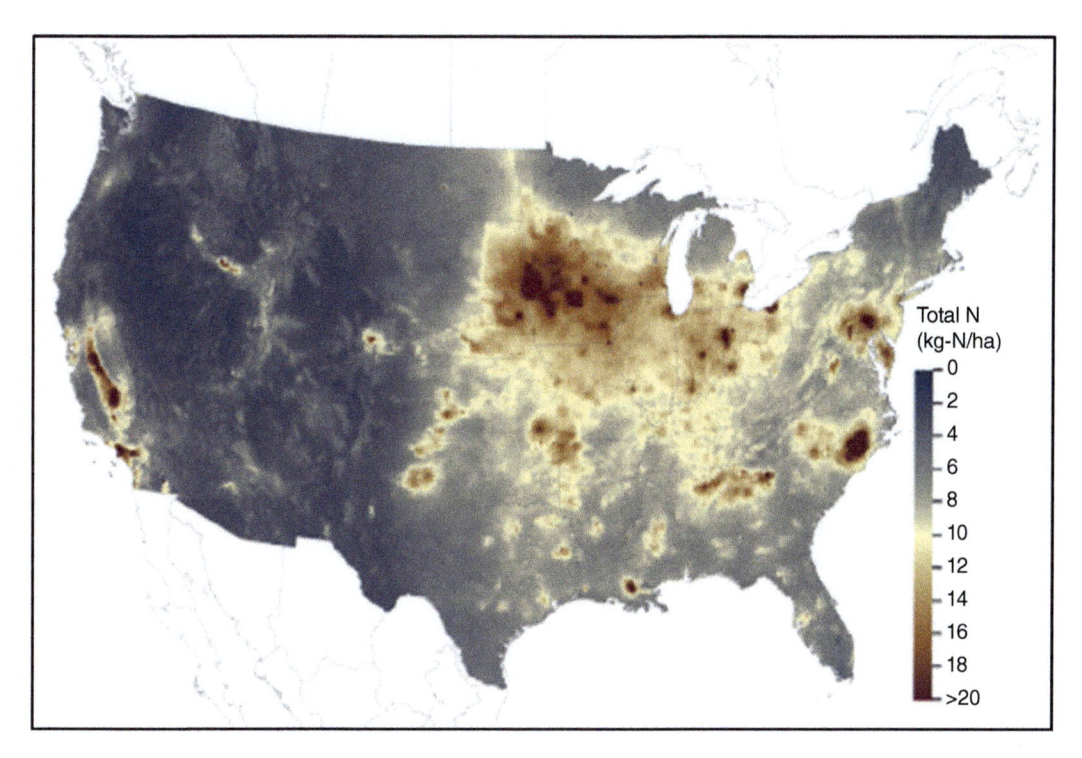

Figure 3.6 Map of the United States showing 2010 total deposition of nitrogen from TDEP model. Darkest areas show highest total nitrogen deposition. *Source*: Modified from colored Figure 6 in Schwede and Lear (2014).

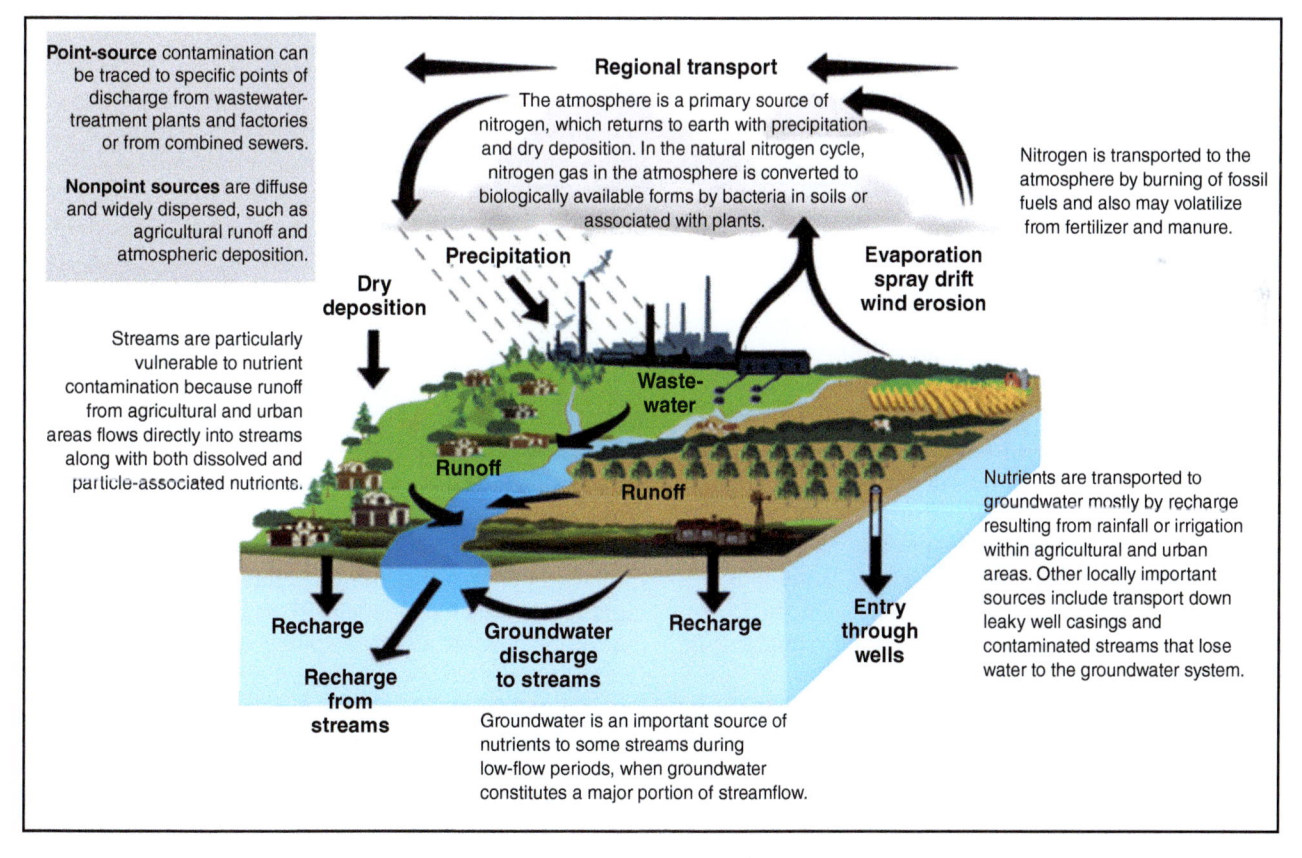

Figure 3.7 Chemical, physical, and biological processes affecting nitrogen transport from nonpoint and point sources throughout the hydrologic system. *Source*: Modified from Munn et al. (2018). Reproduced with permission of United States Geological Survey, public domain.

Reactive nitrogen can exchange between groundwater and streams (see Chapter 8 for more information about interactions between groundwater and surface water systems). Nitrogen can enter streams in inland and coastal areas via groundwater discharge. Also, during high flow conditions in streams and rivers, nitrogen can enter the groundwater system from bank storage or flow reversals into springs along rivers.

Naturally occurring factors affecting the amount and timing of nitrogen transport include soil type (drainage characteristics, mineralogy, and organic matter content), geology, precipitation, temperature, and slope of the land. In agricultural areas, nitrogen transport also is affected by other factors such as water withdrawal, artificial drainage, and best management practices (discussed in more detail in Chapter 11).

3.4. LARGE NATIONAL AND MULTINATIONAL PROGRAMS AND NITROGEN SOURCE CONTRIBUTIONS

Large-scale programs in the United States and Europe have generated an extensive amount of information on sources of reactive nitrogen in surface waters and groundwater. These programs have focused on major sources of reactive nitrogen from fertilizers, manure, atmospheric deposition, and sewage. The U.S. Geological Survey implemented the National Water-Quality Assessment (NAWQA) program in 1991 to provide information on the quality of streams and groundwater in the United States to national, regional, state, and local agencies involved in water-quality management and policy (http://water.usgs.gov/nawqa). Since 1991 the NAWQA Program has collected information on water chemistry, physical characteristics, stream habitat, and aquatic life. During 1991–2001, the program completed interdisciplinary assessments and established a baseline understanding of water-quality conditions in 51 of the river basins and principal aquifers in the United States (http://water.usgs.gov/nawqa/studyu.html). The focus of several NAWQA studies of nutrients was the fate of agricultural chemicals and their effects on aquatic ecosystems. The NAWQA program findings have significantly advanced the understanding of how human activities affect water quality, sources of contaminants, the transport of those contaminants through the hydrologic system, and the potential effects of contaminants on humans and aquatic ecosystems.

Commercial fertilizer was the largest single nonpoint source of nitrogen, with more than 10–12 million tons (9.1–10.9 Tg N/yr) of nitrogen applied each year in the United States (Dubrovsky et al., 2010). The use of commercial fertilizers is crop-dependent, with the highest amounts and areal rates applied to corn (particularly in the Midwest United States). The use of nitrogen fertilizers increased substantially (tenfold) between about 1950 and 1980 in the United States. The applications of fertilizers in the United States have remained relatively constant since about 1980, which have been attributed to increased fertilizer costs, growing environmental issues, and changing agricultural practices. Most of the fertilizer applications are for agricultural usage; however, about 2–4% of the total nitrogen fertilizer applied is used in urban areas, such as city parks, golf courses, and residential lawns. In comparison, manure contributes about 6 million tons of nitrogen per year (5.4 Tg N/yr) in the United States, whereas atmospheric deposition contributes about 2.5 million tons per year (2.3 Tg N/yr) (Dubrovsky et al., 2010). Fertilizers applied to lawns in residential areas (outside of densely populated areas) typically contribute less than 1760 kg/km^2 N/yr.

Data on major sources of nitrogen (farm and nonfarm fertilizers, manure, atmospheric deposition, and sewage) were compiled by county throughout the United States. Different sources of nitrogen were found to be predominant in different regions of the country (Dubrovsky et al., 2010). For example, commercial fertilizers were found to be the largest source of nitrogen in the United States. Applications of commercial fertilizers were highest in agricultural areas of the upper Midwest United States (>7000 kg/km^2 over an extensive area from Indiana to Nebraska), parts of the Great Plains, the Northwest (southern Idaho), California (San Joaquin Valley), and the southeast Atlantic Coastal Plain (particularly in parts of Florida). In parts of the South and Southeast United States, manure sources of nitrogen were dominant as well as in western rangelands. Atmospheric deposition rates of nitrogen increase from west to east and are highest in the eastern part of the United States.

In some agricultural areas in the United States, manure-derived nitrogen was a dominant source where there are large populations of confined livestock, such as in the White River Basin in Arkansas, the Susquehanna River Basin in Pennsylvania, and parts of the South and Southeast (Dubrovsky et al., 2010). In these areas, nitrogen loading rates can exceed 7000 kg/km^2. Atmospheric deposition of nitrogen is the largest nonpoint source of nitrogen in undeveloped watersheds in the eastern part of the country where deposition rates are highest (e.g., the Connecticut River Basin). Atmospheric deposition also is a major source of nitrogen near the Great Lakes, and in mostly sparsely populated areas in the arid and mountainous West.

In urban watersheds in the United States, commercial nitrogen fertilizers also were found to be the major source of nitrogen (Dubrovsky et al., 2010), although some urban watersheds have significant contributions of nitrogen from wastewater-treatment plants. For example, in parts of Arizona, Nevada, and southern California, the main source of nitrogen in streams is effluent from wastewater treatment plants.

The European Nitrogen Assessment Program, described previously in Chapter 1 of this book, (Sutton et al., 2011) has provided extensive quantitative information on sources of reactive nitrogen in Europe and global estimates. In 2004, 25 European countries produced 21% of the global meat production and 20% of global cereal production (Erisman et al., 2011). For Europe, BNF for these crops was estimated to be 3.9–5 Tg N/yr (Erisman et al., 2011; Velthof et al., 2009), compared to the global level of 30 Tg N/yr (Galloway et al., 2004) (Fig. 3.8). The increase in fertilizer application in Europe followed the global increase, with a large drop in 1990–1992 followed by a slight increase and decrease during the mid to late 1990s (Fig. 3.9). Fluctuations in fertilizer usage were related to changes in economic conditions and agricultural policies. Erisman et al. (2011) noted that Europe accounts for about 10% of the global use of nitrogen fertilizers.

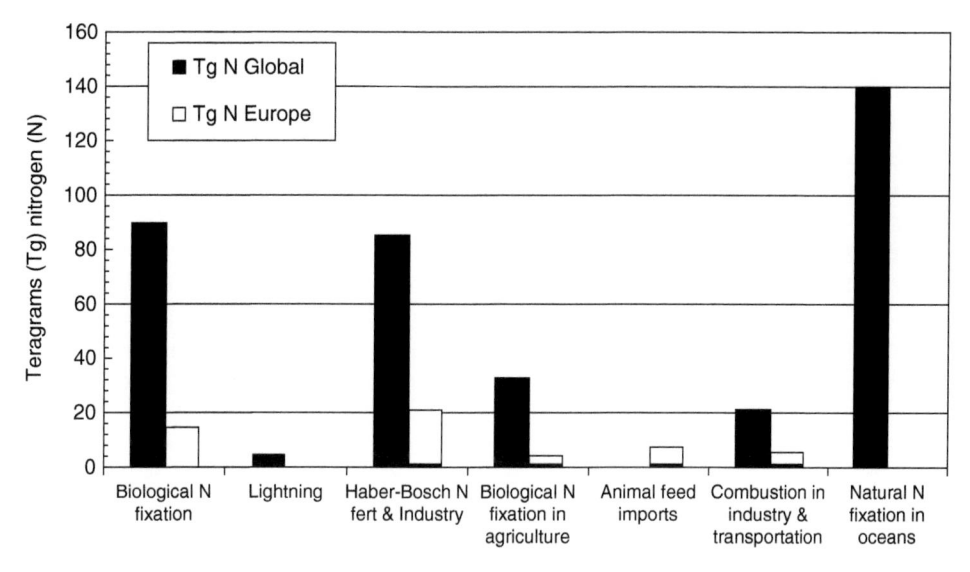

Figure 3.8 Reactive nitrogen produced by various processes in Europe and globally. *Source*: Modified from Erisman et al. (2015), with permission. Reproduced with permission of World Wildlife Fund.

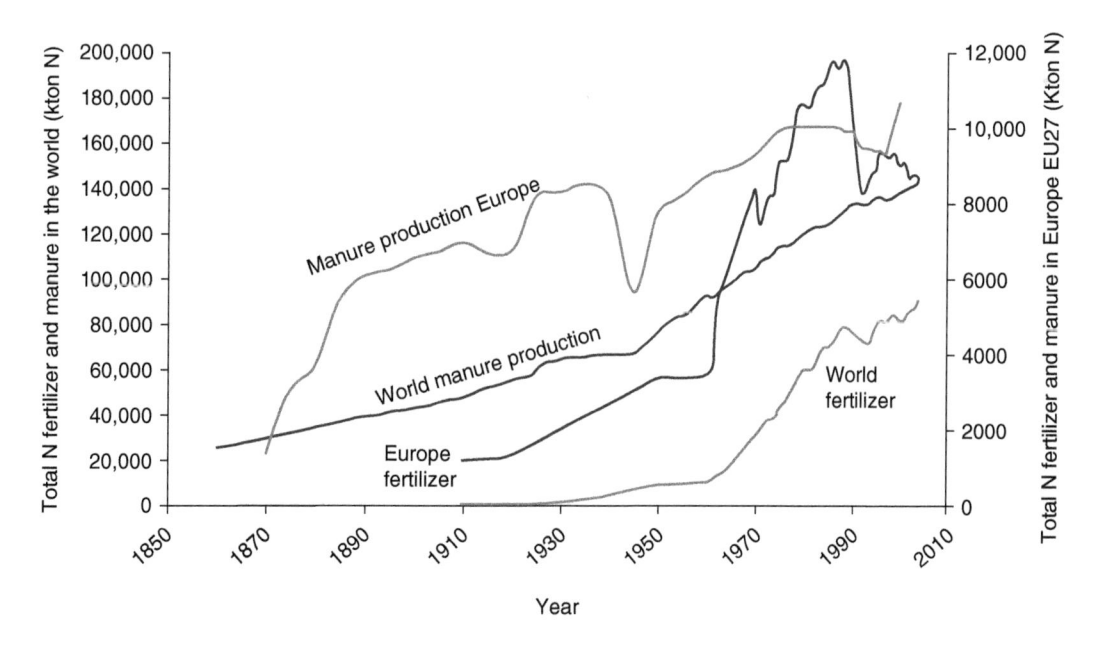

Figure 3.9 European and global reactive nitrogen inputs from livestock manure and fertilizer nitrogen use, worldwide and in Europe (EU27, European Union 27 Countries). *Source*: Modified from Erisman et al. (2011). Reproduced with permission of Elsevier.

Europe has a large proportion of livestock production compared to global numbers: 8.9% poultry, 18% pig, 7.1% cattle, and 8.3% small ruminant (data from Erisman et al., 2011). The increases in manure from livestock production in Europe and globally are shown in Figure 3.9. Although livestock production does not lead directly to the creation of new reactive nitrogen, it does contribute to a concentration of reactive nitrogen in some areas (confined animal feeding operations), and a transfer of reactive nitrogen in meats and cereals over various regions and globally. In Europe, total reactive nitrogen application from manure is approximately equal to the use of mineral fertilizer (Oenema et al., 2007). In comparison, the amount of manure applied globally in the past was an order of magnitude higher than that applied in fertilizer. However, currently it has decreased to a factor of two, as more agricultural land in Europe is fertilized (Erisman et al., 2011).

DeVries (2011) presented detailed nitrogen budgets for 27 European countries in agricultural and other terrestrial ecosystems for the year 2000. Based on the results from various models (INTEGRATOR, IDEAg, MITERRA, and IMAGE) for agricultural systems, it was estimated that a total nitrogen input in European agriculture was 23.3–25.7 Mton N/yr. Estimates of nitrogen uptake ranged from 11.3 to 15.4 Mton N/yr, which resulted in a total N surplus that ranged from 10.4 to 13.2 Mton N/yr. Estimated emissions of ammonia ranged from 2.8 to 3.1 Mton N/yr and emissions of N_2O ranged from 0.33 to 0.43 Mton N/yr. In nonagricultural systems (forests and seminatural vegetation), DeVries (2011) estimated a total N input of approximately 3.2 Mton N/yr, with a net nitrogen uptake of approximately 1.1 Mton N/yr. This resulted in an estimated nitrogen surplus of approximately 2.1 Mton N/yr. In comparison to agricultural systems, the estimated nitrogen fluxes in nonagricultural systems were about five times lower for N_2O emissions and 10 times lower for NO_x and ammonia emissions and for the sum of nitrogen leaching and runoff. Based on a comparison of results from the different models, it was cautioned that there the largest model uncertainties are flux values for nitrogen leaching and runoff followed by N_2O emissions from agricultural ecosystems.

Based on data from Erisman et al. (2011), for global nitrous oxide (NO_x) emissions (year 2000), North

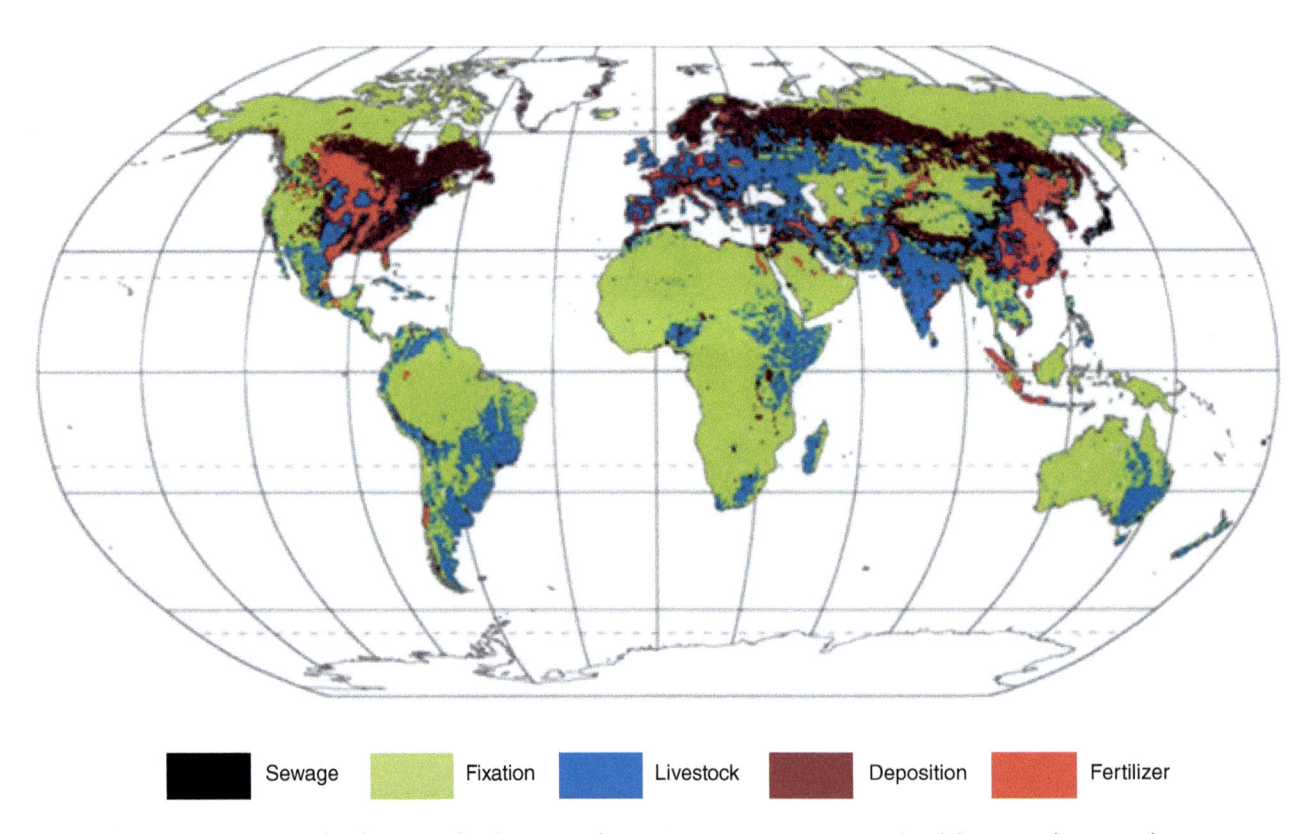

Figure 3.10 Nitrogen loading onto land areas and aquatic systems as a source for delivery to the coastal zone; providing a measure of potential water pollution. Total and inorganic nitrogen loads as deposition, fixation, fertilizer, livestock loads, human loads, and total distributed nitrogen to the land and aquatic system. GEMSTAT database. Map prepared by Water Systems Analysis Group, University of New Hampshire. *Source*: Map taken from Erisman et al. (2011). Reproduced with permission of Cambridge University Press.

America contributed about 17% of global NO_x emissions. This was slightly higher than Europe (14%) and Asia (12% for S/SE Asia and 14% for East Asia, respectively). Most of the European NO_x emissions originate from mobile combustion sources, which contribute an estimated 62% (road transport 30.7%, other mobile sources 16.2%, and international shipping 15%), followed by stationary combustion (15.7% from large combustion plants and 4.5% residential and commercial combustion). The data from the EDGAR inventory typically do not include natural and biogenic sources of emissions, which generally are very low compared to anthropogenic sources.

The Global Environment Monitoring System for Water (GEMS/Water) collects worldwide freshwater quality data for supporting scientific assessments of ground and surface waters and decision-making processes at global, regional, and local levels (for more information please see, https://www.unenvironment.org/explore-topics/water/what-we-do/monitoring-water-quality; http://gemstat.org/about/#gemstat, accessed 29 March 2018). GEMS/Water was established in 1978 as an interagency program under the auspices of the United Nations through the United Nations Environment Program (UNEP), the WHO, the World Meteorological Organization (WMO), and the United Nations Educational, Scientific and Cultural Organization (UNESCO). GEMStat (https://gemstat.org/, accessed 29 March 2018) is hosted, operated, and maintained by the International Centre for Water Resources and Global Change (ICWRGC) in Koblenz, Germany, within the framework of the GEMS/Water Program (UNEP), and in cooperation with the Federal Institute of Hydrology. GEMS/Water also offers training, advice, and other tools to developing countries interested in creating monitoring programs and for conducting water-quality assessments. One of the products from GEMS is a nitrogen loading indicator, which provides a spatial assessment of potential water pollution by mapping the extent of natural and anthropogenic reactive nitrogen loading to the land and aquatic systems (Fig. 3.10). The map shows the results of a mass balance assessment of nitrogen loads to the landscape that takes into account nitrogen sources, uptake, transport, and leakages to terrestrial and riverine systems at global, continental, regional, and coastline-specific areas. As shown in Figure 3.10, in North America, nitrogen fertilizers are the dominant source of nitrogen pollution; whereas in Europe, nitrogen pollution is contributed by livestock production followed by nitrogen fertilizer production. In southern parts of Latin America and India, livestock production is the dominant contributor to nitrogen pollution. In most other areas, BNF is the dominant source of reactive nitrogen.

To reiterate, this chapter has documented the numerous anthropogenic and natural sources that contribute reactive nitrogen to the atmosphere, hydrosphere, and biosphere. Impacts from reactive nitrogen sources are described in more detail in subsequent chapters. Chapter 5 discusses adverse human health effects. Chapter 6 presents information on terrestrial biodiversity and surface water impacts from reactive nitrogen. Chapters 7 and 8 provide information on reactive nitrogen contamination of groundwater and spring systems, respectively.

REFERENCES

Adelana, S. M. A., Bale, R. B., & Wu, M. C. (2003). Hydrochemistry of domestic solid waste area in the vicinity of Lagos, Nigeria. In *Proceedings of the first international workshop on aquifer vulnerability and risk*, 28–30 May (Vol. 2, pp. 43–60). Salamanca, Mexico.

Allègre, C., Manhés, G., & Lewin, E. (2001). Chemical composition of the Earth and the volatility control on planetary genetics, Earth and Planet. *Science Letters*, 185, 46–69.

American Society of Civil Engineers (ASCE). (2017). *America's infrastructure report card: Wastewater*. Retrieved from https://www.infrastructurereportcard.org/cat-item/wastewater/, accessed 29 September 2019.

Beal, C., Menzies, N., & Gardner, E. A. (2005). Process, performance, and pollution potential: A review of septic tank-soil absorption systems. *Australian Journal of Soil Research*, 43(7). https://doi.org/10.1071/SR05018

Billen, G., Silvestre, M., Grizzetti, B., Leip, A., Garnier, J., Voss, M., et al. (2011). Nitrogen flows from European watersheds to coastal marine waters. Ch. 13. In M. A. Sutton, C. M. Howard, J. W. Erisman, G. Billen, A. Bleeker, P. Grennfelt, H. van Grinsven, & B. Grizzetti (Eds.), *The European nitrogen assessment: Sources, effects, and policy perspectives*. Cambridge, UK: Cambridge University Press, 664 p.

Böhlke, J. K. (2002). Groundwater recharge and agricultural contamination. *Hydrogeology Journal*, 10(1), 153–179.

Bosman, C. (2009). The hidden dragon: Nitrate pollution from open pit mines a case study from the Limpopo Province, South Africa. In *Abstracts of the International Mine Water Conference*, 19–23 October 2009 Proceedings, Pretoria, South Africa. ISBN Number 978-0-9802623-5-3.

Bouraoui, F., Grizzetti, B., & Aloe, A. (2009). *Nutrient discharge from river to seas for year 2000*. EC-JRC (Report EUR 24002 EN). Luxembourg, 70 p.

Bradley, J. (2008). New Zealand. In R. J. LeBlanc, P. Matthews, & R. P. Richard (Eds.), *Global atlas of excreta, wastewater sludge, and biosolids management: Moving forward the sustainable and welcome uses of a global resource* (pp. 447–454). Nairobi: United Nations Human Settlements Programme (UN-HABITAT).

Carey, R. O., Hochmuth, G. J., Martinez, C. J., Boyer, T. H., Dukes, M. D., Toor, G. S., & Cisar, J. L. (2013). Evaluating nutrient impacts in urban watersheds: Challenges and research opportunities. *Environmental Pollution*, 173, 138–149. https://doi.org/10.1016/j.envpol.2012.10.004

Carey, R. O., Hochmuth, G. J., Martinez, C. J., Boyer, T. H., Nair, V. D., Dukes, M. D., et al. (2012). A review of turfgrass fertilizer management practices: Implications for urban water quality. *HortTechnology*, *22*(3), 280–291.

Carey, R. O., & Miglaccio, K. W. (2009). Contribution of wastewater treatment plant effluents to nutrient dynamics in aquatic systems: A review. *Environmental Management*, *44*, 205–217. https://doi.org/10.1007/s00267-009-9309-5

Chesapeake Bay Program. 2011. *Summary report to the Chesapeake executive council, a message to the executive council from Jim Edward*, Acting Director, 11 July 2011, 10 p.

Cook, E. M., Hall, S. J., & Larson, K. L. (2012). Residential landscapes as social-ecological systems: A synthesis of multi-scalar interactions between people and their home environment. *Urban Ecosystems*, *15*, 19–52.

Davidson, E. A., David, M. G., Galloway, J. N., Goodale, C. L., Haeuber, R., Harrison, J. A., et al. (2012). Excess nitrogen in the U.S. environment: Trends, risks, and solutions. *Ecological Society of America, Issues in Ecology*, *15*, 1–16.

De, M., & Toor, G. S. (2015). Fate of effluent-borne nitrogen in the mounded drainfield of an onsite wastewater treatment system. *Vadose Zone Journal*, *14*(12). https://doi.org/10.2136/vzj2015.07.0096

Degnan, J. R., Böhlke, J. K., Pelham, K., Langlais, D. M., & Walsh, G. J. (2016). Identification of groundwater nitrate contamination from explosives used in road construction: Isotopic, chemical, and hydrologic evidence. *Environmental Science and Technology*, *50*(2), 593–603. https://doi.org/10.1021/acs.est.5b03671

DeVries, W. (2011). Geographical variation in terrestrial nitrogen budgets across Europe. In M. A. Sutton, C. M. Howard, J. W. Erisman, G. Billen, A. Bleeker, P. Grennfelt, et al. (Eds.), *The European nitrogen assessment: Sources, effects, and policy perspectives* (pp. 317–344). Cambridge, UK: Cambridge University Press.

Divers, M. T., Elliott, E. M., & Bain, D. J. (2013). Constraining nitrogen inputs to urban streams from leaking sewers using inverse modeling: Implications for dissolved inorganic nitrogen (DIN) retention in urban environments. *Environmental Science and Technology*, *47*, 1816–1823.

Divers, M. T., Elliott, E. M., & Bain, D. J. (2014). Quantification of nitrate source to an urban stream using dual nitrate isotopes. *Environmental Science and Technology*, *48*(18), 10580–10587. https://doi.org/10.1021/es404880

Dubrovsky, N.M., Burow, K.R., Clark, G.M., Gronberg, J.M., Hamilton P.A., Hitt, K.J., et al. (2010). The quality of our Nation's waters—*Nutrients in the Nation's streams and groundwater, 1992–2004*. U.S. Geological Survey Circular 1350, 174 p. http://water.usgs.gov/nawqa/nutrients/pubs/circ1350, accessed 19 September 2019.

Erisman, J. W., van Grinsven, H., Grizzetti, B. Bouraoui, F., Powlson, D., Sutton, M. A., Bleeker, A., & Reis, S. (2011). The European nitrogen problem in a global perspective. Ch. 2. In M.A. Sutton, C.M. Howard, J.W. Erisman Billen, G., Bleeker, A., Grennfelt, P., H. van Grinsven, & B. Grizzetti (Eds.), *The European nitrogen assessment: Sources, effects, and policy perspectives* (pp. 9–31). Cambridge, UK: Cambridge University Press.

Erisman, J. W., Galloway, J. N., Dice, N. B., Sutton, M. A., Bleeker, A., Grizzetti, B., Leach, A. M., & de Vries, W. (2015). *Nitrogen, too much of a vital resource*. Science Brief. Zeist, The Netherlands: World Wildlife Fund.

Felix, J. D., Elliott, E. M., Gish, T., Magrihang, R., Clougherty, J., & Cambal, L. (2014). Examining the transport of ammonia emissions across landscapes using nitrogen isotope ratios. *Atmospheric Environment*, *95*, 563–570. https://doi.org/10.1016/j.atmosenv.2014.06.061

Follett, R. F., & Walker, D. J. (1989). Ground water quality concerns about nitrogen. In R. F. Follett (Ed.), *Nitrogen management and ground water protection* (pp. 1–22). Amsterdam: Elsevier Science Publishers.

Fowler, D., Coyle, M., Skiba, U., Sutton, M. A., Cape, J. N., Reis, S., et al. (2013). The global nitrogen cycle in the twenty-first century. *Philosophical Transactions of the Royal Society B*, *368*, 1621. https://doi.org/10.1098/rstb.2013.0164

Fowler, D., Steadman, C. E., Stevenson, D., Coyle, M., Rees, R. M., Skiba, U. M., et al. (2015). Effects of global change during the 21st century on the nitrogen cycle. *Atmospheric Chemistry and Physics*, *15*, 13849–13893. https://doi.org/10.5194/acp-15-13849-2015

Galloway, J. N., Aber, J. D., Erisman, J. W., Seitzinger, S. P., Howarth, R. W., Cowling, E. B., & Cosby, B. J. (2003). The nitrogen cascade. *BioScience*, *53*(4), 341–356. https://doi.org/10.1641/0006-3568(2003)053[0341:TNC]2.0.CO;2

Galloway, J. N., Dentener, F. J., Capone, D. G., Boyer, E. W., Howarth, R. W., Seitzinger, S. P., et al. (2004). Nitrogen cycles: past, present and future. *Biogeochemistry*, *70*, 153–226.

Galloway, J. N., Townsend, A. R., Erisman, J. W., Bekunda, M., Cai, Z., Freney, J. R., et al. (2008). Transformation of the nitrogen cycle: Recent trends, questions, and potential solutions. *Science*, *320*, 889–892.

Gascoyne, M., & Thomas, D. A. (1997). Impact of blasting on groundwater composition in a fracture in Canada's Underground Research Laboratory. *Journal of Geophysical Research*, *102*(B1), 573–584.

Groffman, P. M., Law, N. L., Belt, K. T., Band, L. E., & Fisher, G. T. (2004). Nitrogen fluxes and retention in urban watershed ecosystems. *Ecosystems*, *7*, 393–403.

Guillard, K., & Kopp, K. L. (2004). Nitrogen fertilizer form and associated nitrate leaching from cool-season lawn turf. *Journal of Environmental Quality*, *33*(5), 1822–1827.

Hertel, O., Reis, S., Skjøth, C. A., Bleeker, A., Harrison, R., Cape, J.N., et al. (2011). Nitrogen processes in the atmosphere. In M. A. Sutton, C. M. Howard, J. W. Erisman, G. Billen, A. Bleeker, P. Grennfelt, H. van Grinsven, & B. Grizzetti (Eds.), *The European nitrogen assessment: Sources, effects, and policy perspectives* (pp. 177–207). Cambridge, UK: Cambridge University Press.

Holloway, J. M., & Dahlgren, R. A. (2002). Nitrogen in rock: Occurrences and biogeochemical implications. *Global Biogeochemical Cycles*, *16*(4), 65-1–65-17.

Holloway, J. M., Dahlgren, R. A., Hansen, B., & Casey, W. H. (1999). Contribution of bedrock nitrogen to high nitrate concentrations in stream water. *Nature*, *395*, 785–788.

Hopkins, K. G., & Bain, D. J. (2018). Research note: Mapping spatial patterns in sewer age, material, and proximity to surface waterways to infer sewer leakage hotspots. *Landscape and Urban Planning*, *170*, 320–324. http://dx.doi.org/10.1016/j.landurbplan.2017.04.011

Ihlenfeld, C., Oates, C. J., Bullock, S., & Van, Z. (2009). Isotopic fingerprinting of groundwater nitrate sources around Anglo

Platinum's rpm Mogalakwena operation (Limpopo Province, South Africa). In *International Mine Water Conference*, Pretoria, South Africa, p. 10.

Jaber, F. H., Shukla, S., Stofella, P. J., Obreza, T. A., & Hanlon, E. A. (2005). Impact of organic amendments on groundwater nitrogen concentrations for sand and calcareous soils. *Compost Science and Utilization, 13*(3), 194–202.

Katz, B. G., Eberts, S. M., & Kauffman, L. J. (2011). Using Cl/Br ratios and other indicators to assess potential impacts on groundwater quality from septic systems: A review and examples from principal aquifers in the United States. *Journal of Hydrology, 397*, 151–166.

Kaushal, S. S., Groffman, P. M., Band, L. E., Elliott, E. M., Shields, C. A., & Kendall, C. (2011). Tracking nonpoint source nitrogen pollution in human-impacted watersheds. *Environmental Science and Technology, 45*, 8225–8232, dx.doi.org. doi:10.1021/es200779e

Kinzig, A. P., Warren, P., Martin, C., Hope, D., & Katti, M. (2005). The effects of human socioeconomic status and cultural characteristics on urban patterns of biodiversity. *Ecology and Society, 10*(1), 23.

Kjeldsen, P., Barlaz, M. A., Rooker, A. P., Baun, A., Ledin, A., & Christensen, T. H. (2002). Present and long-term composition of MSW landfill leachate: A review. *Critical Reviews in Environmental Science and Technology, 32*, 297–336.

Knowles, O. A., Robinson, B. H., Contangelo, A., & Clucas, L. (2011). Biochar for the mitigation of nitrate leaching from soil amended with biosolids. *Science of the Total Environment, 409*, 3206–3210.

LaPointe, B. E., Herren, L. W., & Paule, A. L. (2017). Septic systems contribute to nutrient pollution and harmful algal blooms in the St. Lucie Estuary, Southeast Florida, USA. *Harmful Algae, 70*, 1–22. https://doi.org/10.1016/j.hal.2017.09.005

Law, N. L., Band, L. E., & Grove, J. M. (2004). Nitrogen input from residential lawn care practices in suburban watersheds in Baltimore County, Maryland. *Journal of Environmental Planning and Management, 47*(5), 737–755.

Li, L., & Davis, A. P. (2014). Urban stormwater runoff nitrogen composition and fate in bioretention systems. *Environmental Science and Technology, 48*(6), 3403–3410. https://doi.org/10.1021/es4055302

Longe, E. O., & Balogun, M. R. (2010). Groundwater quality assessment near a municipal land fil, Lagos, Nigeria. *Research Journal of Applied Sciences, Engineering and Technology, 2*(1), 39–44.

Lowe, K. S., Rothe, N. K., Tomaras, J. M. B., DeJong, K., Tucholke, M. B., Drewes, J. E., McCray, J. E., & Munakata-Marr, J. (2007). *Influent constituent characteristics of the modern waste stream from single sources: Literature review* (Tech Report 04 DEC 1a). Water Environment Research Foundation.

Lu, C., Bowman, D., Rufty, T., & Shi, W. (2015). Reactive nitrogen in turfgrass systems: Relations to soil physical, chemical, and biological properties. *Journal of Environmental Quality, 44*(1), 210–218.

Lu, Q., Zhenli, L. H., & Stoffella, P. J. (2012). Land application of biosolids in the USA: A review. *Applied and Environmental Soil Science, 2012*, 201462. http://dx.doi.org/10.1155/2012/201462

Lusk, M. G., Toor, G. S., & Inglett, P. W. (2018). Characterization of dissolved organic nitrogen in leachate from a newly established and fertilized turfgrass. *Water Research, 131*, 52–61. https://doi.org/10.1016/j.watres.2017.11.040

Lusk, M. G., Toor, G. S., Yang, Y., Mechtensimer, S., De, M., & Obreza, T. A. (2017). A review of the fate and transport of nitrogen, phosphorus, pathogens, and trace organic chemicals in septic systems. *Critical Reviews in Environmental Science and Technology, 47*(7), 455–541.

Maupin, M. A., & Ivahnenko, T. (2011). Nutrient loadings to streams of the continental United States from municipal and industrial effluent. *Journal of the American Water Resources Association, 47*(5), 950–964. https://doi.org/10.1111/j.1752-1688.2011.00576.x

Milesi, C., Running, S. W., Elvidge, C. D., Dietz, J. B., Tuttle, B. T., & Nemani, R. R. (2005). Mapping and modeling the biogeochemical cycling of turf grasses in the United States. *Environmental Management, 36*(3), 426–438.

Montross, G. G., McGlynn, B. L., Montross, S. N., & Gardner, K. K. (2013). Nitrogen production from geochemical weathering of rocks in southwest Montana, USA. *Journal of Geophysical Research, 118*(3), 1068–1078.

Morford, S., Houlton, B. Z., & Dahlgren, R. A. (2011). Increased forest nitrogen and carbon storage from nitrogen-rich bedrock. *Nature, 477*, 78–81.

Munn, M.D., Frey, J.W., Tesoriero, A.J., Black, R.W., Duff, J.H., Lee, K.E., et al. (2018). *Understanding the influence of nutrients on stream ecosystems in agricultural landscapes*. U.S. Geological Survey Circular 1437, 80 p.

Oenema, O., Oudendag, D., & Velthof, G. L. (2007). Nutrient losses from manure management in the European Union. *Livestock Science, 112*, 261–272.

O'Reilly, A. M., Wanielista, M. P., Chang, N. B., Xuan, Z., & Harris, W. G. (2012). Nutrient removal using biosorption activated media: Preliminary biogeochemical assessment of an innovative stormwater infiltration basin. *Science of the Total Environment, 432*, 227–242.

Panno, S. V., Kelley, W. R., Hackley, K. C., & Weibel, C. P. (2007). Chemical and bacterial quality of aeration-type wastewater treatment system discharge. *Ground Water Monitoring and Remediation, 27*(2), 71–76.

Pionke, H. B., Sharma, M. L., & Hirschberg, K. J. (1990). Impact of irrigated horticulture on nitrate concentrations in groundwater. *Agricultural Ecosystems and Environment, 32*, 119–132.

Puckett, L. J. (1995). Identifying the major sources of nutrient water pollution. *Environmental Science and Technology, 29*(9), 408A–414A. https://doi.org/10.1021/es00009a743

Reis, S., Pinder, R. W., Zhang, M., Lijie, G., & Sutton, M. A. (2009). Reactive nitrogen in atmospheric emission inventories. *Atmospheric Chemistry and Physics, 9*, 7657–7677.

Renou, S., Givaudan, J. G., Poulain, S., Dirassouyan, F., & Moulin, P. (2008). Landfill leachate treatment: Review and opportunity. *Journal of Hazardous Materials, 150*, 468–493.

Reynolds, J. H., & Barrett, M. H. (2003). A review of the effects of sewer leakage on groundwater quality. *Water and Environment Journal, 17*(1), 34–39.

Risch, E., Gutierrez, O., Roux, P., Boutin, C., & Corominas, L. (2015). Life cycle assessment of urban wastewater systems: Quantifying the relative contribution of sewer systems. *Water Research, 77*, 35–48.

Rueedi, J., Cronin, A. A., & Morris, B. L. (2009). Estimation of sewer leakage to urban groundwater using depth-specific hydrochemistry. *Water and Environment Journal*, *23*(2), 134–144.

Schlesinger, W. H. (1997). *Biogeochemistry – An analysis of global change*. San Diego, CA: Academic Press, 588 p.

Schueler, T. (2003). *Impacts of impervious cover on aquatic systems* (Watershed Protection Research Monograph No. 1). Ellicot City, MD: Center for Watershed Protection.

Schwede, D. B., & Lear, G. G. (2014). A novel hybrid approach for estimating total deposition in the United States. *Atmospheric Environment*, *92*, 207–220.

Simpson, D., Aas, W., Bartnicki, J., Berge, H., Bleeker, A., Cuvelier, K., et al. (2011). Atmospheric transport and deposition of reactive nitrogen in Europe. In M. A. Sutton, C. M. Howard, J. W. Erisman, G. Billen, A. Bleeker, P. Grennfelt, H. van Grinsven, & B. Grizzetti (Eds.), *The European nitrogen assessment: Sources, effects, and policy perspectives* (pp. 298–316). Cambridge, UK: Cambridge University Press.

Skinner, K.D., & Maupin, M.A. (2019). *Point-source nutrient loads to streams of the conterminous United States, 2012*. U.S. Geological Survey Data Series 1101, 13 p. https://doi.org/10.3133/ds1101.

State-EPA Nutrient Innovations Task Group. (2009). *An urgent call to action*. Report to EPA, 27 August 2009, 170 p.

Sullivan, D. M., Fransen, S. C., Cogger, C. G., & Bary, A. I. (1997). Biosolids and dairy manure as nitrogen sources for prairiegrass on a poorly drained soil. *Journal of Production Agriculture*, *10*(4), 589–596.

Sutton, M. A., Howard, C. M., Erisman, J. W., Billen, G., Bleeker, A., Grennfelt, P., et al. (2011). *The European nitrogen assessment: Sources, effects, and policy perspectives*. Cambridge, UK: Cambridge University Press, 664 p.

Taylor, G. D., Fletcher, T. D., Wong, T. H. F., Breenc, P. F., & Duncan, H. P. (2005). Nitrogen composition in urban runoff—Implications for stormwater management. *Water Research*, *39*, 1982–1989.

Taylor, R. G., Cronin, A. A., Lerner, D. N., Tellam, J. H., Bottrell, S. H., Rueedi, J., & Barrett, M. H. (2006). Hydrochemical evidence of the depth of penetration of anthropogenic recharge in sandstone Aquifers underlying two mature cities in the UK. *Applied Geochemistry*, *21*, 1570–1592.

Telenko, D. E. P., Shaddox, T. W., Unruh, J. B., & Trenholm, L. E. (2015). Nitrate leaching, turf quality, and growth rate of "Floratam" St. Augustinegrass and common centipedegrass. *Crop Science*, *55*, 1320–1328.

Toor, G. S., Lusk, M., & Obreza, T. (2017). *Onsite sewage treatment and disposal systems: An overview*. University of Florida Institute of Food and Agricultural Sciences Publication SL347, 10 p.

Tucker, W. A., Diblin, M. C., Mattson, R. A., Hicks, R. W., & Wang, Y. (2014). Nitrate in shallow groundwater associated with residential land use in central Florida. *Journal of Environmental Quality*, *43*, 639–643. https://doi.org/10.2134/jeq2012.0370

U.S. Department of Agriculture (USDA). 2012. *Census of agriculture*. U.S. Summary and State Data. Retrieved from http://www.agcensus.usda.gov/Publications/2012/, accessed 29 September 2019.

U.S. Environmental Protection Agency. (2002b). *2000 National water quality inventory*. Washington, DC. Retrieved from https://www.epa.gov/nps/urban-runoff-national-management-measures: U.S. Environmental Protection Agency, Office of Water, accessed 30 March 2018

U.S. Environmental Protection Agency. (2008). *Municipal solid waste in the United States: 2007 facts and figures*. EPA530-R-08-010. Washington, DC. http://www.epa.gov/epawaste/nonhaz/municipal/pubs/msw07-rpt.pdf

Velthof, G., Oudendag, D., Witzke, H. P., Asman, W. A. H., & Klimont, Z. (2009). Integrated assessment of nitrogen losses from agriculture in EU-27 using MITERRA-EUROPE. *Journal Environmental Quality*, *38*, 402–417.

Wakida, F. T., & Lerner, D. N. (2005). Non-agricultural sources of groundwater nitrate: A review and case study. *Water Research*, *39*(1), 3–16. https://doi.org/10.1016/j.watres.2004.07.026

Wang, S., Tang, C., Song, X., Yuan, R., Han, Z., & Pan, Y. (2016). Factors contributing to nitrate contamination in a groundwater recharge area of the North China Plain. *Hydrological Processes*, *30*, 2271–2285. https://doi.org/10.1002/hyp.10778

Wang, S., Tang, C., Song, X., Yuan, R., Wang, Q., & Zhang, Y. (2013). Using major ions and $\delta^{15}N$-NO_3 to identify nitrate sources and fate in an alluvial aquifer of Baiyangdian lake watershed, North China Plain. *Environmental Science: Processes and Impacts*, *15*, 1430–1443.

Ward, B. B. (2012). The global nitrogen cycle. In A. H. Knoll, D. E. Canfield, & K. O. Konhauser (Eds.), *Fundamentals of geomicrobiology* (pp. 36–48). Chichester, UK: Wiley-Blackwell.

Zhang, Q., Sun, J., Liu, J., Huang, G., Lu, C., & Zhang, Y. (2015). Driving mechanism and sources of groundwater nitrate contamination in the rapidly urbanized region of south China. *Journal of Contaminant Hydrology*, *182*, 221–230.

Zhang, Q. H., Yang, W. N., Ngo, H. H., Guo, W. S., Jin, P. K., Dzakpasu, M., et al. (2016). Current status of urban wastewater treatment plants in China. *Environment International*, *92–93*, 11–22. https://doi.org/10.1016/j.envint.2016.03.024

Zhang, W., Tian, Z., Zhang, N., & Li, X. (1996). Nitrate pollution of groundwater in northern China. *Agriculture, Ecosystems, and Environment*, *59*, 223–231.

Zhou, W., Troy, A., & Grove, M. (2008). Modeling residential lawn fertilization practices: Integrating high resolution remote sensing with socioeconomic data. *Environmental Management*, *41*, 742–752.

FURTHER READING

Arnade, L. J. (1999). Seasonal correlation of well contamination and septic tank distance. *Ground Water*, *37*, 920–923.

Barton, L., & Colmer, T. D. (2006). Irrigation and fertilizer strategies for minimizing nitrogen leaching from turfgrass. *Agricultural Water Management*, *80*, 160–175.

Beaulieu, J., Tank, J., Hamilton, S., Wollheim, W., Hall, R., Mulholland, P., et al. (2011). Nitrous oxide emission from denitrification in stream and river networks. *Proceedings of the National Academy of Science*, *108*, 214–219. https://doi.org/10.1073/pnas.1011464108

Candela, L., Fabregat, S., Josa, A., Suriol, J., Vigues, N., & Mas, J. (2007). Assessment of soil and groundwater impacts by treated urban wastewater reuse: A case study: Application in a golf course (Girona Spain). *Science of the Total Environment, 374*, 26–35.

Costa, J. E., Heufelder, G., Foss, S., Milham, N. P., & Howes, B. (2002). Nitrogen removal efficiencies of three alternative septic system technologies and a conventional septic system. *Environment Cape Cod, 5*(1), 15–24.

Dubeux, J. C. B., Jr., Sollenberger, L. E., Mathews, B. W., Scholberg, J. M., & Santos, H. Q. (2007). Nutrient cycling in warm-climate grasslands. *Crop Science, 47*, 915–928.

EDGAR. (2010). *Integrated project of climate change and impact research: The Mediterranean environment* (Project No. 036961–CIRCE). Retrieved from http://www.mnp.nl/edgar/model/v32ft2000edgar/, accessed 29 September 2019.

Grove, J. M., Cadenasso, M. L., Burch, W. R., Pickett, S. T. A., Schwarz, K., O'Neil-Dunne, J., & Wilson, M. (2006). Data and methods comparing social structure and vegetation structure of urban neighborhoods in Baltimore, Maryland. *Society and Natural Resources, 19*, 117–136.

Hertel, O., Skjoth, C. A., Lofstrom, P., Geels, C., Frohn, L. M., Ellermann, T., et al. (2006). Modelling nitrogen deposition on a local scale: A review of the current state of the art. *Environmental Chemistry, 3*, 317–337.

Jordan, M. J., Nadelhoffer, H. J., & Fry, B. (1997). Nitrogen cycling in forest and grass ecosystems irrigated with ^{15}N-enriched wastewater. *Ecological Applications, 7*(3), 864–881.

Katz, B. G., Sepulveda, A. A., & Verdi, R. J. (2009). Estimating nitrogen loading to ground water and assessing vulnerability to nitrate contamination in a large karstic springs basin, Florida. *Journal of the American Water Resources Association, 45*, 3.

Kaushal, S. S., McDowell, W. H., & Wollheim, W. M. (2014). Tracking evolution of urban biogeochemical cycles: Past, present, and future. *Biogeochemistry, 121*, 1–21. https://doi.org/10.1007/s10533-014-0014-y

Liao, L., Green, C. T., Bekins, B. A., & Bohlke, J. K. (2012). Factors controlling nitrate fluxes in groundwater in agricultural areas. *Water Resources Research, 48*, 1–18.

McCray, J. E., Kirkland, S. L., Siegrist, R. L., & Thyne, J. D. (2005). Model parameters for simulating fate and transport of on-site wastewater nutrients. *Groundwater, 43*, 628–639.

Mueller, D.K., & Helsel, D.R. (1996). *Nutrients in the Nation's waters: Too much of a good thing?* U.S. Geological Survey Circular 1136, 31 p.

Obour, A. K., Silveira, M. L., Vendramini, J. M. B., Adjei, M. B., & Sollenberger, L. E. (2010). Evaluating cattle manure application strategies on phosphorus and nitrogen losses from a Florida spodosol. *Agronomy Journal, 102*(5), 1511–1520.

Paramasivam, S., & Alva, A. K. (1997). Leaching of nitrogen forms from controlled-release nitrogen fertilizers. *Communications in Soil Science and Plant Analysis, 28*(17&18), 1663–1674.

Poiani, K. A., Bedord, B. L., & Merrill, M. D. (1996). A GIS-based index for relating landscape characteristics to potential nitrogen leaching to wetlands. *Landscape Ecology, 11*, 237–255.

Qiu, J. (2018). Safeguarding China's water resources. *National Science Review, 5*, 102–107. https://doi.org/10.1093/nsr/nwy007

Ruddy, B.C., Lorenz, D.L., & Mueller, D.K. (2006). *County-level estimates of nutrient inputs to the land surface of the conterminous United States, 1982–2001*. U.S. Geological Survey Scientific Investigations Report 2006–5012.

Sprague, L.A., & Gronberg, J.M. (2013). *Estimation of anthropogenic nitrogen and phosphorus inputs to the land surface of the conterminous United States—1992, 1997, and 2002*. U.S. Geological Survey Scientific Investigations Report 2012–5241.

U.S. Environmental Protection Agency (U.S. EPA). (2001). *National Pollutant Discharge Elimination System permit regulation and effluentlimitations guidelines and standards for Concentrated Animal Feeding Operations (CAFOs)*. Federal Register, 66, 2960–3145.

U.S. Environmental Protection Agency. (2002a). *Onsite wastewater treatment systems manual* (EPA/625/R-00/008). Washington, DC: Office of Water.

Wang, L., Butcher, A. S., Stuart, M. E., Gooddy, D. C., & Bloomfield, J. P. (2013). The nitrate time bomb: A numerical way to investigate nitrate storage and lag time in the unsaturated zone. *Environmental Geochemistry and Health, 35*, 667–681.

White-Leech, R., Liu, K., Sollenberger, L. E., Woodard, K. R., & Interrante, S. M. (2013a). Excreta deposition on grassland patches. I. Forage harvested, nutritive value, and nitrogen recovery. *Crop Science, 53*, 688–695.

White-Leech, R., Liu, K., Sollenberger, L. E., Woodard, K. R., & Interrante, S. M. (2013b). Excreta deposition on grassland patches. II. Spatial pattern and duration of forage responses. *Crop Science, 53*, 696–703.

Wolf, L., Klinger, J., Hoetzl, H., & Mohriok, U. (2007). Quantifying mass fluxes from urban drainage systems to the urban soil-aquifer system. *Journal of Soils and Sediments, 7*(2), 85–95. https://doi.org/10.1065/jss2007.02.207

4
Methods to Identify Sources of Reactive Nitrogen Contamination

Previously, in Chapter 3, we learned about the numerous point and nonpoint sources of reactive nitrogen that have contributed to contamination of surface waters and groundwater in many places around the world. Widespread nitrate contamination of groundwater and surface waters (lakes, streams, rivers, estuaries, and oceans) has resulted from the extensive use of synthetic fertilizers in agriculture, animal wastes, human wastewater (effluent from septic tanks and wastewater treatment facilities), and atmospheric deposition of reactive nitrogen from fossil fuel combustion and biomass burning. Elevated nitrate concentrations have caused eutrophication of aquatic systems, harmful algal blooms (HABs), loss of biodiversity, acidification of soils and surface waters, and adverse human health issues. It is critical to have detailed information on sources of nitrate contamination and the processes that control the fate and transport of nitrate and other reactive nitrogen species. This information will lead to more effective resource management to protect water quality from further degradation and to develop more informed remediation and restoration plans for sites that are contaminated. However, identifying the sources of nitrate and its fate in environmental systems can be complex, as our understanding of sources and fate can be confounded by mixtures of nitrate from multiple sources, temporal and spatial variability of the components of the nitrogen cycle, and our ability to quantify the amount and rates of nitrate attenuation (principally from denitrification) in groundwater, rivers, oceans, and other systems.

Over the past four decades, research scientists have developed and used many innovative methods for identifying sources of nitrate contamination. Various isotopic and other chemical indicators have been used to assess nitrate contamination sources and the fate of reactive nitrogen in diverse environmental systems (e.g., Badruzzaman et al., 2012; Kendall et al., 2007; Pasten-Zapata et al., 2014; Verstraeten et al., 2005; Xue et al., 2009; and other references follow). The following sections describe the various isotopic and other chemical techniques and methods for identifying sources of nitrate contamination and processes that control the transport and fate of nitrate in environmental systems. It is important to keep in mind that while these isotopic and other chemical methods are useful tools for assessing sources of contamination, their effectiveness is dependent on a thorough understanding of the hydrologic system (e.g., groundwater flow patterns, recharge processes, groundwater surface-water interactions, river dynamics during baseflow and high-flow conditions, submarine groundwater discharge, ocean transport processes).

4.1. NITROGEN ISOTOPES FOR SOURCE IDENTIFICATION

Stable isotopes (one or more forms of a chemical element with a different number of neutrons in the atom's nucleus that do not decay, as do radioactive isotopes of elements) of nitrogen, oxygen, and other elements (e.g., carbon, hydrogen, boron, and sulfur) have been used extensively in water-quality and ecological research studies as an effective tool for determining sources and sinks of nitrate contamination. Stable isotopes are very useful tools because typically there is a large difference between the abundance of the common and rare isotopes of nitrogen and other elements (e.g., oxygen, hydrogen, carbon, and boron). For example, 99.64% of atmospheric N is made up of the abundant ^{14}N isotope compared to the rare isotope, ^{15}N, which makes up only 0.36%. Likewise, for stable oxygen isotopes, the distribution of oxygen isotopes is: ^{16}O is 99.76%; ^{17}O: 0.04%; and ^{18}O: 0.20%). Isotopic ratios for measured samples are reported in terms of delta (δ) values for isotopic ratios: ^{15}N/^{14}N (δ^{15}N), ^{18}O/^{16}O (δ^{18}O) and ^{17}O/^{16}O (δ^{17}O), which are expressed (reported) as parts per thousand differences (‰) from isotope ratios in accepted standards (for nitrogen, the atmosphere; for oxygen, the Vienna Standard Mean Ocean Water).

Nitrogen Overload: Environmental Degradation, Ramifications, and Economic Costs, Geophysical Monograph 250, First Edition. Brian G. Katz.
© 2020 American Geophysical Union. Published 2020 by John Wiley & Sons, Inc.

The following equation is used to determine isotopic compositions (parts per thousand) of samples:

$$\delta^{15}N_{sample} \text{ or } \delta^{18}O_{sample} = ((R_{sample} - R_{standard})/(R_{standard}))*1000 \quad (4.1)$$

Where R_{sample} and $R_{standard}$ represents the isotopic ratio for samples and a standard. Theoretical information regarding isotopic tracers, basic isotope chemistry, and fractionation effects are beyond the scope of this book and are discussed in detail elsewhere (e.g., Aravena et al., 1993; Kendall & Aravena, 2000; Kendall & McDonnell, 1998; Sulzman, 2007; Xue et al., 2009).

Early studies on differentiating sources of nitrate focused on the use of nitrogen isotopes in groundwater to provide information mainly on inorganic versus organic sources of nitrogen (Heaton, 1986; Kohl et al., 1971; Kreitler, 1975; Kreitler & Browning, 1983; Mayer et al., 2002; Peterson & Fry, 1987). Different nitrogen sources and sinks exhibit a range of nitrogen isotope values as shown in Figure 4.1. There are several nitrogen isotope fractionation processes that can modify nitrogen isotope ratios. These processes can result in enrichment in the heavier or lighter isotope relative to each other (e.g., volatilization, denitrification, and nitrification) leading to a large range of values that complicates the use of nitrogen isotopes as a source indicator. Nitrogen that originates from inorganic sources, such as ammonium fertilizer and nitrate fertilizer (which are produced by fixation of atmospheric N_2), shows a relatively small range in $\delta^{15}N$ values, typically between −6 and +6‰ (Flipse & Bonner, 1985).

Nitrogen isotope values in atmospheric deposition tend to show a large range (−13 to +13‰) due to various sources of nitrogen emissions and various chemical reactions in the atmosphere and atmospheric sources such fossil fuel combustion and biomass burning (Kendall, 1998; Kendall et al., 2007; Xue et al., 2009). Organic nitrogen sources, such as manure and sewage, tend to be enriched in the ^{15}N relative to ^{14}N. This enrichment results from volatilization of ammonia during storage, treatment, and land application of manure and human wastes. Values of $\delta^{15}N$ of nitrate originating from manure typically range from +5 to +25‰, and those originating from sewage range from +4 to +19‰ (Xue et al., 2009). Volatilization during waste storage or disposal results in a large enrichment of $\delta^{15}N$ in the residual ammonium and the isotopically enriched ammonium is converted into $\delta^{15}N$-enriched nitrate. Soil nitrogen typically has $\delta^{15}N$ values that range from 0 to +8‰ that are related to the relative rate of mineralization (ammonification) and nitrification, soil depth, vegetation, climate, and land-use history (Kendall, 1998). Nitrate contamination in surface water and groundwater typically originates from agricultural activities, human waste disposal, and some industrial activity (atmospheric emissions and deposition) and $\delta^{15}N$ values generally range from −4 to +15‰. Once nitrate enters surface water or groundwater, most studies have assumed that nitrate moves conservatively unless it encounters reducing conditions (Böhlke & Denver, 1995; Mariotti et al., 1981). However, given the large amount of overlap for $\delta^{15}N$ values among the various nitrogen sources, and biogeochemical processes

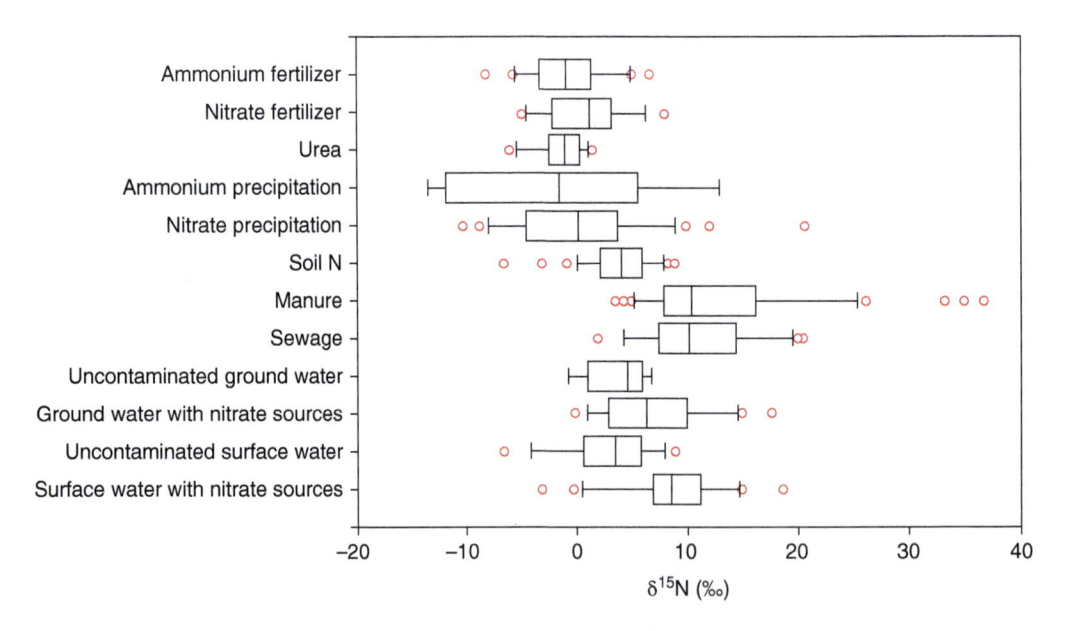

Figure 4.1 Ranges of $\delta^{15}N$ values of nitrate for various sources and sinks (From Xue et al., 2009, based on data from the many references contained therein). Box plots show the 25th, 50th, and 75th percentiles, the whiskers indicate the 10th and 90th percentiles, and the circles indicate outlier values. *Source*: Reproduced with permission of Elsevier.

that change $\delta^{15}N$ values, additional isotopic and other chemical tracers have been increasingly used to resolve sources of nitrate contamination.

4.2. DUAL ISOTOPES OF NITRATE (NITROGEN AND OXYGEN) FOR SOURCE IDENTIFICATION

During the past 20 years, the combined use of both nitrogen ($\delta^{15}N$) and oxygen ($\delta^{18}O$) isotopes of nitrate have provided useful additional information that can account for fractionation processes and thus address some of the problems with overlapping nitrogen isotopic values from different sources (e.g., Badruzzaman et al., 2012; Deutsch et al., 2006; Fenech et al., 2012; Kendall, 1998; Kendall et al., 2007; Kendall & Aravena, 2000; Xue et al., 2009). These new approaches have been increasingly used to study the impact of agricultural sources on groundwater and surface water and to investigate nitrogen transformations in agricultural, urban, and forested watersheds that ranged in size from small to very large basins (Kendall et al., 2007). Many applications of nitrate isotopes in different environmental and ecological settings (e.g., small forested catchments, urban streams, small agricultural rivers, large river basins, wetlands, groundwater, and coastal and estuarine environments) are summarized by Kendall et al. (2007).

Several analytical techniques have been developed for determining both $\delta^{15}N-NO_3$ and $\delta^{18}O-NO_3$ in water samples. Descriptions of these methods, along with the advantages and disadvantages of each method are discussed in detail by Kendall et al. (2007) and Xue et al. (2009). Here, we present a brief summary of the three methods that are most widely used. An ion-exchange or silver nitrate ($AgNO_3$) method was developed by Chang et al. (1999) and Silva et al. (2000). Nitrate is purified and concentrated by passing samples through cation and then anion exchange resin columns. Nitrate is then eluted, neutralized with silver oxide and then filtered to remove the silver chloride (AgCl). A second method for determining $\delta^{15}N-NO_3$ and $\delta^{18}O-NO_3$ in water samples is referred to as the bacterial denitrification method (Casciotti et al., 2002; Rock & Ellert, 2007; Sigman et al., 2001). $\delta^{15}N-NO_3$ and $\delta^{18}O-NO_3$ are determined simultaneously from analysis by mass spectrometry of N_2O, which is produced from the conversion of nitrate by denitrifying bacteria. These bacteria produce N_2O because they lack N_2O-reductase activity, the enzyme that catalyzes the reduction reaction of N_2O to N_2. A third method involves the conversion of nitrate to N_2O by a two-step chemical reduction procedure (McIlvin & Altabet, 2005). This cadmium-reduction method or azide method reduces nitrate to nitrite, which is followed by reduction to N_2O using sodium azide. The N_2O is analyzed using mass spectrophotometric techniques. These methodological advances (especially the bacterial denitrification method) have enabled researches to evaluate environmental variability of nitrogen contamination sources and deduce patterns of contamination.

4.2.1. Ranges of $\delta^{15}N$ and $\delta^{18}O$ Values of Nitrate for Various Sources

Figure 4.2 contains a plot showing typical ranges of $\delta^{15}N$ and $\delta^{18}O$ values of nitrate representing various nitrogen sources and environmental processes (Kendall et al., 2007). Atmospheric data are divided into the ranges observed for samples analyzed using the denitrifier and $AgNO_3$ (nondenitrifier) methods. The two arrows in Figure 4.2 represent generalized slopes for data indicating denitrification of nitrate starting with initial values of $\delta^{15}N-NO_3$ (+6‰) and $\delta^{18}O-NO_3$ (−9‰). Soil waters tend to have higher $\delta^{18}O-NO_3$ values, and a larger range of $\delta^{18}O-NO_3$ values than groundwater because of the higher $\delta^{18}O$ values of O_2 and/or water in soils (from Kendall & Aravena, 2000; Kendall et al., 2007). Although there is still considerable overlap of nitrate isotopic values for some sources (Fig. 4.2), the additional information from $\delta^{18}O-NO_3$ clearly allows for distinguishing between fertilizer versus soil nitrate or atmospheric nitrate versus soil nitrate that would not be possible using $\delta^{15}N$ alone.

Figure 4.3 shows box plots of $\delta^{18}O-NO_3$ values that are useful for distinguishing between nitrification, nitrate in precipitation, and fertilizers containing nitrate (Xue et al., 2009) even though nitrification of ammonium and/or organic-N in fertilizer, precipitation, and organic waste can have a wide range of $\delta^{15}N$ and $\delta^{18}O$ values. The large range of $\delta^{18}O-NO_3$ values in precipitation results from a combination of complex atmospheric processes with $\delta^{18}O-NO_3$ values that vary spatially and temporally. Also, factors that can affect $\delta^{18}O-NO_3$ values in precipitation include the type of combustion source (e.g., coal-fired power plants, fossil fuel combustion emissions from different vehicles) and isotopic fractionation during nitrate formation caused by thunderstorms, incorporation of the $\delta^{18}O$ of reactive oxygen in the atmosphere, and other isotopic fractionation processes as nitrate reacts with other compounds in the atmosphere (Kendall, 1998; Kendall et al., 2007; Xue et al., 2009). As a result, the $\delta^{18}O$ of atmospheric nitrate is isotopically enriched relative to atmospheric oxygen (+23.5‰) and ranges from +25 to +75‰ (Xue et al., 2009). The $\delta^{15}N$ values of atmospheric nitrate and ammonia generally are in the range of −15 to +15‰, relative to atmospheric nitrogen gas ($\delta^{15}N$ = 0). Also, $\delta^{15}N$ and $\delta^{18}O$ values in wet deposition and dry deposition can show considerable variations seasonally and spatially (e.g., depending on proximity to marine-derived nitrogen) (Kendall et al., 2007).

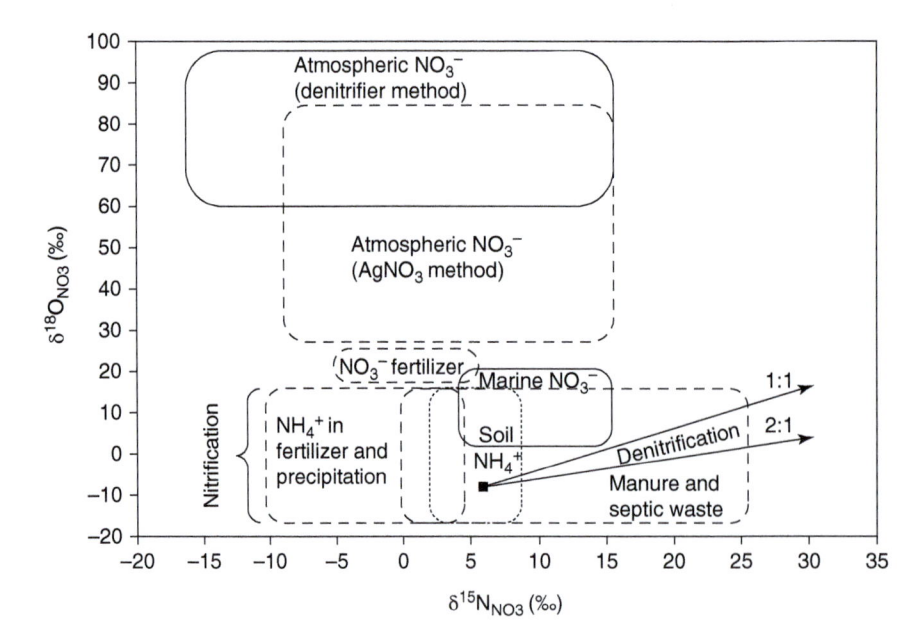

Figure 4.2 Typical ranges of $\delta^{15}N$ and $\delta^{18}O$ values of nitrate representing various sources and processes. *Source*: From Kendall et al. (2007). From a John Wiley & Sons publication.

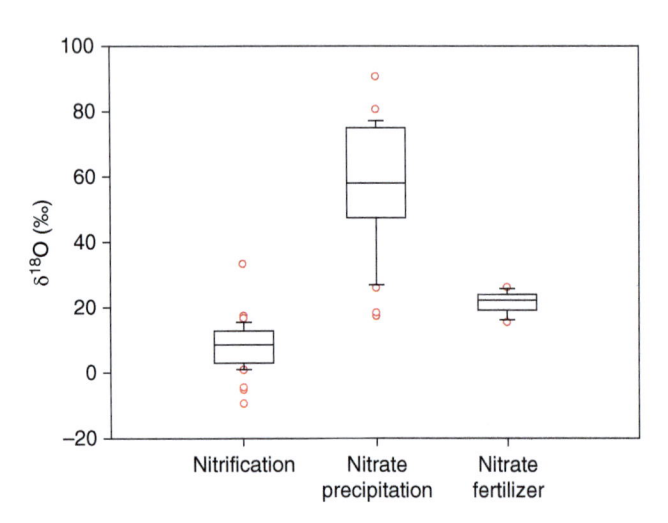

Figure 4.3 Box plots showing the ranges of $\delta^{18}O$ values of nitrate from nitrification, nitrate precipitation, and nitrate fertilizer. Plots show the 25th, 50th, and 75th percentiles, the whiskers indicate the 10th and 90th percentiles, and the circles indicate outlier values. *Source*: From Xue et al. (2009), a summary based on data from the many references contained therein. Reproduced with permission of Elsevier.

There is a wide variety of fertilizer types containing varying forms and amounts of nitrogen that are added to different types of crops and turfgrass, as mentioned in Chapter 3. $\delta^{15}N$ values for fertilizers generally are in the range of −4 to +4‰, although Kendall et al. (2007) reported a range of fertilizer values from −8 to +7‰. The following ranges for $\delta^{18}O\text{-}NO_3$ have been reported for inorganic fertilizers: synthetic (+17 to 25‰) and natural fertilizers derived from Chilean nitrate deposits (+46 to 58‰) (Böhlke et al., 2003).

As can be seen in Figure 4.2, there can be considerable overlap of the $\delta^{18}O\text{-}NO_3$ composition for nitrate derived from soil organic matter, nitrification of ammonium fertilizers, and animal wastes. Some researchers have plotted $\delta^{18}O\text{-}NO_3$ versus $\delta^{18}O\text{-}H_2O$ to assess nitrification in contact with ambient water (McMahon & Böhlke, 2006; Wankel et al., 2006). Although using $\delta^{15}N$ values, Kendall et al. (2007) noted that soil nitrate produced from fertilizers (average $\delta^{15}N$ value is $+4.7 \pm 5.4$‰) generally can be distinguished from soil nitrate produced from animal waste (average $\delta^{15}N$ value is $+14.0 \pm 8.8$‰). Septic tank and animal wastes tend to have overlapping signatures. However, Fogg (1998) extracted nitrate from 218 core samples collected below natural (soil organic matter) fertilizers, onsite sewage disposal systems (septic tank effluent), and animal sources located in Salinas and Sacramento Valleys, California. They reported that $\delta^{15}N$ values of animal sources varied from about +8 to +20‰ and were dependent on site and animal source. Furthermore, the $\delta^{15}N$ of on-site sewage disposal sources varied from about +2 to +12‰, and the mean was significantly different from that of animal sources at a 90% confidence level.

During denitrification, nitrate isotopes ($\delta^{15}N$ and $\delta^{18}O$) vary systematically, causing the $\delta^{15}N\text{-}NO_3$ and $\delta^{18}O\text{-}NO_3$ in the remaining pool of nitrate to increase in a predictable way. The isotopic enrichment of $\delta^{15}N\text{-}NO_3$ and $\delta^{18}O\text{-}NO_3$ increases in approximately a 2:1 ratio or slope shown on Figure 4.2. However, recent work involving pure cultures of denitrifying bacteria has indicated a slope equal to 1.0 for the respiratory process of denitrification (Sigman et al., 2001). Even with various slopes reported in denitrification studies, the range of observed denitrification slopes fall within the two slope lines on Figure 4.2, which still allows for the effective determination of the denitrification process when plotting data for $\delta^{15}N\text{-}NO_3$ and $\delta^{18}O\text{-}NO_3$. If a mixture of two sources of nitrate is suspected, a plot of $\delta^{15}N\text{-}NO_3$ or $\delta^{18}O\text{-}NO_3$ data versus 1/NO_3 will result in a straight line. Several other studies have analyzed the $\delta^{15}N$ of N_2 gas produced during denitrification to estimate the initial nitrate concentration to study the denitrification reaction progress (e.g., Böhlke et al., 2002; McMahon et al., 2008a).

4.2.2. Limitations of the Dual Nitrate Isotope Method for Determining Nitrate Sources of Contamination

The general ranges of nitrate isotopic values shown in Figure 4.2 can be used as a first step for differentiating nitrate sources; however, the complex processes involved in nitrogen cycling in environmental systems can result in measured isotopic values that are outside these ranges. Source differentiation becomes even more difficult when there is a mixed pool of nitrate from different sources (including cycling through an organic matter pool) or when there are seasonal or storm-related cycles of denitrification and nitrification. Thus, as Kendall et al. (2007) noted, "nitrate isotope techniques are not a panacea for source identification; they are merely one set of tools that hydrogeologists and biogeochemists can use when assessing nitrogen cycling processes in groundwater, surface water and in various ecosystems." To reiterate, the main limitations for using isotope methods include the overlapping isotopic signatures of different sources, spatial and temporal variations in isotopic composition of sources, and isotopic fractionations in environmental systems that can mask or prevent distinctive isotopic compositions for sources. In freshwater systems, the use of $\delta^{18}O\text{-}NO_3$ values to identify sources can be affected by various processes including evaporation, seasonal changes of $\delta^{18}O\text{-}NO_3$ in rainfall, changes in the proportion of oxygen from H_2O or O_2 sources, microbial respiration, and nitrification occurring at the same time from both heterotrophic and autotrophic microbial pathways (Kendall et al., 2007).

Studies over the past two decades have pointed out various limitations when using the dual nitrate isotope method for nitrogen source identification. For example, based on lysimeter field experiments in agricultural soils, Mengis et al. (2001) found that the $\delta^{18}O\text{-}NO_3$ values were affected by microbial immobilization and subsequent mineralization and nitrification to nitrate. The original oxygen isotope ratio was masked by this process during summer months when microbial activity was high, but not affected during winter months when microbial activity was low. A soil incubation experiment in another study (Kool et al., 2011) indicated that oxygen exchange between oxygen from nitrate and H_2O affected the isotopic signature of nitrate. This exchange would result in an overestimation of microbial nitrification as a source of nitrate and underestimate the contribution from other sources such as fertilizer and atmospheric deposition. Also, the exchange of oxygen may lead to an underestimation of denitrification progress.

In a recent study, Granger and Wankel (2016) used a numerical model to resolve inconsistencies from experimental studies of nitrifying and denitrifying bacterial cultures. Their models indicated that there are two main driving factors for the observed isotopic overprinting, and they are related to $\delta^{18}O$ of ambient water and the relative flux of nitrate production under net denitrifying conditions (catalyzed either aerobically or anaerobically). Slopes less than 1 for denitrification trends in freshwater systems, such as in anoxic aquifers, resulted from concurrent nitrate production by anammox, which is a process that has been widely overlooked in these systems. In marine systems, where slopes for denitrification trends for dual isotope trajectories are greater than 1.0, these typically result from elevated rates of nitrite (NO_2) re-oxidation relative to nitrate reduction and to the $\delta^{18}O$ value of seawater (Granger & Wankel, 2016).

The targeting of the actual source(s) of nitrate contamination in water bodies and ecosystems is critical for effective remediation of contaminated sites to reduce adverse effects to public health and ecosystems. Given these aforementioned limitations for using nitrate isotopes for source identification, other novel approaches have been developed. These approaches are discussed in the following sections and typically involve the combined use of isotopes along with other chemical markers.

4.3. OTHER CHEMICAL INDICATORS AND TRACERS OF NITRATE CONTAMINATION SOURCES

As is evident from Figures 4.1 and 4.3, to reiterate, it is difficult to use information from nitrogen isotopes or the dual nitrate isotopes to distinguish between animal waste (manure) sources and sewage-derived nitrates, as these two sources have overlapping isotopic signatures. As a way of enhancing the usefulness of the nitrate isotope approach for differentiating between these and other sources of nitrate contamination, Kendall et al. (2007) noted four innovative and effective methods that have been used. These include (a) nitrogen isotopic analysis of the N_2 gas produced by denitrification used to "correct" for the fractionating effects of denitrification, which can help determine the initial $\delta^{15}N$ of nitrate (and therefore a more accurate assessment of its source) (Böhlke et al., 2002; McMahon & Böhlke, 2006), (b) methods for age-dating groundwater recharged during the past 50+ years (with precisions of 1–3 years) using chlorofluorocarbons (CFCs), sulfur hexafluoride, and helium-3/tritium have led to elucidating the timescales of agricultural nitrogen inputs and fate (e.g., Böhlke & Denver, 1995; Katz, 2004; Katz et al., 1999, 2001, 2004, 2005; Knowles et al., 2010; McMahon & Böhlke, 2006), (c) the dual nitrate isotope analyses of biota, for example, incorporating the $\delta^{15}N$ of algae and fish as "proxies" for assessing the $\delta^{15}N$ of nitrate contributed by different land uses (Finlay & Kendall, 2007; and described in more detail in a later section in this chapter), and (d) combination of dual nitrate isotopes with other isotopic and chemical tracers (see following sections for more information and references).

Therefore, to enhance the applicability of isotopic methods, it is important to combine them with an approach that includes other biogeochemical data and detailed hydrologic information about the study area. There has been a considerable amount of research that has used nitrate isotopes along with other tracers (including other isotopes, ion ratios, artificial sweeteners, bacteriophage occurrence, viruses, boron, and pharmaceutical compounds). Many of these approaches are described in detail in several review papers (Badruzzaman et al., 2012; Fenech et al., 2012; Field & Samadpour, 2007; Nestler et al., 2011; Pasten-Zapata et al., 2014). Several examples of these innovative biogeochemical indicators including multi-tracer approaches are described below.

4.3.1. Boron, Boron Isotopes and Other Stable Isotopes

Several research studies have used boron isotopes ($^{11}B/^{10}B$) to help identify nitrate sources in water, particularly as a tracer for sewage contamination in groundwater (Bassett et al., 1995; Eisenhut & Heumann, 1997; Vengosh et al., 1994; Verstraeten et al., 2005; Widory et al., 2004, 2005) and surface water (Chetelat & Gaillardet, 2005; Vengosh et al., 1999). The two stable isotopes of boron, ^{11}B and ^{10}B, have natural abundance of approximately 80 and 20%. An advantage of using boron and boron isotopes is that boron is not affected by transformation processes, mainly denitrification (Bassett et al., 1995; Vengosh et al., 1994). Boron migrates with nitrate and therefore can be used as a tracer of wastewater input to surface waters and groundwater. Boron is also present as a trace or minor constituent in many water types (Bassett et al., 1995; Chetelat & Gaillardet, 2005) due to its relatively high solubility and its use in many soaps and detergents. Boron also is present in fertilizers; however, fertilizer $\delta^{11}B$ generally has a large range of values (−10 to 60‰), which differ from isotope ratios in sewage (−2 to +2‰), detergents (0–10‰) (which can end up in sewage), and manure (5–50‰) (Briand et al., 2013; Widory et al., 2005). Using boron isotopes, Widory et al. (2004) was able to distinguish between two types of sewage, which would not have been possible using $\delta^{15}N$ alone. Also, from plots of $\delta^{11}B$ versus 1/B, Widory et al. (2004) showed some differentiation among animal sources of boron including sewage, cattle, hogs, and poultry. Another study by Widory et al. (2005) used both nitrate and boron isotopes to identify nitrate contamination from wastewater treatment plant effluent in a study of subsurface alluvial groundwater. They were able to distinguish between animal and human waste sources based on differences in boron isotopic composition because $\delta^{11}B$ values for cattle manure ranged from +11.6 to 19.2‰, whereas wastewater effluent values were −1.8 ± 0.1‰.

Saccon et al. (2013) used a multi-isotope approach ($\delta^{15}N$, $\delta^{18}O$, and $\Delta^{17}O$ of nitrate; $\delta^{34}S$ and $\delta^{18}O$ of sulfate; and boron isotopes) in the Marano Lagoon (NE Italy) and parts of its catchment area to identify nitrate pollution. They found that nitrate contamination in the lagoon was originating not only from agricultural activities but also from urban waste water, in situ nitrification, and atmospheric deposition. The combination of nitrate isotopes and stable isotopes of sulfate also can provide useful information about nitrate sources. For example, several other studies have used analyses of $\delta^{34}S$ and $\delta^{18}O$ of sulfate to determine atmospheric sources of acidic rain, fertilizers, animal and human wastes, and biogeochemical processes in the water column or sediments (Kendall et al., 2007).

The stable isotopes of water have been used to provide important information on different sources of nitrate associated with rivers, groundwater, and oceans. Fractionation effects mainly from evaporation are used to distinguish the origin of waters from different geographical areas, mixing of waters, and interactions between groundwater and surface water. Detailed reviews of these topics are provided by Kendall and McDonnell (1998) and Aggarwal et al. (2005).

Desaulty and Millot (2015) used lithium isotopes to trace wastewater sources in a small river basin near Orléans in France. Lithium is used in various industrial applications including production of batteries for both mobile devices (computers, tablets, smartphones, etc.) and electric vehicles, and also in pharmaceutical formulations. Based on the measured lithium isotopic composition (δ^7Li) in river water samples collected upstream and downstream after the release from a wastewater treatment plant connected to an hospital, the values in upstream river water were around −0.5 ± 1‰ ($n = 7$), but wastewater had a $\delta^7Li = +4‰$.

If phosphate is present along with nitrate in surface water and potentially groundwater, analysis of the $\delta^{18}O$ of phosphate can be used to differentiate between various sources. For example, M. Young et al. (2009) reported the following ranges of $\delta^{18}O$ of phosphate in the following sources: sewage (+7 to +12‰), detergents (+13 to +19‰), and fertilizers (+16 to +23‰). Another study reported $\delta^{18}O$ of phosphate of +16 to +19‰ for sewage (Gruau et al., 2005), which could make it difficult to distinguish between wastewater, detergents, and fertilizer sources in some areas.

4.3.2. Ion Ratios and Selected Ion Indicators for Nitrate-Source Identification

Ion ratios have been used effectively in studies to differentiate among sources of nitrate contamination. Anions, such as chloride, bromide, and iodine, are conservative as they travel in groundwater systems and have minimal interactions with the environment. Panno et al. (2006a) plotted the ratio of iodine to sodium (I/Na) against bromide, and from distinct ranges of values for various sources, they were able to differentiate between manure, sewage, and landfill sources affecting groundwater quality (Fig. 4.4). Using analyses of major ions, NO_3/Cl, along with $\delta^{15}N-NH_4^+$, $\delta^{15}N-NO_3$, and $\delta^{18}O-NO_3$, in surface water and groundwater in the Guizhou Province, southwest China, Liu et al. (2006) found large variations of nitrate sources in groundwater between winter and summer seasons. During the summer season, they identified a mixture of multiple sources of nitrate (fertilizer and nitrification of organic nitrogen materials) due to the rapid response of groundwater to rainfall or surface water inflows in the karst area.

Several studies have used the Cl/Br mass ratio as an indicator of wastewater contamination (Alcala & Custodio, 2008; S.N. Davis et al., 2004; Katz et al., 2011; Panno et al., 2006a). Based on a review of data from previous studies and current information from the U.S. Geological Survey National Water Quality Assessment Program, Katz et al. (2011) assessed influences from septic tank effluent on groundwater quality in major aquifers in the United States by developing several end-member fields and mixing lines for dilute groundwater, septic tank leachates, treated sewage effluent, animal manure and urine, and road salt. They found that Cl/Br ratios were useful in evaluating potential impacts from septic systems on groundwater quality and were most useful in shallow monitoring wells and shallow domestic wells. However, other indicators (such as boron, sulfate, nitrate isotopes, and organic wastewater compounds (OWCs)) would help to reduce uncertainty in discriminating between septic tank effluent and other sources of groundwater contamination. The Cl/Br ratio can vary from one area to another as a result of local factors such as proximity to coastal areas that affect ratios in precipitation (S.N. Davis et al., 2004), wastewater treatment methods (Vengosh & Pankratov,

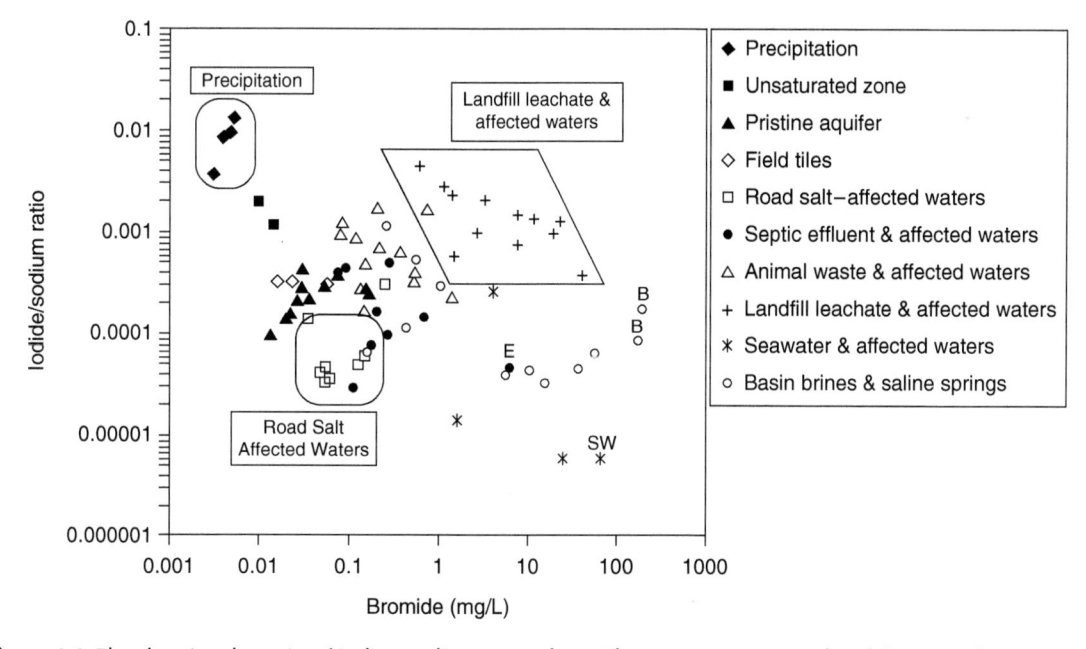

Figure 4.4 Plot showing the ratio of iodine-sodium versus bromide concentrations used to differentiate between precipitation, landfill leachate and leachate contaminated water, and road-salt contaminated water from other sources. Samples designated as end-members are indicated with the following letters: B, basin brine; E, commercially available salt for water softeners; SW, sea water. *Source*: From Panno et al. (2006b). From a Wiley publication.

1998), types of landfill leachate, and differences in animal waste composition and agrichemical usage (Panno et al., 2006a). Although, Cl/Br ratios that plot within ranges for domestic wastewater/animal waste affected areas tend to have higher nitrate concentrations than those from precipitation, pristine groundwater, or halite dissolution, Pasten-Zapata et al. (2014) noted that these results from Cl/Br ratios generally are compatible with nitrate isotope data.

Katz et al. (2017) compared nitrogen isotope data with Cl/Br ratios for water samples from 20 monitoring wells in the Upper Floridan aquifer as part of an ongoing study of the transport and fate of nitrate in a karst spring basin in northern Florida (Fig. 4.5). The study area is dominated by agricultural activities (row crops, hay, and dairies); however, there are a few monitoring wells located near residential areas with septic tanks and a small sprayfield receiving treated municipal wastewater. They found there was a statistically significant correlation between the Cl/Br mass ratio and δ^{15}N-NO$_3$ concentrations (Spearman's rho = 0.53; $p < 0.0001$) and Cl/Br mass ratios less than 400 were consistent with an inorganic source of nitrate as indicated by δ^{15}N-NO$_3$ values less than 5 ppt. They proposed that Cl/Br could be used as a proxy for nitrate source identification (inorganic versus organic) if end-member information is available along with other confirmatory information.

At an agricultural study area in North Carolina, Spruill et al. (2002) used statistically based classification-tree models that incorporated various ion ratios to distinguish among five nitrate sources (fertilizer on crops, fertilizer on golf courses, irrigation spray from hog wastes, septic tank leachate, and poultry litter leachate) with better than an 80% degree of success. By evaluating 32 variables and selecting four primary indicator variables (δ^{15}N, nitrate to ammonium ratio, sodium to potassium ratio, and zinc), they reported the following source differentiation: δ^{15}N-NO$_3$ and potassium greater than 18.2 mg/L indicated animal sources; a potassium concentration less than 18.2 mg/L indicated inorganic or soil organic N. Agricultural crops were indicated by a nitrate to ammonium ratio greater than 575; whereas a ratio less than 575 indicated nitrate from golf course. Septic system wastes were indicated by a sodium to potassium ratio (Na/K) greater than 3.2; whereas a Na/K ratio less than 3.2 indicated spray wastes from hog lagoons, and a Na/K ratio <2.8 indicated poultry wastes. Zinc values greater than 2.8 µg/L indicated spray wastes from hog lagoons, whereas values less than 2.8 µg/L indicated poultry wastes.

Other studies have used various combinations of isotopes, selected anions, and other indicators to evaluate impacts from septic tanks on shallow groundwater quality at various site-specific or local scales. For example, Nishikawa et al. (2003) used nitrate/chloride ratios to determine that effluent from septic systems accounted for the increase in nitrate concentrations in an area where imported water was artificially recharged to groundwater. Landon et al. (2008) found that elevated concentrations of chloride (>30 mg/L), sulfate (90–220 mg/L) and boron (>50 µg/L) in water from wells in a shallow unconfined aquifer were associated with initial (δ^{15}N-nitrate values greater than 8.6‰ and were related to numerous upgradient septic-tank systems). Several studies have used a multiple-tracer approach (various combinations of chemical and microbiological indicators) to investigate the migration of effluent from on-site wastewater systems to groundwater

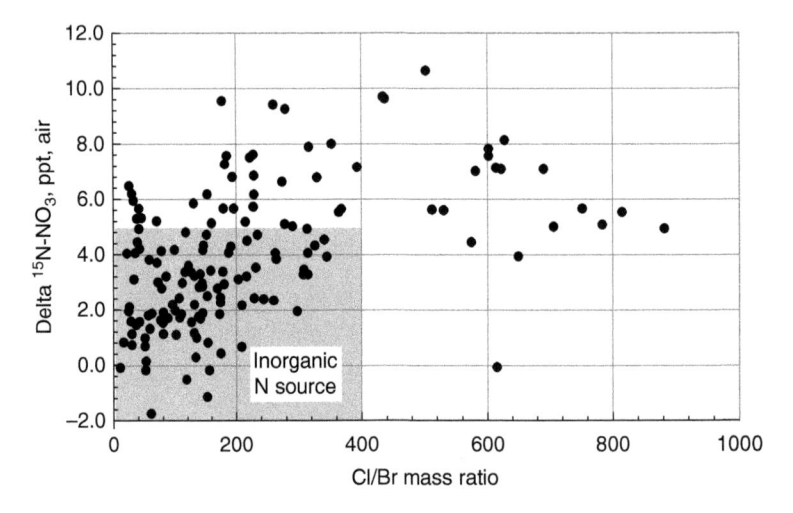

Figure 4.5 Plot showing the relation between the Cl/Br mass ratio and the δ^{15}N-NO$_3$ for water samples from monitoring wells in a karst spring basin in northern Florida. Cl/Br values less than 400 are consistent with an inorganic nitrogen source, as indicated by nitrogen isotope values less than 5 ppt.

(e.g., Godfrey et al., 2007; Phillips et al., 2015; Szabo et al., 2004). As a way of reducing some of the uncertainties associated with nitrate source identification and the fate of nitrate in groundwater, Minet et al. (2017) used sodium (Na$^+$) to differentiate nitrogen-rich organic effluents discharged from septic tanks and farmyards from synthetic fertilizers applied in their study area in southeast Ireland. The high mobility of Na$^+$ in the subsurface and its relatively low affinity for soil exchange sites makes it useful as a chemical tracer to delineate plumes of contamination from human organic effluents (Aravena & Robertson, 1998; Robertson et al., 1991; Weil & Brady, 2017). By combining information on Na$^+$ contamination thresholds and nitrate isotopes, Minet et al. (2017) found that denitrification only occurred in samples where organic point source contamination was characterized or suspected and was related to a high dissolved organic carbon (DOC) concentration in groundwater.

4.3.3. Nitrate Isotopes in Combination with Other Isotopic Indicators to Determine Sources of Nitrate

Stable isotopes (e.g., $\delta^{18}O$, δ^2H of water) along with other chemical constituents also have been used to assess sources of nitrate contamination, especially for urban systems. For example, Grimmeisen et al. (2017) used information from stable isotopes of water ($\delta^{18}O$, δ^2H), chloride concentrations, and nitrate isotopes to quantify surface and sub-surface mixing processes in an urban groundwater system in Jordan. They determined that 25% of nitrate contamination in groundwater originated from leaky sewer systems in the urban area. The combined tracer data (particularly using data from $\delta^{18}O$ and δ^2H of water) also provided a quantification of artificial recharge from water main losses that carry imported water from distant surface water and groundwater sources (river, lake, and wells). Urban stormwater runoff can contain reactive nitrogen from multiple sources. Yang and Toor (2016) collected stormwater runoff samples from over 25 stormwater events and analyzed these samples for $\delta^{18}O\text{-}NO_3^-$ and $\delta^{15}N\text{-}NO_3^-$, along with $\delta^{18}O$ and hydrogen (δD) of water to assess sources of contamination and nitrogen transformations. They found that the dominant nitrate sources in residential stormwater runoff were atmospheric deposition (range 43–71%) and nitrogen fertilizers (range <1–49%), although sources were continuously changing during stormwater events. Also, mixtures of multiple nitrate sources and nitrification processes were the main factors affecting nitrate transport in stormwater runoff from the residential catchment. Kaushal et al. (2011) investigated the fate and transport of nonpoint sources of nitrogen in urbanized, forested, and agricultural watersheds at the Baltimore Long-Term Ecological Research site. Based on information on stable isotopes combined with watershed nitrogen balances, they found negative (inverse) correlations between $\delta^{15}N\text{-}NO_3$ and $\delta^{18}O\text{-}NO_3$ in urban watersheds, which likely indicated mixing between atmospheric deposition and wastewater. Contributions of nitrogen related to storm magnitude, with atmospheric sources contributing approximately 50% at peak storm nitrogen loads. Their findings of positive correlations between $\delta^{15}N\text{-}NO_3$ and $\delta^{18}O\text{-}NO_3$ in urban and agricultural watersheds likely indicated that denitrification was removing nitrate from septic systems and agricultural activities. However, they found that removal of nitrate from leaking sewer lines was less susceptible to denitrification. The types of nitrogen sources in stormwater runoff varied with differences in storms and they attributed transformations of nitrogen species in storm drains to an increase in organic carbon inputs.

Nitrogen transport and transformation were studied beneath stormwater infiltration basins in two areas in central Florida using extensive hydrologic measurements (groundwater levels, rainfall, and soil moisture) along with a multi-tracer approach that included major and trace elements, nutrients, organic carbon, and dissolved gases; $\delta^{18}O$ and hydrogen (δ^2H) of water, $\delta^{18}O\text{-}NO_3^-$ and $\delta^{15}N\text{-}NO_3$, soil mineralogy and chemistry, and nitrite-reductase gene density using real-time polymerase chain reaction (O'Reilly et al., 2012a, 2012b). Based on data from water samples collected from ponded stormwater, suction lysimeters, and shallow wells near the water table, the researchers found contrasting conditions involving nitrate transformations at the two sites (SO and HT). Denitrification occurred at the SO site (Fig. 4.6) where wet soil conditions persisted (along with sulfate reduction and methanogenesis). In contrast, ammonification and nitrification occurred at the HT site (Fig. 4.6), where dry soil moisture conditions occurred and nitrate leaching to groundwater was observed (O'Reilly et al., 2012a, 2012b).

To differentiate between nitrogen sources in stream water an urban watershed near Pittsburg, Pennsylvania, Divers et al. (2013) used dual nitrate isotopes and mixing model analyses to quantify nitrogen inputs from sewage and atmospheric deposition. Results from mixing models using Bayesian techniques indicated that up to 94% of nitrate in stream water originated from sewage sources during baseflow conditions. During storms, sewage-derived nitrate was still the dominant source (66%), although the contribution of nitrate in stream samples from atmospheric deposition was 34%. Based on fractionation effects from nitrate isotope data, Divers et al. (2013) also noted that denitrification was evident in up to 19% of the sewage-derived samples.

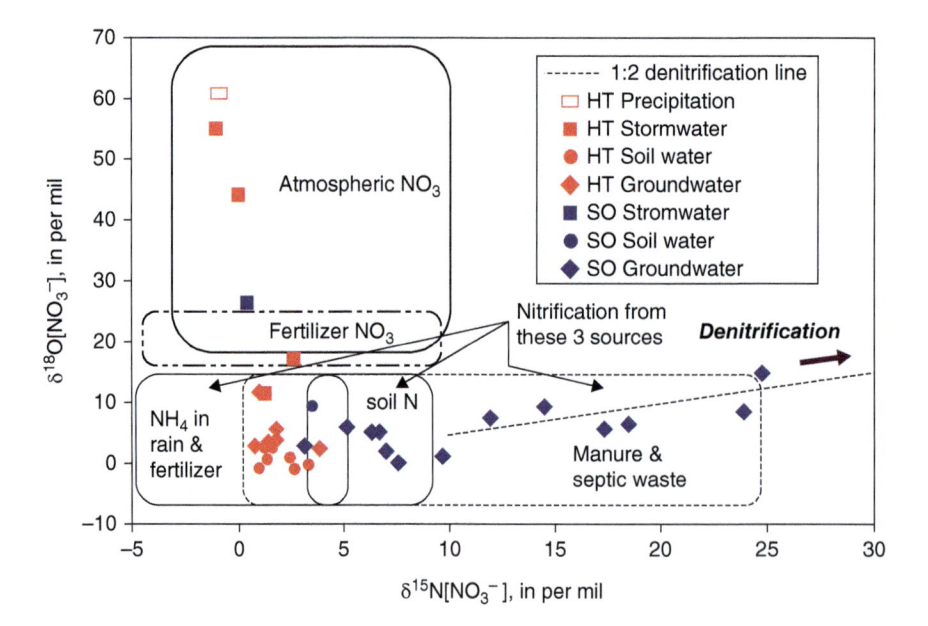

Figure 4.6 $\delta^{15}N$ and $\delta^{18}O$ of nitrate (NO_3^-) in precipitation, stormwater, soil water, and groundwater at the SO and HT sites showing samples plotted relative to typical source ranges. Results indicate that denitrification is naturally occurring in the subsurface at the SO basin but not at the HT basin. *Source*: From https://fl.water.usgs.gov/projects/oreilly_stormwaternitrogen/index.html, accessed 20 April 2018. From a USGS website.

4.4. SOURCES AND IDENTIFICATION OF ANTHROPOGENIC AMMONIUM

4.4.1. Groundwater

Anthropogenic ammonium (NH_4^+) has been found in groundwater contamination plumes in anoxic aquifers contaminated by landfill leachate and wastewater disposal practices (such as septic systems and the land application of treated wastewater). Given that ammonium can cause degradation of groundwater and surface waters, there have been relatively few studies of the complex ammonium transport and reaction processes in aquifer systems (Fig. 4.7).

In a study of a contamination plume originating from the artificial recharge of treated wastewater for almost 60 years at a study site on Cape Cod, Massachusetts, Böhlke et al. (2006) measured the distribution of ammonium, nitrate, and N_2, gas. They combined this information with isotopic analyses of coexisting aqueous ammonium, nitrate, N_2, and adsorbed NH_4^+ along with groundwater age data, natural gradient $^{15}NH_4^+$ tracer tests and numerical simulations of $^{15}NH_4^+$, $^{15}NO_3$, and $^{15}N_2$ breakthrough data. In the anoxic core of the ammonium contamination plume, they found no evidence for nitrification of anammox reactions that would consume NH_4^+. At the oxic boundary of the plume, however, they found that nitrification (conversion of ammonium to nitrate) was occurring. They projected that the main mass of NH_4^+ in the plume would probably reach a discharge area

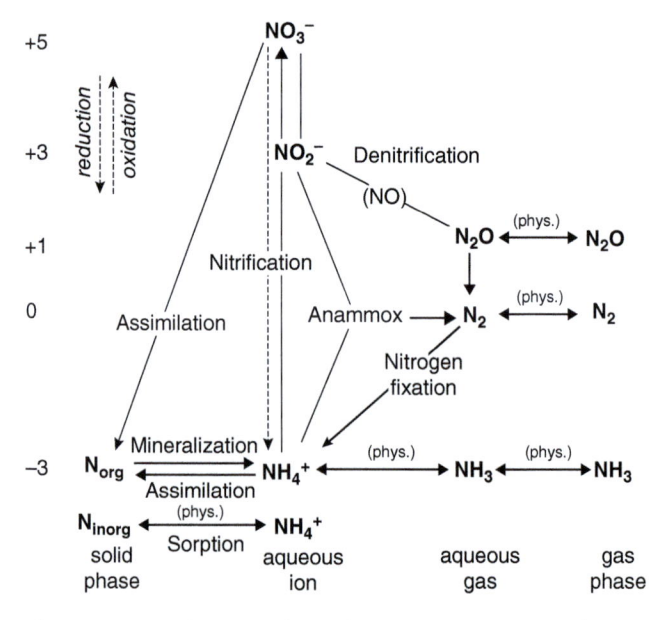

Figure 4.7 Nitrification, denitrification, anammox, and NH_4^+ exchange with aquifer solids were the main processes studied in the transport of ammonium in a wastewater contamination plume. *Source*: Böhlke et al. (2006). From WRR (AGU/John Wiley & Sons publication).

(although at a low transport rate) without substantial reactions and could result in nitrate contamination of surface water depending on redox conditions near the discharge area.

4.4.2. Atmospheric Deposition

Felix et al. (2014) reported that the proportion of reactive nitrogen in wet deposition attributable to ammonium (NH_4^+) has increased over the last three decades in the United States resulting from a continual increase in NH_3 emissions along with reductions in NO_x emissions. They used the nitrogen isotopic composition of ammonia, $NH_3(\delta^{15}N\text{-}NH_3)$, to characterize the transport of NH_3 emissions at the landscape-scale. $\delta^{15}N\text{-}NH_3$ values in ambient NH_3 was sampled and analyzed from a variety of landscapes including conventionally managed cornfield, tallgrass prairie, concentrated animal feeding operation (CAFO), dairy operation, and an urban setting. Felix et al. (2014) noted that volatilized fertilizer is a primary contributor to ambient NH_3 after fertilizer application; however, during periods of low or no fertilization, vehicle NH_3 emissions were a substantial contributor to ambient NH_3 over cornfields next to roadways. At the CAFO site, modeled NH_3 deposition flux contributed a substantial amount of nitrogen to the landscape and $\delta^{15}N\text{-}NH_3$ values were useful in tracing the contributing livestock source. In the urban area, fossil fuel-based emissions were the dominant source of ammonia; however, it was noted that there were large spatial variations in ammonia concentrations.

Ammonia emissions to the atmosphere have increased during the past several decades and have resulted in the formation of particulate matter through reactions with nitric and sulfuric acids. For example, NH_3 air concentrations and NH_4 in precipitation have shown increasing trends (+7.0% per year) over a large area of the United States from 2008 to 2015 (Butler et al., 2016). These harmful compounds in the atmosphere can be dispersed over large areas of the world resulting in eutrophication and acidification of terrestrial, freshwater, and marine habitats (Bobbink et al., 2010; Erisman et al., 2015; Felix et al., 2014). Also, deposition of ammonia can damage sensitive vegetation, particularly bryophytes and lichens that consume most of their nutrients from the atmosphere (Sheppard et al., 2011). Sala et al. (2000) noted that certain biomes (particularly northern temperate, boreal, arctic, alpine, grassland, savannah, and Mediterranean) are very sensitive to reactive nitrogen deposition because of the limited availability of nitrogen in these systems under natural conditions.

4.5. ORGANIC WASTEWATER COMPOUNDS AND EMERGING CONTAMINANTS AS INDICATORS OF NITRATE CONTAMINATION SOURCES

Over the past two decades, there has been a plethora of research studies on the occurrence and fate of trace organic chemicals (e.g., compounds in pharmaceuticals and personal care products, fire retardants, hormones, veterinary drugs, and pesticides used around the home) in sewage effluents, treated wastewaters, and receiving waters (rivers, lakes, estuaries, and groundwater). These compounds, commonly have been referred to as emerging contaminants, emerging substances of concern (ESOCs), OWCs and other terms and acronyms, have been used in numerous studies as tracers of wastewater and associated nitrate contamination in rivers, lakes, estuaries, oceans, groundwater, and other waterbodies. Sources of these compounds include domestic wastewater (septage, leaky sewer lines, and treated wastewater disposal), biosolids, and reclaimed wastewater. These compounds can also come from agricultural animal husbandry, landfills, and industrial activities. Many of these compounds pass through wastewater treatment processes in trace concentrations (parts per billion).

To be able to use pharmaceutical compounds or other wastewater indicators to differentiate between sewage and manure, three important considerations are detection frequency, marker specificity, and analytical detection capability for extremely low concentrations. A persistent compound would likely have a higher detection frequency, as well as its use in the study area. For example, pharmaceutical compounds most detected would be those that are dispensed and used at the highest levels in the study area (Kasprzyk-Hordern et al., 2008).

In a review of the literature on pharmaceuticals found in surface waters (rivers, lakes, and oceans) in the United States, Deo (2014) reported a total of 93 compounds detected, including 27 antibiotics, 15 antidepressants, 9 antihypertensives, 7 analgesics, 7 anticonvulsants, 6 antilipidemics (to reduce serum lipid levels), 3 contraceptives, 3 stimulants, and 2 each of antihistamines, blood thinners, disinfectants, antacids, antitussives, anti-anxiety, anti-inflammatory, and diuretic agents. Fenech et al. (2012) listed 30 pharmaceutical compounds that have been used as indicators of sewage contamination based on information from seven different studies (Benotti & Brownawell, 2007; Buerge et al., 2006a, 2006b, 2008; Glassmeyer et al., 2005; Kasprzyk-Hordern et al., 2008; Kolpin et al., 2002; Nakada et al., 2008; T.A. Young et al., 2008). Also, Fenech et al. (2012) summarizes concentration data (maximum, mean, detection frequency) for 17 potential markers of sewage contamination in fresh surface waters based on information from dozens of environmental studies listed therein. Likewise, they also list concentrations for 17 potential chemical markers of manure contamination in fresh surface waters based on data from dozens of studies listed therein. Other researchers have identified numerous other pharmaceutical compounds that can be used to trace wastewater inputs to surface waters and groundwaters (e.g., Fenech et al., 2012; Foster et al., 2012; Kolpin et al., 2002; Meffe

& de Bustamante, 2014; Sui et al., 2015; Valcárcel et al., 2011). Badruzzaman et al. (2012) reviews several additional compounds that are potential markers of nitrate contamination including carbamazepine (antiepileptic drug), galaxolide (synthetic musk), and gadolinium (magnetic tracer used in magnetic resonance imaging facilities) (Verplanck et al., 2005).

Trace organic chemicals also have been used as tracers of nitrate contamination in groundwater (Barnes et al., 2004). Sui et al. (2015) presents a detailed list of pharmaceuticals and personal care products that have been detected (number of samples, detection frequency, and concentration ranges) in groundwater during 2012–2014 from throughout the world. In Florida, where 90% of the population relies on groundwater for their water supply, Seal et al. (2016) used sucralose (an artificial sweetener in a variety of food products) as a tracer to infer statewide percentages of groundwater resources affected by wastewater sources. Sucralose tends to bypass wastewater treatment processes and is not removed in biosolids during treatment (Soh et al., 2011). Based on a 2012 probabilistic survey of groundwater in wells, Seal et al. (2016) reported sucralose detection rates of 18% in unconfined aquifers and 2% in confined aquifers. In a 2014 sampling of the unconfined aquifer network, sucralose was detected in 25% of 116 wells and 12% of these wells also had detections of three pharmaceutical compounds (acetaminophen, carbamazepine, and primidone). These results confirm the vulnerability of unconfined aquifers in Florida to contamination and as potential reservoirs for storage of sucralose and other pharmaceutical compounds. Katz et al. (2010) used multiple tracers (nitrate isotopes, pharmaceutical compounds, and microbiological indicators) to investigate the movement of nitrate from septic tanks in the Woodville Karst Plain of northern Florida. Samples of septic tank effluent, soil moisture from lysimeters, and groundwater from background and drainfield wells were collected at three sites during dry and wet periods. Attenuation of nitrogen ranged from 25 to 40% at the three sites. Most pharmaceutical compounds were attenuated beneath septic tank drainfields; however, caffeine, acetaminophen, and sulfamethoxazole were detected in groundwater samples.

Fram and Belitz (2011) reported detections of pharmaceutical compounds in 2.3% of 1231 samples of groundwater used for public drinking-water supply in California. They found that pharmaceutical detection frequencies were much lower than those for pesticides (33%), trihalomethanes (28%), and volatile organic compounds not including trihalomethanes (23%) in the same 1231 samples. Oppenheimer et al. (2011, 2012) used sucralose and other trace organic compounds as indicators of wastewater loading to surface waters in urbanized areas. Spoelstra et al. (2017) used four artificial sweeteners to study the movement of wastewater from septic systems to groundwater in a rural are in Ontario, Canada. About 30% of groundwater samples had detectable levels of one or more artificial sweeteners, which indicated the presence of water derived from septic system effluent. They did not find a relationship between the acesulfame and nitrate concentrations in the aquifer, which indicated that other sources likely were contributing nitrate.

4.6. STABLE ISOTOPIC MONITORING OF BIOTA IN ECOSYSTEMS

Finlay and Kendall (2007) provided a comprehensive review stable isotopes (primarily $\delta^{13}C$ and $\delta^{15}N$, and to a lesser extent, $\delta^{34}S$) used in ecosystem studies, with an emphasis on nutrient and energy flow at the base of aquatic food webs. They focused on the environmental basis for variations in $\delta^{13}C$, $\delta^{15}N$, and $\delta^{34}S$ of organic matter at the base of food webs in freshwaters (mainly rivers) and their connections with surrounding watersheds. Finlay and Kendall (2007) concluded that spatial and temporal isotopic variation among organic matter sources is extensive and is not as well understood as other factors that affect use of stable isotope approaches, such as trophic fractionation. They also recommended the importance of incorporating measurements of multiple source tracers, consumer growth rates, movement rates of predators and prey, and modeling approaches. However, the principles, patterns, biogeochemical predictors, and applications of stable isotope measurements in river environments discussed by Finlay and Kendall (2007) can be transferred to studies of other types of ecosystems.

For example, studies have used nitrogen isotopic ratios of macrophytes and associated periphyton to identify sources in subtropical spring-fed streams (De Brabandere et al., 2007) and in macroalgae to identify sources of nitrogen to a bay in Kauai, Hawaii (Derse et al., 2007). Albertin et al. (2012) used dual isotope analyses of nitrate along with analyses of $\delta^{15}N$ of algal tissue during a 3-year study to detect point sources of nitrate in synoptic sampling of 15 large springs and along spring-fed river runs in northern Florida. Focusing on four of these springs, they found a correlation between the $\delta^{15}N$ composition of nitrate and algae at both the regional scale samples of springs and at individual spring-run sites. However, they concluded that the large range of fractionation factors for algae uptake of nitrate precluded the use of algal $\delta^{15}N$ to accurately infer the $\delta^{15}N$ of nitrate in spring water and hence identification of nitrate sources. Other studies in nutrient-rich environments have reported that the $\delta^{15}N$ of algae can closely track the $\delta^{15}N$ of nitrate in river water (with an isotope fractionation of 4–5‰) in large river basins, such as the Mississippi Basin (Battaglin et al., 2001a, 2001b) and the San Joaquin River basin (Kratzer et al., 2004).

LaPointe et al. (2015) investigated the sources contributing to nutrient-fueled algal blooms in the 251 km long Indian River Lagoon (IRL) in eastern Florida. During 2011–2012, they conducted three sampling events at 20 sites for dissolved nutrients (dissolved inorganic nitrogen (DIN), soluble reactive phosphorus (SRP), total dissolved nitrogen (TDN), total dissolved phosphorus (TDP)) and chlorophyll a, and isotopic analyses ($\delta^{15}N$, $\delta^{13}C$) of benthic macroalgae to identify potential nutrient sources. The researchers documented widespread occurrences of HABs during the study and enriched $\delta^{15}N$ values in macroalgae throughout the IRL (+6.3‰), which were similar to values reported for macroalgae from other sewage-polluted coastal waters. Their findings indicated that nonpoint source nitrogen enrichment from septic tanks (about 300,000 in the area) likely is a significant nitrogen source to the IRL. LaPointe and Herren (2016) also conducted 1-year study of the St. Lucie Estuary, also on the east coast of Florida, to investigate the interactions between septic tanks (on-site treatment and disposal systems), groundwater, and surface water in the estuary and nearshore reefs. The study included analyses of dissolved nutrients, stable isotopes, sucralose and acetaminophen, and analysis of macroalgae and phytoplankton for stable isotopes ($\delta^{15}N$, $\delta^{13}C$). Stable nitrogen isotopic ratios indicated a predominantly wastewater source in the surface water and in macroalgae and phytoplankton collected during the study. Detections of sucralose also indicated that wastewater was reaching surface waters of the estuary and the nearshore reefs north and south of St. Lucie Inlet.

4.7. MICROBIAL INDICATORS OF NITRATE SOURCE CONTAMINATION

Human enteroviruses and Hepatitis A viruses (HAVs) have been used in recent years to successfully determine the source of fecal contamination (Griffin et al., 1999, 2000, 2003). Two other groups of viruses, noroviruses, and adenoviruses also have been used along with enteroviruses and HAV for source tracking in marine ecosystems. These three groups of viruses have been utilized to determine whether human wastewater exists as continuous source and whether it responds to storm surge and/or precipitation events. Also, these viruses can adapt and live in shallow aquatic sediments and terrestrial soils resulting in consistently elevated concentrations in near shore waters. The major advantage of using these viruses is that they are only shed in human feces and, if detected, identify human wastewater as the source of contamination, which is not the case for standard bacterial indicators (fecal coliforms and enterococci) that are found in the feces of humans and other animals. Viral source tracking is library independent and, therefore can be more accurate than bacterial source tracking assays.

Microbial indicators, such as bacteriophages, have been used in studies to identify human fecal sources in springs (Griffin et al., 2000) and for tracing the movement of nitrate in groundwater from the land application of treated wastewater (Katz & Griffin, 2008; Katz et al., 2009) and septic tank effluent (Katz et al., 2010). McMinn et al. (2017) describe bacteriophages as viruses that are dependent on a bacterial host for replication. Bacteriophages can infect fecal indicator bacteria or other commensal intestinal species (e.g., Bacteroides) and subsequently are shed by host organisms and typically follow similar routes of wastewater entry into the environment as enteric viral pathogens (Leclerc et al., 2000). To investigate sources of nitrate pollution in the Sava River Basin, Slovenia, Vrzel et al. (2016) used a combination of nitrate isotope measurements ($\delta^{15}N_{NO3}$ and $\delta^{18}O_{NO3}$) along with a microbial source tracking (MST) method. They found that nitrate contamination in the river originated from continuous discharge of untreated human wastewaters.

In a karstic aquifer located in an agricultural area in southwest France, Briand et al. (2013) used a multi-tracer combination of nitrate and boron isotopes along with microbial indicators (genotyping methods for bacteriophages and *Bacteroidales* (a bacteria type present in gastrointestinal tract of warm-blooded animals including humans, ruminants, and pigs)). They were able to determine that nitrate contamination originated from an animal waste source and that there was rapid movement (2–3 days) of nitrate from surface water into groundwater in the karstic limestone aquifer.

4.8. COMBINING NITRATE SOURCE IDENTIFICATION WITH GROUNDWATER AGE AND RESIDENCE TIME INFORMATION

During the past 50 years, several chemical and isotopic substances have been released to the atmosphere from human activities. These substances dissolve in precipitation and enter the subsurface recharging groundwater systems. The concentrations of environmental tracers in groundwater, such as CFCs, tritium (^3H), sulfur hexafluoride, and other chemical and isotopic substances, have been used to trace the flow of young water (water recharged within the past 50 years) and to determine the time elapsed since recharge (https://pubs.usgs.gov/fs/FS-134-99/, accessed 15 May 2018).

Combining information on the age of groundwater and groundwater residence time with isotopes and other chemical tracers can provide a set of tools for effectively estimating rates of recharge (Plummer, 2004), determining the fate of nitrate associated with biogeochemical processes, such as denitrification (e.g., Böhlke & Denver, 1995; Böhlke, 2002; Green et al., 2008; Tesoriero et al., 2007), assessing vulnerability of groundwater to contamination (Eberts et al., 2013), and predicting lag times for remediation (e.g., McMahon et al., 2008b; Puckett et al., 2011). Other studies have used groundwater age estimated for calibrating groundwater flow models (Eberts et al., 2012) and for estimating fractions of young and old water mixtures in public supply wells and springs (Eberts et al., 2013; Jurgens et al., 2016; McMahon et al., 2008a). More information on groundwater age dating and estimation of groundwater residence time is presented in Chapter 8 for the discussion on nitrate contamination in spring waters.

4.9. USING NITRATE ISOTOPES AND OTHER INDICATORS TO ASSESS LEGACY REACTIVE NITROGEN IN SUBSURFACE STORAGE

As mentioned briefly in Chapter 1, there are large amounts of reactive nitrogen, mostly in the form of nitrate, that can leach below the root zone in soils to the unsaturated (vadose) zone (the material between the base of the soil to the water table at the top of the saturated zone) and eventually to groundwater (aquifers) where nitrate can be stored (Wang et al., 2013). This is especially prevalent in agricultural areas where there has been a surplus of nitrogen fertilizers added to cropland. Based on global patterns of nitrate storage in the vadose zone assessed by using estimates of groundwater depth and nitrate leaching for the period 1900–2000, Ascott et al. (2017) estimated a peak global storage of 605–1814 Tg of nitrate in the vadose zone, with the highest storage per unit area in North America, China, and Europe (areas with thick vadose zones and extensive historical agricultural activities).

Several recent studies have shown that groundwater nitrogen applied decades ago can still be found in aquifers (Alley et al., 2002; Basu et al., 2012; McMahon et al., 2006; Meals et al., 2010; Puckett et al., 2011; Sanford and Pope, 2013; Sebilo et al., 2013). The large quantities of nitrate stored in aquifers can be released to surface waters slowly over time, as a function of the groundwater residence time, which can range from weeks to several thousands of years (Alley et al., 2002). This continual release of nitrate over time has significant implications contamination of deeper aquifers used for drinking water supplies, and eutrophication of surface water bodies (Schlesinger, 2009).

An important unknown in these groundwater systems and in the unsaturated zone is the potential for denitrification and the rate of denitrification if it is occurring. Kendall (1998) noted the difficulties in measuring denitrification in groundwater systems and the large variability in measured rates (5–40%). Higher fractionation factors occur in aquifers with longer groundwater residence times; therefore, it is important to use age-dating tracers and possibly groundwater flow modeling to estimate travel times in these aquifers with large nitrate loadings. Panno et al. (2006b) used $\delta^{18}O$-NO_3^- and $\delta^{15}N$-NO_3 to investigate the degree of denitrification in tile drain discharge to the Mississippi River in agricultural areas of east-central Illinois, USA. They found that most of the nitrate in the river was derived from synthetic fertilizers; however, denitrification ranged from 0 to 55% in some tile drains prior to discharge into the Mississippi River. In a forested headwater catchment, based on fluctuations of $\delta^{18}O$-NO_3^-, $\delta^{15}N$-NO_3, and nitrate, Osaka et al. (2010) concluded that nitrification and denitrification were occurring concurrently in groundwater; however, denitrification was more important on net nitrogen transformation in groundwater than nitrification because nitrate concentration was lower and $\delta^{15}N$-NO_3 was higher in groundwater than in soil water. They also noted that the amount of denitrification in groundwater was mainly controlled by groundwater residence time.

Green et al. (2008) investigated natural attenuation of nitrate contamination in four intensive agricultural settings in the United States (San Joaquin watershed, California, the Elkhorn watershed, Nebraska, the Yakima watershed, Washington, and the Chester watershed, Maryland). Based on chemical analyses of major ions, dissolved gases, nitrogen and oxygen stable isotopes, and estimates of recharge date, field measurements, and flow and transport modeling for monitoring well transects, they found zero-order rates of denitrification ranged from 0 to 0.14 µmol N/L/d, which were comparable to results from other studies using the same methods. However, they noted that many denitrification rates reported in the literature were several orders of magnitude higher, which they attributed to possible method limitations and bias for selection of sites with rapid denitrification. An important conclusion from this study was that denitrification is limited in extent in the shallow aquifers below the studied agricultural fields, and likely would require residence times of decades or longer to mitigate present-day nitrate contamination.

Eschenbach et al. (2015) investigated in situ denitrification rates using 28 "push–pull" ^{15}N tracer field tests in two sandy Pleistocene aquifers in northern Germany. They reported in situ denitrification rates that ranged from 0 to 51.5 µg N /kg/d. Their study demonstrated that the microbial community in the nitrate-free zone just below the nitrate contaminated zone could be adapted to denitrification by nitrate injections into wells for an extended period. After preconditioning with nitrate, Eschenbach et al. (2015) found that in situ denitrification rates were 30–65 times higher than prior to any injections. Results from this study demonstrated that preconditioning with nitrate is critical for measuring in situ denitrification rates in deeper nitrate-free aquifer material that are below upper zones with nitrate contamination.

In addition to using $\delta^{18}O\text{-}NO_3^-$, $\delta^{15}N\text{-}NO_3$, several have measured noble gas concentrations in groundwater samples to determine excess nitrogen produced during denitrification. To quantify the magnitude and variability of denitrification in the Upper Floridan aquifer in Florida, USA, Heffernan et al. (2012) used noble gas tracers, neon (Ne) and argon (Ar), along with nitrate isotopes and other chemical data to predict nitrogen gas concentrations for 112 observations from 61 springs. More information about this aquifer system is presented in Chapter 8. They found that excess N_2 (from denitrification) was spatially highly variable and was inversely correlated with dissolved oxygen. Also, inverse relationships between O_2 and $\delta^{15}N\text{-}NO_3$ across a larger dataset of 113 springs, well-constrained isotopic fractionation coefficients, and strong $^{15}N:^{18}O$ covariation indicated that denitrification was occurring in this organic-matter-depleted aquifer system. Although they found low average rates, denitrification accounted for 32% of estimated aquifer N inputs across all sampled springs from the Upper Floridan aquifer. Nitrate contamination of springs is discussed in more detail in Chapter 8. Einsiedl and Mayer (2006) also have documented that denitrification had occurred in the porous rock matrix of a karst aquifer in Germany.

Interactions between groundwater and surface water systems affect the transformations of reactive nitrogen and have important environmental and ecological implications. Chapter 7 explores these interactions in detail, such as groundwater flow through hyporheic zones beneath streams, riparian zones, and submarine groundwater discharge to oceans.

In summary, numerous point and nonpoint sources of reactive nitrogen have caused widespread nitrate contamination of groundwater and surface waters (lakes, streams, rivers, estuaries, and oceans). Elevated nitrate concentrations have caused eutrophication of aquatic systems, HABs, loss of biodiversity, acidification of soils and surface waters, and adverse human health issues. It is critical to have detailed information on sources of nitrate contamination and the processes that control the fate and transport of nitrate and other reactive nitrogen species. This information will lead to more effective resource management to protect water quality from further degradation and to develop more informed remediation and restoration plans for sites that are contaminated.

Fortunately, over the past two decades, substantial analytical advances have been made in being able to differentiate among various sources of reactive nitrogen (synthetic fertilizers in agriculture, animal wastes, human wastewater, and atmospheric deposition of reactive nitrogen (fossil fuel combustion and biomass burning). Our understanding of sources can be confounded by mixtures of nitrate from multiple sources. However, the combination of the dual nitrate isotope technique along with complementary tracers (such as boron isotopes, oxygen and hydrogen isotopes, trace organic chemicals (pharmaceuticals and personal care products), and ion ratios, has helped substantially in our understanding of the sources and fate of nitrate in environmental systems.

REFERENCES

Aggarwal, P. K., Gat, J. R., & Froehlich, K. F. (Eds.). (2005). *Isotopes in the water cycle: Past, present and future of a developing science*. The Netherlands: Springer, 382 p.

Albertin, A. R., Sickman, J. O., Pinowska, A., & Stevenson, R. J. (2012). Identification of nitrogen sources and transformations within karst springs using isotope tracers of nitrogen. *Biogeochemistry, 108*, 219–232.

Alcala, F. J., & Custodio, E. (2008). Using the Cl/Br ratio as a tracer to identify the origin of salinity in aquifers in Spain and Portugal. *Journal of Hydrology, 359*, 189–207.

Alley, W. M., Healy, R. W., LaBaugh, J. W., & Reilly, T. E. (2002). Flow and storage in groundwater systems. *Science, 296*, 1985–1990.

Aravena, R., Evans, M. L., & Cherry, J. A. (1993). Stable isotopes of oxygen and nitrogen in source identification of nitrate from septic systems. *Ground Water, 31*(2), 180–186.

Aravena, R., & Robertson, W. D. (1998). Use of multiple isotope tracers to evaluate denitrification in ground water: Study of nitrate from a large-flux septic system plume. *Ground Water, 36*, 975–982.

Ascott, M. J., Gooddy, D. C., Wang, L., Stuart, M. E., Lewis, M. A., Ward, R. S., & Binley, A. M. (2017). Global patterns of nitrate storage in the vadose zone. *Nature Communications*, 1–7. https://doi.org/10.1038/s41467-017-01321-w

Badruzzaman, M., Pinzon, J., Oppenheimer, J., & Jacangelo, J. G. (2012). Sources of nutrients impacting surface waters in Florida: A review. *Journal of Environmental Management, 109*, 80–92. https://doi.org/10.1016/j.envman2012.04.040

Barnes, K. K., Christenson, S. C., Kolpin, D. W., Focazio, M. J., Furlong, E. T., Zaugg, S. D., et al. (2004). Pharmaceuticals and other waste water contaminants within a leachate plume downgradient of a municipal landfill. *Ground Water Monitoring and Remediation, 24*(2), 119–126.

Bassett, R. L., Buszka, P. M., Davidson, G. R., & Chong-Diaz, D. (1995). Identification of groundwater solute sources using boron isotopic composition. *Environmental Science and Technology, 29*, 2915–2922.

Basu, N. B., Jindal, P., Schilling, K. E., Wolter, C. F., & Takle, E. S. (2012). Evaluation of analytical and numerical approaches for the estimation of groundwater travel time distribution. *Journal of Hydrology, 475*, 65–73.

Battaglin, W. A., Kendall, C., Chang, C. C. Y., Silva, S. R., & Campbell, D. H. (2001a). *Chemical and isotopic composition of organic and inorganic samples from the mississippi river and its tributaries*, 1997–98. U.S. Geological Survey Water Resources Investigation Report 01-4095, 57 p.

Battaglin, W. A., Kendall, C., Chang, C. C. Y., Silva, S. R., & Campbell, D. H. (2001b). Chemical and isotopic evidence of nitrogen transformation in the Mississippi River, 1997–98. *Hydrological Processes*, *15*, 1285–1300.

Benotti, M. J., & Brownawell, B. J. (2007). Distributions of pharmaceuticals in an urban estuary during both dry- and wet-weather conditions. *Environmental Science and Technology*, *41*(16), 5795–5802.

Bobbink, R., Hicks, K., Galloway, J., Spranger, T., Alkemade, R., Ashmore, M., et al. (2010). Global assessment of nitrogen deposition effects on terrestrial plant diversity: A synthesis. *Ecological Applications*, *20*, 30–59.

Böhlke, J. K. (2002). Groundwater recharge and agricultural contamination. *Hydrogeology Journal*, *10*, 153–179. https://doi.org/10.1007/s10040-001-0183-3

Böhlke, J. K., & Denver, J. M. (1995). Combined use of groundwater dating, chemical, and isotopic analyses to resolve the history and fate of nitrate contamination in two agricultural watersheds, Atlantic coastal plain, Maryland. *Water Resources Research*, *31*, 2319–2339.

Böhlke, J. K., Mroczkowski, S. J., & Coplen, T. B. (2003). Oxygen isotopes in nitrate: New reference materials for $^{18}O:^{17}O:^{16}O$ measurements and observations on nitrate-water equilibration. *Rapid Communications in Mass Spectrometry*, *17*, 1835–1846.

Böhlke, J. K., Smith, R. L., & Miller, D. N. (2006). Ammonium transport and reaction in contaminated groundwater: Application of isotope tracers and isotope fractionation studies. *Water Resources Research*, *42*, W05411. https://doi.org/10.1029/2005WR004349

Böhlke, J. K., Wanty, R., Tuttle, M., Delin, G., & Landon, M. (2002). Denitrification in the recharge area and discharge area of a transient agricultural nitrate plume in a glacial outwash sand aquifer, Minnesota. *Water Resources Research*, *38*(7). https://doi.org/10.1029/200!WR000663

Briand, C., Plagnes, V., Sebilo, M., Louvat, P., Chesnot, T., Schneider, M., et al. (2013). Combination of nitrate (N, O) and boron isotopic ratios with microbiological indicators for the determination of nitrate sources in karstic groundwater. *Environmental Chemistry*, *10*, 365–369. http://dx.doi.org/10.1071/EN13036

Buerge, I. J., Buser, H., Poiger, T., & Mueller, M. D. (2006a). Occurrence and fate of the cytostatic drugs cyclophosphamide and ifosfamide in wastewater and surface waters. *Environmental Science and Technology*, *40*(23), 7242–7250.

Buerge, I. J., Kahle, M., Buser, H., Mueller, M. D., & Poiger, T. (2008). Nicotine derivatives in wastewater and surface waters: Application as chemical markers for domestic wastewater. *Environmental Science and Technology*, *42*(17), 6354–6360.

Buerge, I. J., Poiger, T., Mueller, M. D., & Buser, H. (2006b). Combined sewer overflows to surface waters detected by the anthropogenic marker caffeine. *Environmental Science and Technology*, *40*(13), 4096–4102.

Butler, T., Vermeylen, F., Lehmann, C. M., Likens, G. E., & Puchalski, M. (2016). Increasing ammonia concentration trends in large regions of the USA derived from the NADP/AMoN network. *Atmospheric Environment*, *146*, 132–140. https://doi.org/10.1016/j.atmosenv.2016.06.033

Casciotti, K. L., Sigman, D. M., Galanter Hastings, M., Bohlke, J. K., & Hilkert, A. (2002). Measurement of the oxygen isotopic composition of nitrate in seawater and freshwater using the denitrifier method. *Analytical Chemistry*, *74*, 4905–4912.

Chang, C. C. Y., Langston, J., Riggs, M., Campbell, D. H., Silva, S. R., & Kendall, C. (1999). A method for nitrate collection for $\delta^{15}N$ and $\delta^{18}O$ analysis from waters with low nitrate concentrations. *Canadian Journal of Fisheries and Aquatic Sciences*, *56*, 1856–1864.

Chetelat, B., & Gaillardet, J. (2005). Boron isotopes in the Seine River, France: A probe of anthropogenic contamination. *Environmental Science and Technology*, *39*, 2486–2493.

Davis, S. N., Fabryka-Martin, J. T., & Wolfsberg, L. E. (2004). Variations of bromide in potable groundwater in the United States. *Ground Water*, *42*(6), 902–909.

Davis, S. N., Whittemore, D. O., & Fabryka-Martin, J. (1998). Uses of chloride-bromide ratios in studies of potable water. *Ground Water*, *36*, 338–350.

De Brabandere, L., Frazer, T. K., & Montoya, J. P. (2007). Stable nitrogen isotope ratios of macrophytes and associated periphyton along a nitrate gradient in two subtropical, spring-fed streams. *Freshwater Biology*, *52*, 1564–1575.

Deo, R. P. (2014). Pharmaceuticals in the surface water of the USA: A review. *Current Environmental Health Report*, *1*, 113–122. https://doi.org/10.1007/s40572-014-0015-y

Derse, E., Knee, K. L., Wankel, S. D., Kendall, C., Berg, C. J., & Paytan, A. (2007). Identifying sources of nitrogen to Hanalei Bay, Kauai, utilizing the nitrogen isotope signature of macroalgae. *Environmental Science and Technology*, *41*, 5217–5223.

Desaulty, A.M., & Millot, R. (2015). *Tracing waste water with Li isotopes*. AGU Fall Meeting - American Geophysical Union, December 2015, San Francisco, CA, United States.

Deutsch, B., Kahle, P., & Voss, M. (2006). Assessing the source of nitrate pollution in water using stable N and O isotopes. *Agronomy for Sustainable Development*, *26*(4), 263–267.

Divers, M. T., Elliott, E. M., & Bain, D. J. (2013). Constraining nitrogen inputs to urban streams from leaking sewers using inverse modeling: Implications for dissolved inorganic nitrogen (DIN) Retention in Urban Environments. *Environmental Science and Technology*, *47*, 1816–1823.

Eberts, S. M., Böhlke, J. K., Kauffman, L. J., & Jurgens, B. C. (2012). Comparison of particle-tracking and lumped-parameter age-distribution models for evaluating vulnerability of production wells to contamination. *Hydrogeology Journal*, *20*, 263–282. http://dx.doi.org/10.1007/s10040-011-0810-6

Eberts, S.M., Thomas, M.A., & Jagucki, M.L. (2013). *The quality of our Nation's waters—Factors affecting public-supply-well vulnerability to contamination—Understanding observed water quality and anticipating future water quality*. U.S. Geological Survey Circular 1385, 120 p.

Einsiedl, F., & Mayer, B. (2006). Hydrodynamic and microbial processes controlling nitrate in a fissured-porous karst aquifer of the Franconian Alb, Southern Germany. *Environmental Science and Technology*, *40*(21), 6697–6702. https://doi.org/10.1021/es061129x

Eisenhut, S., & Heumann, K. G. (1997). Identification of ground water contaminations by landfills using precise boron isotope ratio measurements with negative thermal ionization mass spectrometry. *Fresenius Journal of Analytical Chemistry, 359,* 375–377.

Erisman, J. W., Galloway, J. N., Dice, N. B., Sutton, M. A., Bleeker, A., Grizzetti, B., Leach, A. M., & de Vries, W. (2015). *Nitrogen, too much of a vital resource.* Science Brief. Zeist, The Netherlands: World Wildlife Fund.

Eschenbach, W., Well, R., & Walther, W. (2015). Predicting the denitrification capacity of sandy aquifers from in situ measurements using push–pull 15N tracer tests. *Biogeosciences, 12,* 2327–2346. https://doi.org/10.5194/bg-12-2327-2015

Felix, J. D., Elliott, E. M., Gish, T., Maghirang, R., Cambal, L., & Clougherty, J. (2014). Examining the transport of ammonia emissions across landscapes using nitrogen isotope ratios. *Atmospheric Environment, 95,* 563–570. https://doi.org/10.1016/j.atmosenv.2014.06.061

Fenech, C., Rock, L., Nolan, K., Tobin, J., & Morrissey, A. (2012). The potential for a suite of isotope and chemical markers to differentiate sources of nitrate contamination. *Water Research, 46,* 2023–2041. https://doi.org/10.1016/j.watres.2012.01.044

Field, K. G., & Samadpour, M. (2007). Fecal source tracking, the indicator paradigm, and managing water quality. *Water Research, 41*(16), 3517–3538.

Finlay, J. C., & Kendall, C. (2007). Stable isotope tracing of temporal and spatial variability in organic matter sources to freshwater ecosystems. In K. Lajtha & R. H. Michener (Eds.), *Stable isotopes in ecology and environmental science,* Ch. 10 (2nd ed., pp. 283–333). Oxford: Blackwell.

Flipse, W. J., & Bonner, F. T. (1985). Nitrogen-isotope ratios of nitrate in ground water under fertilized fields, Long Island, New York. *Ground Water, 23,* 59–67.

Fogg, G. E., Rolston, D. E., Decker, D. L., Louie, D. T., & Grismer, M. E. (1998). Spatial variation in nitrogen isotope values beneath nitrate contamination sources. *Ground Water, 36,* 418–426.

Foster, A.I., Katz, B.G., & Meyer, M. (2012). *Occurrence and potential transport of selected pharmaceuticals and other organic wastewater compounds from wastewater-treatment plant influent and effluent to groundwater and canal systems in Miami-Dade County, Florida.* U.S. Geological Survey Scientific Investigations Report 2012–5083, 64 p.

Fram, M. S., & Belitz, K. (2011). Occurrence and concentrations of pharmaceutical compounds in groundwater used for public drinking-water supply in California. *Science of the Total Environment, 409,* 3409–3417.

Glassmeyer, S. T., Furlong, E. T., Kolpin, D. W., Cahill, J. D., Zaugg, S. D., Werner, S. L., et al. (2005). Transport of chemical and microbial compounds from known wastewater discharges: Potential for use as indicators of human fecal contamination. *Environmental Science and Technology, 39*(14), 5157–5169.

Godfrey, E., Woessner, W. W., & Benotti, M. J. (2007). Pharmaceuticals in on-site sewage effluent and ground water, western Montana. *Ground Water, 45*(3), 263–271.

Granger, J., & Wankel, S. D. (2016). Isotopic overprinting of nitrification on denitrification as a ubiquitous and unifying feature of environmental nitrogen cycling. *Proceedings of the National Academy of Sciences U.S.A, 113,* E6391–E6400.

Green, C. T., Puckett, L. J., Böhlke, J. K., Bekins, B. A., Phillips, S. P., Kauffman, L. J., et al. (2008). Limited occurrence of denitrification in four shallow aquifers in agricultural areas of the United States. *Journal of Environmental Quality, 37,* 994–1009.

Griffin, D. W., Donaldson, K. A., Paul, J. H., & Rose, J. B. (2003). Pathogenic human viruses in coastal waters. *Clinical Microbiology Reviews, 16*(1), 129–143.

Griffin, D. W., Gibson, C. J., III, Lipp, E. K., Riley, K., Paul, J. H., & Rose, J. B. (1999). Detection of viral pathogens by RT-PCR and of microbial indicators using standard methods in the canals of the Florida Keys. *Applied and Environmental Microbiology, 65*(9), 4118–4125.

Griffin, D. W., Stokes, R., Rose, J. B., & Paul, J. H., III (2000). Bacterial indicator occurrence and the use of an F(+) specific RNA coliphage assay to identify fecal sources in Homosassa Springs, Florida. *Microbial Ecology, 39,* 56–64.

Grimmeisen, F., Lehmann, M. F., Liesch, T., Goeppert, N., Klinger, J., Zopfi, J., & Goldscheider, N. (2017). Isotopic constraints on water source mixing, network leakage and contamination in an urban groundwater system. *Science of the Total Environment.* http://dx.doi.org/10.1016/j.scitotenv.2017.01.054

Gruau, G., Legeas, M., Riou, C., Gallacier, E., Martineau, F., & Hénin, O. (2005). The oxygen isotopic composition of dissolved anthropogenic phosphates: A new tool for eutrophication research? *Water Research, 39,* 232–238.

Heaton, T. H. E. (1986). Isotopic studies of nitrogen pollution in the hydrosphere and atmosphere: A review. *Chemical Geology (Isotope GeoScience Section), 59,* 87–102.

Heffernan, J. B., Albertin, A. R., Fork, M. L., Katz, B. G., & Cohen, M. J. (2012). Denitrification and inference of nitrogen sources in the karstic Floridan Aquifer. *Biogeosciences, 9,* 1671–1690.

Jurgens, B. C., Böhlke, J. K., Kauffman, L. J., Belitz, K., & Esser, B. K. (2016). A partial exponential lumped parameter model to evaluate groundwater age distributions and nitrate trends in long-screened wells. *Journal of Hydrology, 543A,* 109–126. http://dx.doi.org/10.1016/j.jhydrol.2016.05.011

Kasprzyk-Hordern, B., Dinsdale, R. M., & Guwy, A. J. (2008). The occurrence of pharmaceuticals, personal care products, endocrine disruptors and illicit drugs in surface water in South Wales, UK. *Water Research, 42*(13), 3498–3518.

Katz, B. G. (2004). Sources of nitrate contamination and age of water in large karstic springs of Florida. *Environmental Geology, 46,* 689–706.

Katz, B. G., Bohlke, J. K., & Hornsby, H. D. (2001). Timescales for nitrate contamination of spring waters, northern Florida. *Chemical Geology, 179,* 167–186.

Katz, B. G., Chelette, A. R., & Pratt, T. R. (2004). Use of chemical and isotopic tracers to assess sources of nitrate and age of ground water, Woodville Karst Plain, USA. *Journal of Hydrology, 289,* 36–61.

Katz, B. G., Copeland, R., Greenhalgh, T., Ceryak, R., & Zwanka, W. (2005). Using multiple chemical indicators to assess sources of nitrate and age of ground water in a karstic spring basin. *Environmental and Engineering Geoscience, XI*(4), 439–453.

Katz, B. G., Eberts, S. M., & Kauffman, L. J. (2011). Using Cl/Br ratios and other indicators to assess potential impacts on groundwater quality from septic systems: A review and examples from principal aquifers in the United States. *Journal of Hydrology, 397*, 151–166.

Katz, B. G., & Griffin, D. W. (2008). Using chemical and microbiological indicators to track the possible movement of contaminants from the land application of reclaimed municipal wastewater in a karstic springs basin. *Environmental Geology, 55*, 801–821.

Katz, B. G., Griffin, D. W., & Davis, H. D. (2009). Groundwater quality impacts from the land application of treated municipal wastewater in a large karstic spring basin: Chemical and microbiological indicators. *Science of the Total Environment, 407*, 2872–2886.

Katz, B. G., Griffin, D. W., McMahon, P. B., Harden, H., Wade, E., Hicks, R. W., & Chanton, J. P. (2010). Fate of effluent-borne contaminants beneath septic tank drainfields overlying a karst aquifer. *Journal of Environmental Quality, 39*, 1181–1195.

Katz, B.G., Hicks, R.W., & Holland, K. (2017). *Implementation of best management practices in the Santa Fe restoration focus area: Nitrate-N concentrations in groundwater, springs, and river water.* In Four-year Progress Report, January 2013–December 2016, Florida Department of Environmental Protection and the Florida Department of Agriculture and Consumer Services, Office of Agricultural Water Policy, 58 p.

Katz, B.G., Hornsby, H.D., Bohlke, J.K., & Mokray, M.F. (1999). *Sources and chronology of nitrate contamination of springwaters, Suwannee River Basin, Florida.* U.S. Geological Survey Water- Resources Investigations Report 99–4252, 54 p.

Katz, B. G., Sepulveda, A. A., & Verdi, R. J. (2009). Estimating nitrogen loading to ground water and assessing vulnerability to nitrate contamination in a large karstic spring basin. *Journal of the American Water Resources Association, 45*, 607–627.

Kaushal, S. S., Groffman, P. M., Band, L. E., Elliott, E. M., Shields, C. A., & Kendall, C. (2011). Tracking nonpoint source nitrogen pollution in human-impacted watersheds. *Environmental Science and Technology, 45*, 8225–8232. https://doi.org/10.1021/es200779e

Kendall, C. (1998). Tracing nitrogen sources and cycling in catchments. In C. Kendall & J. J. McDonnell (Eds.), *Isotope tracers in catchment hydrology* (pp. 519–576). Amsterdam: Elsevier Science B.V.

Kendall, C., & Aravena, R. (2000). Nitrate isotopes in groundwater systems. In P. G. Cook & A. L. Herczeg (Eds.), *Environmental tracers in subsurface hydrology* (pp. 261–297). Boston, MA: Kluwer Academic Publishers.

Kendall, C., Elliott, E. M., & Wankel, S. D. (2007). Tracing anthropogenic inputs of nitrogen to ecosystems. In K. Lajtha & R. H. Michener (Eds.), *Stable isotopes in ecology and environmental science* (2nd ed., pp. 375–449). Oxford: Blackwell.

Kendall, C., & McDonnell, J. J. (Eds.). (1998). *Isotope tracers in catchment hydrology.* Amsterdam: Elsevier Science B.V, 839 p.

Knowles, L., Jr., Katz, B. G., & Toth, D. J. (2010). Using multiple chemical indicators to characterize and determine the age of groundwater from selected vents of the Silver Springs Group, central Florida, USA. *Hydrogeology Journal, 18*, 1825–1838.

Kohl, D. H., Shearer, G. B., & Commoner, B. (1971). Fertilizer nitrogen: Contribution to nitrate in surface water in a corn belt watershed. *Science, 174*, 1331–1334.

Kolpin, D. W., Furlong, E. T., Meyer, M. T., Thurman, E. M., Zaugg, S. D., Barber, L. B., & Buxton, H. T. (2002). Pharmaceuticals, hormones, and other organic wastewater contaminants in U.S. Streams, 1999-2000: A national reconnaissance. *Environmental Science and Technology, 36*(6), 1202–1211.

Kool, D. M., Wrage, N., Oenema, O., Van Kessel, C., & Van Groenigen, J. W. (2011). Oxygen exchange with water alters the oxygen isotopic signature of nitrate in soil ecosystems. *Soil Biology and Biochemistry, 43*, 1180–1185. https://doi.org/10.1016/j.soilbio.2011.02.006

Kratzer, C.R., Dileanis, P.D., Zamora, C., Silva, S.R., Kendall, C., Bergamaschi, B.A., & Dahlgren, R.A. (2004). *Sources and transport of nutrients, organic carbon, and chlorophyll-a in the San Joaquin River upstream of Vernalis, California, during summer and fall, 2000 and 2001.* U.S. Geological Survey Water Resources Investigations Report 03–4127, 113 p.

Kreitler, C. W. (1975). Determining the source of nitrate in groundwater by nitrogen isotope studies. In *Bureau of economic geology report of investigations 83.* Austin, TX: University of Texas, 57 p.

Kreitler, C. W., & Browning, L. A. (1983). Nitrogen-isotope analysis of groundwater nitrate in carbonate aquifers: Natural sources versus human pollution. *Journal of Hydrology, 61*, 285–301.

Landon, M.K., Clark, B.R., McMahon, P.B., McGuire, V.L., & Turco, M.J. (2008). *Hydrogeology, chemical-characteristics, and transport processes in the zone of contribution of a public-supply well in York, Nebraska.* US Geological Survey Scientific Investigations Report 2008-5050, 149 p.

LaPointe, B.E., & Herren, L.W. (2016). *Martin County watershed to reef septic tank study final report.* Prepared for Martin County Board of County Commissioners, 4 March 2016, 71 p.

LaPointe, B. E., Herren, L. W., Debortoli, D. D., & Vogel, M. A. (2015). Evidence of sewage-driven eutrophication and harmful algal blooms in Florida's Indian River Lagoon. *Harmful Algae, 43*, 82–102.

Leclerc, H., Edberg, S., Pierzo, V., & Delattre, J. M. (2000). Bacteriophages as indicators of enteric viruses and public health risk in groundwaters. *Journal of Applied Microbiology, 88*, 5–21.

Liu, C., Li, S., Lang, Y., & Xiao, H. (2006). Using $\delta^{15}N$ and $\delta^{18}O$ values to identify nitrate sources in karst groundwater, Guiyang, southwest China. *Environmental Science and Technology, 40*(22), 6928–6933. https://doi.org/10.1021/es0610129

Mariotti, A., Germon, J., Hubert, P., Kaiser, P., Letolle, R., Tardieux, A., & Tardieux, P. (1981). Experimental determination of nitrogen kinetic isotope fractionation: Some principles;

illustration for the denitrification and nitrification processes. *Plant Soil, 62*(3), 413–430.

Mayer, B., Boyer, E. W., Goodale, C., Jaworski, N. A., Breeman, N. V., Howarth, R. W., et al. (2002). Sources of nitrate in rivers draining sixteen watersheds in the northeastern U.S.: Isotopic constraints. *Biogeochemistry, 57/58*, 171–197.

McIlvin, M. R., & Altabet, M. A. (2005). Chemical conversion of nitrate and nitrite to nitrous oxide for nitrogen and oxygen isotopic analysis in freshwater and seawater. *Analytical Chemistry, 77*, 5589–5595.

McMahon, P. B., & Böhlke, J. K. (2006). Regional patterns in the isotopic composition of natural and anthropogenic nitrate in groundwater, High Plains, USA. *Environmental Science and Technology, 40*(9), 2965–2970.

McMahon, P. B., Dennehy, K. F., Bruce, B. W., Bohlke, J. K., Michel, R. L., Gurdak, J. J., & Hurlbut, D. B. (2006). Storage and transit time of chemicals in thick unsaturated zones under rangeland and irrigated cropland, High Plains, United States. *Water Resources Research, 42*(3), W03413. https://doi.org/10.1029/2005WR004417

McMahon, P. B., Böhlke, J. K., Kauffman, L. J., Kipp, K. L., Landon, M. K., Crandall, C. A., et al. (2008a). Source and transport controls on the movement of nitrate to public supply wells in selected principal aquifers of the United States. *Water Resources Research, 44*, W04401, 17 p., doi: https://doi.org/10.1029/2007WR006252

McMahon, P. B., Burow, K. R., Kauffman, L. J., Eberts, S. M., Böhlke, J. K., & Gurdak, J. J. (2008b). Simulated response of water quality in public-supply wells to land-use change. *Water Resources Research, 44*, W00A06. https://doi.org/10.1029/2007WR006731

McMinn, B. R., Ashbolt, N. J., & Korajkic, A. (2017). Bacteriophages as indicators of faecal pollution and enteric virus removal. Review Article. *Letters in Applied Microbiology, 65*, 11–26. https://doi.org/10.1111/lam.12736

Meals, D. W., Dressing, S. A., & Davenport, T. E. (2010). Lag time in water quality response to best management practices: A review. *Journal of Environmental Quality, 39*, 85–96. https://doi.org/10.2134/jeq2009.0108

Meffe, R., & de Bustamante, I. (2014). Emerging organic contaminants in surface water and groundwater: A first overview of the situation in Italy. *Science of the Total Environment, 15*(481), 280–295. https://doi.org/10.1016/j.scitotenv.2014.02.053

Mengis, M., Walther, U., Bernasconi, S. M., & Wehrli, B. (2001). Limitations of using $\delta^{18}O$ for the source identification of nitrate in agricultural soils. *Environmental Science and Technology, 35*, 1840–1844. https://doi.org/10.1021/es0001815

Minet, E. P., Goodhue, R., Meier-Augenstein, W., Kalin, R. M., Fenton, O., Richards, K. G., & Coxon, C. E. (2017). Combining stable isotopes with contamination indicators: A method for improved investigation of nitrate sources and dynamics in aquifers with mixed nitrogen inputs. *Water Research, 124*, 85–96. http://dx.doi.org/10.1016/j.watres.2017.07.041

Musgrove, M., Opsahl, S. P., Mahler, B. J., Herrington, C., Sample, T. L., & Banta, J. R. (2016). Source, variability, and transformation of nitrate in a regional karst aquifer: Edwards aquifer, central Texas. *Science of the Total Environment, 568*, 457–469.

Nakada, N., Kiri, K., Shinohara, H., Harada, A., Kuroda, K., Takizawa, S., & Takada, H. (2008). Evaluation of pharmaceuticals and personal care products as water-soluble molecular markers of sewage. *Environmental Science and Technology, 42*(17), 6347–6353.

Nestler, A., Berglund, M., Accoe, F., Duta, S., Xue, D., Boeckx, P., & Taylor, P. (2011). Isotopes for improved management of nitrate pollution in aqueous resources: Review of surface water field studies. *Environmental Science and Pollution Research, 18*(8), 519–533.

Nishikawa, T., Densmore, J.N., Martin, P., & Matti, J. (2003). *Evaluation of the source and transport of high nitrate concentrations in ground water, Warren Subbasin, California.* US Geological Survey Water-Resources Investigations Report 03-4009, 118 p.

O'Reilly, A. M., Chang, N. B., & Wanielista, M. P. (2012a). Cyclic biogeochemical processes and nitrogen fate beneath a subtropical stormwater infiltration basin. *Journal of Contaminant Hydrology, 133*, 53–75. https://dx.doi.org/10.1016/j.jconhyd.2012.03.005

O'Reilly, A. M., Wanielista, M. P., Chang, N. B., Xuan, Z., & Harris, W. G. (2012b). Nutrient removal using biosorption activated media: Preliminary biogeochemical assessment of an innovative stormwater infiltration basin. *Science of the Total Environment, 432*, 227–242. https://dx.doi.org/10.1016/j.scitotenv.2012.05.083

Oppenheimer, J. A., Badruzzaman, M., & Jacangelo, J. G. (2012). Differentiating sources of anthropogenic loading to impaired water bodies utilizing ratios of sucralose and other microconstituents. *Water Research, 46*, 5904–5916.

Oppenheimer, J. A., Eaton, A. D., Badruzzaman, M., Haghani, A. W., & Jacangelo, J. G. (2011). Occurrence and suitability of sucralose as an indicator compound of wastewater loading to surface waters in urbanized regions. *Water Research, 45*, 4019–4027.

Osaka, K., Ohte, N., Koba, K., Yoshimizu, C., Katsuyama, M., Tani, M., et al. (2010). Hydrological influences on spatiotemporal variations of $\delta^{15}N$ and $\delta^{18}O$ of nitrate in a forested headwater catchment in central Japan: Denitrification plays a critical role in groundwater. *Journal of Geophysical Research, 115*, G02021. https://doi.org/10.1029/2009JG000977

Panno, S. V., Hackley, K., Hwang, H., Greenberg, S., Krapac, I., Landsberger, S., & O'Kelly, D. (2006a). Characterization and identification of Na–Cl sources in ground water. *Ground Water, 44*, 176–187.

Panno, S. V., Hackley, K., Kelly, W. R., & Hwang, H. (2006b). Isotopic evidence of nitrate sources and denitrification in the Mississippi River, Illinois. *Journal of Environmental Quality, 35*(2), 495–504.

Pasten-Zapata, E., Ledesma-Ruiz, R., Harter, T., Ramirez, A. I., & Mahlknecht, J. (2014). Assessment of sources and fate of nitrate in shallow groundwater of an agricultural area by using a multi-tracer approach. *Science of the Total Environment, 470–471*, 855–864.

Peterson, B. J., & Fry, B. (1987). Stable isotopes in ecosystem studies. *Annual Review of Ecology, Evolution and Systematics, 18*, 293–320.

Phillips, P. J., Schubert, C. E., Argue, D. M., Fisher, I. J., Furlong, E. T., Foreman, W. T., et al. (2015). Concentrations of hormones, pharmaceuticals and other micropollutants in groundwater affected by septic systems in New England and New York. *Science of the Total Environment, 512–513*, 43–54. https://doi.org/10.1016/j.scitotenv.2014.12.067

Plummer, L. N. (2004). Dating of young groundwater. In P. K. Aggarwal, J. R. Gat, & K. F. O. Froelich (Eds.), *Isotopes in the water cycle—Past, present, and future of a developing science* (pp. 193–220). Dordrecht, The Netherlands: Springer.

Puckett, L. J., Tesoriero, A. J., & Dubrovsky, N. M. (2011). Nitrogen contamination of surficial aquifers - A growing legacy. *Environmental Science and Technology, 45*, 839–844.

Robertson, W. D., Cherry, J. A., & Sudicky, E. A. (1991). Ground-water contamination from two small septic systems on sand aquifers. *Ground Water, 29*, 82–92.

Rock, L., & Ellert, B. H. (2007). Nitrogen-15 and oxygen-18 natural abundance of potassium chloride extractable soil nitrate using the denitrifier method. *Soil Science Society of America Journal, 71*, 355–361.

Saccon, P., Leis, A., Marca, A., Kaiser, J., Campisi, L., Bottcher, M. E., et al. (2013). Multi-isotope approach for the identification and characterization of nitrate pollution sources in the Marano lagoon (Italy) and parts of its catchment area. *Applied Geochemistry, 34*, 75–89. https://doi.org/10.1016/j.apgeochem.2013.02.007

Sala, O. E., Chapin, F. S., III, Armesto, J. J., Berlow, E., Bloomfield, J., Dirzo, R., et al. (2000). Global biodiversity scenarios for the year 2100. *Science, 287*, 1770–1774.

Sanford, W. E., & Pope, J. P. (2013). Quantifying groundwater's role in delaying improvements to Chesapeake Bay water quality. *Environmental Science and Technology, 47*, 13330–13338.

Schlesinger, W. H. (2009). On the fate of anthropogenic nitrogen. *Proceedings of the National Academy of Sciences of the USA, 104*, 203–208.

Seal, T., Woeber, N. A., & Silvanima, J. (2016). Using tracers to infer potential extent of emerging contaminants in Florida's groundwater. *Florida Scientist, 79*(4), 279–289.

Sebilo, M., Mayer, B., Nicolardot, B., Pinay, G., & Mariotti, A. (2013). Long-term fate of nitrate fertilizer in agricultural soils. *Proceedings of the National Academy of Sciences, 110*(45), 18185–18189.

Sheppard, L. J., Leith, I. D., Mizunuma, T., Cape, J. N., Crossley, A., Leeson, S., et al. (2011). Dry deposition of ammonia gas drives species change faster than wet deposition of ammonium ions: Evidence from a long-term field manipulation. *Global Change Biology, 17*(12), 3589–3607.

Sigman, D. M., Casciotti, K. L., Andreani, M., Barford, C., Galanter, M., & Bohlke, J. K. (2001). A bacterial method for the nitrogen isotopic analysis of nitrate in seawater and freshwater. *Analytical Chemistry, 73*, 4145–4153.

Silva, S. R., Kendall, C., Wilkison, D. H., Ziegler, A. C., Chang, C. C. Y., & Avanzino, R. J. (2000). A new method for collection of nitrate from fresh water and the analysis of nitrogen and oxygen isotope ratios. *Journal of Hydrology, 228*, 22–36.

Soh, L., Connors, K. A., Brooks, B. W., & Zimmerman, J. (2011). Fate of sucralose through environmental and water treatment processes and impact on plant indicator species. *Environmental Science and Technology, 45*, 1365–1369.

Spoelstra, J., Senger, N. D., & Schiff, S. L. (2017). Artificial sweeteners reveal septic system effluent in rural groundwater. *Journal of Environmental Quality.* https://doi.org/10.2134/jeq2017.06.0233

Spruill, T. B., Showers, W. J., & Howe, S. S. (2002). Application of classification-tree methods to identify nitrate sources in ground water. *Journal of Environmental Quality, 31*, 538–1549.

Sui, Q., Cao, X., Lu, S., Zhao, W., Qiu, Z., & Yu, G. (2015). Occurrence, sources and fate of pharmaceuticals and personal care products in the groundwater: A review. *Emerging Contaminants, 1*, 14–24.

Sulzman, E. W. (2007). Stable isotope chemistry and measurement: A primer. In K. Lajtha & R. H. Michener (Eds.), *Stable isotopes in ecology and environmental science* (2nd ed., pp. 1–21). Oxford: Blackwell.

Szabo, Z., Jacobsen, E., & Reilly, T. (2004). Preliminary evaluation of organic wastewater contaminants in septic tanks for possible use as effluent tracers in shallow ground water. In *Proceedings of the 4th international conference on pharmaceuticals and endocrine disrupting chemicals in water.* Westerville, OH: National Ground Water Association.

Tesoriero, A. J., Saad, D. A., Burow, K. R., Frick, E. A., Puckett, L. J., & Barbash, J. E. (2007). Linking ground-water age and chemistry data along flow paths: Implications for trends and transformations of nitrate and pesticides. *Journal of Contaminant Hydrology, 94*, 139–155.

Valcárcel, Y., González, A. S., Rodriguez-Gil, J. L., Gil, A., & Catalá, M. (2011). Detection of pharmaceutically active compounds in the rivers and tap water of the Madrid region (Spain) and potential ecotoxicological risk. *Chemosphere, 84*, 1336–1348. https://doi.org/10.1016/j.chemosphere.2011.05.014

Vengosh, A., Barth, S., Heumann, K. G., & Eisenhut, S. (1999). Boron isotopic composition of freshwater lakes from Central Europe and possible contamination sources. *Acta Hydrochimica Hydrobiologica, 27*, 416–442.

Vengosh, A., Heumann, K. G., Juraske, S., & Kasher, R. (1994). Boron isotope application for tracing sources of contamination in groundwater. *Environmental Science and Technology, 28*, 1968–1974.

Vengosh, A., & Pankratov, L. (1998). Chloride/bromide and chloride/fluoride ratios of domestic sewage effluents and associated contaminated groundwater. *Ground Water, 36*(5), 815–824.

Verplanck, P. L., Taylor, H. E., Nordstrom, D. K., & Barber, L. B. (2005). Aqueous stability of gadolinium in surface waters receiving sewage treatment plant effluents, Boulder, Colorado. *Environmental Science and Technology, 39*(18), 6923–6929.

Verstraeten, I. M., Fetterman, G. S., Meyer, M. T., Bullen, T., & Sebree, S. K. (2005). Use of tracers and isotopes to evaluate vulnerability of water in domestic wells to septic waste. *Ground Water Monitoring and Remediation, 25*(2), 107–117.

Vrzel, J., Vukovic-Gacid, B., Kolarevic, S., Gačić, Z., Kračun-Kolarević, M., Kostić, J., et al. (2016). Determination of the sources of nitrate and the microbiological sources of pollution in the Sava River Basin. *Science of the Total*

Environment, *573*, 1460–1471. https://doi.org/10.1016/j.scitotenv.2016.07.213

Wang, L., Butcher, A., Stuart, M., Gooddy, D., & Bloomfield, J. (2013). The nitrate time bomb: A numerical way to investigate nitrate storage and lag time in the unsaturated zone. *Environmental Geochemistry and Health*, *35*, 667–681.

Wankel, S. D., Kendall, C., Francis, C. A., & Paytan, A. (2006). Nitrogen sources and cycling in the San Francisco Bay Estuary: A nitrate dual isotope approach. *Limnology and Oceanography*, *51*, 1654–1664.

Weil, R. R., & Brady, N. C. (2017). The colloidal fraction: Seat of soil chemical and physical activity. In D. Fox & A. Gilfillan (Eds.), *The nature and properties of soils*, Ch. 8 (15th ed.). Essex, England: Pearson Education, Inc.

Widory, D., Kloppmann, W., Chery, L., Bonnin, J., Rochdi, H., & Guinamant, J. L. (2004). Nitrate in groundwater, an isotope multi-tracer approach. *Journal of Contaminant Hydrology*, *72*, 165–188.

Widory, D., Petelet-Giraud, E., Négrel, P., & Ladouche, B. (2005). Tracking the sources of nitrate in groundwater using coupled nitrogen and boron isotopes: A synthesis. *Environmental Science and Technology*, *39*, 539–548.

Xue, D., Botte, J., De Baets, B., Accoe, F., Nestler, A., Taylor, P., et al. (2009). Present limitations and future prospects of stable isotope methods for nitrate source identification in surface- and groundwater: Review. *Water Research*, *43*, 1159–1170. https://doi.org/10.1016/jwatres.2008.12.048

Yang, Y., & Toor, G. S. (2016). $\delta^{15}N$ and $\delta^{18}O$ reveal the sources of nitrate-nitrogen in urban residential stormwater runoff. *Environmental Science and Technology*, *50*(6), 2881–2889. https://doi.org/10.1021/acs.est.5b05353

Young, M., McLaughlin, K., Kendall, C., Stringfellow, W., Rollog, M., Elsbury, K., et al. (2009). Characterizing the oxygen isotopic composition of phosphate sources to aquatic ecosystems. *Environmental Science and Technology*, *43*(14), 5190–5196. https://doi.org/10.1021/es900337q

Young, T. A., Heidler, J., Matos-Perez, C. R., Sapkota, A., Toler, T., Gibson, K. E., et al. (2008). Ab Initio and in situ comparison of caffeine, triclosan, and triclocarban as indicators of sewage-derived microbes in surface waters. *Environmental Science and Technology*, *42*(9), 3335–3340.

FURTHER READING

Bassett, R. L. (1990). A critical evaluation of the available measurements for the stable isotopes of boron. *Applied Geochemistry*, *5*, 541–554.

Böhlke, J. K., Jurgens, B. C., Uselmann, D. J., & Eberts, S. M. (2014). Educational webtool illustrating groundwater age effects on contaminant trends in wells. *Ground Water*, *52*(S1), 8–9. (plus 19 p Appendix). http://dx.doi.org/10.1111/gwat.12261

Burns, D. A., Boyer, E. W., Elliott, E. M., & Kendall, C. (2009). Sources and transformations of nitrate from streams draining varying land uses: Evidence from dual isotope analysis. *Journal of Environmental Quality*, *38*, 1149–1159.

Carey, R. O., Hochmuth, G. J., Martinez, C. J., Boyer, T. H., Dukes, M. D., Toor, G. S., & Cisar, J. L. (2013). Evaluating nutrient impacts in urban watersheds: Challenges and research opportunities. *Environmental Pollution*, *173*, 138–149. http://dx.doi.org/10.1016/j.envpol.2012.10.004

Cey, E. E., Rudolph, D. L., Aravena, R., & Parkin, G. (1999). Role of the riparian zone in controlling the distribution and fate of agricultural nitrogen near a small stream in southern Ontario. *Journal of Contaminant Hydrology*, *37*, 45–67.

Chang, C. C. Y., Kendall, C., Silva, S. R., Battaglin, W. A., & Campbell, D. H. (2002). Nitrate stable isotopes: Tools for determining nitrate sources among different land uses in the Mississippi River Basin. *Canadian Journal of Fisheries and Aquatic Sciences*, *59*, 1874–1885. https://doi.org/10.1139/F02-153

Davis, H. D., Katz, B. G., & Griffin, D. W. (2010). *Nitrate-N movement in groundwater from the land application of treated municipal wastewater and other sources in the Wakulla Springs Springshed, Leon and Wakulla Counties, Florida, 1966-2018*. Tallahassee, FL: U.S. Geological Survey Scientific Investigations Report 2010-5099, 80 p.

Deutsch, B., Voss, M., & Fischer, H. (2009). Nitrogen transformation processes in the Elbe River: Distinguishing between assimilation and denitrification by means of stable isotope ratios in nitrate. *Aquatic Science Research Across Boundaries*, *71*, 228–237.

Dubrovsky, N.M., Burow, K.R., Clark, G.M., Gronberg, J.M., Hamilton P.A., Hitt, K.J., et al. (2010). *The quality of our Nation's waters—Nutrients in the Nation's streams and groundwater, 1992–2004*. U.S. Geological Survey Circular 1350, 174 p. Retrieved from http://water.usgs.gov/nawqa/nutrients/pubs/circ1350, 25 September 2019.

Fukada, T., Hiscock, K. M., & Dennis, P. F. (2004). A dual-isotope approach to the nitrogen hydrochemistry of an urban aquifer. *Applied Geochemistry*, *19*, 709–719.

Grizzetti, B., Bouraoui, F., Billen, G., van Grinsven, H., Cardoso, A.C., Thieu, V., et al. (2011). Nitrogen as a threat to European water quality. Ch. 17. In M. A. Sutton, C. M. Howard, J. W. Erisman, G. Billen, A. Bleeker, P. Grennfelt, H. van Grinsven, & B. Grizzetti (Eds.), *The European nitrogen assessment: Sources, effects, and policy perspectives* (pp. 379–404). Cambridge, UK: Cambridge University Press.

Henson, W. R., Huang, L., Graham, W. D., & Ogram, A. (2017). Nitrate reduction mechanisms and rates in an unconfined eogenetic karst aquifer in two sites with different redox potential. *Journal of Geophysical Research: Biogeosciences*, *122*, 1062–1077.

Hinkle, S. R., & Tesoriero, A. J. (2014). Nitrogen speciation and trends, and prediction of denitrification extent, in shallow U.S. groundwater. *Journal of Hydrology*, *509*, 343–353.

Kasprzyk-Hordern, B., Dinsdale, R. M., & Guwy, A. J. (2007). Multiresidue method for the determination of basic/neutral pharmaceuticals and illicit drugs in surface water by solid-phase extraction and ultra-performance liquid chromatography-positive electrospray ionization tandem mass spectrometry. *Journal of Chromatography A*, *1161*(1-2), 132–145.

Katz, B. G. (2012). Nitrate contamination in karst groundwater. In W. B. White & D. C. Culver (Eds.), *Encyclopedia of caves* (pp. 564–568). Boston, MA: Academic Press, Elsevier. https://doi.org/10.1016/B978-0-12-383832-2.00146-8

Kendall, C., & Coplen, T. B. (2001). Distribution of oxygen-18 and deuterium in river waters across the United States. *Hydrological Processes*, *15*, 1363–1393.

Kolpin, D. W., Skopec, M., Meyer, M. T., Furlong, E. T., & Zaugg, S. D. (2004). Urban contribution of pharmaceuticals and other organic wastewater contaminants to streams during differing flow conditions. *Science of the Total Environment*, *328*, 119–130.

Nolan, B. T., & Hitt, K. J. (2006). Vulnerability of shallow groundwater and drinking-water wells to nitrate in the. *United States: Environmental Science and Technology*, *40*(24), 7834–7840.

Panno, S. V., Kelly, W. R., Hackley, K. C., Hwang, H.-H., & Martinsek, A. T. (2008). Sources and fate of nitrate in the Illinois River Basin, Illinois. *Journal of Hydrology*, *359*, 174–188.

Pope, C. A., Burnett, R. T., Thun, M. J., Calle, E. E., Krewski, D., Ito, K., & Thurston, G. D. (2002). Lung cancer, cardiopulmonary mortality, and long-term exposure to fine particulate air pollution. *Journal of the American Medical Association*, *287*, 1132–1141.

Puckett, L. J., & Cowdery, T. K. (2002). Transport and fate of nitrate in a glacial outwash aquifer in relation to groundwater age, land use practices, and redox processes. *Journal of Environmental Quality*, *31*, 782–796.

Scavia, D., & Bricker, S. B. (2006). Coastal eutrophication assessment in the United States. *Biogeochemistry*, *79*(1–2), 187–208.

Sebilo, M., Billen, G., Mayer, B., Billiou, D., Grably, M., Garnier, J., & Mariotti, A. (2006). Assessing nitrification and denitrification in the Seine River and estuary using chemical and isotopic techniques. *Ecosystems*, *9*, 564–577.

Seiler, R. L. (2005). Combined use of ^{15}N and ^{18}O of nitrate and ^{11}B to evaluate nitrate contamination in groundwater. *Applied Geochemistry*, *20*, 1626. https://doi.org/10.1016/J.APGEOCHEM.2005.04.007

Tesoriero, A. J., Liebscher, H., & Cox, S. E. (2000). Mechanism and rate of denitrification in an agricultural watershed—Electron and mass balance along groundwater flow paths. *Water Resources Research*, *36*(6), 1545–1559.

U.S. Environmental Protection Agency. (2009). *An urgent call to action*. Report of the State-EPA Nutrient Innovations Task Group, 41 p.

Walvoord, M. A., Phillips, F. M., Stonestrom, D. A., Evans, R. D., Hartsough, P. C., Newman, B. D., & Striegl, R. G. (2003). A reservoir of nitrate beneath desert soils. *Science*, *302*(5647), 1021–1024.

Wang, L., Stuart, M. E., Lewis, M. A., Ward, R. S., Skirvin, D., Naden, P. S., et al. (2016). The changing trend in nitrate concentrations in major aquifers due to historical nitrate loading from agricultural land across England and Wales from 1925 to 2150. *Science of the Total Environment*, *542*, 694–705.

Wassenaar, L. I., Hendry, M. J., & Harrington, N. (2006). Decadal geochemical and isotopic trends for nitrate in a transboundary aquifer and implications for agricultural beneficial management practices. *Environmental Science and Technology*, *40*, 4626–4632.

Wick, K., Heumesser, C., & Schmid, E. (2012). Groundwater nitrate contamination: Factors and indicators. *Journal of Environmental Management*, *111*, 178–186.

5
Adverse Human Health Effects of Reactive Nitrogen

Previously, in Chapter 3, we saw that numerous point and nonpoint sources have released forms of reactive nitrogen that has contributed to widespread air and water pollution all over the world. This pollution has not only contributed to adverse human health effects but has also have caused eutrophication of aquatic systems, harmful algal blooms (HABs), loss of biodiversity, and acidification of soils and surface waters.

The widespread alteration of the nitrogen cycle from the production and use of synthetic fertilizers has had both positive and negative consequences for human health. More than 15 years ago, Townsend et al. (2003) showed schematically how the net public health benefit from human fixation of nitrogen gas to produce fertilizers and the use of reactive nitrogen has peaked and declined because of an exponential increase in air and water pollution and ecological feedbacks to disease (Fig. 5.1). However, one certainly cannot discount the enormous benefits from the use of nitrogen fertilizer to grow crops in developing countries, which has led to an increase in food production and substantial reductions in malnutrition. With better nutrition, healthier diets can lead to more efficient immune response to parasitic and infectious diseases (Nesheim, 1993). Unfortunately, shortages of food and malnutrition still exist in many large areas of the world. This chapter focuses on some of the adverse human health effects from reactive forms of nitrogen associated with air and water pollution, and with the degradation of ecosystems.

5.1. HEALTH EFFECTS ASSOCIATED WITH REACTIVE NITROGEN IN AIR POLLUTION

In Chapter 3, we learned about emissions of reactive nitrogen compounds (ammonia, NO_x, and nitrate) to the atmosphere in gaseous and particulate forms mainly from fossil fuel combustion and agricultural activities. Nitrogen oxides (NO_x) actually represent seven oxidized nitrogen compounds including N_2O, nitrous oxide; NO, nitric oxide; N_2O_2, dinitrogen dioxide; N_2O_3, dinitrogen trioxide; NO_2, nitrogen dioxide; N_2O_4, dinitrogen tetroxide; N_2O_5, dinitrogen pentoxide. The U.S. EPA and other agencies regulate nitrogen dioxide (NO_2) as a surrogate for this family of compounds because it is the most prevalent form of NO_x in the atmosphere that is generated by anthropogenic (human) activities. Both ammonia and gaseous NO_x tend to form aerosol phase compounds (ammonium and nitrate) and these compounds can be transported over large distances. Over the past 70 years, global ammonia emissions have increased by more than a factor of two, from 23 to 60 Tg/yr (Bauer et al., 2016).

5.1.1. Ammonia, NO_x, and Particulate Air Pollution

Anthropogenic emissions of ammonia can form inorganic aerosols including ammonium sulfate, ammonium nitrate, and ammonium bisulfate. Inorganic aerosols are the main constituents of anthropogenic pollution of fine particulate matter (PM) that is less than 2.5 μm in diameter (referred to as $PM_{2.5}$), which is 3% of the diameter of a human hair. Studies also have investigated PM_{10}, which is PM less than 10 μm in diameter. Peel et al. (2013) noted that fine PM ($PM_{2.5}$) pollution related to atmospheric emissions of nitrogen (N) and other pollutants can cause premature death and a variety of serious health effects. Due to their minute size, particles smaller than 2.5 μm can bypass the nose and throat and move deep into the lungs and some may even enter the circulatory system. These harmful compounds in the atmosphere can be dispersed over large areas of the world and long-term exposures to $PM_{2.5}$ has contributed to significant health hazards including cardiopulmonary mortality, lung cancer, other respiratory ailments, and lowering of personal resistance. The U.S. Environmental Protection Agency (EPA) and the European Union have established ambient standards (annual average) for $PM_{2.5}$ of 12 and 25 μg/m³, respectively. The World Health Organization (WHO)

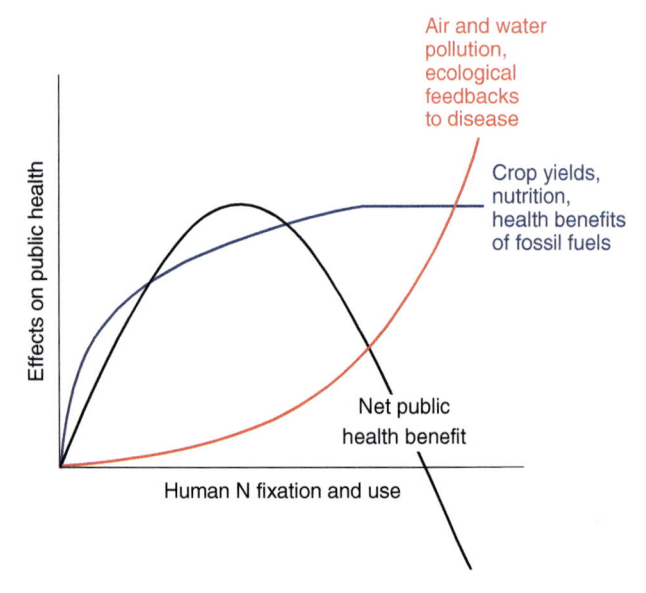

Figure 5.1 Conceptual diagram showing detrimental effects on human health from increasing amounts of excess reactive nitrogen from human fixation. *Source*: From Townsend et al. (2003). Reproduced with permission of John Wiley & Sons.

advises concentrations below 10 µg/m³ (WHO Regional Office for Europe, 2006).

A large national cohort study involving 2.5 million people in Canada investigated associations between cause-specific mortality and ambient concentrations of fine particulate matter ($PM_{2.5}$), ozone, and nitrogen dioxide (Crouse et al., 2015). Hazard ratios for each pollutant were estimated separately over a 16-year period (1991–2006). The investigators found that exposure to $PM_{2.5}$, NO_2, and ozone was positively associated with several common causes of nonaccidental death and cause-specific mortality. Higher NO_2 concentrations were reported to be related to higher PM mortality effect estimates in single-city studies (Dominici et al., 2002; Deguen et al., 2015; Faustini et al., 2016). In multicity studies, cities with higher mean NO_2 levels and mean NO_2/PM_{10} ratios had higher risk estimates of mortality (0.44% for cities with higher mean NO_2 levels compared to 0.17% for cities with lower mean NO_2 levels). Also, cities with higher NO_2/PM_{10} ratios had higher risk estimates of mortality (42% compared to 17% for cities with lower NO_2/PM_{10} ratios (Katsouyanni et al., 2001)). It has been suggested that long-term exposure to higher NO_2 concentrations may deteriorate lung function and increase vulnerability to short-term effects of PM (Faustini et al., 2016).

A study in India showed that there was an association between air pollution with cardiopulmonary diseases (Liu et al., 2013). Regarding human health effects of air pollution and their underlying mechanisms of cellular action, Kampa and Castanas (2008) noted that humans are exposed to different air pollutants such as NO_x primarily through inhalation and ingestion, whereas dermal contact is a minor route of exposure. It was reported that nitrogen oxides increase the susceptibility to respiratory infections, and emphysema-like lesions have been observed in mice exposed to nitrogen dioxide (NO_2).

It is important to note that PM in the atmosphere typically can be a complex mixture of different components including NO_x, ammonia, nitrate, organic carbon compounds, sulfate, and trace elements. There can be considerable heterogeneity in reported adverse health effects from PM in air pollution. This variability from one area to another is related to differences in chemical constituents from different pollution sources. Other important factors include differences in the characteristics of populations (e.g., the proportion of the elderly population and people with preexisting diseases), exposure profile (e.g., air pollution concentrations and composition), and regional climate (Tian & Sun, 2017). For example, Katsouyanni et al. (2001, 2009) pointed out that the short-term mortality effect of an increase of PM_{10} per 10 µg/m³ was 0.29% in cities with relatively cold climate compared to a higher rate of 0.82% for cities in warm climate based on data from the Air Pollution and Health: A European Approach Phase 2 (APHEA-2) study that included 32 European cities. A study of 207 cities in the United States also found a higher association between long-term $PM_{2.5}$ exposure and mortality in warmer cities using data from more than 35 million Medicare enrollees during 2000–2010. (Kioumourtzoglou et al., 2016). They calculated annual $PM_{2.5}$ averages measured at ambient central monitoring sites in each city and found elevated estimates of mortality in the Southeastern, Southern, and Northwestern United States. Based on a literature review of 29 studies in the United States, Stanek et al. (2011) found that PM from motor vehicle emissions were associated with cardiovascular mortality, whereas PM from coal combustion was associated with total mortality. Several recent epidemiological studies have found strong causal relationships between long-term exposure to $PM_{2.5}$ and premature mortality associated with heart disease, stroke, respiratory diseases, and lung cancer. Apte et al. (2015) noted that the Global Burden of Disease Study of Comparative Risk Assessment (Lim et al., 2012) estimated that there were about 3.2 million deaths worldwide attributable to air pollution from $PM_{2.5}$. Yang et al. (2017) reported that based on epidemiological studies, there is a positive association of cardiopulmonary morbidity and mortality, and lung cancer risk with exposure to airborne PM. They noted that oxidative stress and inflammation are the main proposed causal factors involved in mediating PM effects on both cardiovascular and pulmonary health outcomes.

In China, urban air quality has been deteriorating substantially over the past decades. Although nitrogen oxide and sulfur dioxide emissions have decreased since 2013, there have not been significant reductions in severe haze episodes in many cities. Zhang et al. (2018) reported that the highest daily average $PM_{2.5}$ measured concentration was more than $500\,\mu g/m^3$, which is twenty times higher than the WHO guideline. Also, they noted that less than 1% of the 500 largest cities in China meets the WHO air-quality guideline values ($10\,\mu g/m^3$) for the annual mean and $25\,\mu g/m^3$ for the daily mean. Ren et al. (2017) studied the temporal and spatial distribution of major pollutants (PM_{10}, $PM_{2.5}$, sulfur dioxide, nitrogen oxide, and ozone) in China cities. They concluded that reduced emissions have led to some improvements in air quality, the Chinese National Ambient Air Quality Standard (CNAAQS) for PM_{10} and $PM_{2.5}$ is still exceeded in many cities in China in 2015. It was reported that high NO_2 levels were found in Beijing-Tiajin-Hebei and the Yangtze River delta region. Ren et al. (2017) attributed the elevated $PM_{2.5}$ levels mainly to secondary particle formation and motor vehicle exhaust. Wu et al. (2016) postulated that ammonia plays a more important role with the spatiotemporal variation of $PM_{2.5}$ levels than emissions of sulfur dioxide and nitrogen oxide. To test this hypothesis, the researchers analyzed in situ concentration data measured in major cities in China along with gridded emission data from remote sensing and inventories. They found that urban $PM_{2.5}$ pollution in China is significantly affected by ammonia emissions. Wu et al. (2016) concluded that more efforts need to be focused on reduction of ammonia emissions, which in turn could result in substantial cobenefits by improving nitrogen use efficiency in farming systems. They also noted that ammonia reduction strategies would help to address China's food security goals, impede biodiversity losses, reduce greenhouse gas emissions, and promote economic savings.

A 5-year study by a group of doctors in Jiangsu Province, China, indicated that atmospheric pollution produces one tenth of congenital birth defects (http://www.asianews.it/news-en/Pollution-causes-many-birth-defects-in-China-14170.html, accessed 19 May 2018). Birth defects increased by 50% in China from 2001 to 2006, which affected approximately 1.2 million newborns. The most widespread defect involved heart disease, which was directly connected to air pollution and was increasing in both urban and rural areas (Hu et al., 2015; Schoolman & Ma, 2012). Tian and Sun (2017) reported that higher NO_2 concentrations were associated with higher estimates of PM mortality effects in single-city studies.

Somewhat encouraging is the decrease in NO_x emissions in 28 European Countries from 1990 to 2014 (European Environmental Agency, 2016). The decrease was mainly attributed to significant reductions reported by the United Kingdom, France, and Poland. In the United Kingdom, NO_x emissions decreased predominantly in energy production and distribution, and in other sectors (commercial, institutional and households, and road transport). In France, highest emission reductions were from the following sectors: commercial, institutional, and household; energy production and distribution; and the road transport. Also, in Poland, the energy production and distribution sector showed the highest decreases, followed by the commercial, institutional, and household sector (European Environmental Agency, 2016).

In the United States, Dubrovsky et al. (2010) reported that nitrogen oxide emissions from vehicles, factories, and other sources decreased 12% from 1993 to 2002. They noted that the largest decreases were in the Northeastern and Western States and the smallest decreases were in the Rocky Mountains and Great Plains States based on 2003 information from the U.S. EPA. Decreases in NO_x emissions largely coincided with widespread stable or decreasing nitrogen inputs from atmospheric deposition from 1993 to 2003. However, a recent study by Jiang et al. (2018) found that NO_x emissions have not been decreasing as expected in recent years (2011–2015) when they compared estimates from satellites and surface NO_x measurements to the trends predicted from the U.S. EPA's emission inventory data. This discrepancy was attributed to a combination of several factors: the growing relative contribution of industrial, area, and off-road mobile sources of emissions; the decreasing relative contribution of on-road gasoline vehicles, and slower than expected decreases in on-road diesel NO_x emissions. Jiang et al. (2018) recommended for future quantitative evaluations of emission changes for NO_x and subsequent effects on ozone and other air pollutants, models will be required and data with finer spatial scales (e.g., for urban and roadway environments).

While NO_x emissions have been decreasing in many areas, emissions of other forms of reactive nitrogen have also been increasing. Ackerman et al. (2019) used the GEOS-Chem Chemical Transport Model to estimate wet and dry deposition of inorganic nitrogen globally at a spatial resolution of $2\times 2.5°$ for 12 individual years in the period from 1984 to 2016. They reported an 8% increase in global inorganic nitrogen (ammonia, ammonium, and nitrate) deposition from 86.6 to 93.6 Tg N/yr. Increases in inorganic nitrogen deposition were found in regions where overall emissions also have increased in recent decades (e.g., East Asia and Southern Brazil). Accompanying this inorganic nitrogen increase was an increase in the percentage (from 30 to 35%) of chemically reduced nitrogen forms (ammonia and ammonium). This trend was attributed to high regional increases in the proportion of chemically reduced nitrogen deposition over the United States, which was consistent with another study (Li et al., 2016). Ackerman et al. (2019) noted that the increased deposition of chemically reduced nitrogen may adversely affect the competitive balance among plants that have differing needs for various forms of nitrogen.

5.1.2. NO_x Driver of Ozone Production

Elevated NO_x in the atmosphere can enhance the production of tropospheric ozone (O_3) at low altitudes and aerosols, which can cause coughs, asthma, and other reactive-airways disease (Brunekreef & Holgate, 2002; Townsend et al., 2003). When NO_x and volatile organic compounds (VOCs) react in the presence of sunlight in the atmosphere, they also can form photochemical smog, which typically increases when solar radiation is higher (e.g., during the summer). Chronic exposure to O_3 has been reported to reduce lung function and may also be responsible for causing emphysema, and even lung cancer (Kampa & Castanas, 2008).

Jhun et al. (2015) investigated the impact of decreases of NO_x concentrations in air on recent ozone trends in the United States by season and time of day. They analyzed hourly O_3, NO_x, and daily VOCs data from the U.S. EPA's Air Quality System and Photochemical Assessment Monitoring Stations along with meteorological data (24-hour temperature, wind speed, and water vapor pressure) from the National Oceanic and Atmospheric Administration's National Climatic Data Center between 1994 and 2010. Sites with both ozone and NO_x measurements included 35 rural, 55 suburban, and 42 urban sites. Jhun et al. (2015) noted that significant reductions in O_3 precursors including NO_x and VOCs have not resulted in proportionate decreases in ozone. They found that hourly O_3 concentrations decreased during the warm season midday but increased during the cold season; although peak O_3 concentrations were mitigated by NO_x reductions. Increases in O_3 concentrations were found across rural, suburban, and urban areas.

According to the U.S. EPA (2012), in 2010, over one third of the U.S. population resided in areas exceeding the national O_3 standard. Based on a review of scientific evidence, the U.S. EPA in April 2018 issued a decision to retain the current national ambient air quality standards for oxides of nitrogen (NO_x). The EPA concluded that the current ambient air quality standards protect the public health, including the at-risk populations of older adults, children, and people with asthma, with an adequate margin of safety. The air quality standards for nitrogen oxides are a 1-hour standard at a level of 100 ppb (parts per billion) based on the 3-year average of 98th percentile of the yearly distribution of 1-hour daily maximum concentrations, and an annual standard at a level of 53 ppb (https://www.epa.gov/no2-pollution/primary-national-ambient-air-quality-standards-naaqs-nitrogen-dioxide, accessed 18 May 2018).

Ozone concentrations have also increased in other parts of the world. For example, Ren et al. (2017) noted that the average annual ozone concentration was increasing in China during 2013–2015, and 16% of the total cities in 2015 did not meet the CNAAQS, indicating that more focus should be given to O_3 pollution.

5.1.3. N_2O Effects on Ozone Depletion and Climate

At high altitudes and in parts of the stratosphere, ozone can be destroyed due to the production of the catalyst NO as ultraviolet light breaks N_2O apart. The destruction of ozone in the stratosphere would allow more ultraviolet light to reach the Earth's surface, possibly causing more skin cancers. More than two decades ago, Kane (1998) noted that reductions in ozone would likely result in a 10–20% increase in ultraviolet-B radiation and this increase could explain a 20–40% rise in skin cancer in the human population since the 1970s. Particulate air pollution has also been linked to cardiovascular diseases and overall mortality (Pope et al., 2002). N_2O also is a potent greenhouse gas with a long residence time in the atmosphere of 120 years (Howarth et al., 2011) and therefore its harmful effects can persist for decades.

5.1.4. Air-Quality-Related Health Damages Associated with Maize Production

As mentioned previously, agricultural activities can produce air pollution that can adversely affect human health and cause harm to ecosystems. This section highlights the results of a recently published study (J. Hill et al., 2019) that investigated the human health effects of air pollution caused by the production of maize in the United States. Maize is an important agricultural crop that is used for human consumption, animal feed, and ethanol biofuel. The investigators used U.S. county-level data on agricultural production practices and productivity (2010–2014) to generate a spatial life-cycle emissions inventory for maize. Based on this inventory, J. Hill et al. (2019) estimated health damages that accounted for atmospheric pollution transport and chemistry, and human exposure to pollution at high resolution spatially. They showed that reductions in air quality resulting from maize production were associated with 4300 premature deaths annually in the United States, along with significant economic costs. They attributed the reduced air quality to increases in concentrations of $PM_{2.5}$, which were driven by ammonia emissions from fertilizer usage. It was also noted that NO_x and volatile organic compounds emissions from maize production also contributed to ground-level ozone formation that also impacts human health. J. Hill et al. (2019) estimated life-cycle greenhouse gas emissions of maize production and found that damages related to climate change also resulted in substantial costs. Furthermore, the study pointed out that about 10% of maize grown in the United States is consumed as sweeteners, starch, cereals, or beverages that require additional processing, transport, storage, and preparation. These processes likely directly or indirectly contribute to additional air pollution. Several strategies were proposed for reducing damages from maize production, such as changes in the type of fertilizers and application methods, improvements in nitrogen use efficiency, and changing to crops that require less fertilizer. This approach used by J. Hill et al. (2019) would be very useful to apply to other food crops and animal agriculture in the United States and other countries to better understand the overall impact of emissions from agricultural activities on human health and ecosystems.

Air pollution from various anthropogenic activities can also contribute to the contamination of food and water, which in several cases results in ingestion as the main route of reactive nitrogen intake. Health effects associated with consumption of nitrate in drinking water are discussed in the next section.

5.2. HEALTH EFFECTS ASSOCIATED WITH NITRATE IN DRINKING WATER

Humans are increasingly exposed to elevated levels of reactive nitrogen species (mainly nitrate) in drinking water. The U.S. EPA has set a maximum contaminant level of 10 mg/L (parts per million) for nitrate-N (expressed as nitrogen) in drinking water, and the WHO has set a maximum contaminant level of 50 mg/L for nitrate (as NO_3) in drinking water (equivalent to a nitrate-N concentration of 11.3 mg N/L). The regulatory limit for nitrate in drinking water was established to protect against infant methemoglobinemia (described below); however, other potential health effects were not considered at that time. The U.S. EPA (2007) also recommended that chronic oral exposure to nitrate for infants should not exceed a level of 1.6 mg/kg/d and a nitrite level of 0.1 mg/kg/d and drinking water advisories for a 4-kg child include an maximum contaminant level goal of no more than 10 mg/L of nitrate-N and 1 mg/L of nitrite-N.

Other forms of reactive nitrogen, such as nitrite and ammonia, can be present in drinking water. Ammonia and or ammonium can enter surface waters via overland runoff or direct discharges from wastewater sources. There is the potential for nitrification of ammonia (ammonium) and nitrite to nitrate in drinking water systems, as noted by the U.S. EPA (https://www.epa.gov/sites/production/files/2015-09/documents/nitrification_1.pdf, accessed 29 April 2019), possibly in areas where human and animal wastes are present, as well as from fertilizers and industrial wastes. Schullehner et al. (2017) investigated the variability of nitrate, nitrite, and ammonia concentrations in public waterworks in Denmark from 2007 to 2016. Drinking water in Denmark comes exclusively from groundwater most often with minimal treatment (e.g., aeration and sand filtration). They found that nitrification of nitrite and ammonium in the 2498 studied distribution systems resulted in very small increases in nitrate concentrations. They concluded that nitrate concentrations collected at the exit of water systems (e.g., consumer's taps) would be appropriate for calculating risk exposures for all consumers connected to that water system.

5.2.1. Methemoglobinemia

Numerous health effects have been associated with elevated nitrate levels in drinking water; in particular, methemoglobinemia has been repeatedly cited. Fewtrell (2004) offered the following explanation for how methemoglobinemia occurs. During the endogenous (internal) bacterial conversion of nitrate (e.g., from drinking water) to nitrite, methemoglobin (MetHb) is formed when nitrite oxidizes the ferrous iron in hemoglobin (Hb) to the ferric form (Fan et al., 1987). Methemoglobinemia results when MetHb cannot bind oxygen, and it is characterized by cyanosis, stupor, and cerebral anoxia. Under normal conditions, less than 2% of the total Hb circulates as MetHb (Fan et al., 1987). When MetHb reaches levels of 10% or more, signs of methemoglobinemia appear and symptoms can include an unusual bluish gray or brownish gray skin color, irritability, and excessive crying in children with moderate MetHb levels and drowsiness and lethargy at higher levels (Fewtrell, 2004). Methemoglobinemia has also been referred to as "blue-baby syndrome." Methemoglobinemia also can be a life-threatening condition when methemoglobin levels exceed about 10% (Ward et al., 2005). Diagnosis of methemoglobinemia has been made based on the observation of chocolate-colored blood or laboratory tests that indicate elevated MetHb levels (Brunning-Fann & Kaneene, 1993). Ward et al. (2018) noted that the U.S. EPA limit of 10 mg/L NO_3-N was set as about one-half the level at which there were no observed cases (Walton, 1951).

5.2.1.1. Reported Cases of Methemoglobinemia

Methemoglobinemia was a serious problem in Eastern Europe, particularly in Hungary until the late 1980s (Hill, 1999). In fact, in 1985, WHO reported that there were greater than 1300 cases of methemoglobinemia (with 21 fatalities) in Hungary over a 5-year period. Ward et al. (2018) noted that a 2002 WHO report on water and health (WHO, 2002) noted that there were 41 cases in Hungary annually, 2913 cases in Romania from 1985 to 1996 and 46 cases in Albania in 1996. Nitrate levels greater than 10 mg/L NO_3-N were usually associated with increased methemoglobin levels; however, clinical methemoglobinemia was not always present (Hoering &

Chapman, 2004; Ward et al., 2018). The most recent U.S. cases of methemoglobinemia related to nitrate in drinking water were reported by Knobeloch et al. (2000). They noted that nitrate concentrations in studied private wells in Wisconsin were two-times the maximum contaminant level (MCL) (10 mg/L, nitrate as N), and bacterial contamination was not a factor. Knobeloch et al. (2000) also discussed another case in the United States in 1999 related to nitrate contamination of a private well and six infant deaths attributed to methemoglobinemia in the United States between 1979 and 1999. They noted that only one of these studies was reported in the literature (C.J. Johnson et al., 1987). Ward et al. (2018) reported that a cross-sectional study conducted in Gaza found elevated methemoglobin levels in infants receiving supplemental feeding with formula made from well water in an area with high nitrate concentrations (highest mean concentration of was 195 mg/L nitrate as NO_3; ranging from 18 to 440) compared to an area with lower nitrate concentrations (mean: 119 mg/L nitrate as NO_3; range 18–244). Another cross-sectional study in Morocco cited by Ward et al. (2018) found a 22% increased risk of methemoglobinemia in infants in an area with drinking water nitrate concentrations greater than 50 mg/L nitrate as NO_3 (equivalent to 11 mg/L for NO_3-N) compared to infants in an area with nitrate levels less than 50 mg/L nitrate as NO_3. Zeman et al. (2011) reported results from a retrospective cohort study in Iowa of persons (aged 1–60 years) that showed a positive relationship between methemoglobin levels in the blood and the amount of nitrate consumption of water from private wells with nitrate levels less than 10 mg/L nitrate as N. As noted by Knobeloch et al. (2000), older cases of methemoglobinemia may have been underreported and only a small proportion of cases may have been thoroughly investigated and described in the literature.

5.2.1.2. Risk Factors and Other Causes Affecting Methemoglobinemia

Several studies have indicated that methemoglobinemia could be related to other causes and risk factors. For example, where illness was attributed to methemoglobinemia, Hanukoglu and Danon (1996) pointed out that many of the cases predated the early 1990s, and they attributed the apparent decline in the incidence of methemoglobinemia to possibly an infectious etiology. Some of the previous studies described by Fewtrell (2004) suggested several different factors for the causes of methemoglobinemia including an association between gastrointestinal illness and symptoms of methemoglobinemia in the absence of exogenous nitrate exposure. It was noted that diarrhea produces an oxidant stress that increases MetHb production resulting in acidosis that harms the MetHb reductase systems. In other studies cited by Fewtrell (2004), nitric oxide (NO) can be produced by several tissues in response to infection and inflammation, which could be a possible mechanism because nitrite is a product of NO metabolism. Fewtrell (2004) further noted there were other complex cofactor relationships (discussed in previous studies) that did not permit a quantitative exposure–response relationship for human exposure to nitrates in food or water and the subsequent development of methemoglobinemia. Further complicating the understanding of the causes of methemoglobinemia were that methemoglobinemia was not a notifiable disease, and the definitions of methemoglobinemia (in terms of the required level of MetHb) have varied considerably in the medical literature. Fewtrell (2004) noted that mean incidence rates of methemoglobinemia varied between 117 and 363 of 100,000 live births (for the full 5 years between 1990 and 1994) in three counties in Transylvania (Romania). However, it was pointed out that these rates were substantially below previously reported levels of 13,000 of 100,000 live births. The decrease in rates was attributed to a decrease in nitrate levels from the lack of access to inexpensive inorganic fertilizers.

L'hirondel et al. (2006) pointed out that although links between nitrate and health risk have been studied for more than 50 years, none of the health claims against dietary nitrate have been substantiated. Based on their analysis, they concluded that if there was not already an established MCL of 10 mg/L for nitrate-N in drinking water in the United States set by U.S. EPA or 50 mg/L set by WHO, it would be very difficult, if not impossible, to justify one based on the extensive results of studies summarized at the time of the published paper. Powlson et al. (2008) also pointed out that the evidence for nitrate as a cause of methemoglobinemia was questionable. It was indicated that some scientists suggested that there was sufficient evidence for increasing the permitted concentration of nitrate in drinking water without increasing risks to human health. However, Powlson et al. (2008) noted that certain groups within a population (such as infants) would be more susceptible than others to adverse health effects of nitrate. They concluded that more comprehensive and independent epidemiological studies need to be conducted to determine if the current nitrate limit for drinking water would be rigorously supported.

However, a more recent review of methemoglobinemia (Richard et al., 2014) re-examined some of the risks of nitrates in drinking water on public health as a cause of methemoglobinemia and other chronic diseases. They concluded that there were multiple causes of methemoglobinemia including nitrites in processed baby foods (especially vegetables such as carrots) and meats treated with sodium nitrite, environmental chemical exposures, commonly prescribed pharmaceuticals, and the endogenous generation of oxides of nitrogen. In their recent review paper, Ward et al. (2018) also noted that other risk factors for infant methemoglobinemia included formula made with water containing high nitrate levels, foods and medications that have high nitrate levels, and enteric infections. Richard et al. (2014) noted that infants under 6 months of age, especially between birth and 4 months of age, likely are predisposed to methemoglobinemia because their immature hepatic cytochrome system cannot produce sufficient circulating levels of methemoglobin reductase. Richard et al. (2014) stated that G6PD deficiency is the most common genetic enzyme deficiency and is especially prevalent in the United States, and may affect approximately 10% of African Americans and Americans of southern Mediterranean descent. They further explain that G6PD is a critical enzyme in the synthesis of reduced nicotinamide dinucleotide phosphate (NADPH), which activates the enzyme NADPH-dependent methemoglobin reductase to quickly reduce methemoglobin levels exceeding normal range (1–3%).

Richard et al. (2014) cited additional recent epidemiological investigations that indicated other possible sources of nitrogenous substance exposures in infants, including protein-based formulas and foods, and the production of nitrate precursors (nitric acid) by bacterial action in the infant gut in response to inflammation and infection. They noted that infants with congenital enzyme deficiencies in glucose-6-phosphate dehydrogenase and methemoglobin reductase are at greater risk of nitrite-induced methemoglobinemia from nitrates in water and food and from exposures to hemoglobin oxidizers. This re-examination by Richard et al. (2014) includes research that refutes the previous conclusions about exogenous nitrate-to-nitrite sources as causes of methemoglobinemia and supports endogenous (internal) nitrite production secondary to genetic abnormalities or NO generation in an inflamed infant gut as a causative mechanism for infantile methemoglobinemia. Based on this information, it was suggested that the current MCL for nitrate-N in drinking water of 10 mg/L (nitrate as nitrogen) may need to be revisited.

5.2.2. Other Adverse Health Effects of Nitrate in Drinking Water

In addition to methemoglobinemia, several other studies noted that nitrate in drinking water has been implicated in a number of other potential health effects. Based on several studies prior to the early 2000s, Fewtrell (2004) noted several potential adverse health effects included cancer (via the bacterial production of N-nitroso compounds), hypertension, increased infant mortality, central nervous system birth defects, diabetes, spontaneous abortions, respiratory tract infections, and changes to the immune system. It was further stated that the role of nitrates has not been proven conclusively in the formation of N-nitroso compounds and nitrite, which have been shown to promote cancer. The International Agency for Research on Cancer (IARC) Working Group reviewed human, animal, and other studies of cancer through mid-2006 and concluded that ingested nitrate and nitrite, under conditions that result in endogenous nitrosation, are probably carcinogenic (IARC, 2010). Other studies during the past 25 years have reported that nitrate-N concentrations, even below the maximum contaminant level of 10 mg/L (as N) in drinking water, can cause other health maladies including non-Hodgkins's lymphoma (Ward et al., 1996, 2005), thyroid disease (Ward et al., 2018), increased risk of bladder and ovarian cancers (Weyer et al., 2001), and colorectal cancer (Schullehner et al., 2018) and neural tube defects (Ward et al., 2018).

Based on a nationwide population-based cohort study in Denmark, Schullehner et al. (2018) investigated the association between long-term drinking water nitrate exposure (1978–2011) and colorectal cancer risk for men and women ages 35 and older. Colorectal cancer is one of the most common forms of cancer in Denmark and the third most frequent worldwide. Their study included nitrate exposure for 2.7 million adults and data for more than 200,000 drinking water quality analyses at public water supplies and small private wells. For their main analyses, 1.7 million individuals with highest exposures (>9.3 mg/L as nitrate) were assessed. They identified 5944 incidents of colorectal cancer risk cases based on 23 million person-years at risk. People exposed to the highest level of drinking water nitrate (mainly in small private water supplies) had a hazard ration of 1.16 for colorectal cancer compared with people exposed to the lowest level. This translates to a 156% greater risk of getting colorectal cancer compared to those who were exposed to nitrate levels less than 1.3 mg/L as nitrate. An important finding from the Schullehner et al. (2018) study was that statistically significant increased risks of colorectal cancer were noted at drinking water nitrate levels above 3.87 mg/L expressed as NO_3, which is considerably below the current WHO drinking water standard of 50 mg/L for nitrate as NO_3. The investigators noted that the health risk is related to the conversion of nitrate into carcinogenic N-nitroso compounds in the body. Although, nitrate concentrations have been reduced at the public waterworks where the majority of people in Denmark get their water, relatively high nitrate concentrations are found in small private wells. These wells are located in areas where local soil and geological conditions promote nitrate leaching from fertilizers to groundwater. Schullehner et al. (2018) concluded that lowering of the drinking water standard needs to be strongly considered and focusing efforts for reducing exposure to nitrate in drinking water from small private wells.

The following selected studies highlight health effects associated with ingestion of nitrate in drinking water. Zeman et al. (2011) examined data for people (ages 1–60 years) consuming drinking water from 150 private domestic wells in Iowa, United States. They selected wells that had nitrate-N concentrations below the 10 mg/L (as N) maximum contaminant level and analyzed health history data and blood samples (for hemoglobin fractions and immunological parameters). Their results indicated positive associations (bivariate fit) between higher nitrate exposure and body mass index, lower recreational activity, perceived poorer health, and perceptions of susceptibility to illness. In addition, they reported a high tumor necrosis factor-beta (TNF-β) expression (bivariate fit, $f = 3.76$, $p = 0.05$). Also, other health complaints in the Iowa study included stomach/intestinal difficulties (heartburn/reflux >50%; $f = 5.274$, $p = 0.0231$) and bone, muscle, and nerve issues (osteoarthritis (rheumatoid excluded) = 47%; $f = 6.0533$, $p = 0.0150$) associated with increasing nitrate exposure. In addition, the Iowa study noted that the in vivo exposures of nitrate-N associated with complaints of bone/joint disorders or with altered ex vivo production of anti-inflammatory factors (TNF-β or Th2/T_{reg} cytokine interleukin-10) had not been previously reported with environmental exposures of nitrate in drinking water.

In Valencia, Spain, Morales-Suarez-Varela et al. (1995) analyzed data for nitrate levels in drinking water from 258 municipalities in association with cancer mortality rate of different cancers. They reported a correlation with prostate and gastric cancer, but not for bladder cancer and colorectal cancers. When these data from Valencia were compared with data from the rest of Spain, Morales-Suarez-Varela et al. (1993, 1995) concluded that there were higher rates of bladder cancer incidence in Valencia for both genders, especially for people consuming drinking water with a nitrate level greater than 50 mg/L as NO_3.

Gulis et al. (2002) conducted a study of 237,000 people in the Trnava District, Slovakia, who were categorized as having low (0–10 mg/L), medium (10.1–20), or high (20.1–50) nitrate-as NO_3 concentrations in drinking water over a period of 20 years. Their study found a direct correlation between the nitrate levels and stomach cancer in women, colorectal cancer and non-Hodgkin lymphoma in both genders, but no associations for renal cell carcinoma (RCC) or bladder cancer. In a study in Hungary, Sandor et al. (2001) show an association between gastric cancer mortality and nitrate (as NO_3) levels in drinking water that were widely variable (as high as 290 mg/L). Conversely, other studies have found either an inverse association between nitrate (as NO_3) levels (up to 91 mg/L) and stomach (gastric) cancer incidence (Van Leeuwen et al., 1999) or no relationship at all (Barrett et al., 1998; Van Loon et al., 1998).

Volkmer et al. (2005) investigated the effect of nitrate levels in the drinking water on the incidence of urological malignancies in Bolcholt, Germany, a community of 70,000 people. Two different waterworks provided the population with drinking water for 28 years, with different nitrate levels: 60 mg NO_3/L for group A (57,253 people) and 10 mg NO_3/L for group B (10,037 people). A total of 527 urological malignancies (urothelial cancer 39.8%, RCC 10.8%, testicular tumors 8.0%, penile carcinoma 1.7%, and prostate cancer 39.7%) were recorded. The incidence of urinary tract tumors per 100,000 people/yr was 33.8 in group A and only 17.1 in group B (relative risk (RR) 1.98, 95% confidence interval (CI) 1.10–3.54). The RR was 0.87 (0.34–2.22) for renal tumors, 0.66 (0.14–2.88) for penile cancer, and 1.06 (0.76–1.48) for prostate cancer. For testicular tumors, there was an inverse association with nitrate level, with a RR of 0.43 (0.21–0.90). Based on these data, it was concluded that there was a trend for a direct association of the dietary intake of nitrate and the incidence of urothelial cancers in both genders (renal pelvis, ureter, urinary bladder, and urethra), but no association for prostate cancer, renal tumors, or penile tumors.

Zhai et al. (2017) reviewed the published data to investigate and quantify the potential risks of groundwater nitrate pollution to human health in China. They considered exposure through both drinking and dermal contact pathways. The study population was divided into four groups to assess impacts of age and gender on the outcome: infants (0–6 months), children (7 months–17 years old), adult males (≥18 years old), and adult females (≥18 years old). They found that there were seven units where groundwater nitrate concentrations exceeded the standard (especially Shaanxi having a serious condition). Health risk levels were highest in infants with risks decreasing in the order: children > adult males > adult females. Zhai et al. (2017) concluded that minors and males were more vulnerable compared with adults and females, respectively. In Shaanzi and Shandong, adult males, children, and infants are exposed to various degrees of health risk. They further suggested that there were no adverse health effects on adult females in the whole country.

In their extensive review of more than 30 epidemiological studies since 2004, Ward et al. (2018) evaluated consumption of drinking water nitrate and links with cancer, adverse reproductive outcomes, and other health effects. They concluded that the strongest evidence for adverse health effects and consumption of nitrate in drinking water (besides methemoglobinemia) were for colorectal cancer, thyroid disease, and neural tube defects. Ward et al. (2018)

cautioned that given the small number of studies for any one outcome, firm conclusions regarding risk are somewhat limited. Multiple studies were summarized that assessed consumption of nitrate in drinking water and adverse pregnancy outcomes. In their earlier review (Ward et al., 2005) of studies prior to 2005 linking maternal exposure to drinking water nitrate and spontaneous, they cited one with a positive association and another that found no association. Their more recent review did not identify any studies on this possible link. Ward et al. (2018) noted that five of six studies conducted since the 1980s found positive associations between higher drinking water nitrate exposure during pregnancy and neural tube defects or central nervous system defects combined. The sixth study did not find a relationship when they compared average drinking water nitrate concentrations between mothers with and without neural tube defect-affected births. However, Ward et al. (2018) stated that accumulated evidence indicates that higher nitrate intake during pregnancy is a risk factor for neural tube defects. Also, they noted that several studies have indicated a positive relation between low birth weight (and small for gestational age births) with maternal prenatal exposure to nitrate concentrations below 10 mg N/L. In terms of linkages between drinking water nitrate and various cancers, there have been periodic follow-ups of the Iowa cohort study of postmenopausal women in Iowa. Ward et al. (2010) reported that thyroid cancer risk was 2.6 times higher for women consuming water for more than 10 years with nitrate concentrations that exceeded 5 mg N/L compared to women who consumed water with less than 5 mg N/L. They also noted that ovarian cancer risk among private well users was also elevated compared to the lowest nitrate quartile of concentrations from public water systems (PWSs). They also discussed how cancer risks were evaluated when accounting for other factors (such as vitamin C intake, smoking, and folate intake). Ward et al. (2018) also reported that bladder cancer risk remained elevated for women who consumed drinking water with nitrate concentrations exceeding 5 mg N/L for at least 4 years based on a follow-up of the cohort study through 2010 (Jones et al., 2016). In terms of other health effects from consumption of nitrate in drinking water, a population-based cohort study in Wisconsin, United States, found a positive association with an increased incidence of early and late age-related macular degeneration with nitrate-N concentrations greater than 5 mg N/L compared to less than 5 mg N/L in private drinking water supplies (Klein et al., 2013; Ward et al., 2018).

As alluded to previously, one common finding by Ward et al. (2018) was that numerous studies had observed increased risks of adverse health effects with ingestion of nitrate levels in drinking water that were below regulatory limits. Several suggestions were offered for future epidemiological studies evaluating exposures to drinking water nitrate including (a) evaluating the amount of drinking water nitrate consumption at home and at work, (b) assessing the data from public supply wells that have been collected during the past decade, (c) evaluating exposure to nitrate along with co-contaminants (such as atrazine and other pesticides), (d) estimating intakes of nitrate from dietary sources, antioxidants, and other inhibitors of endogenous nitrosation, (e) investigating the interaction of nitrate ingestion and nitrosatable drugs (e.g., those containing secondary and tertiary amines or amides), (f) studying the extent of N-nitroso compounds (NOC) formation following ingestion of drinking water with nitrate concentrations below the MCL along with the reduction of nitrate to nitrite, and (g) focusing on populations of private well users particularly in agricultural regions that typically have the highest exposure to nitrate in drinking water.

5.2.3. Nitrate Levels and Violations of Standards in Public Water Systems

PWSs in many countries can contain elevated levels of nitrate that might be of health concern. For example, Pennino et al. (2017) analyzed trends for the U.S. EPA in nitrate in drinking water from U.S. PWSs from 1994 to 2016. They defined a public water system as any system that provides water for human consumption with at least 15 service connections or that regularly serves an average of 25 or more individuals daily at least 60 d/yr. There are more than 150,000 active PWSs in the United States that can be publicly or privately owned systems. Approximately 66% of the population served by a public water system gets their drinking water from surface water sources, whereas 34% of the PWS population is served by a groundwater source. They found that nitrate violations (those exceeding the maximum contaminant level for nitrate-N in drinking water of 10 mg N/L) significantly increased from 0.28 to 0.42% between 1994 and 2009, and then decreased from 0.32% by 2016. The number of people served by system with nitrate violations decreased from 1.5 million in 1997 to 200,000 in 2014. It was noted that surface water systems have been improving over time; however, groundwater systems in violation of the nitrate MCL and the average duration of violations are increasing. Pennino et al. (2017) pointed out that there is a potential for underreporting of nitrate violations, so the reported violations in their study are conservative. In other parts of the world, community systems have been contaminated with high levels of nitrate. For example, Vitoria Minana et al. (1991) reported that nitrate concentrations in the Valencia region of Spain exceeded the WHO guideline level (50 mg/L as NO_3) in 95 towns, with 18 municipalities reporting nitrate levels (as NO_3) greater than 150 mg/L.

5.2.4. Nitrate Levels in Groundwater and Private Domestic Wells

In the United States, Dubrovsky et al. (2010) estimated that 10–20% of groundwater sources may exceed 10 mg/L for nitrate-N. Sutton et al. (2011) estimated that about 3% of the population in 15 European countries that relies on groundwater as a drinking water source is exposed to nitrate concentrations exceeding the WHO drinking water standard of 50 mg NO_3/L (11.2 mg N/L). In addition, it was estimated that 5% of the population using groundwater is chronically exposed to nitrate concentrations exceeding 5.6 mg N/L, which could double the risk of colon cancer for people that consume above median amounts of meat (Sutton et al., 2011). In other parts of the world, nitrate-N concentrations in many aquifers exceed the maximum contaminant level set by the WHO. For example, large aquifers in China have been contaminated with elevated nitrate-N concentrations (Wang et al., 2016). About 90% of China's shallow groundwater is polluted, and nitrate is considered one of the main pollutants of main concern (Qiu, 2011).

Privately owned wells (especially those with shallow depths) may contain water with high nitrate concentrations. A survey of private well owners in Minnesota found that 5–10% of drinking water wells have nitrate concentrations that exceed health standards (Lewandowski et al., 2008). About 43 million people, or 15% of the population of the United States, use drinking water from private wells, which are not regulated by the Federal Safe Drinking Water Act. Nolan et al. (2006) found that 9% of the privately owned domestic wells in the United States sampled for nitrates during the period 1993–2000 exceeded the MCL of 10 mg/L nitrate-N, compared to only 2% for publicly owned groundwater access wells. Based on a more recent extensive study of 2132 private drinking water wells sampled for nitrate in principal aquifers of the United States (Fig. 5.2), DeSimone (2009) reported that 4.4% exceeded the maximum contaminant level of 10 mg N/L. Wells that exceed the MCL typically were located in or near agricultural areas in the following aquifer systems: certain basin-fill aquifers in the Southwest and California, glacial aquifers in the upper Midwest, and some coastal-plain and crystalline rock aquifers in the Central Appalachian Region (with more than 10% of wells >10 mg/L as N). Using data available at the time, Fewtrell (2004) estimated that 2 million household wells would exceed the nitrate drinking water standard of 10 mg N/L, assuming an exceedance rate of 13% (based on a survey of 5500 wells in nine Midwestern states (CDC, 1998)). Using birth rates from a study by Knobeloch et al. (2000), Fewtrell (2004) estimated that 40,000 infants under 6 months of age would be expected to be living in homes with high-nitrate drinking water.

5.3. HEALTH EFFECTS OF NITRATE AND NITRIC OXIDE IN FOODS

It is important to keep in mind that drinking water is not the only source of nitrate intake by infants, children, and adults. Nitrate consumed from foods can have positive and negative health consequences. Several human health benefits for adults have been reported from the intake of foods containing nitrate and nitrite. For example, bioactive NO is generated endogenously in humans and other mammals from food containing nitrate and nitrite and this compound (NO) is the most important molecule in regulating blood pressure and maintaining homeostasis (keeping blood vessels in their relaxed state) (Sareer et al., 2015). Studies have shown that a diet rich in nitrate can produce a source of bioactive NO that can benefit endothelial (cells that line the interior surface of blood vessels and lymphatic vessels) function.

Nitrate naturally is present in many foods, and some green leafy vegetables and root vegetables tend to contain the highest levels (Ward et al., 2018). The IARC (2010) estimated that average daily intake of nitrate from food ranges from 30 to 130 mg/d as NO_3 (equivalent to 7–29 mg N/d). Due to the inhibition of NOC formation by ascorbic acid, phenols, and other compounds present in elevated levels in most vegetables, nitrate intake from vegetables possibly would not result in significant NOC formation (IARC, 2010). However, the IARC also noted that nitrate and nitrite consumption in diets under conditions that result in endogenous nitrosation are probably carcinogenic.

Recently, Kerley et al. (2018) tested the effect of daily nitrate intake compared with placebo in subjects with uncontrolled hypertension (HTN) using a 7-day, double-blind, randomized, placebo-controlled, cross-over trial. Their results supported existing information that suggested an anti-hypertensive effect of dietary nitrate in people with treated yet uncontrolled hypertension. Also, they concluded that targeted dietary strategies appear to be an effective way to control blood pressure.

Ward et al. (2018) noted that nitrate and nitrite are strong nitrosating agents and the amount of nitrosatable precursors is a key factor in the formation of NOC. Most NOCs are known carcinogens and teratogens. Particularly important factors in the formation of NOC included the amount of dietary consumption of red and processed meat, as increased consumption of red meat (600 vs 60 g/d), but not white meat, was found to cause a threefold increase in fecal NOC levels. It was also reported that studies have shown that heme iron stimulated endogenous nitrosation, which provided a possible explanation for the differences in colon cancer risk between red and white meat consumption. Ward et al. (2018) pointed out that the link between meat consumption and colon cancer

PRINCIPAL AQUIFER ROCK OR SEDIMENT TYPE

- Basin-fill and other nonglacial sand and gravel
- Glacial sand and gravel (discontinuous)
- Coastal plain Semiconsolidated sand
- Sandstone
- Sandstone and carbonate
- Carbonate
- Basalt
- Crystalline

EXPLANATION

Nitrate, in milligrams per liter as N

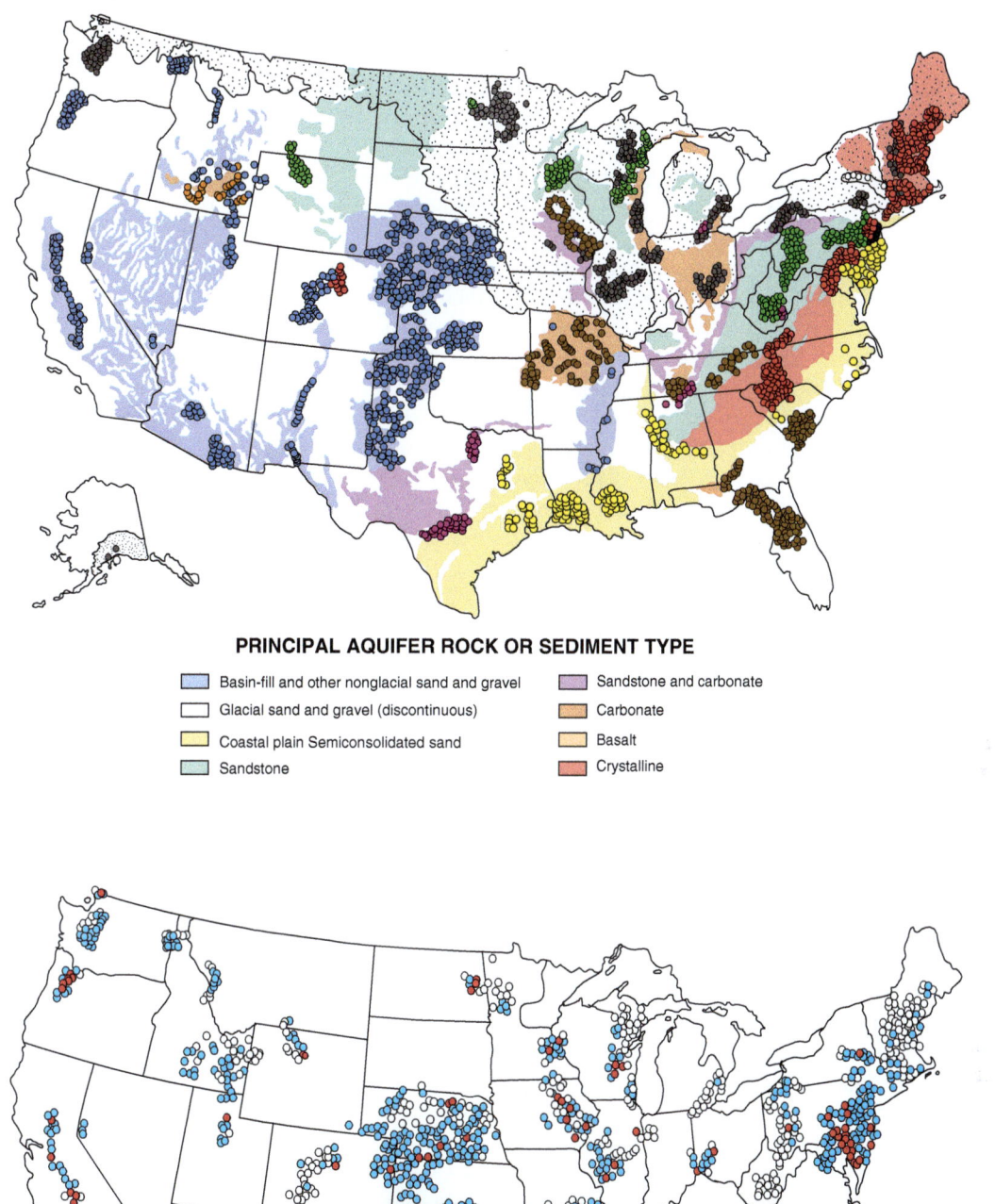

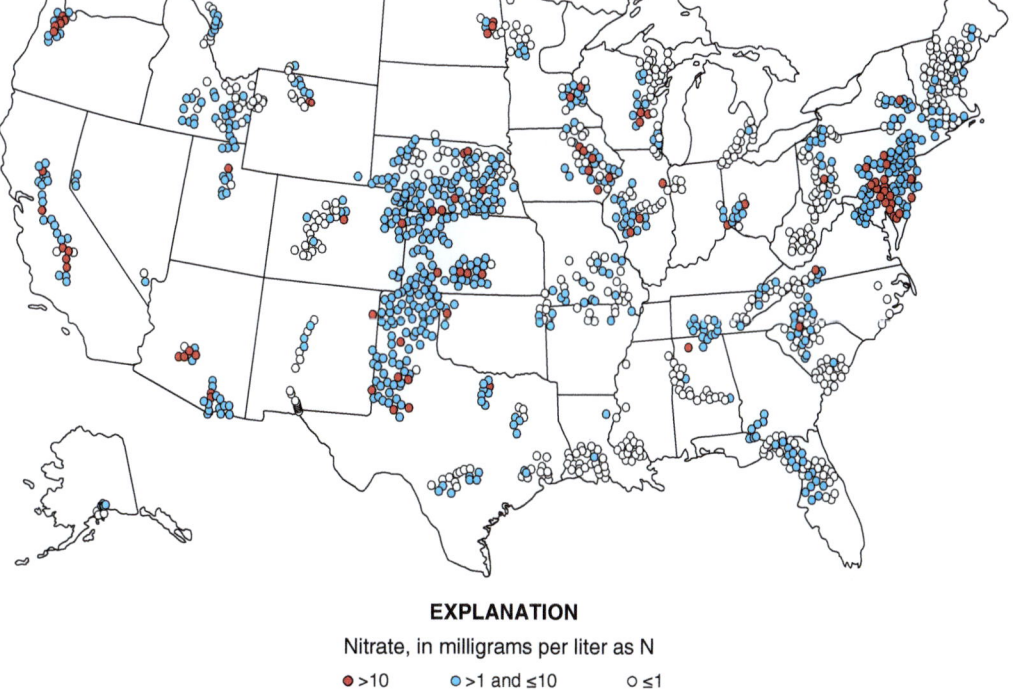

- >10
- >1 and ≤10
- ≤1

Figure 5.2 Upper diagram shows private drinking water wells sampled as part of the U.S. Geological Survey National Water Quality Assessment Program. Modified from colored maps in DeSimone et al. (2009) and DeSimone et al. (2014). Lower diagram shows concentrations of nitrate, in milligrams per liter as nitrogen, in sampled domestic wells. *Source*: From United States Geological Survey, public information.

risk is even greater for nitrite-preserved processed meat than for fresh meat. This was noted in an IARC review that concluded processed meat is carcinogenic to humans.

5.4. ECOSYSTEM DAMAGE AND POTENTIAL IMPACTS ON HUMAN HEALTH

Atmospheric deposition of reactive nitrogen compounds on the earth's surface has resulted in widespread eutrophication and acidification of terrestrial, freshwater, and marine habitats (Bobbink et al., 2010; Erisman et al., 2015). In particular, deposition of ammonia can damage sensitive vegetation, particularly bryophytes and lichens that consume most of their nutrients from the atmosphere (Sheppard et al., 2011). Sala et al. (2000) noted that certain biomes (particularly northern temperate, boreal, arctic, alpine, grassland, savannah, and Mediterranean) are very sensitive to reactive nitrogen deposition because of the limited availability of nitrogen in these systems under natural conditions.

5.4.1. Exploring Links Between Ecosystem Degradation and Human Diseases

Ongoing interdisciplinary research has been investigating possible links between ecosystem damages and human diseases. More than a decade ago, Townsend et al. (2003) hypothesized that an increase in nutrient availability can favor opportunistic, disease-causing organisms and could lead to changes in the epidemiology of human diseases. For example, they noted that certain vectors for infectious diseases (such as malaria) might be affected by changes in availability of reactive nitrogen. Townsend et al. (2003) listed several studies that have shown a positive correlation between inorganic nitrogen concentrations in surface water and larval abundance for malarial mosquitoes, along with carriers of encephalitis, and West Nile virus (Camargo & Alonso, 2006). Although, it was pointed out that ecological responses to changing nitrogen concentrations are complex, and these reactions vary based on the organism's direct response, its food sources, and its reactions with parasitic species.

P.T.J. Johnson et al. (2010) investigated the consequences of nutrient pollution on directly transmitted, vector-borne, complex life cycle, and noninfectious pathogens, including West Nile virus, malaria, HABs, coral reef diseases, and amphibian malformations. The complexities associated with elucidating the effects of environmental nutrient enrichment on disease were emphasized, as these effects vary depending on the type of pathogen, host species and condition, attributes of the ecosystem, and the degree of enrichment (some pathogens increase in abundance, while others decline or disappear). P.T.J. Johnson et al. (2010) referred to available evidence that shows that ecological changes related to nutrient enrichment often aggravate infection and disease caused by "generalist" parasites with direct or simple life cycles. They indicated that observed pathways include changes in host/vector density, host distribution, infection resistance, pathogen virulence or toxicity, or the direct supplementation of pathogens. P.T.J. Johnson et al. (2010) cautioned that these pathogens likely would be extremely dangerous due to their persistence (they can continue to cause mortality even as their hosts decline), which would potentially result in continued epidemics or chronic pathology. It was concluded that tropical and subtropical regions could be especially susceptible to diseases associated with nutrient enrichment due to forecasted increases in nutrient application along with the abundance of infectious pathogens.

5.4.2. Harmful Effects of Algae on Humans

The worldwide increase in the eutrophication of coastal and marine ecosystems has resulted in a proliferation of HABs. Some toxins associated with cyanobacterial HABs have been linked to neurological, amnesic, paralytic, and or diarrheic shellfish poisoning (Townsend et al., 2003). Studies also have reported associations between algal blooms, inorganic nutrient concentrations, and abundances of mosquitoes that can carry pathogenic microorganisms (Camargo & Alonso, 2006).

Algae proliferation due to elevated nutrient (N and P) concentrations is not just harmful to ecosystems but can severely impact human health, in addition to being aesthetically disgusting. A species of algae called *Lyngbya wollei* can produce toxins that are harmful to humans (Zanchett & Oliveira-Filho, 2013). Hoagland et al. (2002) noted that there were more than 60,000 incidents of human exposure to algal toxins annually in the United States, and this resulted in approximately 6500 deaths. In 2002, state park officials in Florida started keeping track of swimmers, kayakers, anglers, and tubers who touched, or brushed their skin up against this algae and then complained of rashes, hives, nausea, itching, and asthma attacks. Since 2002, there have been over 140 reported incidents (Pittman, 2012). Another study from Florida, reported that there was an increase in emergency room visits in Sarasota County to treat respiratory illnesses associated with algal blooms (Hoagland et al. (2009). Recently, reports of attacks to the immune systems of sport fisherman in Lake Erie were linked to algal blooms (Flesher & Kastanis, 2017). Figure 5.3 shows coastal areas around the world where hypoxic and eutrophic conditions have been prevalent (Erisman et al., 2015).

HABs, particularly cyanobacteria, can also cause serious health problems to people living near affected bodies of water. For example, some types of cyanobacteria known to

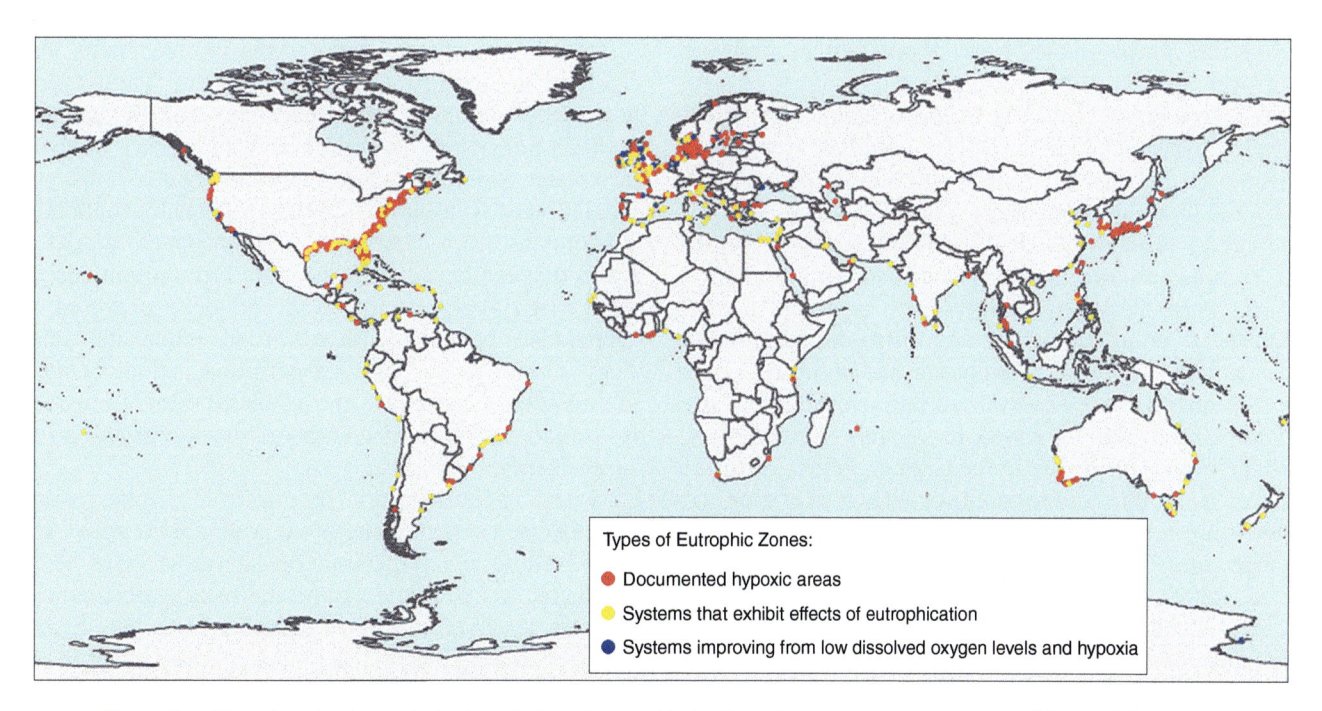

Figure 5.3 Map showing hypoxic (red circles) and eutrophic (yellow circles) coastal areas around the world. *Source*: Modified from the colored figure 4 in Erisman et al. (2015). Reproduced with permission of World Wildlife Fund.

live in Lake Okeechobee in Florida can produce harmful algal toxins under certain conditions. Research scientists from the University of Miami and the Institute for Ethnomedicine were featured in a recent film that documents the connections between chronic human exposure to toxic algae and increased risks of amyotrophic lateral sclerosis, Alzheimer's disease, and Parkinson's disease related to a toxin called BMAA (β-N-methylamino-L-alanine) (Cox et al., 2016) produced by some algal blooms (https://www.toxicpuzzle.com/, accessed 22 July 2018). There are other cyanotoxins that are found in Florida waters that can pose other health problems, such as gastrointestinal and respiratory problems as well as liver damage (toxins have been linked to an increased risk of terminal liver failure and liver cancer). Scientists are warning that any *Microcystis aeruginosa* bloom should be avoided due to its potential toxicity. This warning also included no swimming, no fish consumption, and avoid breathing the toxin that can become airborne (http://flylifemagazine.com/lets-go-fishing-its-only-cyanobacteria/, accessed 22 July 2018). Cyanobacterial blooms of *Microcystis* have occurred on the west coast of Florida, around Cape Coral. The National Centers for Coastal Ocean Science provided a grant to Florida Gulf Coast University to determine if toxins and cell particles are present in air near the blooms and to identify particle sizes (https://coastalscience.noaa.gov/news/study-explores-airborne-health-risks-from-cyanobacteria-blooms-in-florida/, accessed 06 May 2019). The researchers found that airborne cyanobacteria toxins can travel more than 1.6 km inland, and the toxins have been measured in particle sizes that can penetrate deep into human lungs. The ongoing study pointed out that more research needs to be conducted to better understand the impacts to human health because these toxins can potentially cause neurodegenerative effects.

5.5. REACTIVE NITROGEN STORAGE IN AQUIFERS AND FUTURE IMPACTS ON DRINKING WATER

Storage of nitrate and other reactive forms of nitrogen in the subsurface could be a continual source of nitrate contamination of drinking water in aquifers for many years to come. As mentioned in earlier chapters, there are large amounts of reactive nitrogen, mostly in the form of nitrate, that leaches below the root zone in soils to the vadose (unsaturated) zone (the material between the base of the soil to the water table at the top of the saturated zone) and eventually to groundwater (aquifers) where nitrate can be stored (Wang et al., 2013). This is especially prevalent in agricultural areas where there has been a surplus of nitrogen added to cropland. Ascott et al. (2017) assessed global patterns of nitrate storage in the vadose zone by using estimates of groundwater depth and nitrate leaching for the period 1900–2000. They estimated a peak global storage of 605–1814 Tg of nitrate in the vadose zone, with the highest storage per unit area in North America, China, and Europe (areas with thick vadose zones and extensive historical agricultural activities).

Several recent studies have shown that groundwater nitrogen applied decades ago can still be found in aquifers (Basu et al., 2012; McMahon et al., 2006; Meals et al., 2010; Puckett et al. 2011; Sanford & Pope, 2013; Sebillo et al., 2013). Aquifers can supply base flow to streams in many parts of the world. This continual discharge of groundwater with elevated nitrate concentrations can cause ecosystem damage over long periods of time. Particularly noteworthy are cases in Eastern European countries where the nitrogen surplus has decreased by half (due to economic and political changes in the early 1990s), however, no improvements in water quality have been observed in streams (Sutton et al., 2011). This is likely due to the large quantities of nitrate stored in aquifers and released very slowly over time, as a function of the groundwater residence time, which can range from weeks to several thousands of years (Alley et al., 2002).

Also, the release of reactive forms of nitrogen from the unsaturated zone, an aforementioned important reservoir for the storage and release of reactive nitrogen (mainly in the form of nitrate) over time, has significant implications for human health associated with contamination of deeper (oxic) aquifers used for drinking water supplies, global warming, and eutrophication of surface water bodies (Schlesinger, 2009). Also, various organic forms of nitrogen can be stored in the bottom sediments of lakes, estuaries, and other waterbodies that could be released to the water column during storms and other weather-related events.

5.6. CONCLUDING REMARKS

As we have learnt in this chapter, there are many ways that anthropogenic releases of reactive nitrogen to the environment can have adverse impacts to human health. Various forms of reactive nitrogen injurious to human health can occur from short- and long-term exposures to air pollution (small particles and aerosols). The Global Burden of Disease Study of Comparative Risk Assessment (Lim et al., 2012) estimated that there were about 3.2 million deaths worldwide attributable to air pollution from $PM_{2.5}$. Apte et al. (2015) assessed several scenarios for reducing the risk of mortality rates related to linear and nonlinear response relationships for $PM_{2.5}$ concentrations. They found that relatively small improvements in $PM_{2.5}$ in less-polluted areas (such as in North America and Europe) would result in large avoided mortality. In contrast, in more polluted regions (such as China and India), it was reported that to keep $PM_{2.5}$ attributable mortality rates constant, $PM_{2.5}$ levels would need to decrease by 20–30% over the next 15 years in order to balance increases in mortality in aging populations from $PM_{2.5}$ pollution.

Degradation of terrestrial and marine ecosystems also can have severe impacts on human health. These effects (e.g., neurological, amnesic, paralytic, and or diarrheic shellfish poisoning) have resulted from contact with surface water containing cyanobacterial algal blooms, and inhalation of windblown aerosols or particles containing cyanotoxins from HABs. P.T.J. Johnson et al. (2010) noted that ecological changes related to nutrient enrichment often aggravate infection and disease caused by "generalist" parasites with direct or simple life cycles. They indicated that observed pathways include changes in host/vector density, host distribution, infection resistance, pathogen virulence or toxicity, or the direct supplementation of pathogens.

Regarding the ingestion of nitrate in drinking water and adverse health effects, Ward et al. (2018) noted that more well-designed epidemiological studies need to be conducted to better understand the risks associated with different levels of nitrate in drinking water along with other interacting factors (e.g., diet, age, preexisting health conditions, exposure to other potential contaminants, and socioeconomic status). So far, the most consistently reported health effects associated with ingestion of nitrate in drinking water have included colorectal cancer, thyroid disease, and central nervous system birth defects. However, many questions remain regarding exposures of populations on public water systems and private wells users to nitrate concentrations below current maximum contaminant levels and health effects. Studies also need to include other important risk factors, such as nitrate-reducing bacteria, dietary exposures, and nitrosation mechanisms. Ward et al. (2018) also stressed the need for additional studies to evaluate increased risk for other health effects (such as cancers of the thyroid, ovary, and kidney) along with adverse reproductive outcomes (e.g., spontaneous abortion, preterm birth, and small gestational births).

REFERENCES

Ackerman, D., Millet, D. B., & Chen, X. (2019). Global estimates of inorganic nitrogen deposition across four decades. *Global Geochemical Cycles*, *33*, 100–107. https://doi.org/10.1029/2018GB005990

Alley, W. M., Healy, R. W., LaBaugh, J. W., & Reilly, T. E. (2002). Flow and storage in groundwater systems. *Science*, *296*, 1985–1990.

Apte, J. S., Marshall, J. D., Cohen, A. J., & Brauer, M. (2015). Addressing global mortality from ambient $PM_{2.5}$. *Environmental Science and Technology*, *49*, 8057–8066. http://dx.doi.org/10.1021/acs.est.5b01236

Ascott, M. J., Gooddy, D. C., Wang, L., Stuart, M. E., Lewis, M. A., Ward, R. S., & Binley, A. M. (2017). Global patterns of nitrate storage in the vadose zone. *Nature Communications*, 1–7. https://doi.org/10.1038/s41467-017-01321-w

Barrett, J. H., Parslow, R. C., & McKinney, P. A. (1998). Nitrate in drinking water and the incidence of gastric, esophageal, and brain cancer in Yorkshire, England. *Cancer Causes Control*, 9, 153–159.

Basu, N. B., Jindal, P., Schilling, K. E., Wolter, C. F., & Takle, E. S. (2012). Evaluation of analytical and numerical approaches for the estimation of groundwater travel time distribution. *Journal of Hydrology*, 475, 65–73.

Bauer, S. E., Tsigardis, K., & Miller, R. (2016). Significant atmospheric aerosol pollution caused by world food cultivation. *Geophysical Research Letters*, 43, 5394–5400. https://doi.org/10.1002/2016GL068354

Bobbink, R., Hicks, K., Galloway, J., Spranger, T., Alkemade, R., Ashmore, M., et al. (2010). Global assessment of nitrogen deposition effects on terrestrial plant diversity: A synthesis. *Ecological Applications*, 20, 30–59.

Brunekreef, B., & Holgate, S. T. (2002). Air pollution and health. *Lancet*, 360, 1233–1242.

Brunning-Fann, C. S., & Kaneene, J. B. (1993). The effects of nitrate, nitrite and N-nitroso compounds on human health: A review. *Veterinary and Human Toxicology*, 35(6), 521–538.

Camargo, J. A., & Alonso, A. (2006). Ecological and toxicological effects of inorganic nitrogen pollution in aquatic ecosystems: A global assessment. *Environment International*, 32, 831–849.

Centers for Disease Control (CDC). (1998). *A survey of the quality of water drawn from domestic wells in nine midwestern states*. Atlanta, GA: Centers for Disease Control and Prevention.

Cox, P. A., Davis, D. A., Mash, D. C., Metcalf, J. S., & Banack, S. A. (2016). Dietary exposure to an environmental toxin triggers neurofibrillary tangles and amyloid deposits in the brain. *Proceedings of the Royal Society, Biological Sciences*, 283(1823), 20152397. https://doi.org/10.1098/rspb.2015.2397

Crouse, D. L., Peters, P. A., Hystad, P., Brook, J. R., van Donkelaar, A., Martin, R. V., et al. (2015). Ambient $PM_{2.5}$, O_3, and NO_2 exposures and associations with mortality over 16 years of follow-up in the Canadian Census Health and Environment Cohort (CanCHEC). *Environmental Health Perspectives*, 123, 1180–1186. http://dx.doi.org/10.1289/ehp.1409276

DeSimone, L.A. (2009). *Quality of water from domestic wells in principal aquifers of the United States, 1991–2004*. U.S. Geological Survey Scientific Investigations Report 2008–5227, 139 p. Available online at http://pubs.usgs.gov/sir/2008/5227, accessed 24 September 2019

DeSimone, L.A., Hamilton, P.A., & Gilliom, R.J. (2009). *The quality of our nation's waters—Quality of water from domestic wells in principal aquifers of the United States, 1991–2004—Overview of major findings*. U.S. Geological Survey Circular 1332, 48 p., Available online at http://pubs.usgs.gov/sir/2008/5227, accessed 19 September 2019.

Dominici, F., Daniels, M., Zeger, S. L., & Samet, J. M. (2002). Air pollution and mortality: Estimating regional and national dose-response relationships. *Journal of the American Statistical Association*, 97(457), 100–111.

Dubrovsky, N.M., Burow, K.R., Clark, G.M., Gronberg, J.M., Hamilton P.A., Hitt, K.J., et al. (2010). *The quality of our Nation's waters—Nutrients in the Nation's streams and groundwater, 1992–2004*. U.S. Geological Survey Circular 1350, 174 p. Retrieved from http://water.usgs.gov/nawqa/nutrients/pubs/circ1350, accessed 24 September 2019

Erisman, J. W., Galloway, J. N., Dice, N. B., Sutton, M. A., Bleeker, A., Grizzetti, B., Leach, A. M., & de Vries, W. (2015). *Nitrogen, too much of a vital resource*. Science Brief. Zeist, The Netherlands: World Wildlife Fund.

European Environmental Agency. (2016). *European Union emission inventory report 1990–2014 under the UNECE convention on Long-range Transboundary Air Pollution (LRTAP)* (EEA Report No. 16/2016).

Fan, A. M., Willhite, C. C., & Book, S. A. (1987). Evaluation of the nitrate drinking water standard with reference to infant methemoglobinemia and potential reproductive toxicology. *Regulatory Toxicology and Pharmacology*, 7(2), 135–148.

Faustini, A., Stafoggiam, M., Renzi, M., Cesaroni, G., Alessandrini, E., Davoli, M., & Francesco, F. (2016). Does chronic exposure to high levels of nitrogen dioxide exacerbate the short-term effects of airborne particles? *Occupational and Environmental Medicine*, 73(11), 772–778. http://dx.doi.org/10.1136/oemed-2016-103666

Fewtrell, L. (2004). Drinking-water nitrate, methemoglobinemia, and global burden of disease: A discussion. *Environmental Health Perspectives*, 112(4), 1371–1374.

Flesher, J., & Kastanis, A. (2017). *Toxic algae: Once a nuisance now a severe nationwide threat*. Associated Press.

Galloway, J. N., Winiwarter, W., Leip, A., Leach, A., Bleeker, A., & Erisman, J. W. (2014). Nitrogen footprints: Past, present, and future. *Environmental Research Letters*, 9. https://doi.org/10.1088/1748-9326/9/11/115003

Gulis, G., Czompolyova, M., & Cerhan, J. R. (2002). An ecologic study of nitrate in municipal drinking water and cancer incidence in Trnava District, Slovakia. *Environmental Research*, 88, 182–187.

Hanukoglu, A., & Danon, P. N. (1996). Endogenous methemoglobinemia associated with diarrheal disease in infancy. *Journal of Pediatrics Gastroenterology and Nutrition*, 23(1), 1–7.

Hill, J., Goodkind, A., Tessum, C., Thakrar, S., Tilman, D., Polasky, S., et al. (2019). Air-quality-related health damages of maize. *Nature Sustainability*, 2, 397–403. https://doi.org/10.1038/s41893-019-0261-y

Hill, M. J. (1999). Nitrate toxicity: Myth or reality? *British Journal of Nutrition*, 81, 343–344. https://doi.org/10.1017/S0007114599000616

Hoagland, P., Anderson, D. M., Kaoru, Y., & White, A. W. (2002). The economic effects of harmful algal blooms in the Unites States: Estimates, assessment issues, and information needs. *Estuaries*, 25, 819–837.

Hoagland, P., Jin, D., Polansky, L. Y., Kirkpatrick, B., Kirkpatrick, G., Fleming, L. E., et al. (2009). The costs of respiratory illnesses arising from Florida Gulf Coast Karenia brevis blooms. *Environmental Health Perspectives*, 117(8), 1239–1243.

Hoering, H., & Chapman, D. (Eds.). (2004). *Nitrate and nitrite in drinking water. WHO Drinking Water Series*. London: IWA Publishing.

Howarth, R., Swaney, D., Billen, G., Garnier, J., Hong, B., Humborg, C., et al. (2011). Nitrogen fluxes from the landscape are controlled by net anthropogenic nitrogen inputs and by climate. *Frontiers in Ecology and the Environment*. https://doi.org/10.1890/100178

Hu, J., Ying, Q., Wang, Y., & Zhang, H. (2015). Characterizing multi-pollutant air pollution in China: Comparison of three air quality indices. *Environment International*, *84*, 17–25. https://doi.org/10.1016/j.envint.2015.06.014. Epub 2015 Jul 18

International Agency for Research on Cancer (IARC). (2010). *IARC monographs on the evaluation of carcionogenic risks to humans: Ingested nitrate and nitrite and cyanobacterial peptide toxins*. Lyon, France: IARC.

Jhun, I., Coull, B. A., Zanobetti, A., & Koutrakis, P. (2015). The impact of nitrogen oxides concentration decreases on ozone trends in the USA. *Air Quality and Atmospheric Health*, *8*(3), 283–292. https://doi.org/10.1007/s11869-014-0279-2

Jiang, Z., McDonald, B. C., Wordena, H., Wordene, J. R., Miyazakif, K., Qu, Z., et al. (2018). Unexpected slowdown of US pollutant emission reduction in the past decade. *Proceeding of the National Academy of Sciences*, *115*(20), 5099–5104. https://doi.org/10.1073/pnas.1801191115

Johnson, C. J., Bonrud, P. A., Dosch, T. L., Kilness, A. W., Senger, K. A., Busch, D. C., & Meyer, M. R. (1987). Fatal outcome of methemoglobinemia in an infant. *Journal of the American Medical Association*, *257*, 2796–2797.

Johnson, P. T. J., Townsend, A. R., Cleveland, C. C., Glibert, P. M., Howarth, R. W., & McKenzie, V. J. (2010). Linking environmental nutrient enrichment and disease emergence in humans and wildlife. *Ecological Applications*, *20*(1), 16–29.

Jones, R. R., Weyer, P. J., DellaValle, C. T., Inoue-Choi, M., Anderson, K. E., Cantor, K. P., et al. (2016). Nitrate from drinking water and diet and bladder cancer among postmenopausal women in Iowa. *Environmental Health Perspectives*, *124*(11), 1751–1758.

Kampa, M., & Castanas, E. (2008). Human health effects of air pollution. *Environmental Pollution*, *151*, 362–367.

Kane, R. P. (1998). Ozone depletion, related UVB changes and increased skin cancer incidence. *International Journal of Climatology*, *18*, 457–472.

Katsouyanni, K., Samet, J. M., Anderson, H., Atkinson, R., Le Tertre, A., Medina, S., et al. (2009). Air pollution and health: A European and North American approach (APHENA). *Health Effects Institute*, *142*, 5–90.

Katsouyanni, K., Touloumi, G., Samoli, E., Gryparis, A., Le Tertre, A., Monopolis, Y., et al. (2001). Confounding and effect modification in the short-term effects of ambient particles on total mortality: Results from 29 European cities within the APHEA2 project. *Epidemiology*, *12*(5), 521–531.

Katz, B. G. (2012). Nitrate contamination in karst groundwater. Ch. 91. In W. B. White, D. C. Culver, & T. Pipan (Eds.), *Encyclopedia of caves* (pp. 756–760). Boston, MA: Academic Press. Elsevier. https://doi.org/10.1016/B978-0-12-383832-2.00146-8

Kerley, C. P., Dolan, E., James, P. E., & Cormican, L. (2018). Dietary nitrate lowers ambulatory blood pressure in treated, uncontrolled hypertension: A 7-d, double-blind, randomised, placebo-controlled, cross-over trial. *British Journal of Nutrition*, *119*(6). https://doi.org/10.1017/S0007114518000144

Kioumourtzoglou, M. A., Schwartz, J., James, P., Dominici, F., & Zanobetti, A. (2016). $PM_{2.5}$ and mortality in 207 U.S. cities: Modification by temperature and city characteristics. *Epidemiology*, *27*(2), 221–227. https://doi.org/10.1097/EDE.0000000000000422

Klein, B. E. K., McElroy, J. A., Klein, R., Howard, K. P., & Lee, K. E. (2013). Nitrate-nitrogen levels in rural drinking water: Is there an association with age-related macular degeneration? *Journal of Environmental Science and Health, Part A*, *48*, 1757–1763.

Knobeloch, L., Salna, B., Hogan, A., Postle, J., & Anderson, H. (2000). Blue babies and nitrate-contaminated well water. *Environmental Health Perspectives*, *108*, 675–678.

Lewandowski, A. M., Montgomery, B. R., Rosen, C. J., & Moncrief, J. F. (2008). Groundwater nitrate contamination costs: A survey of private well owners. *Journal of Soil and Water Conservation*, *63*(3), 153–161. https://doi.org/10.2489/jswc.63.3.153

L'hirondel, J. L., Avery, A. A., & Addiscott, T. (2006). Dietary nitrate: Where is the risk? *Environmental Health Perspectives*, *114*(8), A458–A459.

Li, Y., Schichtel, B. A., Walker, J. T., Schwede, D. B., Chen, X., Lehmann, C. M., et al. (2016). Increasing importance of deposition of reduced nitrogen in the United States. *Proceedings of the National Academy of Sciences*, *113*(21), 5874–5879. https://doi.org/10.1073/pnas.1525736113

Lim, S. S., Vos, T., Flaxman, A. D., Danaei, G., Shibuya, K., Adair-Rohani, H., et al. (2012). Comparative risk assessment of burden of disease and injury attributable to 67 risk factors and risk factor clusters in 21 regions, 1990–2010: A systematic analysis for the Global Burden of Disease Study 2010. *Lancet*, *380*, 2224–2260.

Liu, H., Bartonova, A., Schindler, M., Sharma, M., Behera, S. N., Katiyar, K., & Dikshit, O. (2013). Respiratory disease in relation to outdoor air pollution in Kanpur, India. *Archives of Environmental and Occupational Health*, *68*(4), 204–217.

McMahon, P. B., Dennehy, K. F., Bruce, B. W., Bohlke, J. K., Michel, R. L., Gurdak, J. J., & Hurlbut, D. B. (2006). Storage and transit time of chemicals in thick unsaturated zones under rangeland and irrigated cropland, High Plains, United States. *Water Resources Research*, *42*(3), W03413. https://doi.org/10.1029/2005WR004417

Meals, D. W., Dressing, S. A., & Davenport, T. E. (2010). Lag time in water quality response to best management practices: A review. *Journal of Environmental Quality*, *39*, 85–96.

Morales-Suarez-Varela, M. M., Llopis-Gonzalez, A., & Tejerizo-Perez, M. L. (1993). Concentration of nitrates in drinking water and its relationship with bladder cancer. *Journal of Environmental Pathology, Toxicology, and Oncology*, *12*, 229–236.

Morales-Suarez-Varela, M. M., Llopis-Gonzalez, A., & Tejerizo-Perez, M. L. (1995). Impact of nitrates in drinking water on cancer mortality in Valencia, Spain. *European Journal of Epidemiology*, *11*, 15–21.

Nesheim, M. C. (1993). Human health needs and parasitic infections. *Parasitology*, *107*, S7–S18.

Nolan, B. T., Hitt, K. J., & Ruddy, B. C. (2006). Probability of nitrate contamination of recently recharged groundwaters in the conterminous United States. *Environmental Science and Technology*, *36*(10), 2138–2145.

Peel, J. L., Haeuber, R., Garcia, V., Russell, A. G., & Neas, L. (2013). Impact of nitrogen and climate change interactions on ambient air pollution and human health. *Biogeochemistry, 114*, 121–134. https://doi.org/10.1007/s10533-012-9782-4

Pennino, M. J., Compton, J. E., & Leibowitz, S. G. (2017). Trends in drinking water nitrate violations across the United States. *Environmental Science and Technology, 51*, 13450–13460. https://doi.org/10.1021/acs.est.7b04269

Pittman, C. (2012). Florida's vanishing springs. *Tampa Bay Times*, 23 November 2012. Retrieved from http://www.tampabay.com/news/environment/water/floridas-vanishing-springs/1262988, accessed 20 April 2018.

Pope, C. A., Burnett, R. T., Thun, M. J., Calle, E. E., Krewski, D., Ito, K., & Thurston, G. D. (2002). Lung cancer, cardiopulmonary mortality, and long-term exposure to fine particulate air pollution. *Journal of the American Medical Association, 287*, 1132–1141.

Powlson, D. S., Addiscott, T. M., Benjamin, N., Cassman, K. G., de Kok, T. M., & van Grinsven, H. (2008). When does nitrate become a risk for humans. *Journal of Environmental Quality, 37*, 291–295. https://doi.org/10.2134/jeq2007.0177

Puckett, L. J., Tesoriero, A. J., & Dubrovsky, N. M. (2011). Nitrogen contamination of surficial aquifers - A growing legacy. *Environmental Science and Technology, 45*, 839–844.

Qiu, J. (2011). China to spend billions cleaning up groundwater. *Science, 334*, 745–745.

Ren, L., Yang, W., & Bai, Z. (2017). Characteristics of major air pollutants in China. *Advances in Experimental Medicine and Biology, 1017*, 7–26. https://doi.org/10.1007/978-981-10-5657-4_2

Richard, A. M., Diaz, J. H., & Kaye, A. D. (2014). Reexamining the risks of drinking-water nitrates on public health. *The Ochsner Journal, 14*, 392–398.

Sala, O. E., Chapin, I. I. I. F. S., Armesto, J. J., Berlow, E., Bloomfield, J., Dirzo, R., et al. (2000). Global biodiversity scenarios for the year 2100. *Science, 287*, 1770–1774.

Sandor, J., Kiss, I., & Farkas, O. (2001). Association between gastric cancer mortality and nitrate content of drinking water: Ecological study based on small area inequalities. *European Journal of Epidemiology, 17*, 443–447.

Sanford, W. E., & Pope, J. P. (2013). Quantifying groundwater's role in delaying improvements to Chesapeake Bay water quality. *Environmental Science and Technology, 47*, 13330–13338.

Sareer, O., Mazahar, S., Khanum, W. M., Akbari, A., & Umar, S. (2015). Nitrogen pollution, plants, and human health. In M. Ozturk, M. Ashraf, A. Aksoy, M. S. A. Ahmad, & K. R. Hakeem (Eds.), *Plants, pollutants, and remediation* (Vol. 404). Dordrecht: Springer Science.

Schlesinger, W. H. (2009). On the fate of anthropogenic nitrogen. *Proceedings of the National Academy of Sciences, 106*, 203–208.

Schoolman, E. D., & Ma, C. B. (2012). Migration, class and environmental inequality: Exposure to pollution in China's Jiangsu province. *Ecological Economics, 75*, 140–151. https://doi.org/10.1016/j.ecolecon.2012.01.015

Schullehner, J., Hansen, B., Thygesen, M., Pederson, C. B., & Sigsgaard, T. (2018). Nitrate in drinking water and colorectal cancer risk: A nationwide population-based cohort study. *International Journal of Cancer, 143*, 73–79. https://doi.org/10.1002/ijc.31306

Schullehner, J., Stayner, L., & Hansen, B. (2017). Nitrate, nitrite, and ammonium variability in drinking water distribution systems. *International Journal of Environmental Research and Public Health, 14*(3), 276. https://doi.org/10.3390/ijerph14030276

Sebillo, M., Mayer, B., Nicolardot, B., Pinay, G., & Mariotti, A. (2013). Long-term fate of nitrate fertilizer in agricultural soils. *Proceedings of the National Academy of Sciences of the United States, 110*(45), 18185–18189.

Sheppard, L. J., Leith, I. D., Mizunuma, T., Cape, J. N., Crossley, A., Leeson, S., et al. (2011). Dry deposition of ammonia gas drives species change faster than wet deposition of ammonium ions: Evidence from a long-term field manipulation. *Global Change Biology, 17*(12), 3589–3607.

Stanek, L. W., Sacks, J. D., Dutton, S. J., & Dubois, J. J. B. (2011). Attributing health effects to apportioned components and sources of particulate matter: An evaluation of collective results. *Atmospheric Environment, 45*(32), 5655–5663. https://doi.org/10.1016/j.atmosenv.2011.07.023

Sutton, M. A., Howard, C. M., Erisman, J. W., Billen, G., Bleeker, A., Grennfelt, P., et al. (2011). *The European nitrogen assessment: Sources, effects, and policy perspectives* (Vol. 664). Cambridge, UK: Cambridge University Press.

Tian, L., & Sun, S. (2017). Comparison of health impact of air pollution between China and other countries. *Advances in Experimental and Medical Biology, 1017*, 215–232. https://doi.org/10.1007/978-981-10-5657-4_9

Townsend, A. R., Howarth, R. W., Bazzaz, F. A., Booth, M. S., Cleveland, C. C., Collinge, S. K., et al. (2003). Human health effects of a changing global nitrogen cycle. *Frontiers in Ecology and the Environment, 1*(5), 240–246.

U.S. Environmental Protection Agency (U.S. EPA). (2007). *Toxicity and exposure assessment for children's health chemical summary: Nitrates and nitrites*. Retrieved from https://archive.epa.gov/region5/teach/web/pdf/nitrates_summary.pdf, accessed 12 May 2018.

U.S. Environmental Protection Agency. (2012). *Our nation's air: Status and trends through 2010* (EPA-454/R-12-001). Research Triangle Park, NC, 32 p.

Van Leeuwen, J. A., Waltner-Toews, D., & Abernathy, T. (1999). Associations between stomach cancer incidence and drinking water contamination with atrazine and nitrate in Ontario (Canada) agroecosystems, 1987–91. *International Journal of Epidemiology, 28*, 836–840.

Van Loon, A. J., Botterweck, A. A., & Goldblohm, R. A. (1998). Intake of nitrate and nitrite and the risk of gastric cancer: A prospective cohort study. *British Journal of Cancer, 78*, 129–135.

Vitoria Minana, I., Brines Solanes, J., Morales-Suarez-Varela, M., & Llopis Gonzalez, A. (1991). Nitrates in drinking water in the Valencia community. Indirect risk of methemoglobinemia in infants. *Anales de Pediatría, 34*(1), 43–50.

Volkmer, B. G., Ernst, B., Simon, J., Kuefer, R., Bartsch, G., Jr., Bach, D., & Gschwend, J. E. (2005). Influence of nitrate levels in drinking water on urological malignancies: A community-based cohort study. *BJU International, 95*, 972–976. https://doi.org/10.1111/j.1464-410X.2005.05450.x

Walton, G. (1951). Survey of literature relating to infant methemoglobinemia due to nitrate-contaminated water. *American Journal of Public Health Nation's Health, 41*, 986–996.

Wang, L., Butcher, A., Stuart, M., Gooddy, D., & Bloomfield, J. (2013). The nitrate time bomb: A numerical way to investigate nitrate storage and lag time in the unsaturated zone. *Environmental Geochemistry and Health, 35*, 667–681.

Wang, L., Stuart, M. E., Lewis, M. A., Ward, R. S., Skirvin, D., Naden, P. S., et al. (2016). The changing trend in nitrate concentrations in major aquifers due to historical nitrate loading from agricultural land across England and Wales from 1925 to 2150. *Science of the Total Environment, 542*, 694–705.

Ward, M. H., de Kok, T. M., Levallois, P., Brender, J., Gulis, G., Nolan, B. T., & VanDerslice, J. (2005). Drinking-water nitrate and health—Recent findings and research needs. *Environmental Health Perspectives, 113*(11), 1607–1614.

Ward, M. H., Jones, R. R., Brender, J. D., de Kok, T. M., Weyer, P. J., Nolan, B. T., et al. (2018). Drinking water nitrate and human health: An updated review. *International Journal of Environmental Research and Public Health., 15*(7), pii: E1557. https://doi.org/10.3390/ijerph15071557

Ward, M. H., Kilfoy, B. A., Weyer, P. J., Anderson, K. E., Folsom, A. R., & Cerhan, J. R. (2010). Nitrate intake and the risk of thyroid cancer and thyroid disease. *Epidemiology, 21*, 389–395.

Ward, M. H., Mark, S. D., Cantor, K. P., Weisenburger, D. D., Correa-Villasenor, A., & Zahm, S. H. (1996). Drinking water nitrate and the risk of Non-Hodgkin's Lymphoma. *Epidemiology, 7*, 465–471.

Weyer, P. J., Cerhan, J., Kross, B. C., Hallberg, G. R., Kantamneni, J., Breuer, G., et al. (2001). Municipal drinking water nitrate level and cancer risk in older women: The Iowa Women's Health Study. *Epidemiology, 12*, 327–338.

World Health Organization. (2002). *Water and health in Europe*. Geneva, Switzerland: World Health Organization.

World Health Organization Regional Office for Europe. (2006). *Air quality guidelines. European Series* (Vol. 23). Copenhagen: WHO Regional Publication.

Wu, Y., Gu, B., Erisman, J. W., Reis, S., Fang, Y., Lu, X., et al. (2016). $PM_{2.5}$ pollution is substantially affected by ammonia emissions in China. *Environmental Pollution, 218*, 86–94. https://doi.org/10.1016/j.envpol.2016.08.027

Yang, D., Yang, X., Deng, F., & Guo, X. (2017). Ambient air pollution and biomarkers of health effect. *Advances in Experimental Medicine and Biology, 1017*, 59–102. https://doi.org/10.1007/978-981-10-5657-4_4

Zanchett, G., & Oliveira-Filho, E. C. (2013). Cyanobacteria and cyanotoxins: From impacts on aquatic ecosystems and human health to anticarcinogenic effects. *Toxins, 5*(10), 1896–1917. https://doi.org/10.3390/toxins5101896

Zeman, C., Beltz, L., Linda, M., Maddux, J., Depken, D., Orr, J., & Theran, P. (2011). New questions and insights into nitrate/nitrite and human health effects: A retrospective cohort study of private well users' immunological and wellness status. *Journal of Environmental Health, 74*(4), 8–19.

Zhai, Y., Lei, Y., Wu, J., Teng, Y., Wang, J., Zhao, X., & Pan, X. (2017). Does the groundwater nitrate pollution in China pose a risk to human health? A critical review of published data. *Environmental Science and Pollution Research International, 24*(4), 3640–3653. https://doi.org/10.1007/s11356-016-8088-9. Epub 2016 Nov 24

Zhang, N., Huang, H., Duan, X., Zhao, J., & Su, B. (2018). Quantitative association analysis between PM2.5 concentration and factors on industry, energy, agriculture, and transportation. *Scientific Reports, 8*, 9461. https://doi.org/10.1038/s41598-018-27771-w

FURTHER READING

Aschebrook-Kilfoy, B., Shu, X. O., Gao, Y. T., Ji, B. T., Yang, G., Li, H. L., et al. (2013). Thyroid cancer risk and dietary nitrate and nitrite intake in the Shanghai women's health study. *International Journal of Cancer, 132*(4), 897–904. https://doi.org/10.1002/ijc.27659

Burow, K. R., Nolan, B. T., Rupert, M. G., & Dubrovsky, N. M. (2010). Nitrate in groundwater of the United States, 1991–2003. *Environmental Science and Technology, 44*(13), 4988–4997.

Deguen, S., Petit, C., Delbarre, A., Kihal, W., Padilla, C., Benmarhnia, T., et al. (2015). Neighbourhood characteristics and long-term air pollution levels modify the association between the short-term nitrogen dioxide concentrations and all-cause mortality in Paris. *PLoS One, 10*(7), e0131463.

Dellavalle, C. T., Xiao, Q., Yang, G., Shu, X. O., Aschebrook-Kilfoy, B., Zheng, W., et al. (2014). Dietary nitrate and nitrite intake and risk of colorectal cancer in the Shanghai Women's Health Study. *International Journal of Cancer, 134*(12), 2917–2926. https://doi.org/10.1002/ijc.28612

Espejo-Herrera, N., Cantor, K. P., Malats, N., Silverman, D. T., Tardón, A., García-Closas, R., et al. (2015). Nitrate in drinking water and bladder cancer in Spain. *Environmental Research, 137*, 299–307. https://doi.org/10.1016/j.envres.2014.10.034

Espejo-Herrera, N., Gracia-Lavedan, E., Boldo, E., Aragones, N., Perez-Gomez, B., Pollan, M., et al. (2016). Colorectal cancer risk and nitrate exposure through drinking water and diet. *International Journal of Cancer, 139*(2), 334–346. https://doi.org/10.1002/ijc.30083

Fowler, D., Steadman, C. E., Stevenson, D., Coyle, M., Rees, R. M., Skiba, U. M., et al. (2015). Effects of global change during the 21st century on the nitrogen cycle. *Atmospheric Chemistry and Physics, 15*, 13849–13893. https://doi.org/10.5194/acp-15-13849-2015

Galloway, J. N., Townsend, A. R., Erisman, J. W., Bekunda, M., Cai, Z., Freney, J. R., et al. (2008). Transformation of the nitrogen cycle: Recent trends, questions, and potential solutions. *Science, 320*, 889–892.

Manassaram, D. M., Backer, L. C., Messing, R., Fleming, L. E., Luke, B., & Monteilh, C. P. (2010). Nitrates in drinking water and methemoglobin levels in pregnancy: A longitudinal study. *Environmental Health, 9*, 60.

Marshall, J. D., Apte, J. S., Coggins, J. S., & Goodkind, A. L. (2016). Blue skies bluer? *Environmental Science and Technology.* http://dx.doi.org/10.1021/acs.est.5b03154

Nolan, B. T., & Hitt, K. J. (2006). Vulnerability of shallow groundwater and drinking-water wells to nitrate in the United

States. *Environmental Science and Technology, 40*(24), 7834–7840.

Phillips, S.W., & B.D. Lindsey. (2003). *The influence of ground water on nitrogen delivery to the Chesapeake Bay.* U.S. Geological Survey Fact Sheet FS-091-03.

Sadeq, M., Moe, C. L., Attarassi, B., Cherkaoui, I., ElAouad, R., & Idrissi, L. (2008). Drinking water nitrate and prevalence of methemoglobinemia among infants and children aged 1–7 years in Moroccan areas. *International Journal of Hygiene and Environmental Health, 211*, 546–554.

Schullehner, J., & Hansen, B. (2014). Nitrate exposure in Denmark over the last 35 years. *Environmental Research Letters, 9*, 095001. http://dx.doi.org/10.1088/1748-9326/9/9/095001

State-USEPA Nutrient Innovations Task Group. (2009). *An urgent call to action.* State-USEPA Nutrient Innovations Task Group Report to USEPA, August 2009, 170 p.

Tesoriero, A. J., Duff, H. H., Saad, D. A., Spahr, N. E., & Wolock, D. M. (2013). Vulnerability of streams to legacy nitrate sources. *Environmental Science and Technology, 47*(8), 3623–3629.

U.S. Environmental Protection Agency. (2009). *An urgent call to action.* Report of the State-EPA Nutrient Innovations Task Group, 41 p.

U.S. Environmental Protection Agency. (2002). *Nitrogen—Multiple and regional impacts.* U.S. Environmental Protection Agency Clean Air Markets Division Report EPA-430-R-01-006, 38 p.

6
Terrestrial Biodiversity and Surface Water Impacts from Reactive Nitrogen

During the past 50 years, excess reactive nitrogen has had adverse consequences for many varied types of environmental systems all over the world. As mentioned earlier, only a small proportion of the Earth's biota can transform N_2 to reactive nitrogen forms, as a result, reactive nitrogen tends to be the limiting nutrient in most natural ecosystems and typically for agricultural systems. Therefore, increased amounts of reactive nitrogen in the environment have had detrimental and irreversible effects on ecosystems (Camargo & Alonso, 2006; Erisman et al., 2011, 2013, 2015; Payne et al., 2013; Townsend et al., 2003). Some of these effects include shifts in plant species composition, decreases in species diversity, habitat degradation, eutrophication, decreases in water transparency, increased biomass of benthic and epiphytic algae, and changes in food-web processes in terrestrial ecosystems and in coastal waters (Howarth et al., 2000). These effects are more prevalent in certain areas, such as the highly affected areas in Central and Western Europe, Southern Asia, the eastern United States, and parts of Africa and South America (Erisman et al., 2015).

Biologically, available nitrogen that enters the terrestrial biosphere has more than doubled in the since the 1950s from various activities, including fertilizer production and use, fossil fuel combustion, and cultivation of leguminous crops (Galloway et al., 2004, 2008). This increase in reactive nitrogen inputs to aquatic ecosystems has had a multitude of harmful effects on biota (flora and fauna) in aquatic ecosystems including loss of biodiversity, fish kills, and harmful algal blooms (HABs). These impacts on many different types of ecosystems have been the subject of numerous journal articles describing case studies at local scales, review papers of regional and global scale impacts, special journal issues dedicated to different aspects of ecosystem alterations (e.g., Biogeosciences, 2013, Eds. Sutton et al., https://www.biogeosciences.net/special_issue93.html, accessed 10 June 2018), and special issue papers (e.g., Excess nitrogen in the U.S. Environment: Trends, Risks, and Solutions, Ecological Society of America, Davidson et al., 2011; Focus on Nitrogen Management Challenges: Synthesis and review: Tackling the nitrogen management challenge: from global to local scales, Reis et al., 2016, Environmental Research Letters, http://iopscience.iop.org/journal/1748-9326/page/Nitrogen%20management%20challenges, accessed 10 June 2018), comprehensive reports and books (European Nitrogen Assessment, Sutton et al., 2011b, http://www.nine-esf.org/files/ena_doc/ENA_pdfs/ENA_pre.pdf, accessed 10 June 2018); U.S. National Water Quality Assessment Program; Dubrovsky et al. (2010); Reactive Nitrogen in the United States: An Analysis of Inputs, Flows, Consequences, and Management Options (US EPA, 2011; U.S. EPA, 2011).

Erisman et al. (2013) summarized how reactive nitrogen has contributed to the exceedance of effect levels for various ecosystems and to global human health problems. For example, it was noted that approximately 10% of freshwater systems have nitrate-N concentrations exceeding 1 mg/L, and 40% of the problem (relative to other freshwater pollution and natural causes) has resulted from anthropogenic reactive nitrogen. They estimated that there has been 50% biodiversity loss (50% of the total area of biodiversity hot spots where nitrogen deposition exceeds 5 kg N/ha/yr), with a 15% contribution of the problem from anthropogenic reactive nitrogen. For coastal dead zones, they estimated that 80% of large marine ecosystems (64 in total) has a reactive nitrogen problem, with 50% of the global coastal zone pollution contributed by anthropogenic reactive nitrogen. They show that for nitrate/nitrite intake, 70% of the global population is exposed to above-recommended levels of either nitrate or nitrite in air, water, or food, and the contribution of reactive nitrogen accounts for 80% to the problem (20% of the exposure to above-recommended

---

Nitrogen Overload: Environmental Degradation, Ramifications, and Economic Costs, Geophysical Monograph 250, First Edition. Brian G. Katz.
© 2020 American Geophysical Union. Published 2020 by John Wiley & Sons, Inc.

reactive nitrogen is related to natural sources). For air pollution (human health), 60% of the global population is exposed to air quality above recommended safe levels, with anthropogenic reactive nitrogen contributing 20% of the formation of fine particles. About 4% of the global crop loss is related to air pollution, of which reactive nitrogen from anthropogenic activities contributes 50%, primarily through tropospheric ozone enrichment. For climate change, 20% of the preindustrial N_2O concentration has been exceeded, with the contribution of reactive nitrogen, net cooling of 15% due to all reactive nitrogen impacts on drivers of radiative forcing. For stratospheric ozone levels, the exceedance was estimated at 20% of the preindustrial N_2O concentration, with reactive nitrogen contributing 40% of all stratospheric ozone.

This chapter highlights some of the aforementioned topics and focuses mainly on recent studies that have investigated impacts of reactive forms of nitrogen to terrestrial and coastal aquatic ecosystems and surface waters. Also, the combined effects of climate change and reactive nitrogen loading are discussed briefly in terms of their potential future impacts on various ecosystems. Several recent articles present more detailed information on the impacts of climate change and reactive nitrogen on aquatic ecosystems (e.g., Baron et al., 2013; Fowler et al., 2015; Greaver et al., 2012; Irvine et al., 2017; Pinder et al., 2012a; E.M. Porter et al., 2012; Reis et al., 2016; Stuart et al., 2011).

6.1. WORLDWIDE TERRESTRIAL BIODIVERSITY IMPACTS FROM REACTIVE NITROGEN DEPOSITION

Erisman et al. (2015) noted that reactive nitrogen is a significant driver of biodiversity loss as a result of acidification and eutrophication. In areas with high inputs of reactive nitrogen, faster-growing species that can rapidly assimilate forms of reactive nitrogen and acid-tolerant species would predominate. Other animals, such as insects or fauna, dependent on those plants and habitats would also be affected. Erisman et al. (2015) presented a map (Fig. 6.1) showing modeled global deposition of reactive nitrogen overlain on the spatial distribution of the world's protected areas as delineated by the UN Environment Programme, World Conservation Monitoring Center (UNEP-WCMC, 2008). It was noted that 40% of all protected areas (11% by area surface) have been projected to receive a nitrogen deposition rate greater than 10 kg/ha/yr by 2030. This deposition rate is considered the largest amount that most natural systems can absorb without affecting the ecosystem and its services. Certain parts of the world have been affected more severely than other areas. For example, highly affected areas noted include Central and Western Europe, southern Asia and the eastern United States, as well as parts of Africa and South America.

Several studies are using or have used an approach that estimates critical loads from nitrogen deposition. The term "critical load" refers to the input or exposure of one or more contaminants (i.e., forms of reactive nitrogen) below which no detrimental ecological effects occur over the long term. In fact, critical load values have been adopted by the European Commission and used in Europe since the 1980s and are an important component in pollution management, mapping pollution impacts, and developing international agreements on transboundary air pollution (Payne et al., 2013). Critical load values are now being used in Canada, the United States, and other countries as a way of implementing reductions in nitrogen deposition. These values also have been used to set emissions standards and provide an effective way to evaluate emissions control programs and the potential impact of new sources of pollution. For example, the critical nitrogen load for grasslands in Europe has been set at 10–15 kg/ha/yr.

Based on a large dataset of European grasslands exposed to a gradient of reactive nitrogen deposition, Payne et al. (2013) found that approximately 60% of change points for various grassland species occurred at or below the range of the currently established critical load range. The change points were based on a statistical procedure that determined a point along a nitrogen deposition gradient where a given species showed a statistically significant decline. Based on this work, they concluded that more informed decisions are needed on how much harm can be tolerated, especially in unpolluted areas where sensitive plant species could be exposed to new pollution sources and harm due to low levels of nitrogen deposition.

Bobbink et al. (2010) noted that research studies during the past 20–30 years in Europe and North America have shown adverse effects on ecosystems from deposition of reactive nitrogen. The severity of effects were shown to be dependent on three main factors: (a) the duration, the total amount, and the form reactive nitrogen; (b) the sensitivity of the plant species; and (c) abiotic conditions in the ecosystem. Additional factors include the soil acid neutralizing capacity, the soil nutrient availability, and other soil characteristics that affect the nitrification potential and nitrogen attenuation. Bobbink et al. (2010) reviewed field studies of reactive nitrogen deposition, effects, and thresholds for terrestrial plant diversity protection across a latitudinal range of main categories of ecosystems from arctic and boreal systems to tropical forests. They concluded that nitrogen accumulation in the environment is the main driver of changes to species composition across the whole range of different ecosystem types by driving the competitive interactions that

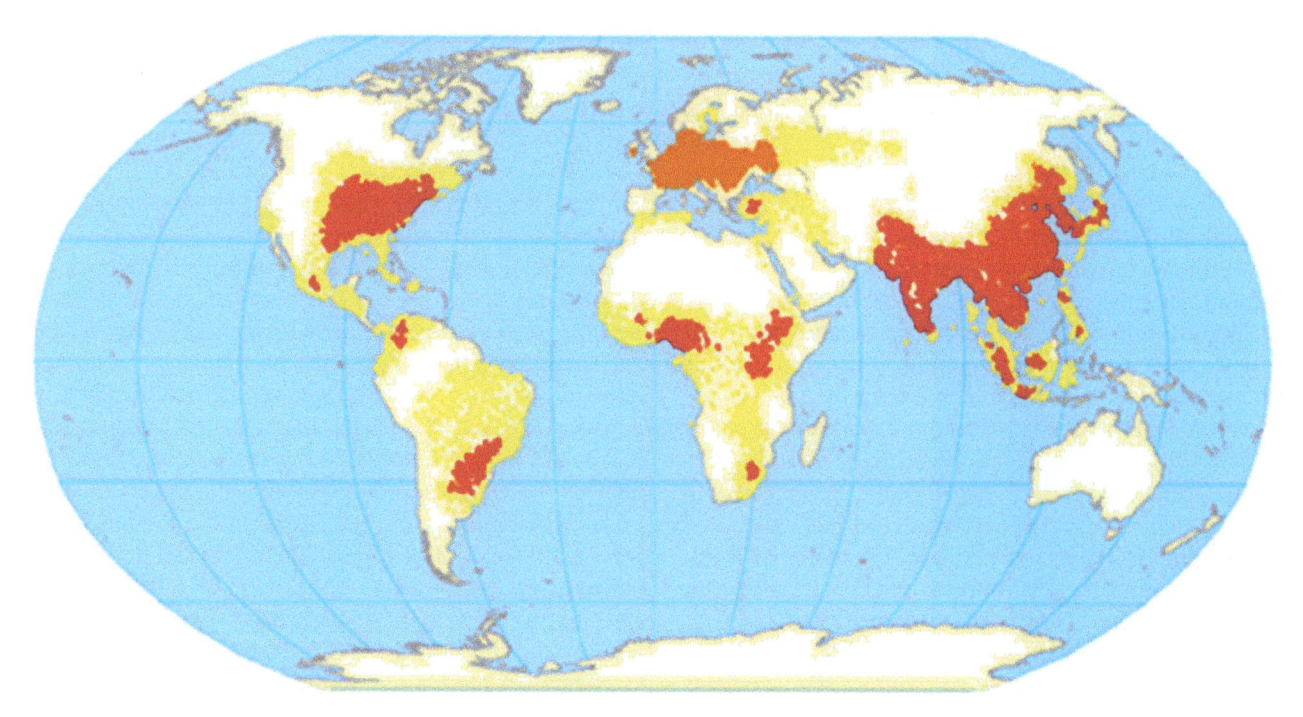

Figure 6.1 Distribution of categories of reactive nitrogen deposition and exceedance of deposition levels in the period 2000–2030 on Protected Areas (PAs) under the Convention on Biological Diversity. Red shaded PAs show an exceedance of 10 kg N/ha/yr and deposition in 2030 higher than in 2000. Orange PAs show a current exceedance, but deposition in 2030 lower than 2000. Yellow PAs might be under threat in the near future since reactive nitrogen deposition exceeds 5 kg N/ha/yr, but is increasing over the period 2000–2030. *Source*: From Erisman et al. (2013, 2015) based on information in Bleeker et al. (2010); with permission.

lead to composition change and/or making conditions unfavorable for some species. Other site-specific ecosystem effects could include direct toxicity from nitrogen gases and aerosols, long-term negative effects of increased ammonium and ammonia availability, soil-mediated effects of acidification, and secondary stress and disturbance. Fowler et al. (2013) noted that plant and soil communities have evolved to sequester and recycle different forms of reactive nitrogen because nitrogen is an essential and often limiting nutrient. Although systems that typically are not nitrogen limited (e.g., tropical and subtropical systems) could be affected in their regeneration phase.

Effects from the higher reactive nitrogen deposition rates have included species composition changes, declines in diversity, and acidification of soils. For example, species that exist in naturally nutrient-poor (oligotrophic) or nutrient-moderate (mesotrophic) and pH neutral habitats are out-competed by nitrophilic (more nitrogen-loving) or acid-tolerant plants (Sutton et al., 2013b). Also, in areas with long-term deposition of reactive nitrogen species, forests can be more susceptible to stresses (e.g., frost damage) and other diseases (Dise et al., 2011). Other studies have shown a linkage between excessive reactive nitrogen and reduced plant species richness in a broad range of European ecosystems (Bobbink et al., 2010; Dise et al., 2011). A general threshold value for adverse effects (annual deposition rate of 5–10 kg N/ha) was proposed (Bobbink et al., 2010), but other studies have shown that effects can occur at even lower levels over the long term (Clark & Tilman, 2008).

In a review of reactive nitrogen impacts on terrestrial biodiversity across major bioclimatic zones throughout Europe, Dise et al. (2011) indicated that vegetation diversity has been harmed by direct foliar damage, eutrophication, acidification, and susceptibility to secondary stress. Particularly susceptible to chronically elevated reactive nitrogen deposition are species and communities that are adapted to low nutrient levels or that exist in areas that are not well buffered against acidic deposition. It was noted that grassland, heathland, peatland, forest, and arctic/montane ecosystems are most vulnerable; however, other ecosystems (Mediterranean, coastal dune, and tundra) have been affected as well. The importance of considering the cumulative load of plant-available nitrogen in the soil as the biodiversity of ecological communities all over Europe have been especially prone to decline after many decades of reactive nitrogen deposition was also emphasized. In fact, Dise et al. (2011) noted that plant species richness in a broad range of European ecosystems has been reduced due to exceedance of critical loads for nitrogen. The following critical loads for nitrogen deposition (in kg N/ha/yr) were summarized by Dise et al. (2011) for various habitats: arctic, alpine and subalpine scrub, 5–15; heathlands, 10–25; peatlands, 5–10; grasslands, 10–30; and forests, 10–20. They concluded that even with reduced deposition of reactive nitrogen full recovery would likely be very slow, particularly for highly impacted ecosystems.

Ammonia emissions to the atmosphere also have increased during the past several decades and can result in the formation of particulate matter through reactions with nitric and sulfuric acids. These harmful compounds in the atmosphere can be dispersed over large areas of the world resulting in eutrophication and acidification of terrestrial, freshwater, and marine habitats (Bobbink et al., 2010; Erisman et al., 2015). Sutton et al. (2013a) noted that emissions to the atmosphere containing high concentrations of ammonia and nitrous oxide especially near direct sources (such as intensive livestock production or industrial facilities) cause extensive foliar damage. Particularly susceptible are lower plants such as lichens and bryophytes that consume most of their nutrients from the atmosphere (Sheppard et al., 2011). Sutton et al. (2013a) provided an example of how *Sphagnum* moss (important for peatland growth and carbon storage) can be suffocated by colonization of the surface with algae, which threatens biodiversity and carbon sinks (Gu et al., 2017). Most of these effects occur within a few kilometers of agricultural sources, although many landscapes that contain a mixture of agricultural and other biota are at risk. Gaseous ammonia (NH_3) can be particularly harmful to vegetation, especially lower plants, through direct foliar damage (Dise et al., 2011). For example, some of the effects they noted were a decrease in nitrophobic epiphytic lichen species and decreased cover of nitrophobic vascular plants in the United Kingdom, a decrease in acidophytic epiphytic lichen species in the Netherlands, an increase in strictly nitrophytic lichen diversity in Italy and Portugal (but a decrease in oligotrophic lichen diversity). Sala et al. (2000) noted that certain biomes (particularly northern temperate, boreal, arctic, alpine, grassland, savannah, and Mediterranean) are very sensitive to reactive nitrogen deposition because of the limited availability of nitrogen in these systems under natural conditions.

In the conterminous United States, Baron et al. (2013) noted that inputs of reactive nitrogen have been increasing over time and total terrestrial reactive nitrogen inputs to the United States in 2002 were 28.5 Tg N/yr (U.S. EPA, 2011). However, the exact amount of reactive nitrogen that leaches to groundwater and deposited from the atmosphere or runs off and enters surface waters has not been well quantified. Various models have given a range of estimates. For example, the U.S. Geological Survey's SPAtially Referenced Regressions on Watershed (SPARROW) attributes model estimated nitrogen inputs to river systems in 2002 to be 4.8 Tg N/yr (Alexander et al., 2008; U.S. EPA, 2011). For North America, Boyer et al. (2006) estimated that approximately 7.0 Tg N/yr is exported via rivers to the coastal zone, inlands, and drylands.

Baron et al. (2011) noted that atmospheric deposition of ammonia and nitrate is an important source of reactive nitrogen to watersheds in mountain areas in the United States. Based on a review of studies in the United States, Compton et al. (2011) showed schematically how increases in reactive nitrogen loading have affected ecosystem services for lakes, forest growth, coastal fish production, soil fertility, and drinkable water (Fig. 6.2).

Critical loads also are being used in the United States to evaluate impacts of nitrogen deposition on ecosystems and to assess eutrophication. Pardo et al. (2011) synthesized current research on the effects of atmospheric nitrogen deposition on major terrestrial and freshwater ecosystems in the United States (Fig. 6.3) and estimated empirical critical loads of reactive nitrogen. Biota receptors included freshwater diatoms, mycorrhizal fungi, lichens, bryophytes, herbaceous plants, shrubs, and trees. The following ecosystem impacts were evaluated: biogeochemical responses (such as increased nitrogen mineralization and nitrification, increased gaseous nitrogen losses, and increased leaching); and individual species, population, and community responses. These latter impacts were evaluated using various parameters (including increased tissue nitrogen, physiological and nutrient imbalances, increased growth, altered root: shoot ratios, increased susceptibility to secondary stresses, altered fire regime, shifts in competitive interactions and community composition, changes in species richness and other measures of biodiversity, and increases in invasive

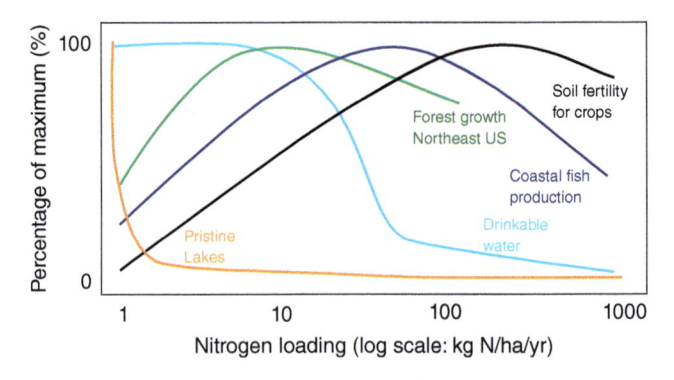

Figure 6.2 A plot showing the relation between increases in nitrogen loading and changes in ecosystem services for lakes, forest growth, coastal fish production, soil fertility, and drinkable water. *Source*: From Compton et al. (2011). Public domain. Reproduced with permission of John Wiley & Sons.

Figure 6.3 Major ecological regions of North America. *Source*: From Pardo et al. (2011).

Legend:
1. Arctic Cordillera
2. Tundra
3. Taiga
4. Hudson Plains
5. Northern Forests
6. Northwestern Forested Mountains
7. Marine West Coast Forests
8. Eastern Temperate Forests
9. Great Plains
10. North American Deserts
11. Mediterranean California
12. Southern Semi-Arid Highlands
13. Temperate Sierras
14. Tropical Dry Forests
15. Tropical Humid Forests

species). Pardo et al. (2011) found that empirical critical nitrogen loads for U.S. ecoregions, inland surface waters, and freshwater wetlands ranged from 1 to 39 kg N/ha/yr. They noted that the empirical critical nitrogen loads tended to increase in the following sequence for biota: diatoms, lichens and bryophytes, mycorrhizal fungi, herbaceous plants and shrubs, and trees.

Recently, a collaborative team of scientists from EPA, U.S. Geological Survey, the U.S. Forest Service, the University of Colorado, and multiple other universities have been investigating the effects of nitrogen deposition on herbaceous plants (those with nonwoody stems such as grass) in a variety of ecosystems across the United States. This research addresses how climatic (precipitation), physical, and chemical factors (e.g., soil acidity, temperature) affect species loss due to nitrogen deposition. More recently, their research has expanded into habitats that have not received much attention, including the forest understory. Simkin et al. (2016) found that 24% of more than 15,000 sites (forest, woodland, shrubland, and grassland) across a wide range of climates, soil conditions, and vegetation types in the United States were susceptible to nitrogen deposition-induced species loss. At extremely low levels of nitrogen deposition, the number of individual plant species tended to increase. However, above a certain threshold level, or "critical load," diversity began to decline. However, grasslands, shrublands, and woodlands were susceptible to species losses at lower loads of nitrogen deposition than forests, and susceptibility to species losses increased in acidic soils. Both grasslands and forests were highly susceptible to nitrogen deposition, with critical loads in the range of current deposition levels. Also, their results showed that forests typically can tolerate slightly higher levels of nitrogen deposition than other ecosystems before showing a negative impact on biodiversity. They hypothesized that forest species living under the canopy can tolerate low-light conditions and therefore are less vulnerable to the effects of increased shade caused by increased nitrogen. Their future studies will attempt to determine the individual plant species that are most at risk and which native and invasive species may increase with elevated nitrogen deposition. These findings are important for the protection of biodiversity and demonstrate the need for establishing more stringent air quality standards.

Clark et al. (2017) reported findings from a workshop that synthesized information on N-induced terrestrial eutrophication from published literature and to relate exceedances in nitrogen critical loads with human beneficiaries by using the STressor–Ecological Production function–final ecosystem Services Framework and the Final Ecosystem Goods and Services Classification System (FEGS-CS). A total of 21 nitrogen critical loads resulted from nitrogen N deposition (ranging from 2 to 39 kg N/ha/yr), which was issued to different beneficiary types through 582 individual pathways in the five ecoregions examined in the United States (Eastern Temperate Forests, Marine West Coast Forests, Northwestern Forested Mountains, North American Deserts, and Mediterranean California). Nitrogen critical load exceedance affected 66 FEGS across a range of final ecosystem service categories (21 categories, such as changes in timber production, fire regimes, and native plant and animal communities) and 198 regional human beneficiaries of distinct types. Various ecosystems were found to have different biological indicators, including grasses and/or forbs (33% of all pathways), mycorrhizal communities (22%), tree species (21%), and lichen biodiversity (11%). Longer exposure times to higher deposition rates in ecoregions resulted in more numerous and varied ecological impacts (e.g., Eastern Temperate Forests, eight biological indicators) in contrast to other ecoregions (e.g., North American Deserts and Marine West Coast Forests each with one biological indicator). Clark et al. (2017) reported that terrestrial eutrophication affected all ecosystems studied, which indicated the widespread nature of terrestrial eutrophication across the United States. Even though the various ecoregions differed by ecological effects from terrestrial eutrophication, the number of FEGS and beneficiaries impacted was similar across the studied ecoregions. Their findings showed which ecosystems and people are most affected according to current information and helped to identify some of the knowledge gaps that would need to be addressed in future research studies.

Another recent review article (Hernández et al., 2016) surveyed approximately 1400 species in the continental United States that were listed as candidate, threatened, or endangered under the U.S. Endangered Species Act (ESA) to evaluate the extent of recognized nitrogen pollution effects on biodiversity in terrestrial and aquatic ecosystems. They found 78 federally listed species recognized as affected by nitrogen pollution. At least nine species (including species of freshwater mussels, amphibians, plants) in their survey were directly affected by toxic or lethal nitrogen effects. Most of the nitrogen-affected species on the ESA list were threatened by eutrophication factors. This was found to be especially prevalent in freshwater ecosystems where the indirect effects of nitrogen deposition resulted in shifts in species composition of primary producers, increased producer biomass and organic matter sedimentation, and reductions in dissolved oxygen, water clarity, and light availability that affected the habitat and trophic dynamics of aquatic species. Also, some native species of plants were harmed by nonnative plant species, and in some cases, native animal species were affected due to exclusion of their food source. It was concluded that thresholds for both atmospheric and aquatic nitrogen inputs should be set in sensitive ecosystems based on an integration of direct observations, experiments, and modeling studies using chronic low nitrogen inputs at multiple scales and management conditions.

6.2. SOIL ACIDIFICATION AND RELATED IMPACTS ON TERRESTRIAL ECOSYSTEMS

Atmospheric deposition of reactive nitrogen also leads to acidification of soils. Bowman et al. (2008) noted that nitrogen deposition rates were above critical loads in many industrially developed countries. Soils in many places around the world are particularly sensitive to nitrogen deposition and with buffering likely transitioning from base cations (Ca, Mg, and K) to nonbase cations (Al and Mn), which could have a toxic impact for terrestrial ecosystems (Bowman et al., 2008; Tian & Niu, 2015). In parts of North America and Europe, Bowman et al. (2008) mentioned previous studies where acid deposition shifted forest and grassland soils into the aluminum buffering range. During 2000–2010, Peñuelas et al. (2013) reported that global land had received more than 50 kg/ha accumulated N deposition. They found that nitrogen additions to the land surface significantly reduced soil pH by 0.26 on average globally. Soil acidification has resulted in the decline of sensitive plant species in high-deposition areas. Previous work has documented patterns of species change in favor of nitrophilic or acid-tolerant species in grasslands, coastal dunes, heathlands, and alpine tundras (Bobbink et al., 2010; Clark & Tilman, 2008; Remke et al., 2009; Stevens, et al., 2004). Significant decreases in forest and grassland productivity have been observed when reactive nitrogen deposition increases above threshold levels. As previously mentioned, these adverse effects can lead to biodiversity changes through the food chain, affecting insects, birds, and other animals that depend on these food sources (Erisman et al., 2013). In cold mountainous regions that contain ecosystems with short growing seasons, there is a high potential for severe impacts of nitrogen deposition on shallow soils in steep terrain. To investigate the impact of elevated nitrogen deposition in mountainous areas, Bowman et al. (2008) conducted an experiment in the Western Tetra Mountains of Slovakia. They reported that increased nitrogen deposition in the region resulted in a depletion of base cations (calcium and magnesium) and soluble aluminum, with an accompanying increase in extractable iron concentrations. As a result, they found a reduction in the biomass of vascular plants induced by nitrogen deposition. They concluded that other sites in Central Europe potentially have reached a more toxic level of soil acidification, where soil acidification has resulted in greater iron release than aluminum into soil water.

6.3. RIVERINE EXPORT OF REACTIVE NITROGEN

Freshwater nitrogen loadings worldwide have increased substantially from 21 million tons in preindustrial times to 40 million tons per year in 2009 (Dodds et al., 2009), while riverine transport of dissolved inorganic nitrogen (DIN) has increased from 2–3 million to 15 million tons. Reactive nitrogen can enter surface water systems via many pathways including stormwater runoff, shallow subsurface flow paths, groundwater flow to streams, submarine groundwater discharge to estuaries and oceans, and direct atmospheric nitrogen deposition. In many large river basins, the two main sources of anthropogenic reactive nitrogen contamination are inorganic fertilizer and animal manure (Baron et al., 2013; Hong et al., 2011; Howarth et al., 2011b; Puckett, 1995). Several studies have shown that approximately 50% of the nitrogen used in agricultural activities is unintentionally lost to the environment with most flowing to freshwaters (Houlton et al., 2012; Howarth et al., 2011b; Sutton et al., 2011a). The excessive nitrogen inputs from agriculture activities have been identified as a major contributor to stream nitrogen loadings (Boyer et al., 2006; Garnier et al., 2010; Hatano et al., 2005). The transport of reactive nitrogen from agricultural activities is related to several factors, including fertilization amount, type, and timing of fertilizer application; the method of fertilizer application to soils; soil properties (e.g., pH, particle size distribution, and organic matter content), crop types and their fertilizer requirements; agronomic practices (e.g., irrigation and dry-land farming); and the extent of animal farming and production (Bechmann, 2014; Bechmann et al., 2014; Kyllmar et al., 2014a, 2014b). In addition, weather conditions (rainfall frequency, amount, and intensity) and river basin land use also affect the amount of nitrogen transport to surface water and groundwater systems (Woli et al., 2008; R. Jiang et al., 2014a; Y. Jiang et al., 2014b; Lawniczak et al., 2016). In urban areas, most of the nitrogen consumed in food is converted to human wastes, which is eventually released as treated municipal wastewater to surface waters or leached to groundwater from septic tank systems.

Globally, DIN and particulate nitrogen (PN) each account for approximately 40% of river transported nitrogen, with dissolved organic nitrogen (DON) comprising approximately 20% (Seitzinger & Harrison, 2008). Throughout most of the world, they concluded that nonpoint sources (natural and/or anthropogenic) are the major contributors to nitrogen export, regardless of the reactive nitrogen form. Fertilizer was estimated to be the single most important source of exported DIN in many north temperate latitude watersheds. Based on a review of previous studies regarding nitrogen inputs to landscapes and the associated delivery of nitrogen to coastal waters around the world, Boyer and Howarth (2008) reported that ammonium fluxes typically are much lower than nitrate fluxes in streams and rivers, however, they could represent a substantial proportion of the increased fluxes resulting from anthropogenic activities (e.g., certain urban areas with wastewater inputs).

Seitzinger et al. (2010) used an integrated modeling approach (Global NEWS model) that included socioeconomic factors and four nutrient management scenarios to predict changes in riverine export of reactive nitrogen (along with other nutrients including phosphorus, silica, and carbon) during 2000–2030. Their modeling results for different areas around the world are shown in

Figure 6.4 and demonstrated substantial spatial variation among continents and various regions. For example, the largest increases in DIN occurred in regions with primarily low- to medium-income countries (e.g., Africa, South America) for 2000–2030 under one model scenario but less under a different scenario. These increases were driven by fast growth in crop and livestock production moderated by improvements in agricultural nutrient efficiency (Conant et al., 2013). Model results indicated that global trends in nutrient export were dominated by South Asia, a region with a large and increasing population. South Asia alone accounted for over half of the global increase during 1970–2000 (note scale differences in Fig. 6.4). There were large changes in the relative contribution of watershed nitrogen sources to export of DIN, transitioning from a pattern in 1970 that was like lesser developed continents (South America and Africa) to a pattern by 2000 that is very similar to Europe and North America. These changes occurred as the contribution from fertilizer and manure increased markedly. Sewage remained a relatively small contributor to riverine export of DIN in all scenarios in South Asia, even though population growth is high.

Other forms of reactive nitrogen, such as ammonia can occur in stream ecosystems in two forms, unionized (NH_3) and ionized (NH_4^+). These forms can occur both naturally (e.g., formed during decomposition of nitrogen-containing compounds such as proteins, livestock wastes, and organic nitrogen) and from anthropogenic sources. The unionized form can be toxic to aquatic plants, invertebrates, and fish at concentrations below 1 mg/L (U.S. EPA, 2002). Based on more than 24,000 samples from 499 stream sites in the United States, Dubrovsky et al. (2010) reported that the concentrations of ammonia rarely exceeded levels thought to adversely affect aquatic life. They analyzed ammonia concentrations along with

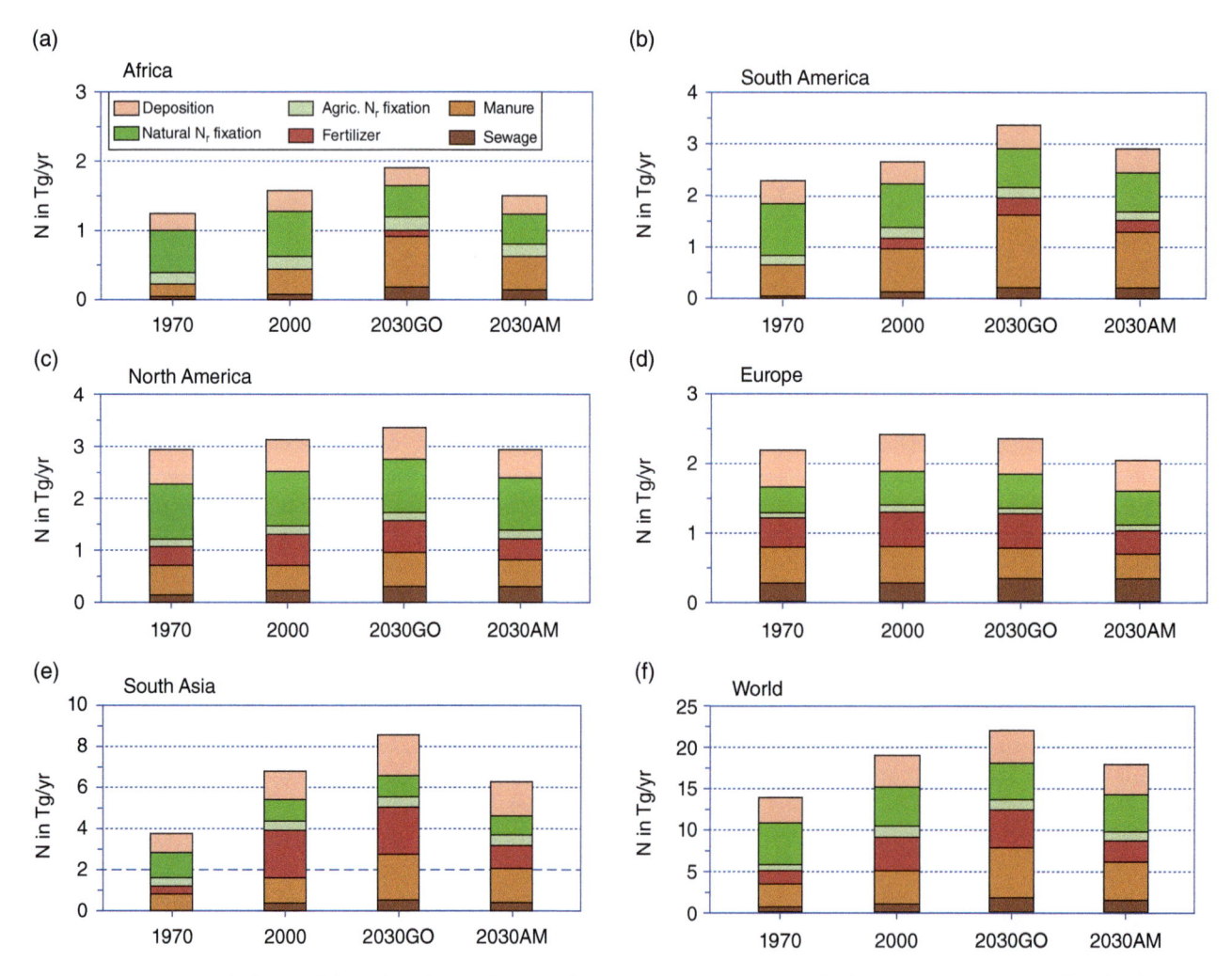

Figure 6.4 (a-f) The model predicted contribution of nitrogen sources in watersheds to riverine export of dissolved sources for various continents, regions, and the world. *Source*: From Seitzinger et al. (2010) (AGU publication). Reproduced with permission of John Wiley & Sons.

pH and temperature with respect to acute and chronic criteria. While acute criteria for ammonia were seldom exceeded, chronic criteria for ammonia were exceeded in 139 samples at 22 sites. These ammonia exceedances of chronic criteria were mainly in urban stream sites or mixed land use areas in the semiarid western United States (most likely from wastewater treatment effluent), although a smaller number were found at agricultural stream sites. The low number of exceedances of the ammonia criteria in agricultural areas most likely indicates that substantial attenuation of ammonia is occurring, either by sorption onto soils, volatilization, nitrification, or uptake by aquatic plants.

Baron et al. (2013) estimated that streams and rivers in the United States denitrify approximately 0.73 Tg N/yr, which they noted is significant relative to nitrogen inputs (roughly 7% of inorganic fertilizer application in the United States. However, the amount of denitrification and nitrogen removal efficiency occurring in streams and rivers is highly variable and is related to several factors including water/sediment contact times, stream temperature, seasonal high/low water conditions, the amount of biogenic nutrients, and respiration rates (Alexander et al., 2009; Boyer et al., 2006). Thus, climate change will have an important effect on nitrogen removal and processing in streams and rivers depending on flow conditions. The amount of denitrification typically increases with increasing nitrate inputs, but the capacity of microorganisms for denitrification might be limited when certain saturation thresholds are reached (Baron et al., 2013).

A recent study (Oelsner & Stets, 2019) determined decadal trends in flow-normalized nutrient and sediment loads for 88 U.S. coastal rivers (95 monitoring stations) for the period 2002–2012. They found that N and P loading from urban watersheds generally decreased during this period. Conversely, in agricultural watersheds, N and P trends in agricultural watersheds were not consistent and indicated disparate improvements in decreasing nutrient loading. Coherent decreases in N loading from agricultural watersheds occurred in the Lake Erie basin, but because P is the primary driver of degradation in the lake, limited benefits are likely. In undeveloped watersheds, the study found that nutrient loading was low, but increased between 2002 and 2012, possibly indicating degradation of coastal watersheds that are minimally affected by human activities. It was found that coastal rivers of the conterminous United States have slightly lower TN yields and slightly higher TP yields compared to global rivers. Oelsner and Stets (2019) noted that recent N:P ratios remained elevated compared to historic values in many areas, even with widespread decreases in N loading in coastal watersheds. They concluded that coastal ecosystems would benefit greatly from additional reductions in N and P loading to U.S. coastal waters, particularly outside of urban areas.

6.4. NUTRIENT-RELATED IMPAIRMENT OF SURFACE WATERS IN THE UNITED STATES

The U.S. EPA lists "nutrient-related" impairment for various water bodies for each state in the "303(d) list," which is compiled from each state's list of impaired and threatened waters (e.g., stream/river segments, lakes). States are required to submit their list every 2 years for approval by the U.S. EPA. The state identifies the pollutant causing the impairment for each impaired water on the list. The EPA list includes waters impaired for nutrients, algal growth, ammonia, noxious aquatic plants, and organic enrichment/oxygen depletion. Impaired waters include those from reporting categories that mostly have a total maximum daily load (TMDL) or need a TMDL. A TMDL (as defined in the 1972 Clean Water Act) describes a plan for restoring impaired waters that identifies the maximum amount of a pollutant that a body of water can receive while still meeting water quality standards. In some cases, the number of assessed waters with a nutrient impairment is underestimated because not all states assess each water for nutrients (https://www.epa.gov/nutrient-policy-data/waters-assessed-impaired-due-nutrient-related-causes#rivers%20, accessed 05 June 2018). According to recent information for U.S. waters, nutrient-related impairment has been reported for 253,210 river kilometers, 19,395 km^2 of lakes and reservoirs area, and 10,152 km of bays and estuaries. According to data from the U.S. Environmental Protection Agency (EPA), states with more than 50% of assessed rivers that have a nutrient-related impairment include Delaware, Florida, Hawaii, Kansas, and Ohio. Agriculture is listed as the source group contributing to the highest numbers of impaired freshwater bodies in the United States. The website (https://ofmpub.epa.gov/waters10/attains_nation_cy.control#imp_water_by_state, accessed 05 June 2018) presents a detailed breakdown of impairment for different types of waterbodies by state, size of assessed waters with causes of impairment, designated uses, approved TMDLs, and other information. Shoda et al. (2019) analyzed water-quality data from 2002 to 2012 at 762 sites in the conterminous United States to determine trends in rivers. They found that the majority of TN and TP trend sites were in exceedance of the U.S. EPA ecoregional criteria. However, where these elevated concentrations of TN and TP were approaching the limit of concern (LOC) from above the LOC, there was a more consistent and more rapid rate of decrease. Many states are spending hundreds of millions of dollars each year to "clean up" HABs in waterbodies caused by excess nutrients. However, there appears to be insufficient efforts to regulate the many activities that contribute nutrients entering these waterbodies. Chapters 10 and 11 discuss in more detail some of the effective strategies and efforts that are being made to reduce nutrient inputs to surface waters and ecosystems.

6.5. IMPACTS TO OTHER SURFACE WATER ECOSYSTEMS

6.5.1. Wetlands

Wetlands (including natural and constructed wetlands) are important sinks for uptake and storage of reactive nitrogen. They also have the highest capacity for nitrogen removal compared to other aquatic ecosystem types (Fig. 6.5) (Baron et al., 2013). In the contiguous United States, Jordan et al. (2011) estimated that wetlands removed 5.8 Tg N/yr, which is more than half the rate of annual inorganic nitrogen fertilizer application in the United States (approximately 11 Tg N/yr, U.S. EPA, 2011). Climate change could have different effects on the capacity of wetland soils and plants to remove nitrogen. In dry periods, wetland soils would have reduced capability to store and remove nitrogen. However, during periods of increased rainfall, flooding could result in an increase in anaerobic sites and more denitrification over larger wetted areas. However, increased denitrification could lead to other adverse environmental effects.

Wetlands also produce the most total N_2O (a potent greenhouse gas), compared to emissions by streams and rivers, reservoirs, lakes, and wastewater treatment facilities (Fig. 6.5). For the entire United States, Baron et al. (2013) estimated an annual aquatic N_2O production rate of 0.58 Tg N/yr for all aquatic system types, which was about 10 times higher than the U.S. greenhouse gas inventory (did not separate out lakes reservoirs, groundwater, streams, or wetlands (U.S. EPA, 2011)). They noted that aquatic N_2O production is almost the same as all terrestrially based N_2O sources in the United States, and concluded that aquatic systems (especially wetlands) constitute an important contributor to the N_2O budget.

Haag and Kaupenjohann (2001) noted that increased nitrogen loads in wetland would contribute to higher nitrogen oxide emission rates due to higher nitrogen availability and conditions promoting denitrification. It was also pointed out that increased CO_2 emissions could also occur due to increased organic matter oxidation in these systems. The increased nitrogen load (both inorganic and organic nitrogen) would adversely affect the ecological status of the wetland (both oligotrophic and mesotrophic wetland systems) and lead to reduced biodiversity and altered structure and function in these ecosystems. Herbert et al. (2015) reported that salinization of inland and coastal wetlands is increasing worldwide at high rates. They attributed salinization of inland wetlands to several factors, including the disposal of wastewater effluent.

Reactive nitrogen removal by aquatic system type
- Streams and rivers 8%
- All reservoirs 9%
- All lakes 19%
- Total wetlands 64%

N_2O production in freshwaters
- Wastewater treatment 14%
- Streams and rivers 8%
- All reservoirs 6%
- All lakes 8%
- Total wetlands 64%

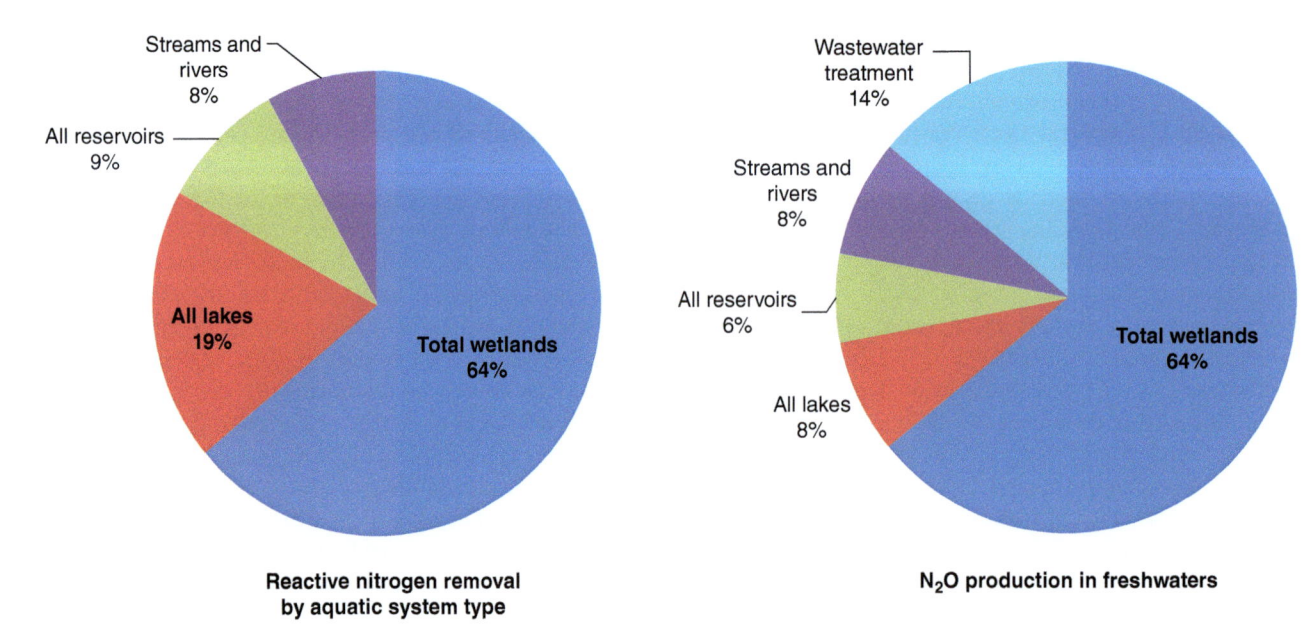

Figure 6.5 Pie diagram on the left shows removal of reactive nitrogen by aquatic system type (through either burial or denitrification) in U.S. freshwater systems. Values are in percent and illustrate the importance of wetlands in removal of reactive nitrogen; pie diagram on the right shows N_2O production in various freshwaters in the United States, in percent. Figure modified more than 20%. *Source*: From Baron et al. (2013). Reproduced with permission of Springer Nature.

6.5.2. Lakes and Reservoirs

Lakes and reservoirs are important sinks for reactive nitrogen. Baron et al. (2013) estimated that lakes and reservoirs in the United States remove approximately 2.6 Tg N/yr. In some lakes and reservoir systems, the nitrogen removal capacity is close to the amount of nitrogen inputs to aquatic systems. Reservoirs are effective sinks for removing nitrogen, and account for more than two thirds of the total nitrogen removed by all lentic waters (still or standing water ecosystems) in the United States even though they only take up about 10% of the lentic surface area in the United States. Nitrogen retention by small lakes and other lentic systems is an important removal process in urban and agricultural areas, especially where increased nitrogen loading has occurred (Seitzinger et al., 2006). However, Jordan et al. (2011) noted there is considerable uncertainty as to the threshold ability of denitrifying microbes to process the increased nitrogen loads.

6.5.3. Coastal Ecosystems, Estuaries, and Oceans

The flux of reactive nitrogen to estuaries and coastal oceans has more than doubled since the industrial and agricultural revolutions (Boyer & Howarth, 2008; Howarth et al., 2011a). Seitzinger et al. (2010) noted that changes in the amount, form (dissolved inorganic, organic, particulate), and ratios in nutrient inputs to coastal ecosystems globally have contributed to numerous adverse human health and environmental impacts, such as loss of habitat and biodiversity, increase in blooms of certain species of harmful algae, red tides, eutrophication, hypoxia, and fish kills (Billen & Garnier, 2007; Diaz & Rosenberg, 2008; Howarth et al., 2000; Rabalais, 2002; Sutton et al., 2013a; Turner et al., 2003). Acidification of surface waters (lakes, streams, and oceans) in many places around the world also has increased and has had detrimental effects on biota (Curtis et al., 2005; Doney et al., 2007). Lefcheck et al. (2018) noted that nutrient pollution is a main threat to nearshore ecosystems, such as marshes, mangroves, kelps, and seagrasses. Studies have shown that the global cover of seagrasses has declined more than 29% during the last century due to nutrient and sediment runoff. Based on 31 years of water-quality data (1984–2014), Lefcheck et al. (2017) reported that seagrass beds have declined substantially since 1991 in Chesapeake Bay. Seagrass beds provide important environmental and ecosystem services, such as a sink for global carbon, habitat for fish and crabs, and filters for water clarity.

Two thirds of estuaries in the United States have been degraded from excess reactive nitrogen (Bricker et al., 2007; U.S. EPA, 2011). The resulting large losses in aquatic habitats have essentially reduced the capability of many ecosystems to assimilate, attenuate, and denitrify reactive nitrogen (Baron et al., 2013). In estuaries and continental shelves, denitrification occurs mainly in the suboxic waters of estuaries (Seitzinger et al., 2006). However, the amount of denitrification is highly dependent on residence time, which is related to river flows, tidal flushing, and geomorphology.

Higher nitrogen deposition rates in the future and climate change likely would lead to an increase in ocean acidification. Acidification of oceans results in considerable damage to calcifying shellfish and crustaceans (Howarth et al., 2011a). Impacts on the nitrogen-cycle from acidification include pH-dependent reductions in nitrification rates and enhancement of open ocean nitrogen-fixation (Beman et al., 2011; Doney et al., 2009; Levitan et al., 2007). Eutrophication increases the vulnerability of coastal ecosystems to ocean acidification through interactions between low oxygen levels and increases in inorganic carbon (carbon dioxide) (Howarth et al., 2011a). Therefore, carbon chemistry changes from ocean acidification would be disproportionately large in hypoxic water bodies. Coastal upwelling along shelves and estuaries where eutrophication has occurred has exhibited partial pressure CO_2 levels in excess of values that were not anticipated to occur in the mean surface ocean until the next century (Feely et al., 2009).

6.6. HARMFUL ALGAL BLOOMS

Excess nitrogen in water bodies can promote the excess growth of microalgae or phytoplankton. When algae grow to levels that are substantially higher than average conditions, this can lead to a bloom. Harmful algae in blooms can include some species of cyanobacteria, some diatoms, and dinoflagellates (Paerl et al., 2016; https://www.whoi.edu/redtide/page.do?pid=9257, accessed 27 April 2019). The proliferation of harmful algal blooms (HABs) has occurred all over the world (Erisman et al., 2015), and HABs have been reported in every state in the United States during the past 10 years (U.S. EPA, 2009). These HABs can produce toxic chemicals, nasty odors when they decompose, and cause massive fish kills.

In addition to fish kills and adverse effects on other aquatic biota, nonaquatic animals also are affected when consuming waters containing algae. Pets and livestock have died after drinking water containing algal blooms (algal toxins), including 32 cattle on an Oregon ranch in July 2017 (Flesher & Kastanis, 2017). It was also noted that more than 100 people fell ill after swimming in Utah's largest freshwater lake.

Toxic algal blooms have contaminated waterways from the Great Lakes to Chesapeake Bay, from the Snake River in Idaho to New York's Finger Lakes and reservoirs in California's Central Valley (Flesher & Kastanis, 2017). As of mid-August 2016, the U.S. EPA noted that states across the United States have reported more than 250 health advisories due to HABs that year. Large algal blooms occurred in the Black Sea in the 1970s and 1980s following the intensive use of fertilizers and livestock production in the Black Sea basin in the 1960s (Bodeanu, 1993; Mee et al., 2005). Most of Europe has a high potential risk of eutrophication of surface freshwaters (Grizzetti et al., 2011) as well as South America, Australia, Southeast Asia, and the United States (Erisman et al., 2015) (Fig. 6.6). The National Aquatic Resource Surveys conducted by the EPA and state and tribal partners in 2012 found that 34% of the lakes surveyed in the United States had high levels of nitrogen associated with harmful ecological impacts. Toxic HABs have expanded to tens of km^2 in areal extent off the coast of the East China Sea and their duration has increased from days to months. These changes have been related to increases in fertilizer use over the past two decades (Li et al., 2009).

Algal blooms have resulted in numerous beach closures in Florida and in 2016, the governor declared a state of emergency in four coastal counties in southeast Florida. Lake Okeechobee in southeastern Florida is the third largest freshwater lake in the United States, and covers an area of about 1900 km^2. Nutrient inputs to the lake have resulted in large algal blooms. Even when algae are not visible on the surface of the lake, cyanobacteria (blue-green algae) have been found in the water column. The Florida governor declared a state of emergency in mid-July 2018 after a NOAA satellite image of the lake from 2 July 2018, showed varying concentrations of cyanobacteria in about 90% of the lake. Some of the types of cyanobacteria known to live in the lake can produce algal toxins under certain conditions. Research scientists from the University of Miami and the Institute for Ethnomedicine were featured in a film that documents he connections between chronic human exposure to toxic algae and increased risks of ALS Alzheimer's disease, and Parkinson's disease related to a toxin called BMAA (β-N-methylamino-L-alanine) (Cox et al., 2016) produced by some algal blooms (https://www.toxicpuzzle.com/, accessed 22 July 2018). Other cyanotoxins found in Florida water pose other health problems, such as gastrointestinal and respiratory problems as well as liver damage. These toxins also have been linked to an increased risk of terminal liver failure and liver cancer. Scientists are warning that any *Microcystis aeruginosa* bloom should be avoided due to its potential toxicity. This warning also included no swimming, no fish consumption, and avoid breathing the toxin that can become airborne (http://flylifemagazine.com/lets-go-fishing-its-only-cyanobacteria/, accessed 22 July 2018).

In the United States, the Interagency Working Group on Harmful Algal Blooms, Hypoxia, and Human Health (IWG-4H), provides Federal coordination for HAB research and response. This group has assessed HABs in freshwater and marine systems. For freshwater systems,

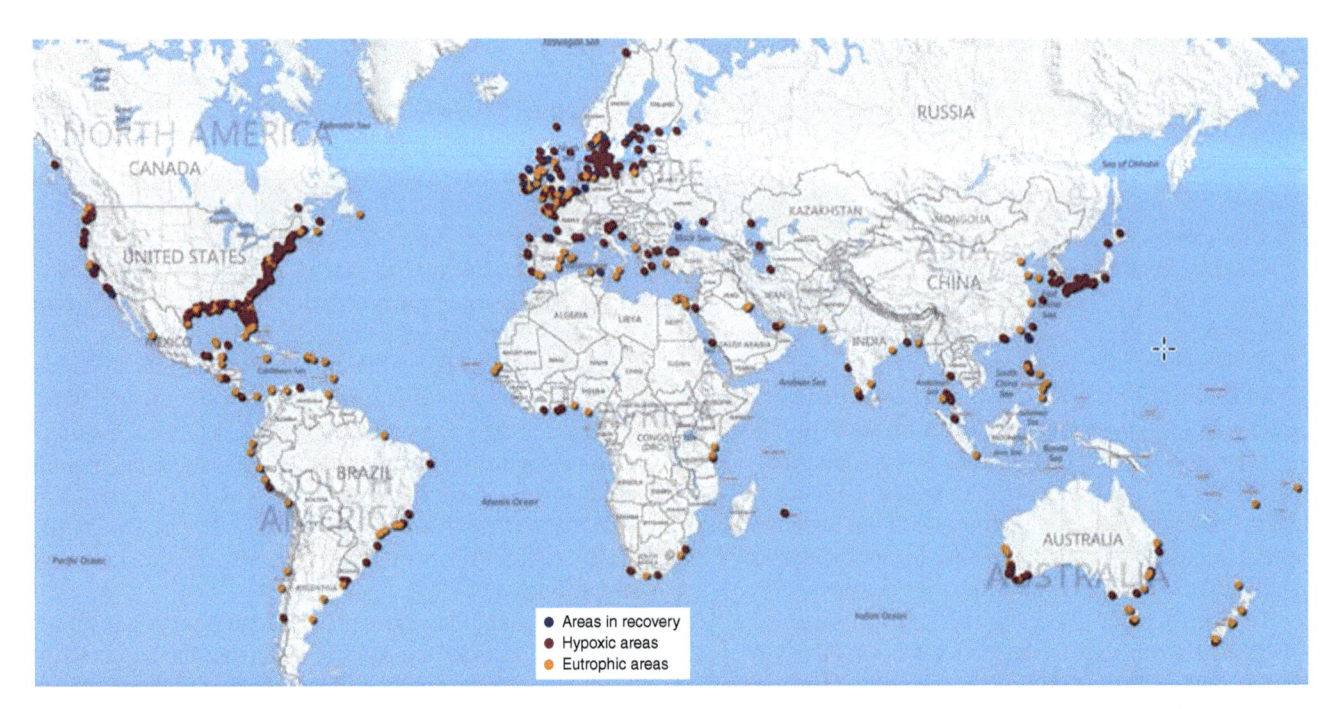

Figure 6.6 Map showing hypoxic and eutrophic coastal areas around the world. *Source*: Modified from the colored figure 4 in Erisman et al. (2015). Reproduced with permission of World Wildlife Fund.

Lopez et al. (2008a) reviewed studies that documented adverse ecosystem impacts stemming from nontoxic high biomass blooms of cyanobacteria. They noted that high biomass blooms have caused low oxygen conditions that and produce toxins that kill fish and bottom dwelling organisms. In addition, their review indicated that dense blooms blocked sunlight penetration, which prevented the growth of other algae and had disruptive impacts on food webs. They indicated that *Prymnesium parvum*, also called "golden algae," was the most severe type, which has caused fish kills in Texas annually since 2001 and has been documented in at least nine other states (Lopez et al., 2008a).

In their assessment of marine HABs in the United States, Lopez et al. (2008b) discussed several topics including the legislative background and report process, the state of the marine HAB problem in the United States, causes and consequences, federal research accomplishments on marine HABs in the United States, and coordination and communication for marine HABs. Briefly, they noted that in the United States, all coastal states have experienced HABs during the 1990s and early 2000s (Fig. 6.7), with new species showing up in some locations that previously did not have known HAB problems. They noted that marine HABs can occur naturally, but human actions that disturb ecosystems (such as increased nutrient loadings and pollution, food web alterations, introduced species, and water flow modifications) have contributed to the increased occurrence of some HAB species.

In the United States two national programs, Ecology and Oceanography of Harmful Algal Blooms (ECOHAB) Program and Monitoring and Event Response for Harmful Algal Blooms (MERHAB) Program, have funded approximately $100 million for marine HAB research since the programs began in 1996 and 2000, respectively. In addition, at least 13 Federal agencies conduct significant marine HAB research as part of other research programs. Urquhart et al. (2017) describe a method using satellite data to quantify cyanobacterial extent in HABs. Other tools have been developed for controlled testing of environmental factors that regulate growth and toxin production of red tide alga. Also, genetic research methods have helped to identify harmful algal genes and improve the understanding of the biological aspects of bloom development. Ongoing research on the role of nutrients has demonstrated links between increased anthropogenic nutrient loadings and prevalence of some species, such as *Pfiesteria* spp., some macroalgae, and Pseudo-nitzschia (can produce a neurotoxin called domoic acid) in the Gulf of Mexico. Models that incorporate biological and physical data in other U.S. coastal areas have explained patterns and spatial variations in shellfish toxicity and have been used to predict weekly bloom status to resource managers since 2005. These model predictions provide critical information to water resource managers for closures of shellfish harvesting areas, which help to reduce economic impacts on local communities in addition to protecting human health.

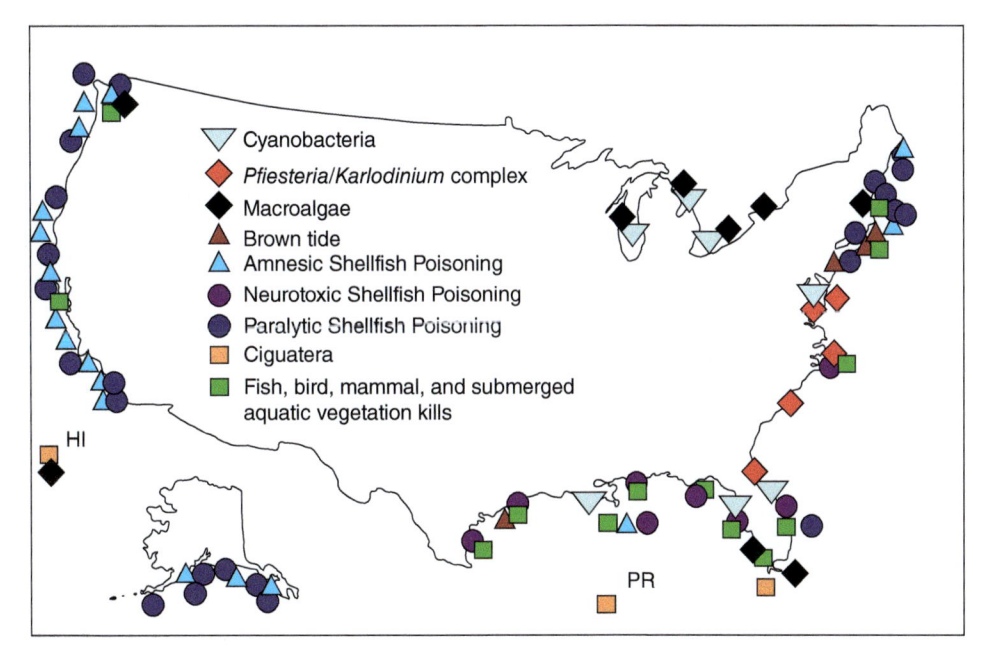

Figure 6.7 Types of harmful algal bloom (HAB) events in coastal areas of the United States. *Source*: From Lopez et al. (2008b).

For stream systems in the United States, the U.S. Geological Survey National Water Quality Assessment Program demonstrated that the effects of nutrient enrichment on algal biomass (as indicated by chlorophyll *a*) were not uniform. Dubrovsky et al. (2010) concluded that regional variations indicated that other factors influence algal communities, such as streamflow, canopy cover, water temperature, and clarity. Based on results from a regression model (using data from 143 sites in 5 different agricultural areas across the United States) they found a weak relation between phosphorus concentrations and algal biomass. Specifically, only 12% of the variability in algal growth was explained by concentrations of phosphorus. The relation was even weaker for nitrogen concentrations. They explained that weak relations in some regions might be related to elevated nutrient concentrations that were greater than the requirements for the plants, therefore, additional increases in nutrients could have had little effect on plant biomass. The strength of the regression model improved (explaining 46% of the algal biomass variability) when stream characteristics (e.g., water temperature and canopy cover) were included in the analysis.

Another severe impact of excessive amounts of reactive nitrogen and the decomposition of algae is the depletion of oxygen (hypoxia), which has led to large dead zones in estuaries and other large water bodies. Since 1950, Breitburg et al. (2018) noted that more than 500 sites in coastal waters worldwide have reported oxygen concentrations ≤2 mg/L (=63 mmol/L), a threshold often used to delineate hypoxia. It was pointed out that less than 10% of these systems were known to have hypoxia before 1950. Recent literature indicates that the number of dead zones could possibly double each decade (Conley et al., 2009; Diaz & Rosenberg, 2008). Figure 6.6 shows a global map of low oxygen zones (2 mg/L of dissolved oxygen or less) and areas of low oxygen zones in the open ocean (Global Ocean Oxygen Network, GO$_2$NE https://en.unesco.org/go2ne, accessed 27 April 2019). These low oxygen zones affect marine ecosystems thereby resulting in less food production from loss of fisheries. Also, in some areas, GO$_2$NE has noted that upwelling of low oxygen waters have caused large fish kills. In the United States, more than 166 dead zones have been identified and have affected large water bodies such as the Chesapeake Bay and the Gulf of Mexico. The dead zone in the Gulf of Mexico grew to approximately 5840 mile^2 in 2013. This extremely large dead zone results from summertime nutrient pollution from the Mississippi River (excess reactive nitrogen most likely from fertilizers), which drains 31 upstream states (https://www.epa.gov/nutrientpollution/sources-and-solutions). Other coastal areas where large increases in dead zones have been noted are the Adriatic Sea, the Black Sea, the Kattegat, and the Baltic Sea (Diaz & Rosenberg, 2008).

6.7. MODELING OF REACTIVE NITROGEN TRANSPORT IN SURFACE WATER SYSTEMS

As noted earlier in this chapter and in Chapter 2, the nitrogen cycle in rivers and other surface water systems has been substantially altered by excess reactive nitrogen from agricultural land uses, urban waste disposal and other activities, and atmospheric deposition. Documenting these alterations across large river networks is a major challenge; therefore, simulation models are important tools for assessing biogeochemical processes that affect reactive nitrogen export in surface water systems and for predicting future changes. Numerous models have been utilized for evaluating nutrient fate and transport at the regional river basin and global scales. Billen et al. (2011) compiled a list and provided a brief description of 10 watershed models including their geographical resolution, data input requirements, basic equations, and principles for representing nutrient transfer, temporal resolution, and other variables. These models all have been validated in a variety of catchment areas and could be used to predict nutrient export from land use and point sources. They all differ in complexity in terms of representing processes and temporal resolution. Some of the earlier riverine nitrogen models concentrated on estimating nitrogen export from large watersheds (Alexander et al., 2002a). More recently, models have focused on biogeochemical processes, such as denitrification, that reduce downstream nitrogen transport (Alexander et al., 2009; Helton et al., 2011; Seitzinger et al., 2010). Also, several models are increasingly being used effectively for identifying critical areas and timing for nitrogen mitigation efforts and their cost-effectiveness. Several mitigation options have been evaluated, such as pollution swapping, considering potential trade-offs in various environmental objectives (e.g., Bouraoui & Grizzetti, 2014; Reis et al., 2016).

This chapter briefly describes selected models used to estimate reactive nitrogen loads as various scales from rivers, watersheds, and globally. Also discussed are models used to assess how future changes in climate may affect reactive nitrogen transport in surface water systems. References and websites are included for the reader to obtain more detailed information about the various models used in the cited studies.

6.7.1. Regional Scale Models of Nitrogen Transport in Watersheds and Large Rivers

Many of the surface water models have been developed and used for simulating nitrogen transport in large-scale watersheds in Europe, the United States, and China. In Europe, the Nitrates Directive (ND) is one of a series of environmental policies that have been implemented in the European Union (EU) to decrease nitrogen emissions and reduce nitrogen runoff and leaching from agriculture. The Water Framework Directive is the main policy device for protecting inland, transitional, and coastal water resources in Europe. The EUROHARP project promoted nine different contemporary methodologies for quantifying diffuse losses of nitrogen and phosphorus in 17 study catchments across gradients in European climate, soils, topography, hydrology, and land use (Bouraoui et al., 2009). These models were applied at the catchment scale and were used to inform policy makers at national and international levels (http://www.eugris.info/index.asp, accessed 12 December 2018). Bouraoui and Grizzetti (2014) reviewed several models for quantifying the effectiveness of mitigation measures to reduce nitrogen losses from agricultural areas to surface and groundwaters. Their evaluation indicated the importance of factoring in the lag time between the implementation of mitigation measures (e.g., fertilizer management, construction of riparian areas and wetlands) and effects on water quality.

Velthof et al. (2014) conducted a study to quantify runoff to surface waters and leaching to groundwater, and on the emissions of ammonia (NH_3), nitrous oxide (N_2O), nitrogen oxides (NO_x), and dinitrogen gas (N_2) to the atmosphere for 27 countries in the European Union (EU-27). They investigated two scenarios, one with and one without implementation of the ND. The model MITERRA-Europe was used to calculate nitrogen emissions on a regional level in the EU-27 for the period from 2000 to 2008. The model estimated a total N loss from agriculture in the EU-27 of 13 million tons (Mton) in 2008, with 53% as N_2, 22% as nitrate, 21% as ammonia, 3% as N_2O, and 1% as NO_x. Overall, there was a slight decrease in nitrogen emissions and leaching in the EU-27 during 2000–2008. With implementation of the ND, the total nitrogen emissions in the EU-27 in 2008 were smaller than without the ND (by 3% for ammonia, 6% for N_2O, 9% for NO_x, and 16% for N leaching and runoff). Large regional differences were found that was related to nitrogen inputs from fertilizers and manures. They concluded that implementation of the ND reduced both nitrogen leaching losses to ground and surface waters, and gaseous emissions to the atmosphere. Also, as more stringent measures are implemented as part of the ND, further decreases in nitrogen emissions would be expected for EU-27.

In the United States, the 1972 Clean Water Act (CWA) established the structure for regulating discharges of pollutants into the waters of the United States and regulating quality standards for surface waters. As part of the CWA, the U.S. EPA has implemented pollution control programs (e.g., setting wastewater standards for industry) and has developed national water quality criteria recommendations for reactive nitrogen and other pollutants in surface waters. Although progress has been made since the enactment of the CWA, elevated levels of reactive nitrogen in large river systems have resulted in widespread environmental damage. For example, the Mississippi River transports large nitrogen loads to the Gulf of Mexico that have resulted in a large dead zone. Other coastal areas in the United States affected by excess nutrients and anoxic zones are shown in Figure 6.8.

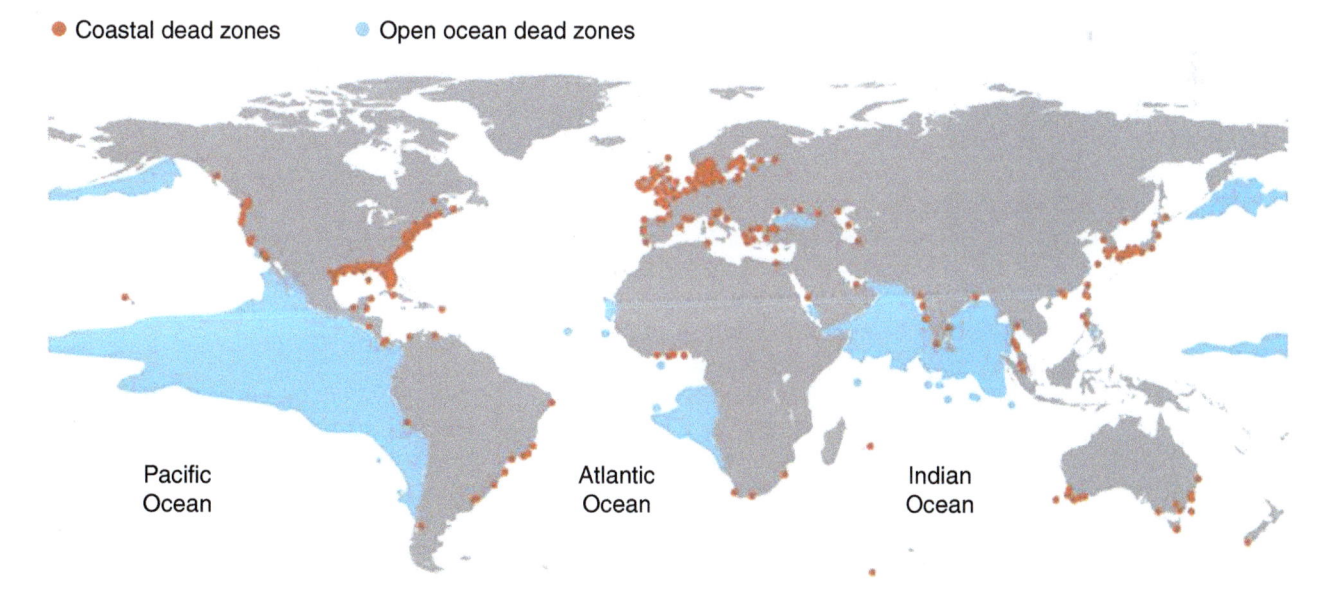

Figure 6.8 Map from Global Ocean Oxygen Network (https://en.unesco.org/go2ne) showing coastal dead zones (red) and open ocean dead zones (blue). Over 500 coastal water bodies now have dissolved oxygen levels below 2 mg/L (Diaz & Rosenberg, 2008). *Source*: Reproduced with permission of American Association of the Advancement of Science.

The U.S. EPA Center for Exposure Assessment Modeling (CEAM) dispenses simulation models and database software designed to quantify movement and concentration of various contaminants in lakes, streams, estuaries, and marine environments (https://www.epa.gov/ceam/surface-water-models-assess-exposures, accessed 15 January 2019). CEAM lists 21 surface water models that were developed by the U.S. EPA in conjunction with other government, academic, and commercial institutions. Many of these models, such as HSPF, can be used to assess biology, compliance, deposition, discharge, environmental effects on estuaries, monitoring, and TMDLs.

As part of the National Water Quality Assessment Program (NAWQA) in the United States, the U.S. Geological Survey developed the model SPARROW attributes (https://pubs.usgs.gov/tm/2006/tm6b3/; and https://water.usgs.gov/nawqa/sparrow/FAQs/faq.html, accessed 15 January 2019). This statistically based regression model uses explanatory variables to obtain correlations of stream water-quality monitoring data with watershed nutrient sources and land use properties to provide empirical estimates of stream export of nutrients (nitrogen and phosphorus) from inland watersheds to larger surface water bodies. Access SPARROW Mappers are interactive tools that can be used to estimate river nutrient loads and yields and the contribution from different sources in a river basin. The user can visualize water quality data using maps and interactive graphs and tables, and relative amounts of nitrogen can be presented for the entire country, state or province, hydrologic unit code, and catchment. SPARROW models have been developed for nitrogen and phosphorus for the conterminous United States and for many basins throughout the United States (https://water.usgs.gov/nawqa/sparrow/#, accessed 15 January 2019). SPARROW models have been used at the U.S. scale to estimate background levels of nitrogen in streams (Smith et al., 2003), which can help in the development of nutrient criteria in streams and asses the effectiveness of agricultural management practices to reduce the amount of nitrogen and phosphorus entering streams (Garcia et al., 2016). They also have been used in other parts of the world (e.g., New Zealand (Alexander et al., 2002b); and Canada (Benoy et al., 2016)) where adequate water-quality, land use, and nutrient source data were available for developing the model. SPARROW also has been used to assess how stream size influences denitrification and attenuation of nitrogen levels with transport downstream (Alexander et al., 2000).

In 2013, the U.S. Geological Survey, the U.S. EPA, and the National River and Stream Assessment collaborated to evaluate stream water quality across the Midwestern United States. This Midwest Stream Quality Assessment (MSQA) covered about 600,000 km² in parts of 11 states (Garrett et al., 2017). The MSQAs comprehensive data base included habitat characterization, weekly samples of water and sediment chemistry, and biological communities (algae, invertebrates, and fish) in 100 small streams (50 shared with NRSA data collection). In the Midwest United States, stream water quality and biological condition in streams have been substantially affected by intensive agricultural practices and urbanization. For the various watersheds, Schmidt et al. (2018) used structural equation modeling (SEM) to hypothesize causal linkages between human development, physical and chemical stressors in streams, and stream ecological health. SEM is a statistical modeling method that explicitly connects empirical data to a theoretical model of interrelated factors.

Schmidt et al. (2018) found that seven environmental variables (agricultural and urban land use, sand content of soils, basin area, percent riparian area as forest, channel erosion, and relative bed stability) were important for all three-community metrics. Their results indicated that the ecological integrity of Midwest streams was affected by both agricultural and urban land uses and by the natural geologic setting, as indicated by the sand content of soils. Different forms of reactive nitrogen had contradictory effects on invertebrate communities. They found a negative correlation between ammonia and invertebrate community condition, although ammonia concentrations were far below toxic levels. Conversely, TN was positively correlated with invertebrate community condition, even though dissolved nitrate concentrations were near acutely toxic thresholds in many streams (Van Metre et al., 2016). It was suggested that interactions with other stressors (e.g., insecticides in Midwest streams affecting invertebrate communities) might explain these contradictory findings. Schmidt et al. (2018) concluded that chemical exposure, particularly to nutrients and pesticides, appears to affect the condition of biological communities. Therefore, they emphasized that reducing nutrient and pesticide loading would have an advantageous effect on stream health at local scales and in downstream receiving waters.

6.7.2. Modeling Nitrogen Attenuation in Large River Systems

One of the uncertainties in modeling nitrogen dynamics in large river systems is the ability to accurately quantify nitrogen retention and distinguish between the temporary and permanent removal of nitrogen. Grizzetti et al. (2015) focused their nitrogen retention study on the large Seine River basin in France (approximately 65,000 km²). They evaluated three widely used models in Europe and elsewhere with different levels of complexity (GREEN, SWAT, and RiverStrahler models) and compared their

Table 6.1 Annual average values (kg N/ha) for modeled nitrogen fluxes and retention in the Seine River basin using the GREEN, RiverStrahler, and SWAT models for the period 1995–2005.

Fluxes/retention	GREEN	RiverStrahler	SWAT
Diffuse sources	93	28	105
Point sources	6	8	8
Diffuse emissions	13	16	18
Basin retention	80		
River retention	5	3	2
Crop uptake and soil retention			87
Riparian retention		12	
Load	14	21	24

Source: From Grizzetti et al. (2015). Reproduced with permission of IOP Publishing.

estimates of water nitrogen retention over an 11-year period (Table 6.1). Nitrogen river retention calculated by the three models varied between 8% and 26% of the total load entering the river. However, the sensitivity of the model parameters was not evaluated or an uncertainty analysis was not performed. They estimated that the error associated with the average annual load was about 10% for the RiverStrahler and SWAT models, but around 40% for the GREEN model. The estimated nitrogen total removal by denitrification in the Seine River network ranged from 0.1 to 3.1 kg N/ha/yr.

Grizzetti et al. (2015) emphasized the challenges in modeling nitrogen retention in large river basins due to the difficulty in distinguishing between permanent and temporary removal of nitrogen. Also, the role of groundwater in storing and releasing nitrogen to the stream water system is highly variable spatially and challenging to quantify. This is due in part to the paucity of data and information about groundwater residence time at the basin scale, and the time lag regarding the delivery of nitrogen through the unsaturated zone to the groundwater system. Models that have addressed some of these groundwater related issues are discussed later in this chapter. It was concluded that models can provide estimates of nitrogen fluxes and pathways that contribute to emissions to rivers and the atmosphere; however, it may not be possible to test the accuracy or reliability of any model against observed data because of the lack of measurements at the river basin scale. A comparison of results from different models (along with uncertainty and sensitivity analyses) provides useful information about where additional investigations are needed.

6.7.3. Global Scale Models of Reactive Nitrogen Transport and Fate

Riddick et al. (2016) developed the Community Land Model (CLM) to assess the magnitude, temporal variability, and spatial heterogeneity of terrestrial nitrogen pathways and ammonia (NH_3) emissions on a global scale from fertilizers and animal manure sources. Their model focused on an explicit climate-dependent approach and included manure management and fertilizer application practices. For the year 2000, they reported approximately 125 Tg N/yr applied as manure and 62 Tg N/yr applied as synthetic fertilizer. The resulting global emissions of ammonia were 21 and 12 Tg N/yr from manure and fertilizer application, respectively. Reactive nitrogen runoff during rain events was estimated at 11 Tg N/yr from manure and 5 Tg N/yr from fertilizer. It was estimated that the remaining nitrogen from manure (93 Tg N/yr) and synthetic fertilizer (45 Tg N/yr) was "captured by" the canopy or transferred to the soil nitrogen pools. Also, they used a transient simulation from 1850 to 2000, which showed that ammonia (NH_3) emissions from manure increased from 14% of nitrogen applied in 1850 (3 Tg NH_3/yr) to 17% of nitrogen applied in 2000 (21 Tg NH_3/yr). Riddick et al. (2016) estimated that an additional 1 Tg NH_3/yr (approximately 3% of manure and fertilizer) would result per degree Celsius of global warming.

As mentioned earlier in this chapter, Seitzinger et al. (2010) used an integrated modeling approach to relate socioeconomic factors and nutrient management options to river export of nitrogen, phosphorus, silica, and carbon based on an updated Global NEWS model. They investigated past trends from 1970 to 2000 and four Millennium Ecosystem Assessment future scenarios (briefly mentioned in this chapter; please see Seitzinger et al. (2010) for more detailed explanation): (a) Global Orchestration, (b) Order from Strength, (c) Technogarden, and (d) Adapting Mosaic. It was recognized that the Global Orchestration scenario depicts a globally connected society that "focuses on global trade and economic liberalization and takes a reactive approach to ecosystem problems, but also takes strong steps to reduce poverty and inequality and to invest in public goods, such as infrastructure and education." The Order from Strength scenario relates to a "regionalized and fragmented world, concerned with security and protection, with the emphasis primarily on regional markets, paying little attention to public goods, and taking a reactive approach to ecosystem problems." The Technogarden scenario describes a "globally connected world, relying strongly on environmentally sound technology, using highly managed, often engineered, ecosystems to deliver ecosystem services, and taking a proactive approach to the management of ecosystems to avoid problems." The Adapting Mosaic scenario portrays regional watershed scale ecosystems as the focus of political and economic activity. Major differences among the four scenarios incorporated in the NEWS model were related to the total crop and livestock production, nutrient use efficiency in agriculture, nutrient releases from sewage, energy use, per capita income, and river discharge especially as it relates to consumptive water use for irrigation.

Seitzinger et al. (2010) found that differences among the various scenarios for nutrient management in agriculture were a major factor controlling the change (magnitude and direction) of future DIN river export. Population changes were an important underlying factor for river export of nitrogen and other nutrients in all scenarios. Large differences were noted among regions around the world. For example, their model show that South Asia alone accounted for over half of the global increase in DIN and dissolved inorganic phosphorus in river export between 1970 and 2000 and in the subsequent 30 years under the Global Orchestration scenario (globally connected with reactive approach to environmental problems). The model results also indicated that DIN river export decreased in the Adapting Mosaic (globally connected with proactive approach) scenario by 2030. It was concluded that risks for coastal eutrophication from riverine DIN loads will likely continue to increase in many regions around the world for the foreseeable future resulting from combined increases in magnitude and changes in nutrient ratios in river nitrogen and phosphorus export.

Fowler et al. (2015) also used an integrative modeling approach and combined information from new and existing models, extensive data analysis, and a review of published literature to assess the responses of the flow of nitrogen through terrestrial and marine ecosystems and the atmosphere to projected changes in climate in the 21st century. They considered that climate change mainly related to the environmental drivers, temperature and rainfall (amount, frequency, seasonal distribution), both of which affected soil environmental conditions, landscape hydrology, vegetation cover, and substrate supply. In addition, they included farming and land management practices that were modified as a result of climate changes. They projected an increase in the terrestrial reactive nitrogen flux from 240 to 328 Tg N/yr from 2010 to 2100. Likewise, their modeling indicated an increase in atmospheric nitrogen flux (wet and dry deposition) from 100 to 120 Tg N/yr during 2010–2100 that would result in more nitrogen transported to oceans and coastal ecosystems. Fowler et al. (2015) concluded that the increased per unit area emissions of reactive nitrogen from farming activities to the atmosphere, soils, and freshwaters are much higher relative to the fluxes involved in biological nitrogen fixation for natural ecosystems and significantly contribute to hot spots for reactive nitrogen emissions and deposition, damage to ecosystems, and adverse effects on human health.

6.7.4. Modeling Reactive Nitrogen in Surface Waters in China

The Chinese Central Government recently announced major policies regarding pollution control and remediation (Han et al., 2016). These include the Water Pollution Prevention and Control Action Plan ("10-Point Water Plan"), which was released in April 2015 (Central People's Government of the People's Republic of China, 2015). This comprehensive policy attempts to address the severe pollution of surface water and groundwater in China. China has a standard 6-class rating system for surface water quality: Class I (pristine water) water quality decreasing to Class VI (not useable). Han et al. (2016) reported that out of 208,000 km of monitored river reaches in China, currently water quality in 31.4% reaches falls into Class IV (industrial water supply) or worse, and therefore unfit for potable use or human contact. Water quality in 14.9% of river reaches is inferior to Class V, indicating complete loss of potential for all consumptive uses or human contact. Of ten major watershed areas, only in the southwest and northwest is water quality in the majority of rivers rated as high to moderate (Classes I–III), while the major northern river systems, for example, the Yellow, Liaohe, and Huaihe Rivers, are rated as Class IV or Class V, and the Haihe River as Class VI. Water in six of nine major coastal bays in China was characterized as "poor" or "very poor" (Class IV or Class V). In the Fall of 2014, the combined coastal area with water quality in Class IV or Class V (unfit for human contact) covered 57,000 km^2. These large river systems drain major industrial and agricultural regions of China that have numerous upstream pollution sources. Han et al. (2016) also pointed out that the quality of water in small tributaries that flow into China's main river systems generally has poorer quality than the larger river systems.

Wang et al. (2014) used a mass-balance model that show nitrogen input to the whole Yangtze River basin had increased by a factor of two over a 20-year period from 1990 to 2010. This likely is related to the increase in fertilizer use in China, which had increased from 0.5 Mt (megatonnes) in the 1960s to 42 Mt in 2010 along with urea applications that increased by a factor of five in the last two decades (Glibert et al., 2014).

6.7.5. Including Groundwater Transport and Lag Times in Surface Water Models for Developing More Effective Management Strategies

Several studies have shown limited progress in achieving water quality improvements with nitrogen reduction goals in large watersheds due to storage of reactive nitrogen and lag times in nitrogen transport to surface waters (e.g., Meals et al., 2010). For example, in the Netherlands, a phased program was designed and started in 1985 to achieve water quality goals for nitrogen and phosphorus in 1985 (Boers, 1996); more than 20 years later, however, only 25% of surface waters in the Netherlands met the standards set for nitrogen and phosphorus (van Puijenbroek et al., 2014). In the United States, Van Meter et al. (2017) noted that challenging goals were set in the 1980s to reduce "controllable" nitrogen and phosphorus loading to the Chesapeake Bay by 40%, and in 2008, the Mississippi River/Gulf of Mexico Waters. For the Gulf of Mexico, the goal was to reduce the size of the summer hypoxic zone to 5000 km² by 2015. In August 2017, the Gulf of Mexico's hypoxic zone was the largest ever measured (22,720 km²) (Fig. 6.9) since mapping began in 1985, and more than four times the Task Force goal of 5000 km² (https://www.noaa.gov/media-release/gulf-of-mexico-dead-zone-is-largest-ever-measured, accessed 25 January 2019). Clearly, the nutrient reduction and water quality improvement goals have not been achieved in both large basins and target dates have been extended up to two decades.

To better understand why nutrient reduction goals have not been achieved for the desired water-quality effects in large river basins, Van Metre et al. (2016) modeled impacts from intensive agriculture throughout the Mississippi River basin. They analyzed long-term soil data (1957–2010) from 2069 sites and showed nitrogen accumulation in the cropland of 25–70 kg/ha/yr, amounting to a total of 3.8 ± 1.8 Mt N/yr at the watershed scale. Their detailed model indicated that the accumulation of soil organic nitrogen (SON) over a 30-year period (142 Tg N) would result in a lag time of 35 years for 99% of legacy SON, even with a complete cessation (no further) of fertilizer applications. Van Meter et al. (2018) estimated that a 60% decrease in watershed nitrogen (N) loading may be necessary to adequately reduce eutrophication in the Gulf of Mexico. However, they noted that little work has been done regarding the effect of nitrogen legacies on achieving water quality goals. In their study involving extensive modeling of future scenarios, Van Meter et al. (2018) demonstrated that even if agricultural N use became 100% efficient, it would take decades to meet target N loads due to legacy nitrogen and release from storage within the Mississippi River basin. They pointed out that there is a substantial amount of nitrogen from runoff that has accumulated in the Mississippi River basin, and even if future agricultural nitrogen inputs are

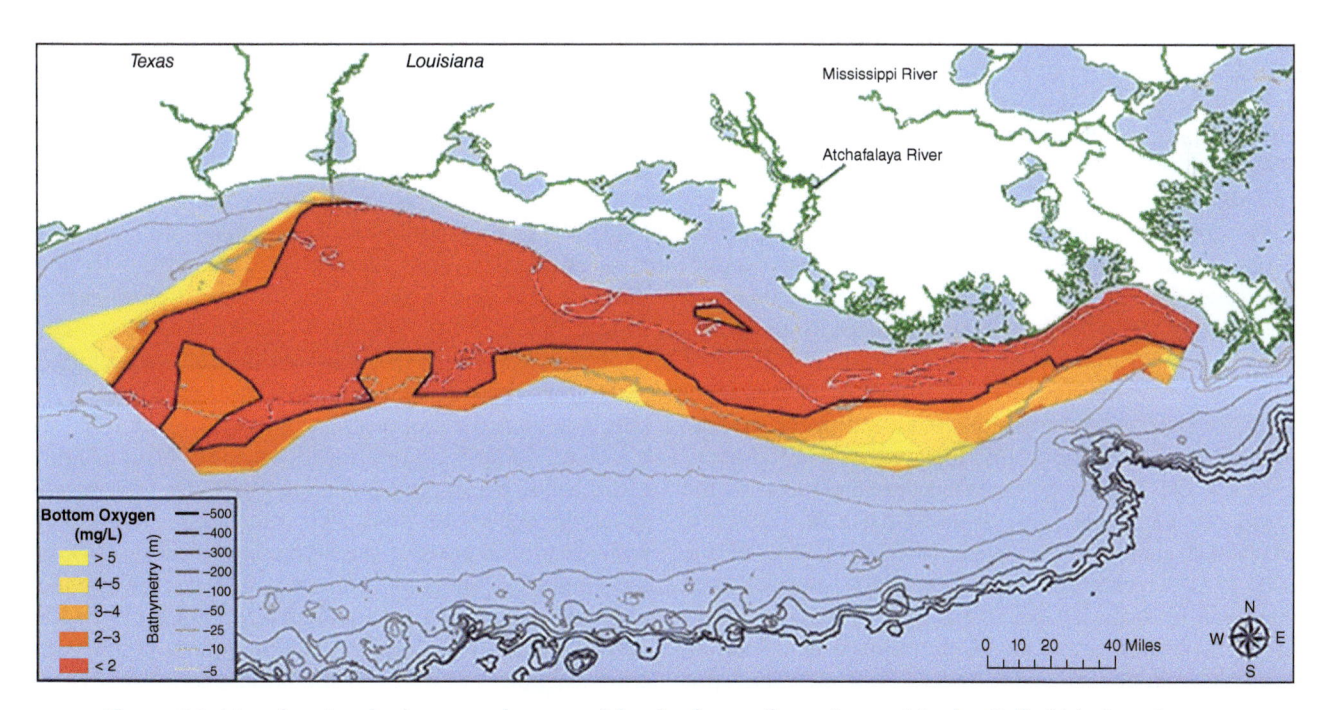

Figure 6.9 Map showing the large areal extent of the dead zone (hypoxic zone) in the Gulf of Mexico. *Source*: From the National Oceanic and Atmospheric Administration (https://www.noaa.gov/media-release/gulf-of-mexico-dead-zone-is-largest-ever-measured, accessed 25 January 2019).

eliminated, it would take approximately 30 years to realize the 60% decrease in load needed to reduce eutrophication in the Gulf. This legacy effect means that a dramatic shift in land-use practices, which may not be compatible with current levels of agricultural production, will be needed to control hypoxia in the Gulf of Mexico. Their combined modeling of reactive nitrogen release from storage and associated time lags provides critical information to water resource managers and decision-makers for exploring and quantifying the tradeoffs between costs to implement conservation measures and the actual amount of time needed to observe the desired concentration reductions. Their results demonstrated that long-term commitments along with large-scale changes in agricultural management practices would be needed to reduce Mississippi River nitrogen loads and to achieve current goals for reducing the size of the hypoxic zone in the Gulf of Mexico.

In another study of lag times in a large watershed (6800 km^2) in southern Ontario, Van Meter and Basu (2017) compared long-term nitrogen input information with stream nitrate concentration data for 16 nested subwatersheds. Their model results showed substantial non-linear hysteresis effects between reactive nitrogen inputs and outputs in streams that indicated lag times on the order of decades (average annual lag time was 24.5 years). Time lags varied seasonally and were negatively correlated with two explanatory variables, namely tile drainage and watershed slope. It was indicated that tile drainage was a dominant control in fall; however, watershed slope was a dominant control during the snowmelt period in the spring.

The Chesapeake Bay watershed in the mid-Atlantic United States contains numerous streams and rivers with elevated nitrate levels that have entered and adversely affected the water quality of the Bay. In response, the U.S. EPA along with the states of Maryland, Delaware, Virginia, and Pennsylvania (Chesapeake Bay Partnership, CBP) has developed a watershed model that is designed to quantify the contribution of nitrogen from various sources and forecast water quality conditions in the Bay. The states have set TMDLs for nitrogen and reduction goals for streams and rivers flowing into the Chesapeake Bay. Sanford and Pope (2013) indicated that the reduction goal for the Maryland and Delaware portions of the Delmarva Peninsula (that contains parts of Delaware, Maryland, and Virginia) is approximately 25% of the 2009 load, which translates to about 5000 tons of total dissolved nitrogen per year by 2020. The CBP model is mostly a surface water model that uses stream loading from the spatial regression model SPARROW but also has a groundwater storage component to simulate transient flow conditions. However, the model does not account for the lag time of nitrate transported from groundwater to surface water. Sanford and Pope (2013) combined a regional groundwater flow model with a nitrogen mass balance regression model to estimate the time required for the effects of nitrogen reductions associated with current and future best management practices on the land surface to reach the Chesapeake Bay through groundwater transport to streams. Their study area included seven watersheds on the Delmarva Peninsula on the eastern side of Chesapeake Bay. Sanford and Pope (2013) found that some nutrient management practices implemented during the past decade have had some positive results but their model predicted a high current load to the Bay (8200 tons N). Their model forecasts indicated that reduced nitrogen loading to groundwater would take decades to be observed in streams. They estimated that a 25% load reduction of nitrogen to the water table would only lead to a 12% load reduction to the Chesapeake Bay. Sanford and Pope (2013) concluded that these decadal time lags relating to delays in groundwater transport of nitrogen loads must be clearly understood for an accurate assessment of the time required for restoring the health of Chesapeake Bay.

Van Meter et al. (2017) also modeled the nutrient dynamics in the Susquehanna River, the largest river that drains into Chesapeake Bay. The Susquehanna River accounts for more than half of the annual nutrient load to the bay (Zhang et al., 2016). After agricultural land use peaked in the early 20th century, increases in urbanization and widespread reforestation have occurred in the Susquehanna River Basin (SRB) (Drummond & Loveland, 2010; Thompson et al., 2013). However, even with the reduced proportion of agricultural land use in the basin, the increased use of commercial fertilizers and elevated levels of atmospheric nitrogen deposition have resulted in increased primary productivity (eutrophication) and hypoxia issues in Chesapeake Bay. Van Meter et al. (2017) developed a modeling framework, called ELEMeNT, that combined information on land use, population, agricultural production data, and biological nitrogen fixation to quantify soil surface nitrogen surplus trajectories during 1800–2015 across the SRB. This information was used to quantify nitrogen legacies and to assess their impacts on water quality. Van Meter et al. (2017) estimated a groundwater nitrate accumulation rate of 23.0 ± 5.4 kg/ha (1990–2013) in the Susquehanna Basin, with a large part coming from the high atmospheric N deposition rates. They noted that even though nonagricultural areas of the SRB comprise about 80% of the watershed, there is no N removal from crop production and N uptake levels are relatively low. Given the low nutrient soils of the region that provide minimal buffering capacity, nitrogen from atmospheric deposition moves quickly through the landscape due to high recharge rates and enters the groundwater system. Their model predicted that groundwater nitrogen in the SRB is increasing at a rate of approximately 11.6 kg/ha/yr. With a total nitrogen accumulation in the SRB of approximately 980,275 kg/ha, which is almost double that in the Mississippi River Basin (508,237 kg/ha).

6.7.6. Modeling Nitrogen and Other Nutrient Impacts on Harmful Algal Blooms

Glibert and Burford (2017) noted that the relationship between nitrogen and phosphorus pollution and HABs was more complex than once thought. They recognized that other factors need to be considered, such as physiological adaptations of the harmful algal species (and or strain), environmental conditions, interactions with co-occurring organisms that can change conditions that may enhance or inhibit HAB occurrence. Therefore, modeling the occurrence/proliferation of HABs is challenging, and it is necessary to have accurate estimates of nutrient loads entering receiving waters. Several models can provide this information at different catchment scales (e.g., the SPARROW model (described above), and the Nutrient Export from Watersheds (NEWS; Seitzinger et al., 2005, 2010)). In their review article that describes continuing challenges with understanding the increasing global occurrence of HABs, Glibert and Burford (2017) concluded that even though other factors are important in controlling HABs, more effective management approaches need to focus on control and regulation of both nitrogen and phosphorus reduction strategies. They noted that HABs can proliferate, survive, and become more toxic at either end of the continuum of nitrogen to phosphorus limitation.

Townhill et al. (2018) used species distribution modeling along with environmental conditions to assess the response of harmful algal species to future climate change. Their study incorporated high resolution, downscaled shelf seas climate projections for the north-west European shelf along with lower resolution global projections, to evaluate changes in the distribution of harmful algae by the mid to end of the 21st century. Their model predictions indicated that the habitat of most species (defined by temperature, salinity, depth, and stratification) would shift north this century, with suitability conditions increasing in the central and northern North Sea. They noted that an increase in more frequent harmful blooms would occur if wind, irradiance, and nutrient levels are also suitable. They concluded that increased monitoring of species in highly susceptible areas would lead to more effective early warning systems for health protection strategies and for aquaculture.

6.8. FURTHER RESEARCH NEEDS AND CONCLUDING REMARKS

Several research studies have identified topics for further investigations regarding transport of reactive nitrogen in surface water systems and impacts on sensitive ecosystems. For example, Bobbink et al. (2010) noted that the critical load concept for nitrogen deposition has helped European governments make progress toward reducing reactive nitrogen loads on sensitive ecosystems. However, they emphasized that more work needs to be done in Europe and North America, especially for the more sensitive ecosystem types, including several ecosystems with high conservation importance. Their detailed assessment of global nitrogen impacts indicated that further studies are needed in other vulnerable ecoregions including eastern and southern Asia (China, India), the Mediterranean ecoregion (California, southern Europe), and several subtropical and tropical parts of Latin America and Africa. Bobbink et al. (2010) concluded that reductions in plant diversity from increased atmospheric nitrogen deposition were likely more widespread than first thought, and there is a need for more targeted studies in areas with low background levels of atmospheric nitrogen deposition. Information also is limited on the effects of reactive nitrogen on faunal diversity, although faunal communities respond to changes in vegetation diversity, composition, or structure. Other uncertainties include the impact on ecosystems from the increased load reactive nitrogen that has accumulated in soils. Estimates are needed to determine the amount of time needed for recovery after reduction in nitrogen deposition occur. Adjustments may need to be made regarding critical loads and their impacts on plant species richness.

Nearshore ecosystems (e.g., marshes, mangroves, kelps, and especially seagrasses) are threatened by excess nutrient inputs. Waycott et al. (2009) noted that the global cover of seagrasses has declined by over 29% in the last century, largely because of nutrient and sediment runoff. This has resulted in substantial losses of ecosystem services, such as habitat and nurseries for commercially important species, shoreline protection, nutrient cycling, and carbon storage (Breitburg et al., 2018; Leslie, 2018). Recovery of sea grasses and other submerged aquatic vegetation (SAV) has been very slow or unsuccessful, but there has been some recovery in small-scale areas over short time periods. However, a recent study has provided an example of successful restoration where effective management of nutrients has led to the successful recovery of submersed aquatic vegetation along thousands of kilometers of coastline in the Chesapeake Bay, United States (Breitburg et al., 2018). Their study effectively showed that 30 years of environmental policy promoting biodiversity conservation can lead toward recovery of coastal systems. Breitburg et al. (2018) noted that other regions have seen seagrass recovery along with improved water clarity (e.g., Tampa Bay, Florida); however, the Chesapeake Bay has had a greater total and proportional recovery than any other SAV restoration project that they know of. The encouraging results of their study can contribute an ecosystem-based management strategy for future ecological restoration in other areas.

There are many unknowns regarding the combined effect of climate change and increasing or episodic reactive nitrogen inputs to surface waters (lakes, rivers, reservoirs, and estuaries) and their associated ecosystems. Several researchers have noted that climate change already has contributed additional stresses on ecosystems by altering the rates of transport and loading of reactive nitrogen to fresh and coastal waters and affecting the processes controlling uptake and removal (denitrification) within aquatic ecosystems. For example, Baron et al. (2013) noted that aquatic ecosystems are important denitrification hotspots, with high per-unit-area denitrification rates in soils; sediments of lakes, reservoirs, small streams; floodplains; and wetlands. Seitzinger et al. (2006) estimated that about 20% of global denitrification occurs in freshwaters (lakes, rivers, groundwater), 44% in the continental shelf; 14% in ocean zones with minimal oxygen levels, 20% in terrestrial soils, and 1% in estuaries. However, these estimates of denitrification rates need further research to better quantify reactive nitrogen attenuation in freshwater and marine systems.

Several recent studies have expressed serious concerns about how climate change might affect HABs. For example, Havens and Paerl (2015) and Paerl et al. (2016) noted that climate change will present major challenges for water resource managers to formulate nutrient-based algal bloom thresholds for controlling HABs. Warmer water temperatures, episodic nutrient loading, and more climate extremes (e.g., droughts, increased storm activity, tropical cyclones) can change phytoplankton communities by promoting bloom-forming cyanobacteria. Furthermore, Paerl et al. (2016) noted that legacy nutrients in sediments can persist for long periods of time, which would continue to provide high internal nutrient loads that could fuel algal blooms. Several physical, chemical, and biotic approaches to control cyanobacterial algal blooms were discussed. However, they indicated that these various approaches have only been successful in small ecosystems, but not in larger lakes and reservoirs. Paerl et al. (2016) concluded that critical research needs would include a clearer understanding of how physical, chemical, and biological control measures would work in the future with the anticipated changes in hydrology, stratification, and nutrient dynamics all related to climate change.

In their review of the literature on the impacts of atmospheric nitrogen deposition on remote freshwater ecosystems, Lepori and Keck (2012) concluded that atmospheric nitrogen contributions to episodic acidification will continue to affect industrialized regions in North America, Western Europe, and Japan for a considerable amount of time. They caution that synergistic effects between nitrogen saturation and climate change will likely worsen these impacts. However, Lepori and Keck (2012) also noted that some effects of climate change could help to reduce impacts due to higher temperatures, which may result in increased mineral weathering, increased nitrogen retention through increased winter uptake, and increased leaching of dissolved organic matter (promotes light limitation and could reduce nutrient limitation). Clearly, more integrative modeling studies are needed to assess these impacts from climate change and inputs of reactive nitrogen.

Helton et al. (2011) also recommended a holistic and mechanistic approach to biogeochemical modeling of rivers that incorporates an accurate representation of the hydrogeomorphic dynamics across large river networks. They noted that these holistic models would have important applications for a clearer understanding of nutrient cycling across a many varied fluvial landscapes and diverse biomes. Helton et al. (2011) concluded that holistic models would also be extremely useful for predicting the responses of biogeochemical cycles to climate change across river networks worldwide.

REFERENCES

Alexander, R. B., Bohlke, J. K., Boyer, E. W., David, M. B., Harvey, J. W., Mulholland, P. J., et al. (2009). Dynamic modeling of nitrogen losses in river networks unravels the coupled effects of hydrological and biogeochemical processes. *Biogeochemistry*, *93*, 91–116.

Alexander, R. B., Elliott, A. H., Shanker, U., & McBride, G. B. (2002b). Estimating the sources and transport of nutrients in the Waikato River Basin, New Zealand. *Water Resources Research*, *38*, 1268–1290. https://doi.org/10.1029/2001WR000878

Alexander, R. B., Johnes, P. J., Boyer, E. W., & Smith, R. A. (2002a). A comparison of models for estimating the riverine export of nitrogen from large watersheds. *Biogeochemistry*, *57*, 29S–339S.

Alexander, R. B., Smith, R. A., & Schwarz, G. E. (2000). Effect of stream channel size on the delivery of nitrogen to the Gulf of Mexico. *Nature*, *403*, 758–761.

Alexander, R. B., Smith, R. A., Schwarz, G. E., Boyer, E. W., Nolan, J. V., & Brakebill, J. W. (2008). Differences in phosphorus and nitrogen delivery to the Gulf of Mexico from the Mississippi River Basin. *Environmental Science and Technology*, *42*, 822–830.

Baron, J. S., Driscoll, C. T., Stoddard, J. L., & Richer, E. E. (2011). Empirical critical loads of atmospheric nitrogen deposition for nutrient enrichment and acidification of sensitive US lakes. *BioScience*, *61*, 602–613.

Baron, J. S., Hall, E. K., Nolan, B. T., Finlay, J. C., Bernhardt, E. S., Harrison, J. A., et al. (2013). The interactive effects of excess reactive nitrogen and climate change on aquatic ecosystems and water resources of the United States. *Biogeochemistry*, *114*, 71–92. https://doi.org/10.1007/s10533-012-9788-y

Bechmann, M. (2014). Loss of nitrogen in surface and subsurface runoff from small agricultural dominated catchments in Norway. *Agriculture, Ecosystems, and Environment*, *198*, 13–24.

Bechmann, M., Blicher-Mathiesen, G., Kyllmar, K., Iital, A., Lagzdins, A., & Salo, T. (2014). Nitrogen application, balances and effect on water quality in the Nordic-Baltic countries. *Agriculture, Ecosystems, and Environment, 198*, 104–113.

Beman, J. M., Chow, C.-E., King, A. L., Feng, Y., Fuhrman, J. A., Andersson, A., et al. (2011). Global declines in oceanic nitrification rates as a consequence of ocean acidification. *Proceedings of the National Academy of Sciences, 108*, 208–213.

Benoy, G. A., Jenkinson, R. W., Robertson, D. M., & Saad, D. A. (2016). Nutrient delivery to Lake Winnipeg from the red—Assiniboine River Basin – A binational application of the SPARROW model. *Canadian Water Resources Journal, 41*(3), 429–447. https://doi.org/10.1080/07011784.2016.1178601

Billen, G., & Garnier, J. (2007). River basin nutrient delivery to the coastal sea: Assessing its potential to sustain new production of non-siliceous algae. *Marine Chemistry, 106*, 148–160. https://doi.org/10.1016/j.marchem.2006.12.017

Billen, G., Silvestre, M., Grizzetti, B., Leip, A., Garnier, J., Voss, M., et al. (2011). Nitrogen flows from European watersheds to coastal marine waters. Ch. 13. In M. A. Sutton, C. M. Howard, J. W. Erisman, G. Billen, A. Bleeker, P. Grennfelt, H. van Grinsven, & B. Grizzetti (Eds.), *The European nitrogen assessment: Sources, effects, and policy perspectives*. Cambridge, UK: Cambridge University Press, 664 p.

Bleeker, A., Hicks, W. K., Dentener, F., Galloway, J., & Erisman, J. W. (2010). N deposition as a threat to the World's protected areas under the Convention on Biological Diversity. *Environmental Pollution, 159*, 2280–2288.

Bobbink, R., Hicks, K., Galloway, J., Spranger, T., Alkemade, R., Ashmore, M., et al. (2010). Global assessment of nitrogen deposition effects on terrestrial plant diversity: A synthesis. *Ecological Applications, 20*, 30–59.

Bodeanu, N. (1993). Microalgal blooms in the Romanian area of the Black Sea and contemporary eutrophication conditions. In T. J. Smayda & Y. Shimizu (Eds.), *Toxic phytoplankton blooms in the sea* (pp. 203–209). Amsterdam: Elsevier.

Boers, P. C. M. (1996). Nutrient emissions from agriculture in the Netherlands, causes and remedies. *Water Science and Technology, 33*, 183–189.

Bouraoui, F., & Grizzetti, B. (2014). Modelling mitigation options to reduce diffuse nitrogen water pollution from agriculture. *Science of the Total Environment, 468–469*, 1267–1277. https://doi.org/10.1016/j.scitotenv.2013.07.066

Bouraoui, F., Grizzetti, B., Adelskold, G., Behrendt, H., de Miguel, I., Silgram, M., et al. (2009). Basin characteristics and nutrient losses: The EUROHARP catchment network perspective. *Journal of Environmental Monitoring, 11*(3), 425–525. https://doi.org/10.1039/b822931g

Bowman, W. D., Cleveland, C. C., Halada, L., Hresko, J., & Baron, J. S. (2008). Negative impact of nitrogen deposition on soil buffering capacity. *Nature Geoscience, 1*. https://doi.org/10.1038/ngeo339

Boyer, E. W., Alexander, R. B., Parton, W. J., Li, C., Butterbach-Bahl, K., Donner, S. D., Skaggs, R. W., & Del Grosso, S. J. (2006). Modeling denitrification in terrestrial and aquatic ecosystems at regional scales. *Ecological Applications, 16*, 2123–2142.

Boyer, E. W., & Howarth, R. W. (2008). Nitrogen fluxes from rivers to the coastal oceans. In D. G. Capone, D. A. Bronk, M. R. Mulholland, & E. J. Carpenter (Eds.), *Nitrogen in the marine environment* (2nd ed., pp. 1565–1587). Boston, MA: Elsevier.

Breitburg, D., Levin, L. A., Oschlies, A., Grégoire, M., Chavez, F. P., Conley, D. J., et al. (2018). Declining oxygen in the global ocean and coastal waters. *Science, 359*, eaam7240.

Bricker, S., Longstaff, B., Dennison, W., Jones, A., Boicourt, K., Wicks, C., & Woerner, J. (2007). *Effects of nutrient enrichment in the nation's estuaries: A decade of change. NOAA Coastal Ocean Program Decision Analysis Series No. 26*. Silver Spring, MD: National Centers for Coastal Ocean Science.

Camargo, J. A., & Alonso, A. (2006). Ecological and toxicological effects of inorganic nitrogen pollution in aquatic ecosystems: A global assessment. *Environment International, 32*, 831–849.

Clark, C. M., Bell, M. D., Boyd, J. W., Compton, J. E., Davidson, E. A., Davis, C., et al. (2017). Nitrogen-induced terrestrial eutrophication: Cascading effects and impacts on ecosystem services. *Ecosphere, 8*(7), e01877. https://doi.org/10.1002/ecs2.1877

Clark, C. M., & Tilman, D. (2008). Loss of plant species after chronic low-level nitrogen deposition to prairie grasslands. *Nature, 451*, 712–715.

Compton, J. E., Harrison, J. A., Dennis, R. L., Greaver, T., Hill, B. H., Jordon, S. J., Walker, H., & Campbell, H. V. (2011). Ecosystem services altered by changes in reactive nitrogen: An approach to inform decision-making. *Ecology Letters, 14*, 804–815. doi:10.1111/j.1461-0248.2011.01631.x

Conant, R. T., Berdanier, A. A., & Grace, P. R. (2013). Patterns and trends in nitrogen use and nitrogen recovery efficiency in world agriculture. *Global Biogeochemical Cycles, 27*, 558–566.

Conley, D. J., Paerl, H. W., Howarth, R. W., Boesch, D. F., Setzinger, S. P., Havens, K. E., Lancelot, C., & Likens, G. E. (2009) Controlling eutrophication: Nitrogen and phosphorus. *Science, 323*(5917), 1014–1015.

Cox, P. A., Davis, D. A., Mash, D. C., Metcalf, J. S., & Banack, S. A. (2016). Dietary exposure to an environmental toxin triggers neurofibrillary tangles and amyloid deposits in the brain. *Proceedings of the Royal Society B, 283*, 20152397. http://dx.doi.org/10.1098/rspb.2015.2397

Curtis, C. J., Botev, I., Camarero, L., Catalan, J., Cogalniceanu, D., Hughes, M., et al. (2005). Acidification in European mountain lake districts: A regional assessment of critical load exceedance. *Aquatic Sciences, 67*, 237–251.

Davidson, E. A., David, M. B., Galloway, J. N., Goodale, C. L., Haeuber, R., Harrison, J. A., et al. (2011). Excess nitrogen in the U.S. environment: Trends, risks, and solutions. *Issues in Ecology, 15*.

Diaz, R. J., & Rosenberg, R. (2008). Spreading dead zones and consequences for marine ecosystems. *Science, 321*, 926–929.

Diaz, R.J., Selman, M., & Chique, C. (2010). *Global eutrophic and hypoxic coastal systems*. World Resources Institute. Eutrophication and Hypoxia: Nutrient Pollution in Coastal Waters. Retrieved from https://www.wri.org/our-work/project/eutrophication-and-hypoxia/interactive-map-eutrophication-hypoxia, accessed 24 September 2019.

Dise, N. B., Ashmore, M., Belyazid, S., Bleeker, A., Bobbink, R., de Vries, W., et al. (2011). Nitrogen as a threat to European terrestrial biodiversity, Ch. 20. In M. A. Sutton, C. M. Howard, J. W. Erisman, G. Billen, A. Bleeker, P. Grennfelt, et al. (Eds.), *The European nitrogen assessment* (pp. 463–494). Cambridge University Press.

Dodds, W. K., Bouska, W. W., Eitzmann, J. L., Pilger, T. H., Pitts, K. L., Riley, A. J., et al. (2009). Eutrophication of U.S. Freshwaters: Analysis of potential economic damages. *Environmental Science and Technology*, *43*(1), 12–19.

Doney, S. C., Fabry, V. J., Feely, R. A., & Kleypas, J. A. (2009). Ocean acidification: The other CO_2 problem. *Annual Reviews in Marine Science*, *1*, 169–191.

Doney, S. C., Mahowald, N., Lima, I., Feely, R. A., Mackenzie, F. T., Lamarque, J.-F., & Rasch, P. J. (2007). Impact of anthropogenic atmospheric nitrogen and sulfur deposition on ocean acidification and the inorganic carbon system. *Proceedings of the National Academy of Sciences USA*, *104*(37), 14580–14585. https://doi.org/10.1073/pnas.0702218104

Drummond, M. A., & Loveland, T. R. (2010). Land-use pressure and a transition to forest-cover loss in the eastern United States. *BioScience*, *60*(4), 286–298. https://doi.org/10.1525/bio.2010.60.4.7

Dubrovsky, N.M., Burow, K.R., Clark, G.M., Gronberg, J.M., Hamilton P.A., Hitt, K.J., et al. (2010). *The quality of our Nation's waters—Nutrients in the Nation's streams and groundwater, 1992–2004*. U.S. Geological Survey Circular 1350, 174 p. Retrieved from http://water.usgs.gov/nawqa/nutrients/pubs/circ1350

Erisman, J. W., Galloway, J. N., Dise, N. B., Sutton, M. A., Bleeker, A., Grizzetti, B., et al. (2015). *Nitrogen, too much of a vital resource*. Science Brief. Zeist, The Netherlands: World Wildlife Fund.

Erisman, J. W., Galloway, J. N., Seitzinger, S., Bleeker, A., Dise, N. B., Petrescu, R., et al. (2013). Consequences of human modification of the global nitrogen cycle. *Philosophical Transactions of the Royal Society*, *368*(1621). https://doi.org/10.1098/rstb.2013.0116

Erisman, J. W., van Grinsven, H., Grizzetti, B. Bouraoui, F., Powlson, D., Sutton, M. A., Bleeker, A., & Reis, S. (2011). The European nitrogen problem in a global perspective. Ch. 2. In M.A. Sutton, C.M. Howard, J.W. Erisman Billen, G., Bleeker, A., Grennfelt, P., H. van Grinsven, & B. Grizzetti (Eds.), *The European nitrogen assessment: Sources, effects, and policy perspectives* (pp. 9–31). Cambridge, UK: Cambridge University Press.

Feely, R. A., Doney, S. C., & Cooley, S. R. (2009). Ocean acidification: Present conditions and future changes in a high-CO_2 world. *Oceanography*, *22*, 36–47.

Flesher, J. & Kastanis. A. (2017). Toxic algae: Once a nuisance now a severe nationwide threat. *Associated Press*, 16 November 2017.

Fowler, D., Coyle, M., Skiba, U., Sutton, M. A., Capel, J. N., Reis, S., et al. (2013). The global nitrogen cycle in the twenty-first century. *Philosophical Transactions of the Royal Society B*, *368*(1621). https://doi.org/10.1098/rtsb.2013.0164

Fowler, D., Steadman, C. E., Stevenson, D., Coyle, M., Rees, R. M., Skiba, U. M., et al. (2015). Effects of global change during the 21st century on the nitrogen cycle. *Atmospheric Chemistry and Physics*, *15*, 13849–13893. https://doi.org/10.5194/acp-15-13849-2015

Galloway, J. N., Dentener, F. J., Capone, D. G., Boyer, E. W., Howarth, R. W., Seitzinger, S. P., et al. (2004). Nitrogen cycles: Past, present and future. *Biogeochemistry*, *70*, 153–226.

Galloway, J. N., Townsend, A. R., Erisman, J. W., Bekunda, M., Cai, Z., Freney, J. R., et al. (2008). Transformation of the nitrogen cycle: Recent trends, questions, and potential solutions. *Science*, *320*, 889–892.

Garcia, A. M., Alexander, R. B., Arnold, J. G., Norfleet, L., White, J. J., Robertson, D. M., & Schwarz, G. (2016). Regional effects of agricultural conservation practices on nutrient transport in the Upper Mississippi River basin. *Environmental Science and Technology*, *50*, 6991–7000. https://doi.org/10.1021/acs.est.5b03543

Garnier, M., Recanatesi, F., Ripa, M. N., & Leone, A. (2010). Agricultural nitrate monitoring in a lake basin in central Italy: A further step ahead towards an integrated nutrient management aimed at controlling water pollution. *Environmental Monitoring and Assessment*, *170*, 273–286.

Garrett, J.D., Frey, J.W., Van Metre, P.C., Journey, C.A., Nakagaki, N., Button, D.T., and Nowell, L.H. (2017). *Design and methods of the Midwest Stream Quality Assessment (MSQA), 2013*. U.S. Geological Survey Open-File Report 2017–1073, 59 p., 4 app., https://doi.org/10.3133/ofr20171073.

Glibert, P. M., & Burford, M. A. (2017). Globally changing nutrient loads and harmful algal blooms: Recent advances, new paradigms, and continuing challenges. *Oceanography*, *30*(1), 58–69. https://doi.org/10.5670/oceanog.2017.110

Glibert, P. M., Maranger, R., Sobota, D. J., & Bouwman, L. (2014). The Haber Bosch–harmful algal bloom (HB–HAB) link. *Environmental Research Letters*, *9*, 105001. https://doi.org/10.1088/1748-9326/9/10/105001

Greaver, T. L., Sullivan, T., Herrick, J., Lawrence, G., Herlihy, A., Barron, J., et al. (2012). Ecological effects of air pollution in the U.S.: What do we know? *Frontiers in Ecology and the Environment*, *10*, 365–372.

Grizzetti, B., Bouraoui, F., Billen, G., van Grinsven, H., Cardoso, A.C., Thieu, V., et al. (2011). Nitrogen as a threat to European water quality. Ch. 17. In M. A. Sutton, C. M. Howard, J. W. Erisman, G. Billen, A. Bleeker, P. Grennfelt, H. van Grinsven, & B. Grizzetti (Eds.), *The European nitrogen assessment: Sources, effects, and policy perspectives* (pp. 379–404). Cambridge, UK: Cambridge University Press.

Grizzetti, B., Passy, P., Billen, G., Bouraoui, F., Garnier, J., & Lassaletta, L. (2015). The role of water nitrogen retention in integrated nutrient management: Assessment in a large basin using different modelling approaches. *Environmental Research Letters*, *10*, 065008. https://doi.org/10.1088/1748-9326/10/6/065008

Gu, B., Ju, X., Wu, Y., Erisman, J. W., Bleeker, A., Reis, S., et al. (2017). Cleaning up nitrogen pollution may reduce future carbon sinks. *Global Environmental Change*, *48*, 55–66.

Haag, D., & Kaupenjohann, M. (2001). Landscape fate of nitrate fluxes and emissions in central Europe: A critical review of concepts, data, and models for transport and retention. *Agriculture Ecosystems and the Environment*, *86*(1), 1–21.

Han, D., Currell, M. J., & Cao, G. (2016). Deep challenges for China's war on water pollution. *Environmental Pollution, 218*, 1222–1233. http://dx.doi.org/10.1016/j.envpol.2016.08.078

Hatano, R., Nagumo, T., & Kuramochi, K. (2005). Impact of nitrogen cycling on stream water quality in a basin associated with forest, grassland, and animal husbandry, Hokkaido, Japan. *Ecological Engineering, 24*, 509–515.

Havens, K. E., & Paerl, H. W. (2015). Climate change at a crossroad for control of harmful algal blooms. *Environmental Science and Technology*. https://doi.org/10.1021/acs.est.5b03990

Helton, A. M., Poole, G. C., Meyer, J. L., Wollheim, W. M., Peterson, B. J., Mulholland, P. J., et al. (2011). Thinking outside the channel: Modeling nitrogen cycling in networked river ecosystems. *Frontiers in Ecology and the Environment, 9*(4), 229–238. https://doi.org/10.1890/080211

Herbert, E. R., Boon, P., Burgin, A. J., Neubauer, S. C., Franklin, R. B., Ardón, M., et al. (2015). A global perspective on wetland salinization: Ecological consequences of a growing threat to freshwater wetlands. *Ecosphere, 6*(10), 1–43.

Hernández, D. L., Vallano, D. M., Zavaleta, E. S., Tzankova, A., Pasari, J. R., Weiss, S., et al. (2016). Nitrogen pollution is linked to U.S. listed species declines. *Bioscience, 66*, 213–222.

Hong, B., Swaney, D., & Howarth, R. W. (2011). A toolbox for calculating net anthropogenic nitrogen inputs (NANI). *Environmental Modeling Software, 26*, 623–633.

Houlton, B. Z., Boyer, E., Finzi, A., Galloway, J., Leach, A., Liptzin, D., et al. (2012). Intentional versus unintentional nitrogen use in the United States: Trends, efficiency and implications. *Biogeochemistry*. https://doi.org/10.1007/s10533-012-9801-5

Howarth, R.W., Anderson, D., Cloern, J., Elfring, C., Hopkinson, C., Lapointe, B., et al. (2000). Nutrient pollution of coastal rivers, bays, and seas. *Issues in Ecology*, no. 7. Retrieved from http://www.epa.gov/watertrain/pdf/issue7.pdf, accessed 24 September 2019.

Howarth, R. W., Chan, F., Conley, D. J., Garnier, J., Doney, S. C., Marino, R., & Billen, G. (2011a). Coupled biogeochemical cycles: Eutrophication and hypoxia in temperate estuaries and coastal marine ecosystems. *Frontiers in Ecology and the Environment, 9*, 18–26.

Howarth, R. W., Swaney, D., Billen, G., Garnier, J., Hong, B., Humborg, C., et al. (2011b). Nitrogen fluxes from the landscape are controlled by net anthropogenic nitrogen inputs and by climate. *Frontiers in Ecology and the Environment*. https://doi.org/10.1890/100178

Irvine, I. C., Greaver, T., Phelan, J., Sabo, R. D., & Van Houtven, G. (2017). Terrestrial acidification and ecosystem services: Effects of acid rain on bunnies, baseball, and Christmas trees. *Ecosphere, 8*(6), e01857. https://doi.org/10.1002/ecs2.1857

Jiang, R., Hatano, R., Zhao, Y., Woli, K. P., Kuramochi, K., Shimizu, M., & Hayakawa, A. (2014a). Factors controlling nitrogen and dissolved organic carbon exports across timescales in two watersheds with different land uses. *Hydrological Processes, 28*, 5105–5121.

Jiang, Y., Frankenberger, J. R., Bowling, L. C., & Sun, Z. (2014b). Quantification of uncertainty in estimated nitrate-N loads in agricultural watersheds. *Journal of Hydrology, 519*(A), 106–116.

Jordan, S. J., Stoffer, J., & Nestlerode, J. A. (2011). Wetlands as sinks for reactive nitrogen at continental and global scales: A metaanalysis. *Ecosystems, 4*, 144–155.

Kyllmar, K., Bechmann, M., Deelstra, J., Iital, A., Blicher-Mathiesen, G., Jansons, V., et al. (2014a). Long-term monitoring of nutrient losses from agricultural catchments in the Nordic-Baltic region: A discussion of methods, uncertainties and future needs. *Agriculture, Ecosystems and Environment, 198*, 4–12.

Kyllmar, K., Stjernman Forsberg, L., Andersson, S., & Mårtensson, K. (2014b). Small agricultural monitoring catchments in Sweden representing environmental impact. *Agriculture, Ecosystems and Environment, 198*, 25–35.

Lawniczak, A. E., Zbierska, J., Nowak, B., Achtenberg, K., Grześkowiak, A., & Kanas, K. (2016). Impact of agriculture and land use on nitrate contamination in groundwater and running waters in central-west Poland. *Environmental Monitoring and Assessment, 188*, 172. doi:10.1007/s10661-016-5167-9

Lefcheck, J.S., Orth, R.J., Dennison, W.C., Wilcox, D.J., Murphy, R.R., Keisman, J., et al. 2018. Long-term nutrient reductions lead to the unprecedented recovery of a temperate coastal region. *Proceedings of the National Academy of Sciences of the USA 115* (14); Retrieved from https://www.pnas.org/content/115/14/3658, accessed 24 September 2019.

Lefcheck, J. S., Wilcox, D. J., Murphy, R. R., Marion, S. R., & Orth, R. J. (2017). Multiple stressors threaten the imperiled coastal foundation species eelgrass (*Zostera marina*) in Chesapeake Bay, USA. *Global Change Biology, 23*, 3474–3483. https://doi.org/10.1111/gcb.13623

Lepori, F., & Keck, F. (2012). Effects of atmospheric nitrogen deposition on remote freshwater ecosystems. *Ambio, 41*(3), 235–246.

Leslie, H.M. 2018. Value of ecosystem-based management. *Proceedings of the National Academy of Sciences of the USA* Retrieved from www.pnas.org/cgi/doi/10.1073/pnas.1802180115, accessed 24 September 2019

Levitan, O., Rosenberg, G., Setlik, I., Setlikova, E., Grigel, J., Lkepetar, J., et al. (2007). Elevated CO_2 enhances nitrogen fixation and growth in the marine cyanobacterium Trichodesmium. *Global Changes Biology, 13*, 531–538.

Li, J., Glibert, P. M., Zhou, M., Lu, S., & Lu, D. (2009). Relationships between nitrogen and phosphorus forms and ratios and the development of dinoflagellate blooms in the East China Sea. *Marine Ecology Progress Series, 383*, 11–26. https://doi.org/10.3354/meps07975

Lopez, C.B., Dortch, Q., Jewett, E.B., & Garrison, D. (2008b). *Scientific assessment of marine harmful algal blooms*. Washington, DC: Interagency Working Group on Harmful Algal Blooms, Hypoxia, and Human Health of the Joint Subcommittee on Ocean Science and Technology, 72 p.

Lopez, C.B., Jewett, E.B., Dortch, Q., Walton, B.T., & Hudnell, H.K. (2008a). *Scientific assessment of freshwater harmful algal blooms*. Washington, DC: Interagency Working Group on Harmful Algal Blooms, Hypoxia and Human Health of the Joint Subcommittee on Ocean Science and Technology, 65 p.

Meals, D. W., Dressing, S. A., & Davenport, T. E. (2010). Lag time in water quality response to best management practices: A review. *Journal of Environmental Quality, 39*, 85.

Mee, L. D., Friedrich, J., & Gomoiu, M.-T. (2005). Restoring the Black Sea in times of uncertainty. *Oceanography, 18*, 32–43.

Oelsner, G. P., & Stets, E. G. (2019). Recent trends in nutrient and sediment loading to coastal areas of the conterminous U.S.: Insights and global context. *Science of the Total Environment, 654*, 1225–1240. https://doi.org/10.1016/j.scitotenv.2018.10.437

Paerl, H. W., Gardner, W. S., Havens, K. E., Joyner, A. R., & McCarthy, M. J. (2016). Mitigating cyanobacterial harmful algal blooms in aquatic ecosystems impacted by climate change and anthropogenic nutrients. *Harmful Algae, 54*, 213–222. http://dx.doi.org/10.1016/j.hal.2015.09.009

Pardo, L. H., Fenn, M. E., Goodale, C. L., Geiser, L. H., Driscoll, C. T., Allen, E. B., et al. (2011). Effects of nitrogen deposition and empirical nitrogen critical loads for ecoregions of the United States. *Ecological Applications, 8*, 3049–3082.

Payne, R. J., Dise, N. B., Stevens, C. J., Gowing, D. J., & Partners, B. (2013). Impact of nitrogen deposition at the species level. *Proceedings of the National Academy of Sciences of the USA, 113*, 984–987.

Peñuelas, J., Poulter, B., Sardans, J., Ciais, P., van der Velde, M., Bopp, L., et al. (2013). Human-induced nitrogen–phosphorus imbalances alter natural and managed ecosystems across the globe. *Nature Communications, 4*, 2934. https://doi.org/10.1038/ncomms3934

Pinder, R.W., Davison, E.A., Goodale, C.L., Greaver, T.L., Herrick, J.D., and Liu, L. 2012a. Climate change impacts of U.S. reactive nitrogen. *Proceedings of the National Academy of Science 109* (20): 7671–7675. Retrieved from www.pnas.org/cgi/doi/10.1073/pnas.1114243109, accessed 24 September 2019.

Porter, E. M., Bowman, W. D., Clark, C. M., Compton, J. E., Pardo, L. H., & Soong, J. L. (2012). Interactive effects of anthropogenic nitrogen enrichment and climate change on terrestrial and aquatic biodiversity. *Biogeochemistry, 114*, 93–120. https://doi.org/10.1007/s10533-012-9803-3

Puckett, L. J. (1995). Identifying the major sources of nutrient water pollution. *Environmental Science and Technology, 29*, 408A–414A.

van Puijenbroek, P. J. T. M., Cleij, P., & Visser, H. (2014). Aggregated indices for trends in eutrophication of different types of fresh water in the Netherlands. *Ecological Indicators, 36*, 456–462.

Rabalais, N. N. (2002). Nitrogen in aquatic ecosystems. *Ambio, 31*, 102–112.

Reis, S., Bekunda, M., Howard, C. M., Karanja, N., Winiwarter, W., Yan, X., et al. (2016). Synthesis and review: Tackling the nitrogen management challenge: From global to local scales. *Environmental Research Letters, 11*(2016), 120205. http://dx.doi.org/10.1088/1748-9326/11/12/120205

Remke, E., Brouwer, E., Kooijman, A., Blindow, I., & Roelofs, J. G. M. (2009). Low atmospheric nitrogen loads lead to grass encroachment in coastal dunes, but only on acid soils. *Ecosystems, 12*(7), 1173–1188.

Riddick, S., Ward, D., Hess, P., Mahowald, N., Massad, R., & Holland, E. (2016). Estimate of changes in agricultural terrestrial nitrogen pathways and ammonia emissions from 1850 to present in the Community Earth System Model. *Biogeosciences, 13*, 3397–3426. https://doi.org/10.5194/bg-13-3397-2016

Sala, O. E., Chapin, F. S., III, Armesto, J. J., Berlow, E., Bloomfield, J., Dirzo, R., et al. (2000). Global biodiversity scenarios for the year 2100. *Science, 287*, 1770–1774.

Sanford, W. E., & Pope, J. P. (2013). Quantifying groundwater's role in delaying improvements to Chesapeake Bay water quality. *Environmental Science and Technology, 47*(23), 13330–13338.

Schmidt, T. S., Van Metre, P. C., & Carlisle, D. M. (2018). Linking the agricultural landscape of the Midwest to stream health with structural equation modeling. *Environmental Science and Technology*. https://doi.org/10.1021/acs.est.8b04381

Seitzinger, S. P., & Harrison, J. A. (2008). Land-based nitrogen sources and their delivery to coastal systems. In D. G. Capone, D. A. Bronk, M. R. Mulholland, & E. J. Carpenter (Eds.), *Nitrogen in the marine environment* (2nd ed., pp. 469–510). Boston, MA: Elsevier.

Seitzinger, S. P., Harrison, J. A., Bohlke, J. K., Bouwman, A. F., Lowrance, R., Peterson, B., Tobias, C., & Van Drecht, G. (2006). Denitrification across landscapes and waterscapes: A synthesis. *Ecological Applications, 16*, 2064–2090.

Seitzinger, S. P., Harrison, J. A., Dumont, E., Beusen, A. H. W., & Bouwman, A. F. (2005). Sources and delivery of carbon, nitrogen and phosphorous to the coastal zone: An overview of global nutrient export from watersheds (NEWS) models and their application. *Global Biogeochemical Cycles, 19*, GB4S01. https://doi.org/10.1029/2005GB002606

Seitzinger, S. P., Mayorga, E., Bouwman, A. F., Kroeze, C., Beusen, A. H. W., Billen, G., et al. (2010). Global river nutrient export: A scenario analysis of past and future trends. *Global Biogeochemical Cycles, 24*, GB0A08. https://doi.org/10.1029/2009GB003587

Sheppard, L. J., Leith, I. D., Mizunuma, T., Cape, J. N., Crossley, A., Leeson, S., et al. (2011). Dry deposition of ammonia gas drives species change faster than wet deposition of ammonium ions: Evidence from a long-term field manipulation. *Global Change Biology, 17*(12), 3589–3607.

Shoda, M. E., Sprague, L. A., Murphy, J. C., & Riskin, M. L. (2019). Water-quality trends in U.S. rivers, 2002-2012: Relations to levels of concern. *Science of the Total Environment, 650*, 2314–2324. https://doi.org/10.1016/j.scitotenv.2018.09.377

Simkin, S.M., Allen, E.B., Bowman, W.D., Clark, C.M., Belnap, J. Brooks, M.L., et al. 2016. Conditional vulnerability of plant diversity to atmospheric nitrogen deposition across the United States. *Proceedings of the National Academy of Sciences 113* (15): 4086–4091., Retrieved from www.pnas.org/cgi/doi/10.1073/pnas.1515241113, accessed 24 September 2019.

Smith, R. A., Alexander, R. B., & Schwarz, G. E. (2003). Estimating the natural background concentrations of nutrients in streams and rivers of the conterminous United States. *Environmental Science and Technology*, *37*, 3039–3047.

Stevens, C. J., Dise, N. B., Mountford, J. O., & Gowing, D. J. (2004). Impact of nitrogen deposition on the species richness of grasslands. *Science*, *303*(5665), 1876–1879.

Stuart, M. E., Gooddy, D. C., Bloomfield, J. P., & Williams, A. T. (2011). A review of the impact of climate change on future nitrate concentrations in groundwater of the UK. *Science of the Total Environment*, *409*, 2859–2873.

Sutton, M. A., Bleeker, A., Howard, C. M., Bekunda, M., Grizzetti, B., et al. (2013b). Our nutrient world: The challenge to produce more food and energy with less pollution. In *Global overview of nutrient management*. Edinburgh, UK: Centre for Ecology and Hydrology on behalf of the Global Partnership on Nutrient Management and the International Nitrogen Initiative, 128 p.

Sutton, M. A., Erisman, J. W., Dentener, F., & Moeller, D. (2008). Ammonia in the environment: From ancient times to the present. *Environmental Pollution*, *156*, 583–604.

Sutton, M. A., Howard, C. M., Erisman, J. W., Billen, G., Bleeker, A., Grennfelt, P., van Grinsven, H., & Grizzetti, B. (Eds.). (2011b). *The European nitrogen assessment: Sources, effects, and policy perspectives*. Cambridge, UK: Cambridge University Press, 664 p.

Sutton, M. A., Howard, C. M., Erisman, J. W., Billen, G., Bleeker, A., Grennfelt, P., van Grinsven, H., & Grizzetti, B. (2011a). Assessing our nitrogen inheritance. The need to integrate nitrogen science and policies. Ch. 1. In M. A. Sutton, C. M. Howard, J. W. Erisman, G. Billen, A. Bleeker, P. Grennfelt, H. van Grinsven, & B. Grizzetti (Eds.), *The European nitrogen assessment: Sources, effects, and policy perspectives*. Cambridge, UK: Cambridge University Press, 664 p.

Sutton, M. A., Reis, S., Riddick, S. N., Dragosits, U., Nemitz, E., Theobald, M. R., et al. (2013a). Towards a climate-dependent paradigm of ammonia emission and deposition. *Philosophical Transactions of the Royal Society B*, *368*. https://doi.org/10.1098/rtsb.2013.0166

Thompson, J. R., Carpenter, D. N., Cogbill, C. V., & Foster, D. R. (2013). Four centuries of change in northeastern United States forests. *PLoS One*, *8*(9), e72540. https://doi.org/10.1371/journal.pone.0072540

Tian, D., & Niu, S. (2015). A global analysis of soil acidification caused by nitrogen addition. *Environmental Research Letters*, *10*, 1–10. https://doi.org/10.1088/1748-9326/10/2/024019

Townhill, B. L., Tinker, J., Jones, M., Pitois, S., Creach, V., Simpson, S. D., et al. (2018). Harmful algal blooms and climate change: Exploring future distribution changes. *ICES Journal of Marine Science*, *75*(6), 1882–1893. https://doi.org/10.1093/icesjms/fsy113

Townsend, A. R., Howarth, R. W., Bazzaz, F. A., Booth, M. S., Cleveland, C. C., Collinge, S. K., et al. (2003). Human health effects of a changing global nitrogen cycle. *Frontiers in Ecology and the Environment*, *1*(5), 240–246.

Turner, R. E., Rabalais, N. N., Justic, D., & Dortch, Q. (2003). Global patterns of dissolved N, P and Si in large rivers. *Biogeochemistry*, *64*, 297–317. https://doi.org/10.1023/A:1024960007569

U.S. Environmental Protection Agency (U.S. EPA). (2002). *Nitrogen—Multiple and regional impacts*. U.S. Environmental Protection Agency Clean Air Markets Division Report EPA-430-R-01-006, 38 p.

U.S. Environmental Protection Agency. (2009). *An urgent call to action*. Report of the State-EPA Nutrient Innovations Task Group, 41 p.

U.S. Environmental Protection Agency. (2011). *Reactive nitrogen in the United States: An analysis of inputs, flows, consequences and management options*. A Report of the US EPA Science Advisory Board (EPA-SAB-11-013). Washington, DC, 172 p.

UNEP-WCMC (2008). *The state of the World's Protected Areas 2007: An annual review of global conservation progress*. Cambridge, UK: UNEP-WCMC.

Urquhart, E. A., Schaeffer, B. A., Stumpf, R. P., Loftin, K. A., & Werdell, P. J. (2017). A method for examining temporal changes in cyanobacterial harmful algal bloom spatial extent using satellite remote sensing. *Harmful Algae*, *67*, 144–152. https://doi.org/10.1016/j.hal.2017.06.001

Van Meter, K. J., & Basu, N. B. (2017). Time lags in watershed-scale nutrient transport: an exploration of dominant controls. *Environmental Research Letters*, *12*(8). https://doi.org/10.1088/1748-9326/aa7bf4

Van Meter, K. J., Basu, N. B., & Van Cappellen, P. (2017). Two centuries of nitrogen dynamics, legacy sources and sinks in the Mississippi and Susquehanna River basins. *Global Biogeochemical Cycles*, *31*(1), 2–23.

Van Meter, K. J., Van Capellen, P., & Basu, N. B. (2018). Legacy nitrogen may prevent achievement of water quality goals in the Gulf of Mexico. *Science*, *360*(6387), 427–430. https://doi.org/10.1126/science.aar4462

Van Metre, P. C., Frey, J. W., Musgrove, M., Nakagaki, N., Qi, S., Mahler, B. J., et al. (2016). High nitrate concentrations in some Midwest United States streams in 2013 after the 2012 drought. *Journal of Environmental Quality*, *45*(5), 1696–1704.

Velthof, G. L., Lesschen, J. P., Webb, J., Pietrzak, S., Miatkowski, Z., Pinto, M., et al. (2014). The impact of the Nitrates Directive on nitrogen emissions from agriculture in the EU-27 during 2000-2008. *Science of the Total Environment*, *468–469*, 1225–1233. https://doi.org/10.1016/j.scitotenv.2013.04.058

Wang, Q., Koshikawa, H., Liu, C., & Otsubo, K. (2014). 30-year changes in the nitrogen inputs to the Yangtze River Basin. *Environmental Research Letters*, *9*, 115005. https://doi.org/10.1088/1748-9326/9/11/115005

Waycott, M., Duarte, C. M., Carruthers, T. J. B., Orth, R. J., Dennison, W. C., Olyarnik, S., et al. (2009). Accelerating loss of seagrasses across the globe threatens coastal ecosystems. *Proceedings of the National Academy of Sciences USA, July 28, 2009*, *106*(30), 12377–12381. https://doi.org/10.1073/pnas.0905620106

Woli, K. P., Hayakawa, A., Kuramochi, K., & Hatano, R. (2008). Assessment of river water quality during snowmelt and base flow periods in two catchment areas with different land use. *Environmental Monitoring and Assessment*, *137*(1–3), 251–260.

Zhang, Q., Ball, W. P., & Moyer, D. L. (2016). Decadal-scale export of nitrogen, phosphorus, and sediment from the Susquehanna River Basin, USA: Analysis and synthesis of temporal and spatial patterns. *Science of the Total Environment*, *563–564*, 1016–1029. https://doi.org/10.1016/j.scitotenv.2016.03.104

FURTHER READING

Ascott, M. J., Gooddy, D. C., Wang, L., Stuart, M. E., Lewis, M. A., Ward, R. S., & Binley, A. M. (2017). Global patterns of nitrate storage in the vadose zone. *Nature Communications*, 1–7. https://doi.org/10.1038/s41467-017-01321-w

Benavides, M., & Voss, M. (2015). Five decades of N_2 fixation research in the North Atlantic Ocean. *Frontiers in Marine Science*, *2*, 40. http://dx.doi.org/10.3389/fmars.2015.00040

Galloway, J. N., Aber, J. D., Erisman, J. W., Seitzinger, S. P., Howarth, R. W., Cowling, E. B., & Cosby, B. J. (2003). The nitrogen cascade. *BioScience*, *53*, 341–356.

Glibert, P. M., Anderson, D. M., Gentien, P., Graneli, E., & Sellner, K. G. (2005). The global complex phenomena of harmful algal blooms. *Oceanography*, *18*, 136–147. https://doi.org/10.5670/oceanog.2005.49

Harrison, J. A., Maranger, R. J., Alexander, R. B., Giblin, A. E., Jacinthe, P. A., Mayorga, E., et al. (2009). The regional and global significance of nitrogen removal in lakes and reservoirs. *Biogeochemistry*, *93*, 143–157.

Havens, K.E. (2018). *The future of harmful algal blooms in Florida inland and coastal waters*. University of Florida IFAS Extension TP-231, 4 p. Retrieved from http://edis.ifas.ufl.edu.

Lopez, C.B., Jewett, E.B., Dortch, Q., Walton, B.T., & Hudnell, H.K. (2008c). *Scientific assessment of freshwater harmful algal blooms*. Washington, DC: Interagency Working Group on Harmful Algal Blooms, Hypoxia, and Human Health of the Joint Subcommittee on Ocean Science and Technology, 78 p.

Pinder, R. W., Bettez, N. D., Bonan, G. B., Greaver, T. L., Wieder, W. R., Schlesinger, W. H., & Davidson, E. A. (2012b). Impacts of human alteration of the nitrogen cycle in the US on radiative forcing. *Biogeochemistry*. https://doi.org/10.1007/s10533-012-9787-z

Porter, S. D., Mueller, D. K., Spahr, N. E., Munn, M. D., & Dubrovsky, N. M. (2008). Efficacy of algal metrics for assessing nutrient and organic enrichment in flowing waters. *Freshwater Biology*, *53*(5), 1036–1054.

Preston, S.D., Alexander, R.B., Woodside, M.D., & Hamilton, P.A. (2009). *SPARROW modeling: Enhancing understanding of the Nation's water quality*. U.S. Geological Survey Fact Sheet 2009–3019, 6 p.

Scavia, D., & Bricker, S. B. (2006). Coastal eutrophication assessment in the United States. *Biogeochemistry*, *79*(1–2), 187–208.

State-USEPA Nutrient Innovations Task Group. (2009). *An urgent call to action. State-USEPA Nutrient Innovations Task Group Report to USEPA*, (August 2009). 170 p

Sutton, M. A., Reis, S., & Butterbach-Bahl, K. (2009). Reactive nitrogen in agro-ecosystems: Integration with greenhouse gas interactions. *Agriculture, Ecosystems and Environment*, *133*, 135–138.

U.S. Environmental Protection Agency. (2013). Final aquatic life ambient water quality criteria for ammonia-Freshwater 2013. *Notice in Federal Register*, *78*(163), 52192–52194.

U.S. Environmental Protection Agency. (2016). *Water quality assessment and TMDL information*. Washington, DC: United States Environmental Protection Agency. Available at https://ofmpub.epa.gov/waters10/attains_index.home, accessed 05 June 2018

7
Groundwater Contamination from Reactive Nitrogen

As discussed in Chapter 3, several point and nonpoint sources of reactive nitrogen at the land surface can infiltrate into the subsurface and result in contamination of groundwater. Over the past several decades, many studies have documented the widespread nitrate contamination of groundwater from the use of synthetic fertilizers in agriculture, mineralization of animal wastes, fixation by crops, land application of treated municipal wastewater and biosolids, effluent from septic tanks, and atmospheric deposition of reactive nitrogen from fossil fuel combustion and biomass burning. Shallow unconfined aquifers are particularly vulnerable to contamination from leaching of reactive nitrogen (mainly nitrate). As a result, there are serious concerns related to human health and nitrous oxide emissions (a potent greenhouse gas) to the atmosphere. Chapter 6 discusses in detail some of the adverse effects on human health related to the consumption of elevated nitrate concentrations in drinking water from aquifers. This chapter discusses groundwater contamination from reactive nitrogen focusing on three main areas: the United States, Europe, and China. Other topics presented in this chapter include interchange of reactive nitrogen between groundwater and surface water, impacts from storage and legacy reactive nitrogen, nitrogen contamination of drinking water, and modeling vulnerability of aquifers to nitrate contamination and nitrogen fate and transport in various groundwater systems.

To better understand how reactive nitrogen can travel through the subsurface and end up as a contaminant in a groundwater system, it would be useful to briefly review several hydrogeologic terms. Recharge water (from precipitation or irrigation water) can transport reactive forms of nitrogen first via leaching through the soil zone and continue to infiltrate through unsaturated (or vadose) zone (Fig. 7.1). The unsaturated zone is the area below the land surface and above the water table that is composed of soil, rocks, and air in pore spaces. Below the water table is the saturated zone, which is composed of saturated permeable materials (rock, gravel, and sand). This zone is an aquifer if it yields useful quantities of water to a well. The top of an aquifer that is unconfined is at the water table and water moves directly from the land surface through the unsaturated zone into the aquifer. In contrast, the top of a confined aquifer occurs at a confining layer, which generally is composed of low permeability material (typically clay) that impedes the flow of water into the aquifer. Occasionally, the confining layer is breached (e.g., by a well or sinkhole) and water can flow from an unconfined aquifer into a confined aquifer.

Nitrogen leaching from soils and the vadose zone to groundwater in agricultural areas depends on several factors: (a) fertilization level, type, and timing of fertilizer application; (b) the method of their application to the soil; (c) properties of soils (i.e., pH, structure, and organic matter content), (d) types of crops and their fertilizer requirements; (e) plant nutrient uptake, method of cultivation, and agronomic practices; and (f) the level of animal production (Bechmann, 2014; Kyllmar et al., 2014a, 2014b). Weather conditions and catchment land use also have a crucial impact on the intensity and quantity of nitrogen leaching (Jiang et al., 2014; Yoon, 2005; Woli et al., 2008).

Uptake of nitrogen by plants in soils is related to several factors including the availability of macroelements (e.g., potassium, phosphorus, carbon, and nitrogen) and microelements (e.g., boron, iron, copper, and molybdenum) in the soil, and the mass ratios between elements (e.g., Szczepaniak et al., 2013). Losses of nitrogen (i.e., leaching to groundwater) can be enhanced by deficiencies in phosphorus or potassium, which limits uptake of nitrogen by plants (Lawniczak et al., 2009). This issue may concern two thirds of the world's agricultural land where potassium deficiency occurs (Römheld & Kirkby, 2010).

The percentage of nitrate leached to groundwater generally ranged from 10 to 50% based on several field studies in agricultural areas conducted by the US Geological Survey (USGS) (Dubrovsky et al., 2010).

Nitrogen Overload: Environmental Degradation, Ramifications, and Economic Costs, Geophysical Monograph 250, First Edition. Brian G. Katz.
© 2020 American Geophysical Union. Published 2020 by John Wiley & Sons, Inc.

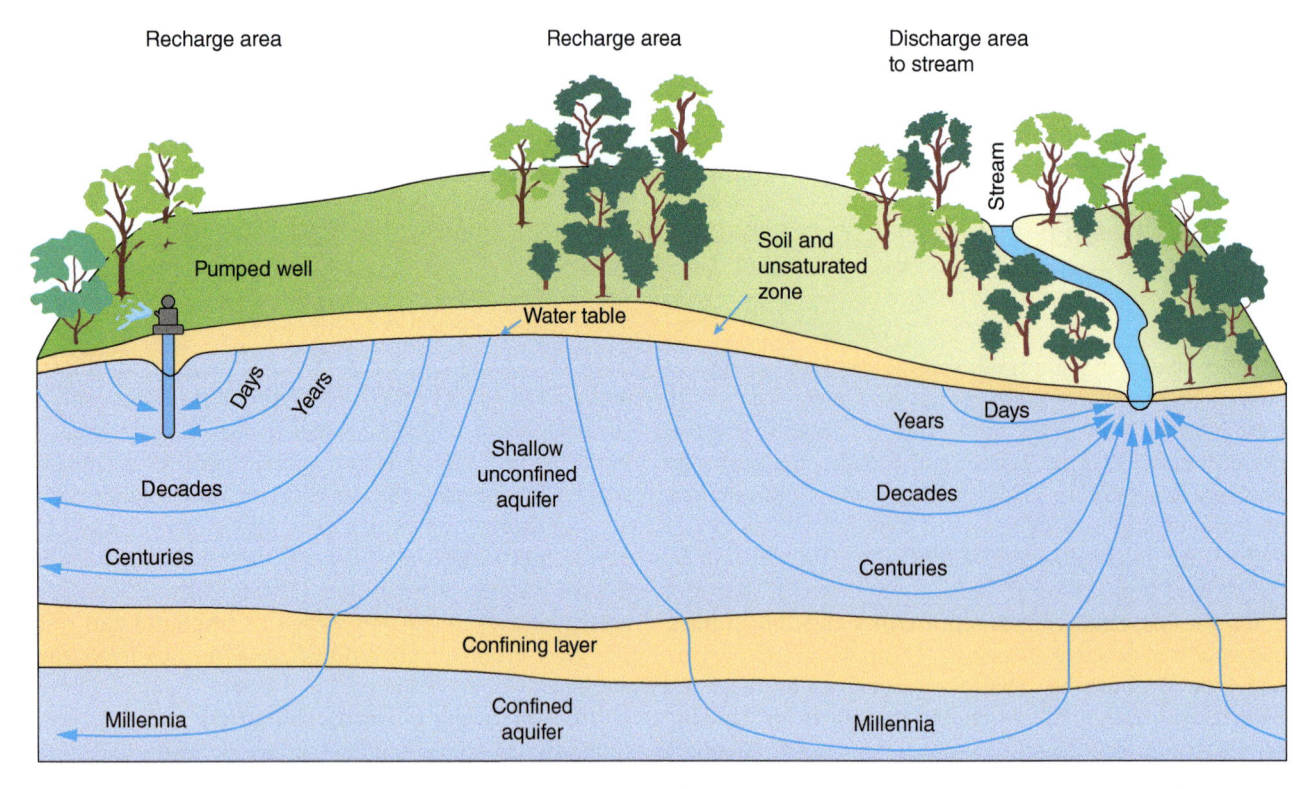

Figure 7.1 Conceptual model showing generalized groundwater flow paths and travel times in recharge and discharge areas for unconfined and confined aquifers (modified after Puckett et al., 2011, ES&T). Groundwater travel times and nitrate concentrations are affected by several factors such as aquifer hydraulic properties, slope, depth, and geochemical (e.g., redox) conditions. If water samples are collected along groundwater flow paths, it would be possible to assess the chronology of nitrate contamination in the aquifer. Public domain. *Source*: Reproduced with permission of American Chemical Society.

They found relatively low amounts of nitrate leaching (less than 20%) in areas where tile drains divert infiltrating groundwater to streams, or where fine-grained sediments impede the vertical movement of water and promote conditions that remove nitrate (denitrification) before it enters the aquifer. These studies found relatively higher nitrate leaching to aquifers that underlie areas with high nitrogen inputs to the land surface (e.g., agricultural cropland), and in areas with coarse-grained (well-drained) soils and rapid rates of vertical movement through the unsaturated zone.

7.1. REACTIVE NITROGEN CONTAMINATION OF GROUNDWATER IN THE UNITED STATES OF AMERICA

Groundwater is an extremely important resource in the United States that is heavily used by agriculture, various industries, and for drinking water (public supply and private domestic wells). In 2015, approximately 85 billion gallons of groundwater were pumped daily from 62 principal aquifers in the United States (Dieter et al., 2018). An estimated 130 million people use groundwater for drinking water, with about 43 million of those people relying on groundwater from domestic (or privately owned) wells for drinking water (DeSimone et al., 2014). As noted by DeSimone et al. (2009; http://info.ngwa.org/GWOL/pdf/091384002.pdf, accessed 27 December 2018), individual homeowners are responsible for maintenance and ensuring the quality of water from domestic wells because drinking water quality from these private wells are not regulated by the Safe Drinking Water Act (as public supply wells [PSWs] are) or, in many cases, by state laws.

The USGS National Water Quality Assessment Program (NAWQA) evaluated nitrate concentrations in water samples from 5101 wells during 1991–2003, that were part of 189 land-use and major aquifer study networks (Dubrovsky et al., 2010). The main emphasis of the NAWQA land-use studies was on shallow groundwater (depths typically less than 6m below the water table) mostly within agricultural and urban land-use settings, although comparisons were made with data from undeveloped areas. The wells sampled included new or existing observation wells or domestic supply wells. Each major aquifer study sampled about 20–30 randomly located wells within each targeted land-use area. Data analyses were based on one sample per well, as water quality in most groundwater systems changes relatively slowly over time compared to surface water systems. Studies also were conducted on how nitrate (and other nutrients) concentrations change along groundwater flow paths from the point of recharge at the land surface to the point of discharge (e.g., to streams, lakes, or pumped water supply well) (Fig. 7.1). These "flow-system studies" included transport of anthropogenic and natural contaminants to drinking-water wells, and transport of agricultural chemicals in groundwater to streams. More detailed information about each major aquifer study and the flow-system studies is summarized in numerous NAWQA reports and journal articles (see http://water.usgs.gov/nawqa/, accessed 20 December 2018).

Based on nitrate-N data from 419 wells in relatively undeveloped areas of the United States, Dubrovsky et al. (2010) reported a background concentration of 1.0 mg/L. The 419 background wells included 320 wells from previous studies (Nolan & Hitt, 2006) and an additional 99 wells sampled in the NAWQA studies. The background concentration for ammonia-N (0.10 mg N/L) was based on samples from 177 wells. Median nitrate concentrations were greater than the groundwater background concentration in 64% of 86 NAWQA shallow aquifer studies in areas underlying agricultural or urban land use. Concentrations of ammonia were not greater than background levels in most areas.

Nitrate-N concentrations were highest in groundwater beneath the agricultural land uses (median of 3.1 mg N/L), compared to the lower median concentration (1.4 mg N/L) in urban land-use area (Dubrovsky et al., 2010) (Fig. 7.2). The overall median nitrate-N concentration in all principal aquifer samples was 0.05 mg N/L. Higher median nitrate-N concentrations were found in unconsolidated sand and gravel (nonglacial origin) and basaltic and volcanic-rock aquifer lithology groups that have higher permeabilities compared to the other aquifer lithologies (Fig. 7.3). Conversely, water from sandstone and semiconsolidated sand and gravel aquifers typically had the lowest nitrate concentrations within each land-use category. The main factors that accounted for large variations in groundwater nitrate concentrations across the United States included differences in (a) the types of nitrogen sources (e.g., fertilizers, manure, atmospheric deposition, and wastewater); (b) physical properties of soils and unsaturated zone material; (c) aquifer lithology and hydrogeology; (d) redox conditions; and (e) groundwater age. Nitrate-N concentrations in shallow groundwater were higher in the northeast, Midwest, and Northwest United States, and like were related to high nitrogen inputs and physical and chemical conditions that enhanced nitrate transport in groundwater. Nitrate concentrations were highest in shallow, oxic groundwater that receives high inputs of nitrogen from fertilizer, manure, and atmospheric deposition. Aquifers with high permeability (lithologies consisting of unconsolidated sand and gravel (nonglacial origin), carbonate rock and basaltic and volcanic-rocks) had relatively high nitrate-N concentrations (greater than 2 mg N/L) in groundwater regardless of land use. Conversely, groundwater in sandstone and semiconsolidated sand and gravel aquifers consistently had the lowest nitrate concentrations within each land-use category.

Dubrovsky et al. (2010) also explored the relationship between groundwater nitrate concentrations and age of groundwater relative to redox conditions. Groundwater age information was obtained from 1885 water samples analyzed for tritium (^{3}H, a radioactive isotope of hydrogen was introduced into the atmosphere following periods of testing of nuclear devices that began in 1952 and reached a maximum in 1963–1964) (referred to as the "bomb peak"; https://www.ldeo.columbia.edu/~martins/isohydro/3h3hedating/tritium_he_dating.html, accessed 26 December 2018). Tritium entered groundwater via recharge from rainfall containing elevated tritium levels and can be used to distinguish between groundwater recharged before 1952 (old groundwater) and after 1952 (young groundwater). Higher nitrate-N concentrations were found in young oxic groundwater compared to groundwater with mixed or reduced redox conditions.

DeSimone et al. (2014) summarized nitrate and other water quality data collected during 1991–2010 from three types of groundwater studies in the United States as part of the NAWQA: (a) parts of 41 principal aquifers used for drinking water supply (from 3669 mostly existing domestic wells) beneath mixed land uses, (b) shallow groundwater from 22 aquifers beneath agricultural land (from 1793 mostly monitoring wells installed for NAWQA), and (c) shallow groundwater from 22 aquifers beneath urban land (from 1158 mostly monitoring wells installed for NAWQA). Typically, there were 20–30 wells randomly selected for the sampling network in each principal aquifer. Drinking water exceedances were summarized for nitrate in water samples collected during 1991–2010 from 41 principal aquifers in the United States that were grouped by eight rock types (basalt and other volcanic rocks, crystalline, carbonate, sandstone and carbonate, sandstone, semiconsolidated sand, glacial, and unconsolidated sand and gravel). They found that nitrate exceedances were highly variable when separated out by the following: (a) parts of aquifers used for drinking water, (b) shallow groundwater beneath agricultural land, and (c) shallow groundwater beneath urban land. For all principal aquifers, 4.14% of samples exceeded the drinking water maximum contaminant level (MCL) for nitrate compared to 22.2% in aquifers beneath agricultural land

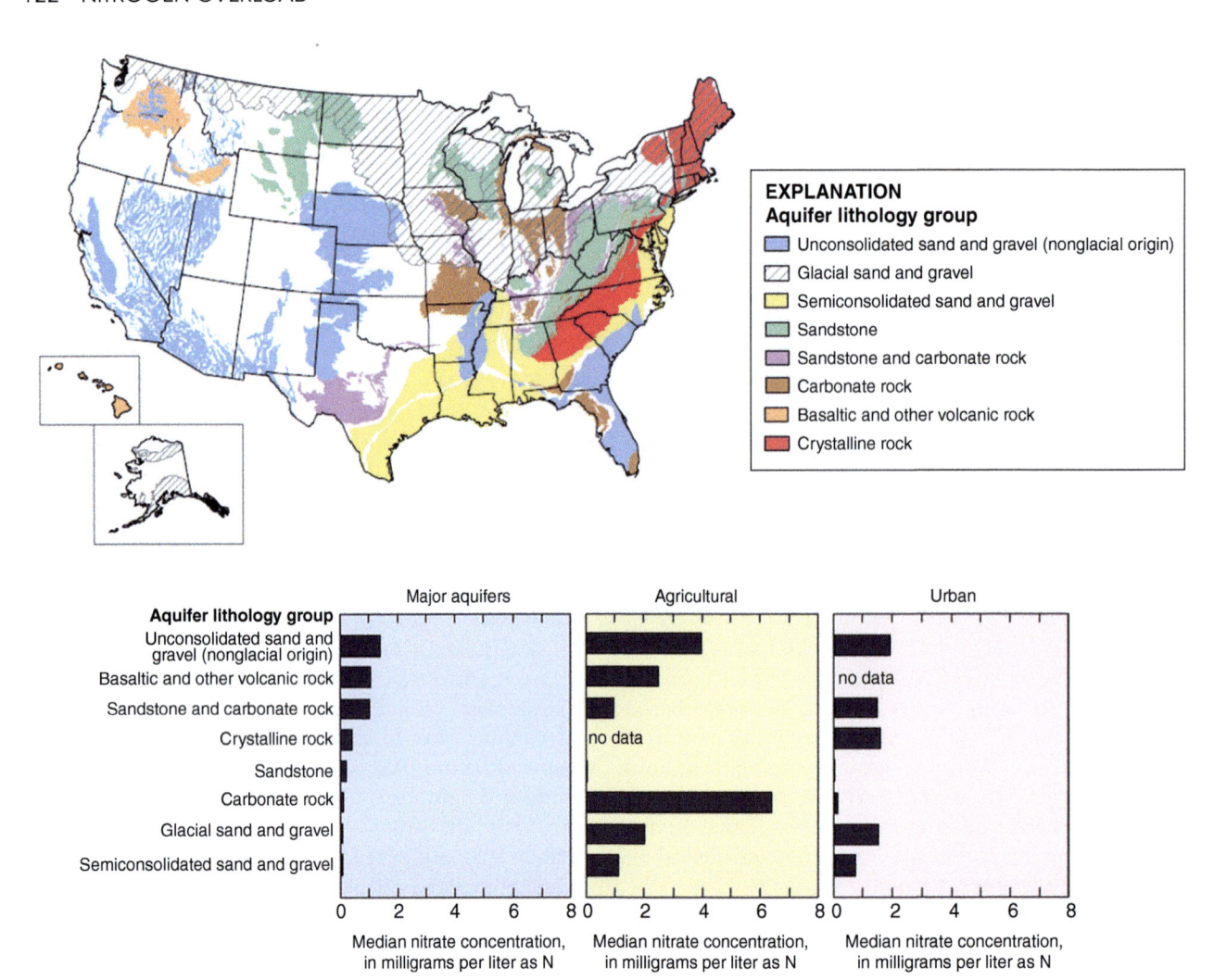

Figure 7.2 Map showing the eight different lithologies for principal aquifers in the United States investigated as part of the NAWQA Program. Graphs show the overall median nitrate-N concentration for each principal aquifer and median nitrate-N concentrations for the various principal aquifer lithologies grouped by agricultural and urban land uses. *Source*: From Dubrovsky et al. (2010), Figure 4–28, p. 81. Public domain. Reproduced with permission of United States Geological Survey.

and 4.2% in aquifers beneath urban land (Table 7.1). The highest percentages of drinking water exceedances by rock type were found in aquifers in unconsolidated sand and gravel and semiconsolidated sand, and carbonate. DeSimone also noted that deep parts of some aquifers are more vulnerable to contamination by nitrate due to irrigation and pumping.

The NAWQA studies found that geology was an important factor in the occurrence and concentration of nitrate-N in principal aquifers in the United States. Nitrate concentrations were relatively high in several unconsolidated sand and gravel aquifers in the western United States and in agricultural areas of the Piedmont and Blue Ridge and the Valley and Ridge carbonate-rock aquifers in the mid-Atlantic region. These aquifers generally have high permeability and oxic conditions that favor rapid infiltration of recharge with elevated nitrate concentrations and the persistence of nitrate in the aquifer. DeSimone et al. (2014) found higher nitrate concentrations in almost all shallow groundwater in agricultural areas compared to deeper groundwater used for drinking-water supply. Also, groundwater nitrate concentrations were higher in shallow wells in urban areas than in deeper groundwater. Concentrations of nitrate in the carbonate-rock aquifers were among the highest in the United States. These aquifers supply a large suburban and rural population in the eastern United States with drinking water.

The highest percentages of nitrate exceedances were in carbonate aquifers beneath agricultural land (DeSimone et al., 2014). These results were consistent with other studies in carbonate-rock aquifers that are highly susceptible to

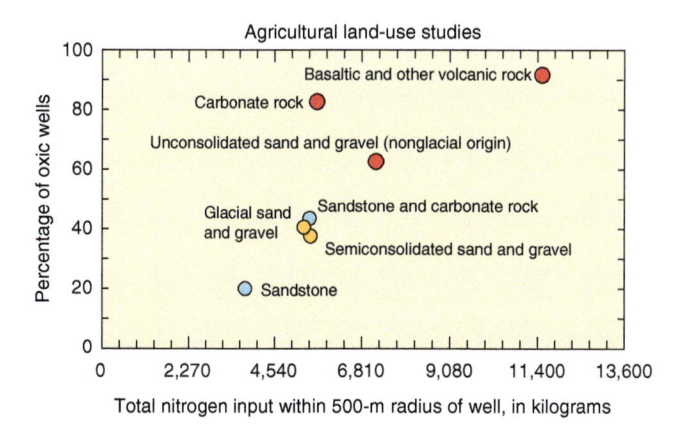

Figure 7.3 Plot showing the percentage of wells in oxic groundwater from agricultural land-use studies with respect to total nitrogen input within a 500-m radius of well. Aquifers with permeable lithologies, such as basaltic and volcanic-rock aquifers, and karst aquifers, are most vulnerable to high nitrate concentrations. *Source*: Modified from Dubrovsky et al. (2010); pounds converted to kilograms. Figure 4–30, p. 82. Public domain. Reproduced with permission of United States Geological Survey.

contamination from human activities at land surface due to the presence of karst (solution) features. Lindsey et al. (2009) studied the water quality in 12 carbonate aquifers across the United States for 1042 samples collected during 1993–2005. They found that nitrate concentrations were significantly higher in unconfined aquifers than in confined aquifers. Aquifers with oxic waters had significantly higher nitrate concentrations than those containing anoxic waters, regardless of land use near the well. Lindsey et al. (2009) found that 5% of sampled wells exceeded the drinking water MCL and most of these were domestic wells in agricultural areas.

Numerous studies have investigated the transport of reactive nitrogen from the disposal of human wastewater sources in the United States. In Florida, treated municipal wastewater typically is applied to the land surface in sprayfields, rapid infiltration basins, absorption fields, and percolation ponds. Due to the mobility of nitrate and the vulnerable karstic Upper Floridan aquifer, studies have documented the occurrence of nitrate contamination of groundwater and springs from the land application of treated municipal water at sprayfields in northern Florida (Davis et al., 2010; Katz & Griffin, 2008; Katz et al., 2010). Ammonium concentrations generally are very low in most groundwater systems; however, elevated ammonium concentrations in groundwater can occur from wastewater sources at the land surface (i.e., treated wastewater and septic tanks). Böhlke et al. (2006) used a combination of methods (e.g., isotopic analyses of coexisting ammonium, nitrate, nitrogen gas, and sorbed ammonium; groundwater age dating; and in situ natural gradient tracer tests with numerical simulations of breakthrough data) to study various processes controlling the movement of ammonium in a treated wastewater plume in a glacial outwash aquifer on Cape Cod, Massachusetts, US. A contaminant plume containing ammonium resulted from local artificial recharge of treated wastewater for more than 50 years (from 1936 to 1995). They noted that the migration and transformation of ammonium was related to redox conditions; for example, in the anoxic core of the plume, no evidence was found for nitrification or anaerobic ammonium oxidation (anammox) reactions that would consume ammonium. Also, the transport of ammonium in the plume was substantially delayed (by a factor of 3–6) with respect to the movement of groundwater and the more mobile constituents. Based on multiple lines of evidence, it was concluded

Table 7.1 Percentages of samples that exceeded the drinking water MCL for nitrate as N for each rock type for principal aquifers in the United States. Number in parentheses is the number of principal aquifers sampled collected 1991–2010 in each of rock type and for wells located in agricultural and urban land settings.

Rock type	Aquifers used for drinking water	Aquifers beneath agricultural land	Aquifers beneath urban land
All aquifers	4.14 (41)	22.23 (26)	4.24 (22)
Basalt and other volcanic rock	0–3.33 (3)	12–24.4 (2)	~
Crystalline	0–6.51 (3)	~	1.96 (1)
Carbonate	0–7.69 (8)	0–60.0 (5)	0 (3)
Sandstone and carbonate	0–4.35 (2)	0 (1)	0 (1)
Sandstone and carbonate	0–5.08 (6)	0–27.3 (3)	5.26–15 (2)
Semiconsolidated sand	0–12.71 (5)	3.7–22.9 (3)	0–5.56 (4)
Glacial	3.33 (1)	22.7 (1)	4.21 (1)
Unconsolidated sand and gravel	0–30.77 (13)	5.56–56.7 (11)	0–20.8 (10)
Total number of samples	**3669**	**1793**	**1187**

that the main mass of ammonium could reach a discharge area without considerable attenuation a long time after the more mobile constituents (e.g., nitrate) is gone.

As noted previously, groundwater tends to move very slowly through most aquifers. Typical flow rates can range from 0.3 m/yr to as low as 0.3 m per decade (Alley et al., 1999). DeSimone et al. (2014) noted that even though groundwater quality changes slowly in most aquifers, indicators of anthropogenic influence on groundwater quality have increased across the United States over relatively short time periods of single decades. To investigate how groundwater quality is changing over time, the NAWQA Program has sampled well networks in groundwater study areas at 10-year intervals. Samples have been collected from 1295 wells in 56 groundwater studies so far. These included principal aquifer studies (which constituted about one third) and land-use studies made up the remainder. Concentrations of nitrate (along with dissolved solids and chloride) in groundwater increased in two thirds of well networks that were sampled at 10-year intervals between the early 1990s and 2010. This was particularly evident in shallow groundwater beneath agricultural and urban areas. Nitrate-N concentrations exceeded human-health benchmarks two to four times more frequently in shallow groundwater beneath agricultural and urban land than in groundwater from the deeper parts of aquifers currently used for drinking water. However, as groundwater moves downward from the shallow part of an aquifer to the deeper parts of an aquifer, nitrate-N concentrations likely will increase. This has several human health and economic ramifications because treatment and restoration of deep contaminated groundwater supplies can be very costly and difficult and may take decades to see improvement. DeSimone et al. (2014) found that contaminant inputs from more than 30 years ago are still being seen in parts of many aquifers across the United States. The USGS NAWQA Program also provides up-to-date information on their website (https://nawqatrends.wim.usgs.gov/Decadal/, accessed 28 December 2018) regarding groundwater quality trends over time for nitrate and other constituents throughout the United States. Legacy nitrogen in groundwater and storage in the vadose zone are discussed in more detail in Section 7.5.

7.2. REACTIVE NITROGEN CONTAMINATION OF GROUNDWATER IN EUROPE

Groundwater is an important resource in Europe, where about two thirds of the population relies on groundwater for domestic supply. Grizzetti et al. (2011) noted that because groundwater is a finite and slowly renewed resource, degradation of groundwater quality used for a drinking water source in Europe has become a serious concern. In many parts of Europe, nitrate concentrations in groundwater have remained high and their persistence in groundwater threatens this important resource. Nitrate in groundwater has been directly linked to the amount of nitrogen applied in agricultural land, and specifically to the nitrogen surplus (Grizzetti et al., 2011). The nitrogen surplus typically is defined as the annual difference between agricultural nitrogen inputs and outputs and represents the amount of nitrogen not directly used in the agricultural production. The surplus of nitrogen for Europe (which has a high risk of leaching to groundwater) was estimated by Bouraoui et al. (2009) at 11.5 Tg for the year 2000 and 10 Tg for 2005. This estimated surplus did not include volatilization from manure, which contributes ammonia to the environment.

In 1991, the European Union introduced the Nitrates Directive, which was designed to reduce groundwater and surface water pollution resulting from agricultural sources of nitrate. The Directive required Member States to promote the use of agricultural action program measures throughout their whole territory or within discrete nitrate vulnerable zones. As summarized in the comprehensive European Nitrogen Assessment (Sutton et al., 2011a), the long-term quality of groundwater and surface water resources are threatened by high reactive nitrogen concentrations in European waterbodies along with increasing nitrate in groundwater. Given the accumulation of nitrogen in soils, sediments, and aquifers coupled with the long expected residence time of nitrate in aquifers, past fertilizer strategies likely will impact the quality of groundwater in Europe for many decades (Grizzetti et al., 2011).

About 3% of the population in 15 European states (EU-15; Austria, Belgium, Denmark, Finland, France, Germany, Greece, Ireland, Italy, Luxembourg, Netherlands, Portugal, Spain, Sweden, and the United Kingdom) that uses drinking water from groundwater resources is potentially exposed to concentrations exceeding the standard for drinking water of 50 mg NO_3/L (11.2 mg N/L), with 5% of the population chronically exposed to concentrations exceeding 25 mg NO_3/L (5.6 mg N/L), which may double the risk of colon cancer for above median meat consumers (Grizzetti et al., 2011).

The Third Assessment Report on the Implementation of the Nitrates Directive (European Commission, 2007; http://ec.europa.eu/environment/water/water-nitrates/reports.html, accessed 16 December 2018) indicated that for the period 2000–2003, about 17% of the wells in EU15 exhibit a nitrate concentration above the limit of 50 mg/L (about 3% of the EU-15 population is potential exposed to concentrations that exceed this limit). An additional 7% of wells were in the range between 40 and 50 mg/L, while about 60% were below 25 mg/L. About 30% of the reported wells showed an improvement in nitrate concentrations based on a comparison of earlier data; however,

36% of wells showed an increasing nitrate trend. The last assessment report for the period 2004–2007 indicated that nitrate contamination of groundwater was still present in 34% of the monitoring sites, with 15% of the wells with nitrate concentration above 50 mg/L. Most of the nitrate contamination of groundwater in Europe originates from anthropogenic sources (mainly agricultural activities), as background (natural) nitrate concentrations are very low (Grizzetti et al., 2011).

The Nitrates Directive has been effective in some areas. Across the 27 EU Member States (EU-27; Austria, Belgium, Bulgaria, Croatia, Republic of Cyprus, Czech Republic, Denmark, Estonia, Finland, France, Germany, Greece, Hungary, Ireland, Italy, Latvia, Lithuania, Luxembourg, Malta, Netherlands, Poland, Portugal, Romania, Slovakia, Slovenia, Spain, Sweden, and the United Kingdom), 39.6% of the area is subject to the implementation of action programs. The new Groundwater Directive (2006; https://www.eea.europa.eu/policy-documents/groundwater-directive-gwd-2006-118-ec, accessed 29 December 2018) states that nitrate concentrations must not exceed the trigger value of 50 mg NO_3/L. Also, several Member States have set their own more stringent limits to improve water quality. Between 2004 and 2007, nitrate concentrations in surface water remained stable or fell at 70% of monitored sites. Nitrate concentrations at 66% of groundwater monitoring points are stable or improving (http://ec.europa.eu/environment/pubs/pdf/factsheets/nitrates.pdf, accessed 17 December 2018).

A comparison of two datasets for groundwater monitoring stations in Europe during 2000–2012 showed an overall slight increase in groundwater nitrate concentrations. Based on a dataset containing 400 stations in 15 countries (1992–2012), the average annual nitrate concentration in 2012 was 4.0 mg N/L. A larger dataset containing 1242 stations in 25 countries (2002–2012) consistently had higher nitrate-N concentrations compared to the one for 15 countries and had an average nitrate concentration in 2012 of 4.3 mg N/L (https://www.eea.europa.eu/data-and-maps/, accessed 8 January 2019).

Schullehner and Hansen (2014) showed that nitrate concentrations in public water supply have been decreasing since the 1970s in Denmark; however, in contrast, nitrate levels have been increasing in water from private wells. They attributed the decrease of nitrate levels in drinking water to structural changes (i.e., more treatment), but they concluded that overall groundwater quality has not improved in Denmark.

A recent report by the European Commission (http://ec.europa.eu/environment/water/water-nitrates/pdf/nitrates_directive_implementation_report.pdf, accessed 15 December 2018) indicated that during the period 2012–2015, 13.2% of groundwater stations (34,901 in 28 European countries, sampled on average twice per year but less than once per year in Denmark, Latvia, Poland, and Sweden to around five times per year in Belgium and Croatia) exceeded 50 mg/L. This represented a slight improvement compared to the previous reporting period where 14.4% of stations exceeded the limit and 5.9% had nitrate concentrations between 40 and 50 mg/L. The report noted that some member states (Ireland, Finland, and Sweden) on average had no groundwater stations exceeding 50 mg/L. However, exceedances in Malta, Germany, and Spain were 71, 28, and 21.5%, respectively. A higher proportion of stations with exceedances of nitrate concentrations were observed for groundwater depths ranging from 5 to 15 m. The report compared groundwater nitrate concentrations from 2012 to 2015 with those from 2008 to 2011. Overall, water quality remained the same or improved in 74% of the stations (42% no change and 32% showed a decrease), but nitrate concentrations increased in 26% of the stations. The highest percentage of stations showing improvements were in Bulgaria (40.9%), Malta (46.3%), and Portugal (43.6%). Areas with the highest percentage showing increasing trends included Estonia (44.4%), Malta (43.9%), and Lithuania (58.5%). In some countries (e.g., Malta), it appears that areas with contaminated groundwater are getting further degraded; whereas, other less-contaminated areas have shown improvements in nitrate concentrations. However, based on these European studies and those conducted in other areas, nitrate concentrations in groundwater likely will be impacted for many decades from past and current anthropogenic activities for many decades. As it was noted in the report, nitrogen surplus continues to remain high in many countries, and there are substantial amounts of reactive nitrogen stored in soils unsaturated zone, and in aquifers. There have been concerns that in Eastern European countries, agricultural activities and fertilization rates may increase; whereas fertilization rates in some countries of Western Europe may stabilize at high levels. Therefore, it is essential to continue to assess the effects on human health and ecosystems health from the elevated reactive nitrogen levels.

The European Environment Agency (EEA) presents a map on their website (https://www.eea.europa.eu/data-and-maps/explore-interactive-maps/nitrates-in-groundwater-by-countries, accessed 08 January 2019) that shows the mean annual concentration of nitrate, expressed as milligrams of NO_3-N per liter of water, as reported by EEA member countries for their groundwater monitoring stations. Additional information on regional and global groundwater resources is available from the International Groundwater Resources Assessment Centre (IGRAC; https://www.un-igrac.org/, accessed 31 December 2018). IGRAC was launched in 2003 as a UNESCO center that works under auspices of

World Meteorological Organization and is supported by the Government of the Netherlands. IGRAC has established a Global Groundwater Monitoring Network (GGMN) that is a web-based network of participatory networks designed to improve the quality and accessibility of groundwater monitoring information. The GGMN portal has web-based software that can provide spatial and temporal analysis of monitoring information. Several products might be of interest to the reader. One of IGRAC's products is the Global Groundwater Information System (GGIS). This system is a web-based interactive Geographic Information System that supports the storage, visualization, analysis, and sharing of groundwater data and information through map-based modules for regional and transboundary-level assessments. Another product is a 2015 report entitled, "A global assessment of nitrate contamination in groundwater" (https://www.un-igrac.org/resource/global-assessment-nitrate-contamination-groundwater, accessed 31 December 2018). This report provided basic information on methods for generating a global map of aquifer vulnerability to nitrate contamination, although the dataset is incomplete, and the vulnerability was only calculated for ten aquifers around the world.

7.3. REACTIVE NITROGEN CONTAMINATION OF GROUNDWATER IN CHINA

Extensive nitrate contamination of groundwater in China has resulted from the substantial increase in socioeconomic development, urbanization, and industrialization since the 1970s (Fu et al., 2007). China has a national water quality assessment program (Ministry of Environmental Protection, MEP) that has been monitoring groundwater quality in 182 main metropolitan areas. The program found that the water quality of 57% (MEP, 2010) of all monitoring wells does not meet the clean groundwater standard (<20 mg N/L nitrate concentration, MEP, 1993). To protect and to more effectively manage groundwater resources, the State Council of China approved the "National Groundwater Pollution Prevention Plan (NGPPP)" in 2011. The NGPPP proposed to invest about 6 billion US dollars to monitor groundwater pollution sources and remediate polluted groundwater before 2020. The plan would also quantify sources of reactive nitrogen loading to groundwater in highly vulnerable areas.

Gu et al. (2013) conducted a detailed study that investigated the changes in the sources of groundwater nitrate on the spatiotemporal scale and the factors impact groundwater nitrate concentrations in China. Their investigation included (a) a literature review and analysis to understand the current status of groundwater nitrate contamination and the effects of land-use type on groundwater nitrate concentrations, (b) an analysis of the contribution from various reactive nitrogen sources based on the coupled human and natural systems (CHANS) approach to identify and quantify specific sources of groundwater nitrate both on temporal (from 1980 to 2008) and spatial (provincial) scales, (c) quantification of the driving forces of changes in groundwater nitrate concentrations related to natural and human factors, (d) an estimate of the spatial distribution of the sources at the county level, which they compared with the results of the literature analysis to assess the accuracy of their analysis of the relative contribution of nitrogen from various sources, (e) a detailed high-resolution map of the relative contribution to groundwater nitrate concentrations from various sources in China, which provides important information for modeling of future scenarios involving groundwater nitrate dynamics. Gu et al. (2013) reported that nitrate was detected in 96% of groundwater samples based on a common detection threshold of 0.2 mg N/L, and 28% of groundwater samples exceeded WHO's MCL (10 mg N/L). Groundwater nitrate concentrations were the highest beneath industrial land (median: 34.6 mg N/L), followed by urban land (10.2 mg N/L), cropland (4.8 mg N/L), and rural human settlement (4.0 mg N/L), with the lowest found beneath natural land (0.8 mg N/L).

During the period 1980–2008, total reactive nitrogen leakage to groundwater in China increased about 1.5 times, from 2.0 to 5.0 Tg N/yr (Gu et al., 2013). They also noted that even though the contribution of cropland to the total amount of reactive nitrogen leakage to groundwater was reduced from 50 to 40% during the past three decades, cropland still was the single largest source (Fig. 7.4). The contribution to nitrogen leakage to groundwater from landfills rapidly increased from 10 to 34%. Their study also found that high amounts of reactive

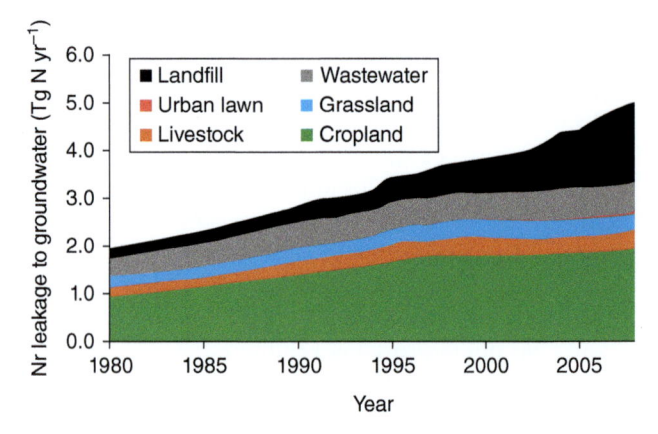

Figure 7.4 Graph showing the increases in reactive nitrogen (Nr) leakage to groundwater in China from various sources. *Source*: From Gu et al. (2013). Reproduced with permission of Elsevier.

N leakage occurred mostly in relatively developed agricultural or urbanized regions with a large population. Gu et al. (2013) concluded that the amount of reactive nitrogen leakage to groundwater was mainly related to and controlled by several anthropogenic factors (such as population, gross domestic product, rate of urbanization, and land-use type). They concluded that their high-resolution maps of reactive nitrogen source contributions could be used for policy development on mitigation of groundwater contamination.

Liu et al. (2017) noted that the Shongdong province was one of the most intensive agricultural areas in the North China Plain (called "China's Granary") and many districts in the province have reported extensive nitrate contamination of groundwater. Their study focused on the concentration and spatial and temporal distribution of groundwater nitrate-N under cropland in Shandong province using statistical and geostatistical techniques. They found that nitrate-N concentration reached a maximum of 184 mg/N and 29.5% of samples had levels in excess of the safety threshold concentration (20 mg N/L). Liu et al. (2017) also found that median nitrate-N concentrations were significantly higher after the rainy season compared to those before the rainy season. Their study also found that groundwater nitrate-N concentrations were significantly higher under vegetable and orchard areas than those under grain. Although groundwater in many districts was heavily contaminated (e.g., Weifang, Linyi in Shandong province), there were no significant trends of nitrate-N for most cities. It was concluded that the high spatial variability of groundwater nitrate concentrations was significantly related to the following variables: vegetable yield per unit area, percentages of orchard area, per capita agricultural production, nitrogen fertilizer application rate per area, livestock per unit area, percentages of irrigation areas, population per unit area, and annual mean temperature.

7.4. INTERCHANGE OF REACTIVE NITROGEN BETWEEN GROUNDWATER AND SURFACE WATER

In many hydrogeologic settings, groundwater and surface water systems are inextricably linked and allow exchange between the two systems. Reactive nitrogen species can discharge from groundwater and contaminate streams, lakes, rivers, coastal areas, and oceans. In some instances, particularly during flood or high-water conditions, surface water can infiltrate into and recharge groundwater systems.

7.4.1. Groundwater Discharge of Reactive Nitrogen to Streams

Groundwater discharge to streams is the principal component of stream base flow. During base-flow conditions, groundwater contributions of reactive nitrogen can be considerable, although they often have been difficult to quantify. Spahr et al. (2010) investigated the proportion of the nitrate load contributed by base flow for 148 relatively small watersheds (each less than 1300 km^2) across the United States. Many streams have a large percentage of their nitrate load contributed by base flow: 66% of streams had more than 37% of the total nitrate load contributed by base flow. Although broad regional patterns are not well defined, small clusters of sites at which similar processes result in similar proportions of nitrate contributions are distinct.

Total leaching of nitrogen to groundwater at the global scale was estimated to be 55 Tg/yr with a contribution of 8 Tg/yr for Europe (Van Drecht et al., 2003). It was estimated that 40% of this leached nitrogen would reach the river outlets. Deep aquifers that have been affected by historical use of fertilizers contribute an estimated 10% of the total load of nitrogen to rivers. These calculations by Van Drecht et al. (2003) were estimated on a global scale and do not account for spatial and temporal variability. However, they are similar when compared with more detailed estimates at regional scales. For example, the groundwater contribution to the total load of nitrogen for the Danube was estimated to be 48% for the period 1998–2000 (Behrendt et al., 2003).

Other estimates of groundwater contribution to total river loads were 38–69% for all German catchments (Schreiber et al., 2003), and 36% in the Po Valley (Palmeri et al., 2005). The large contribution of groundwater nitrate to the total basin load is strongly related to the amount of nitrogen (particularly the surplus amounts) applied to agricultural land, even with differences in agriculture intensity and hydrogeology across Europe.

At the local scale, several factors can affect the transport of nitrate to streams from groundwater discharge (Fig. 7.5): (a) loss through denitrification in the aquifer if sufficient organic carbon is present (reducing zones); (b) uptake by plants in the riparian zone; (c) direct transfer to streams in water via tile drains and ditches; and (d) dilution by low nitrate groundwater that may have recharged through the riparian zone, or by deeper, older groundwater that recharged prior to extensive use of fertilizers.

7.4.1.1. Hyporheic Zone Interactions

Interactions between groundwater and stream water can occur in the hyporheic zone (Fig. 7.6), located beneath and alongside a stream bed. This zone, where mixing occurs between shallow groundwater and surface water, is important for processes affecting stream water quality and ecosystems. In small streams, nitrate attenuation can occur as hyporheic flow brings stream water into contact with the reactive surface of sediment and periphyton where

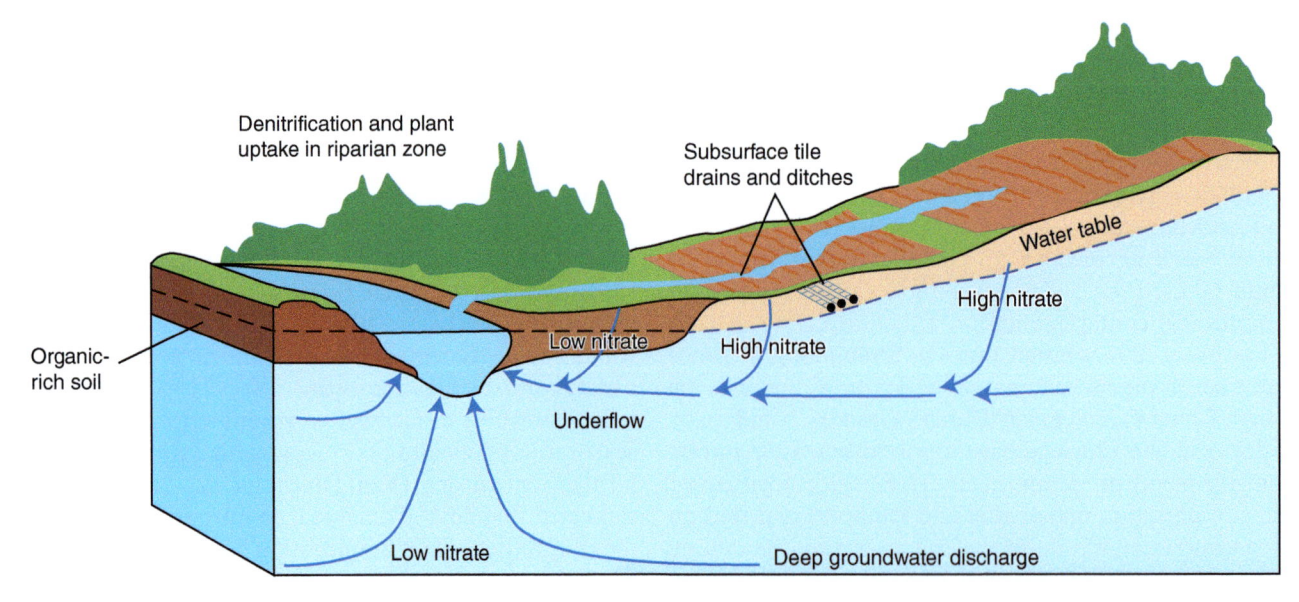

Figure 7.5 Conceptual diagram showing various processes affecting nitrate fate and transport from groundwater to a stream. *Source*: Modified from United States Geological Survey (public domain).

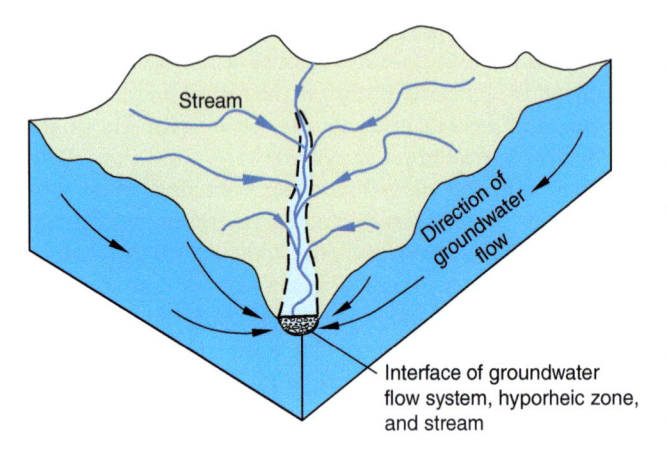

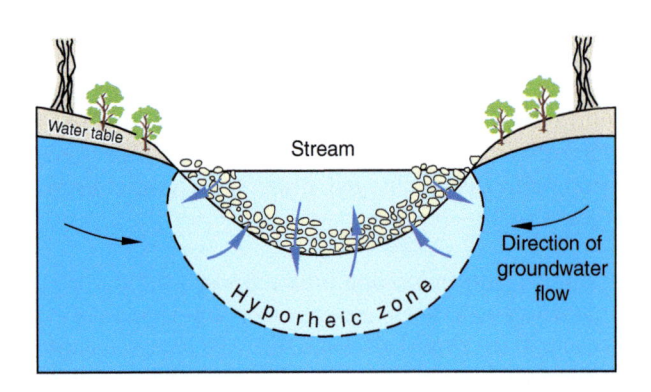

Figure 7.6 Conceptual diagram showing the interface of the groundwater flow system, the hyporheic zone, and the stream. Mixing of shallow groundwater and surface water occurs in the hyporheic zone. Public domain from USGS Circular. *Source*: Reproduced with permission of United States Geological Survey.

microbial activity and redox conditions are favorable for denitrification. Other conditions that may enhance denitrification in the hyporheic zone include the delivery of organic carbon (in dissolved form or as fine particulates); algal communities attached to sediments, rocks, or epiphyton layers; oxic/anoxic interfaces or anoxic microzones in sediments (e.g., Böhlke et al., 2009; Zarnetske et al., 2012). Dubrovsky et al. (2010) reported that based on studies at five sites around the United States, hyporheic processes removed 45–75% of groundwater nitrate before it discharged to low-gradient streams in agricultural areas.

To better understand and quantify denitrification rates in the hyporheic zone, Harvey et al. (2013) conducted a detailed study of first- and second-order reaches of a headwater stream in northwestern Indiana, which is part of the Iroquois, Illinois, and Mississippi River basins. They compared and measured whole-stream denitrification with in situ hyporheic zone denitrification in shallow and deeper flow paths in different geomorphic units. They found that hyporheic denitrification accounted for between 1 and 200% of whole-stream denitrification. The denitrification reaction rate constant was affected by the hyporheic exchange rate (greater substrate delivery), concentrations of substrates dissolved organic carbon (DOC) and nitrate, abundance of microbial denitrifiers, and measures of granular surface area and the presence of anoxic microzones. It was concluded that enhancing biogeochemical reactions in hyporheic zones of streams could lead to more effective stream restoration efforts and improvements in water quality.

7.4.1.2. Riparian Zones Along Streams

Riparian zones are areas located along rivers and other bodies of surface water and typically contain a characteristic assemblage of plants that have adapted to the shallow water table (Fig. 7.5). These areas also are important places where interactions between shallow groundwater and surface water provide important environmental benefits to streams, groundwater, and downstream land areas. Riparian areas also have the potential to remove substantial amounts of nitrate from groundwater and shallow subsurface water before it enters rivers and lakes (e.g., Vidon & Hill, 2004; Vidon et al., 2010). Plant uptake and denitrification are the primary processes that can remove nitrate in riparian zones. Although, plant uptake would temporarily immobilize reactive N that can be returned by flows to surface water bodies.

Riparian zones are particularly important in agricultural areas where large amounts of nitrate may enter streams or other surface water bodies. Anderson et al. (2014) measured in situ groundwater denitrification rates in two riparian zones adjacent to an intensive dairy farm located in the headwaters of the Susquehanna River. Based on monthly samples over a 1-year period, they determined denitrification rates that ranged from 0 to 4177 µg N/kg soil/d (mean, 830 ± 193 µg N/kg soil/d). They found that denitrification varied seasonally, with highest rates in the spring and summer related to higher temperatures and lower dissolved oxygen levels. The study estimated an annual nitrogen loss from denitrification of 470 ± 116 kg/yr/ha of riparian zone with the potential for greater than 20% of this amount occurring as N_2O. This amount of denitrification from shallow groundwater in the riparian zone was equal to 32% of the amount of nitrogen in manure spread on the adjacent upland field. Their study confirmed the importance of riparian zones in agricultural landscapes in removing substantial nitrogen loads entering downstream waters. In urban areas, riparian zones along with wetlands can be very effective in protecting water quality and the river channel by providing hotspots for denitrification and for filtering out other pollutants from stormwater runoff and other sources of nitrogen (e.g., Carey et al., 2013; Groffman et al., 2002; Kaushal et al., 2011).

7.4.2. Groundwater Seepage and Discharge to Lakes and Wetlands

Lakes can receive reactive nitrogen from various sources including wet and dry atmospheric deposition, fixation of nitrogen gas (N_2) by cyanobacteria, internal microbial decomposition of nitrogen stored in biomass and lake sediments, and groundwater flow or seepage. The water quality and ecology of lakes that receive most or all of their water from groundwater seepage or discharge are particularly susceptible to contamination from nitrogen sources from various land uses in the groundwater contributing area. Nutrient-rich groundwater that discharges to seepage lakes can disturb the biodiversity, macrophyte abundance, plant distribution, or benthic algae composition, and trophic state (e.g., Kidmose et al., 2015; and references therein). Stoliker et al. (2016) investigated several factors controlling nitrogen cycling processes in lake sediments at three sites with contrasting hydrologic regimes at a lake on Cape Cod, MA. The factors studied included water chemistry, seepage rates and direction of groundwater flow, and the abundance and potential rates of activity of N-cycling microbial communities. Based on genes coding, they identified denitrification, anammox, and nitrification at all sites regardless of flow direction or groundwater dissolved oxygen concentrations. They noted that the potential for nitrogen attenuation (denitrification in the sediments) was controlled to a large degree by the groundwater flow direction. Rates of denitrification were related to the supply of labile organic matter and varied from 6 to 4500 picomoles N/g/hour from the inflow to the outflow side of the lake. In areas where oxic lake water was migrating downward, potential nitrification rates were found to be substantial. The researchers found that rates of anammox, denitrification, and nitrification likely were linked to rates of organic N-mineralization, which would increase N-mobility and transport downgradient. They found that the direction of groundwater flow into or out of the lake and the chemistry of the water (dissolved oxygen and carbon) control the processing of different forms of nitrogen. Where dissolved oxygen concentrations were low, the microbes removed harmful forms of nitrogen flowing into the lake. In contrast, in an area where the groundwater contained high levels of dissolved oxygen and less flow into the lake, there was little or no removal of nitrogen. When lake water flowed down through the sediments and into the aquifer, nitrification was occurring. During nitrification, nitrogen trapped in the sediments, primarily resulting from decaying algae, was converted into nitrate that was transported out of the lake by flowing groundwater. The researchers concluded that lakes on Cape Cod can act to remove nitrogen that flows into them from groundwater; however, they also can be a source of nitrogen (mainly nitrate) to downstream aquifers. Durand et al. (2011) noted that transformations of nitrogen in shallow lakes is related to turnover of nitrogen, which is controlled by inorganic and organic nitrogen uptake by aquatic plants and organic nitrogen decomposition by the microbial community. In deeper lakes, they mentioned that nitrogen dynamics are controlled by the water transparency, macronutrient availability, sediment–water interactions, the length of contact time between nitrogen inputs to the lake and the biota, and the abundance and type of planktonic communities.

Freshwater wetlands are diverse ecosystems that are capable of reducing or "buffering" the flux of inorganic reactive nitrogen species from groundwater to surface water systems. This buffering of nitrogen fluxes results from storage and transformation of nitrogen by several physical, chemical, and biological processes that are controlled by the hydrologic residence time (Durand et al., 2011). Constructed wetlands also can be very effective in removing nitrate and other contaminants from groundwater via plant uptake and various biogeochemical processes (e.g., adsorption, microbial reactions, and sedimentation). In agricultural areas, constructed wetlands have been used to remove nitrate-enriched tile water before discharging into a stream or other surface water body. Also, constructed wetlands have been effective in removing nitrogen from wastewater prior to infiltrating to groundwater or discharging to surface waters (Knight et al., 2000; Vymazal, 2010). Lee et al. (2009) presented a detailed review of nitrogen removal in constructed wetland systems, with an emphasis on the nitrogen removal technology, environmental factors related to nitrogen removal, and the operation and management of the wetlands. A recent study by Collins and Gillies (2013) investigated the removal effectiveness and cost efficiency of using a constructed wetland treatment system to reduce nitrate discharges into a surface stream in an agricultural area in West Virginia. They found that the constructed wetland reduced stream concentrations of nitrate during the growing season by about 0.14 mg/L at mean streamflow (17% reduction). They found that about 80 kg of nitrate-N were removed annually by the constructed wetland at a cost of about US$30/kg N. It was noted that this per unit cost was at the low range of small wastewater treatment plant costs for nitrate removal, but higher than the costs associated with reduced fertilizer application.

Durand et al. (2011) mentioned some cautionary issues and uncertainties regarding the use of wetlands and constructed wetlands for nitrate mitigation. These include their variable effectiveness and side effects such as increased dissolved organic nitrogen emissions to adjacent open waters, nitrous oxide emissions to the atmosphere, and potential losses of biodiversity. Also, as natural wetland areas are converted to agricultural and or urban use, nutrients stored in the wetland would be transported to the environment and these important ecosystems would be degraded or lost.

7.4.3. Submarine Groundwater Discharge

Submarine groundwater discharge (SGD) can consist of fresh groundwater, recirculated seawater (or a combination of the two), and connate water (water incorporated into rock pores when the rocks form) (Fig. 7.7). SGD can contribute substantial amounts of reactive nitrogen to coastal areas and the sea (e.g., Burnett et al., 2006; Corbett et al., 2002; Kroeger & Charette, 2008). This reactive nitrogen flux has resulted in ecological impacts, such as eutrophication, changes in microbial communities (Garces et al., 2011), the growth of algal bloom-forming phytoplankton (Lecher et al., 2015), coastal dead zones, and red tide (Hu et al., 2006). SGD occurs in many coastal areas, but it can be challenging to accurately quantify especially at continental scales. Bratton (2010) emphasized the need to consider SGD and flow at three spatial scales to improve the design and reporting of results from field and modeling studies: (a) nearshore, which extends about 0–10 m offshore and includes the unconfined surficial aquifer; (b) the embayment, which extends from about 10 m to

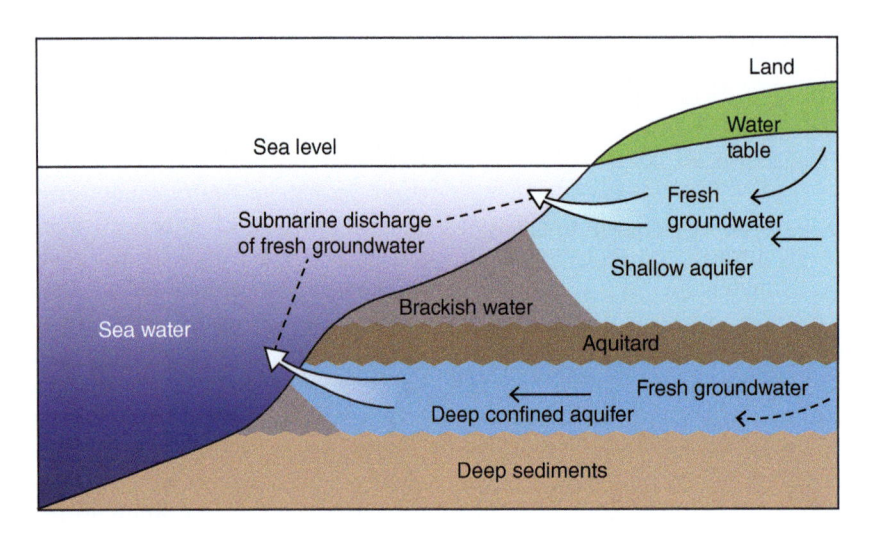

Figure 7.7 Conceptual diagram showing areas where submarine groundwater discharge occurs. *Source*: From Burnett et al. (2006). Quantifying submarine groundwater discharge in the coastal zone via multiple methods. Science of the Total Environment 367 (2–3): 498–543. Figure 1; p. 502, Reproduced with permission of Elsevier.

aboutv10km offshore and includes the first confined submarine aquifer extent; and (c) the shelf, which encompasses the width and thickness of the aquifers of the continental shelf from the base of the first confined aquifer down to the basement, and included geothermal convection and glacioeustatic change in sea level. A recent study by Sawyer et al. (2016) concluded that 12% of the US coastline is especially vulnerable to contamination from SGD, including parts of the northern Gulf Coast, the Pacific Northwest, and the northern Atlantic coast. These vulnerable areas result from high rates of recharge and seepage that coincide with densely populated areas. This study by Sawyer et al. (2016) was unique in that it used high-resolution topographic models of riverbeds, streams, and coastlines across the United States. This information was combined with historical climate data on local rainfall and snowmelt to estimate the contribution from subterranean groundwater flow and surface waters. Their analysis provided detailed spatial information on hotspots for contaminant discharge to coastal and marine waters and saltwater intrusion into coastal aquifers. Other selected studies of reactive nitrogen transport from SGD are presented below.

Montiel et al. (2018) presented an approach to identify and quantify four forms of groundwater discharge in a karstic groundwater system in southern Spain (Maro-Cerro Gordo) that includes an ecologically protected coastal area. They noted the four ways in which groundwater discharged to the sea: (a) groundwater-fed creeks; (b) coastal springs; (c) diffuse groundwater seepage through seabed sediments; and (d) submarine springs. Their study quantified groundwater discharge using a combination of tracer techniques (salinity, ^{224}Ra, and ^{222}Rn) and direct measurements (seepage meters and flowmeters). Groundwater discharge via submarine springs was the most difficult to assess due to their depth (up to 15m) and extensive development of the springs conduits. Montiel et al. (2018) found that the total groundwater discharge for the 16km of shoreline in the study area was at least $11 \pm 3 \times 10^3$ m^3/d for the aforementioned four types of discharge. In a highly populated and farmed section of their study area, they stated that the groundwater-derived nitrate (as NO_3^-) flux was 641 ± 166 mol/d, or approximately 75% of the total NO_3^- loading in the study area.

From January 2005 through January 2006, a harmful algal bloom (*Karenia brevis*) affected coastal waters (depths less than 50m) off west-central Florida. During this bloom, Hu et al. (2006) reported that there was a sustained chorophyll anomaly of 1 mg/m^3 over an area of up to 67,500 km^2. They also noted that red tides occur in this same general area almost every year; however, the extensive 2005 bloom caused a widespread hypoxic zone (dissolved oxygen <2 mg/L) that resulted in mortalities of benthic communities, fish, turtles, birds, and marine mammals. It was hypothesized that SGD provided the missing nutrients and could initiate and sustain the recurrent red tides off west-central Florida. Hu et al. (2006) found that SGD inputs of dissolved inorganic nitrogen (DIN) in Tampa Bay in west-central Florida alone constitute 35% of the nitrogen discharged by all central Florida rivers that drain to the west coast. To promote their hypothesis, they proposed that the atypically large number of hurricanes in 2004 resulted in high runoff, and in higher than normal SGD emerging along the west Florida coast throughout 2005, initiating and feeding the persistent HAB. It was further speculated that this mechanism may explain recurrent red tides in other coastal regions of the Gulf of Mexico.

Kroeger and Charette (2008) investigated SGD by studying the transport, chemical speciation, and attenuation of nitrogen load in a seepage zone at the head of Waquoit Bay in western Cape Cod, Massachusetts, US. They collected 328 nearshore groundwater samples and assessed the distribution and isotopic signature (δ^{15}N) of nitrate and ammonium to estimate nutrient fluxes from terrestrial and marine groundwater sources. In the freshwater zone, there were groundwater plumes containing nitrate and ammonium; however, the deep salinity transition zone (STZ) carried almost exclusively ammonium. Their calculated terrestrial fluxes for nitrate and ammonium were 304 and 67 mmol/m/d, respectively. Marine nitrate and ammonium fluxes were much lower, 0 and 27 mmol/m/d. They reported sharp interfaces between water masses of distinct oxidation: reduction potential, which indicated relatively fast rates of microbial nitrogen transformations compared to dispersive mixing processes. Kroeger and Charette (2008) found that both nitrate and ammonium were removed in freshwater and the deep STZ and postulated several mechanisms including heterotrophic or autotrophic denitrification, coupled nitrification/denitrification, anammox, or Mn oxidation of NH_4^+. The short residence time of water in the shallow STZ most likely prevented any loss of nitrogen. It was concluded that the nearshore aquifer and subterranean estuary are biogeochemically active zones for attenuation of nitrogen even with low levels of organic carbon. They noted that the extent of nitrogen attenuation was controlled by the degree of mixing of water masses with different biogeochemical conditions, which likely occurs at the fresh-saline groundwater interface at most SGD zones.

Lecher et al. (2015) quantified groundwater discharge and associated nutrient fluxes to Monterey Bay, California, during the wet and dry seasons using excess ^{224}Ra as a tracer. In addition, bioassay incubation experiments were conducted to assess the response of bloom-forming phytoplankton to nutrient inputs from SGD.

Study results showed that the high nutrient content (nitrate and silica) in groundwater stimulated the growth of bloom-forming phytoplankton. Local land use likely contributes nonpoint anthropogenic sources of elevated nitrate concentrations in groundwater around Monterrey Bay from agriculture, landfill, and rural housing. Their findings showed that these sources along with SGD provide a continual source of nutrients that stimulate bloom-forming phytoplankton in Monterrey Bay.

7.5. LEGACY REACTIVE NITROGEN STORAGE IN THE VADOSE ZONE AND IN AQUIFERS

A persistent and growing problem concerns the large amounts of reactive nitrogen (mostly in the form of nitrate) that are stored in the vadose zone (Ascott et al., 2016, 2017) and aquifers (Meals et al., 2010; Puckett et al., 2011; Wang et al., 2013). Nitrate from multiple sources (fertilizer, animal wastes, septic tanks, land application of treated wastewater, and biosolids) can leach below the root zone in soils move into the vadose (unsaturated) zone (the material between the base of the soil to the water table at the top of the saturated zone) and eventually travel to the water table and deeper into aquifers. Ascott et al. (2017) assessed global patterns of nitrate storage in the vadose zone by using estimates of groundwater depth and nitrate leaching for the period 1900–2000. They estimated a peak global storage of 605–1814 Tg of nitrate in the vadose zone, with the highest storage per unit area in North America, China, and Europe (areas with thick vadose zones and extensive historical agricultural activities).

Ascott et al. (2017) noted that global-scale nitrogen budgets that quantify anthropogenic impacts on the nitrogen cycle do not explicitly look at nitrate stored in the unsaturated (vadose) zone. It was pointed out that most of the global nitrogen budgets typically consider only "steady-state" conditions. That is, by balancing the various nitrogen inflows and outflows in the environment (e.g., atmospheric nitrogen fixed in soils by leguminous plants, fertilizer application to soils, nitrogen loss from soils, and nitrogen removed during crop harvesting), these budgets have not accounted for the changes in nitrogen storage in soils, the vadose zone, groundwater, or riverine sediment. To estimate storage of nitrogen, Ascott et al. (2017) used estimates of groundwater depth and nitrate leaching for the period 1900–2000 and estimated that the peak global storage of nitrate in the unsaturated (vadose) zone was 605–1814 Tg. Therefore, it was concluded that nitrate stored in the vadose zone should be considered in future budgets for effective policymaking. Using basin-scale and countrywide estimates and observed groundwater nitrate data, they found that nitrate storage (per unit area) was greatest in North America, China, and Europe particularly where there are thick vadose zones and extensive historical agriculture. The impact of changes in agricultural practices in these areas could be substantially delayed (long lag times) due to the long travel times in the vadose zone. This is especially prevalent in agricultural areas where there has been a surplus of nitrogen added to cropland (for more information: http://www.fondriest.com/news/hidden-underground-nitrate-pollution-threatens-groundwater-worldwide.htm, accessed 12 January 2019).

Several studies have documented that large amounts of reactive nitrogen, mostly in the form of nitrate, leaches below the root zone in soils to the vadose (unsaturated) zone (the material between the base of the soil to the water table at the top of the saturated zone) and eventually to groundwater (aquifers) where nitrate can be stored (Walvoord et al., 2003; Wang et al., 2013). This is especially prevalent in agricultural areas where there has been a surplus of nitrogen added to cropland. Ascott et al. (2017) assessed global patterns of nitrate storage in the vadose zone by using estimates of groundwater depth and nitrate leaching for the period 1900–2000. They estimated a peak global storage of 605–1814 Tg (10^{12} g) of nitrate in the vadose zone, with the highest storage per unit area in North America, China, and Europe (areas with thick vadose zones and extensive historical agricultural activities).

There is considerable uncertainty in watershed and global-scale nitrogen budgets in terms of the fate of surplus nitrogen. This excess nitrogen may be lost from the landscape due to denitrification or retained within watersheds as nitrate or organic nitrogen; however, the relative magnitudes of nitrogen fluxes and pools are poorly understood. Van Meter and Basu (2015) used models to assess catchment-scale time lags between implementation of nitrogen conservation measures and improvements in water quality. Their models accounted for both hydrologic and biogeochemical legacies and spatial patterns of landscape conversion on reductions in nitrogen at the watershed outlet. It was noted that nitrogen legacies develop in agricultural watershed due to the long-term application of fertilizers and the relation between soil nitrogen levels and multidecadal nitrogen surpluses. Their parsimonious model incorporated a MODFLOW model and particle tracking simulations (described in more detail in Section 7.7) to determine exponential groundwater travel time distributions in the watershed. Nitrogen attenuation was accounted for in biogeochemical processes that included sorption of organic nitrogen within the root zone, an overall denitrification rate constant, and a legacy nitrogen depletion rate constant. Their model accurately predicted time lags observed in an agricultural watershed in Iowa that had experienced a 41% conversion of row crop areas to native prairie. Model-simulated time lags were dependent on the desired nitrogen concentration reduction in the receiving water. Van Meter and Basu (2015) showed that the biogeochemical legacy could increase the time by a factor of two needed to see substantial concentration reductions at the catchment scale.

Tesoriero et al. (2013) assessed the vulnerability of streams to legacy nitrate sources in groundwater at seven study sites in the United States that had a range of base flow index (BFI) values. The BFI is defined as the ratio of base flow to total streamflow volume. Streams with high BFI values (more groundwater contribution) had high nitrate loads and had significantly higher dissolved oxygen concentrations in streambed pore water. It was noted that nitrate transport through the streambed from groundwater is enhanced by these oxic conditions. A groundwater–surface water interaction study was conducted at one stream site with a high BFI and results indicated that decades-old nitrate-enriched water was discharging into this stream. It was concluded that high nitrate levels in this stream are likely to continue for decades irrespective of current agricultural practices.

Several other studies have shown that groundwater nitrogen applied decades ago can still be found in shallow and deep aquifers (Basu et al., 2012; McMahon et al., 2006; Meals et al., 2010; Puckett et al., 2011; Sanford & Pope, 2013; Sebillo et al., 2013). In 20 study areas in the United States, Puckett et al. (2011) found that denitrification was minimal in recently recharged groundwater that was less than 20 years old. However, in older groundwater (recharged before 1983), a median of 65% of the original nitrate was still present. Their results indicated that shallow zones in many aquifers have limited potential for natural bioremediation from denitrification. It was also noted that their findings regarding limited denitrification do not support the recent global denitrification estimates (Seitzinger et al., 2010) that assumed a 30–40% leaching rate and a 2-year half-life for nitrate in shallow groundwater. A detailed review of nitrate attenuation in groundwater (Rivett et al., 2008) noted that the most important limiting factors controlling denitrification in aquifers are oxygen and electron donor concentration and availability, as denitrifying bacteria are ubiquitous in the surface. However, Heffernan et al. (2012) found that denitrification was an important nitrate attenuation process in the Upper Floridan aquifer, which is a uniquely an organic-matter-poor system. Their study indicated that with even very low average rates, denitrification accounted for 32% of estimated aquifer nitrogen inputs in most sampled large karstic springs that discharge water from relatively deep parts of the Upper Floridan aquifer. Rivett et al. (2008) noted that other processes, such as dissimilatory nitrate reduction to ammonium and assimilation of nitrate into microbial biomass, essentially were not important influences on denitrification in aquifers.

As young groundwater with elevated nitrate concentrations in shallow zones migrates deeper into the aquifer, nitrate concentrations likely would increase in deep zones that are typically used for supplying drinking water. It is important to note that deep groundwater has long residence times in aquifers that results in long lag times as deep groundwater slowly discharges to surface water – the phenomenon of long groundwater residence times. Puckett et al. (2011) concluded that nitrate can persist for long periods of time under oxic conditions in aquifers and nitrate stored could take decades to centuries before discharging to a receiving water body.

This legacy nitrate and its persistence in aquifers have been particularly evident in other areas around the world. Although substantial decreases in fertilizer applications have been noted in the Netherlands, Denmark, and Germany (where the nitrogen surplus is back to the level of that of 1970), concentrations of nitrate-N in groundwater have not responded to this decrease of nitrogen surplus (Sutton et al., 2011a). In addition, in Eastern European countries where the nitrogen surplus has decreased by half (due to economic and political changes in the early 1990s), no improvements in water quality have been observed in streams (Sutton et al., 2011a). This is likely due to the large quantities of nitrate stored in aquifers and released very slowly over time, as a function of the groundwater residence time, which can range from weeks to several thousands of years (Alley et al., 2002).

Van Meter et al. (2016) synthesized data from the Mississippi and Susquehanna River Basins in the United States using a parsimonious, process-based model, referred to as Exploration of Long-trM Nutrient Trajectories (ELEMeNT). They used this model to develop a 214-year (1800–2014) trajectory of nitrogen inputs to the land surface of the continental United States. The model further paired the nitrogen inputs over the two centuries with a travel-time approach that was used to simulate transport and retention of nitrogen along pathways in the subsurface. Van Meter et al. (2016) found that there was substantial nitrogen loading to these basins above baseline levels before the widespread use of commercial fertilizers, mainly from the conversion of forest and grassland to agriculture dominated by row crops. Their model results show that since preindustrial times, the nitrogen loads increased by a factor of 7 and 14 for the Mississippi River Basin and the Susquehanna River Basin, respectively. When quantifying the amount of legacy nitrogen in soil and groundwater pools, they found that approximately 55% of the current annual nitrogen loads in the Mississippi River Basin were older than 10 years of age. Also, for the Susquehanna River Basin, 18% of the current annual nitrogen loads were older than 10 years of age. The contribution of legacy nitrogen likely accounts for the delays in achieving goals for reducing the size of the hypoxic zone in the Gulf of Mexico, and for reducing nitrogen loads to Chesapeake Bay. In terms of remediation strategies involving legacy nitrogen sources, it was suggested that riparian buffers and wetlands should be considered in areas where there are significant amounts of nitrogen in soils and groundwater discharging to streams and other surface water bodies. Van Meter et al. (2016) concluded that legacy nitrogen storage is a substantial source of nutrient loading to coastal areas throughout the world; therefore, more emphasis needs to be put on investigating the magnitudes and spatial distribution of legacy nitrogen to meet nutrient reduction goals.

In arid environmental systems, an increased reliance on dryland ecosystems may result in flushing of naturally occurring nitrate in the unsaturated zone into groundwater in regional basins due to irrigation and vegetation change (Robertson et al., 2017). Thus, the unsaturated zone, an important reservoir for the storage and release of reactive nitrogen (mainly in the form of nitrate) over time, has significant implications for global warming, contamination of deeper aquifers used for drinking water supplies, and eutrophication of surface water bodies (Schlesinger, 2009).

Studies have shown that nitrate concentrations in many watersheds, particularly in the United States and Canada, have increased despite the large decreases in atmospheric nitrogen deposition coupled with higher agricultural nitrogen use efficiency (Van Meter & Basu, 2017). Van Meter et al. (2016, 2017) have done extensive investigations on modeling time lags related to legacy nitrogen accumulation and delayed outputs to surface water systems. Based on study of nitrate data for 16 nested subwatersheds in a 6800 km^2 watershed in southern Ontario, they found that the mean annual lag time was 24.5 years (large seasonal variations) between nitrogen inputs and stream nitrate outputs to these subwatersheds. More information about nitrogen dynamics between groundwater and surface water in large basins is discussed in the following section on interactions between groundwater and surface water systems.

In many European countries (particularly the Netherlands, Denmark, and Germany), Grizzetti et al. (2011) noted that there have been substantial decreases in the nitrogen surplus, where the nitrogen surplus levels approach those in 1970. However, groundwater nitrate concentrations have not reflected the decrease in nitrogen surplus. For example, in Eastern European countries where the nitrogen surplus decreased by 50% (due to economic and political changes in the early 1990s), there have not been improvements in stream water quality in response to these changes. As found in many other places, this most likely is related to large amounts of nitrogen being stored in shallow and deep aquifers that is slowly released depending on soil drainage properties, the groundwater residence time, and the lag time between reductions in fertilizer use and the occurrence in stream water. Groundwater residence times can vary from weeks to several thousands of years, depending on aquifer properties, slope, and variations in recharge and climate (Schlesinger, 2009; Wriedt & Bouraoui, 2009). This is particularly evident in Lithuania, where nitrate concentrations in the Nemunas River have increased due to large amounts of nitrogen stored in soils and groundwater from extensive fertilization during the Soviet period (Sileika et al., 2006).

7.6. NITRATE CONTAMINATION IN DRINKING WATER FROM PUBLIC SUPPLY AND DOMESTIC WELLS

Humans are increasingly exposed to elevated levels of reactive nitrogen species (mainly nitrate) in drinking water. Several worldwide and national agencies have set the regulatory limit for nitrate in drinking water to protect against infant methemoglobinemia. However, based on an updated review of more than 30 epidemiological studies since 2005, Ward et al. (2018) found that there was a strong relationship between nitrate in drinking water and other adverse health effects, including colorectal cancer, thyroid disease, and neural tube defects. They noted that many of these studies found increased health risks with ingestion of drinking water nitrate levels that were below regulatory limits. Shukla and Saxena (2018) summarized the status of nitrate contamination of groundwater as a drinking-water source in many countries around the world. They noted that approximately 118 million people in India are drinking water with nitrate (as NO_3) concentrations greater than 100 mg/L (equivalent to 23 mg N/L). Other countries identified as having areas with groundwater nitrate concentrations greater than World Health Organization permissible limits included Afghanistan, Morocco, Niger, Nigeria, Senegal, Pakistan, Japan, Lebanon, Brazil, South Korea, Serbia, the Gaza strip, Australia, Spain, and Germany (Ward et al., 2018).

In 2015, the US EPA reported that 183 community water systems in the United States exceeded allowable levels of nitrate-N in drinking water. Sutton et al. (2011a) estimated that about 3% of the population in EU-15 that relies on groundwater as a drinking water source is exposed to nitrate concentrations exceeding the drinking water nitrate standard of 50 mg NO_3/L (equivalent to nitrate as N of 11.2 mg N/L). In addition, Sutton et al. (2011a) estimated that 5% of the population using groundwater is chronically exposed to nitrate concentrations exceeding 5.6 mg N/L, which could double the risk of colon cancer for people that consume more than median amounts of meat. In China, many large aquifers have been contaminated with elevated nitrate-N concentrations (Wang et al., 2016b). About 90% of China's shallow groundwater is polluted, and nitrate is considered one of the main pollutants of serious health concerns (Qiu, 2011). In the United States, several areas have been delineated where groundwater nitrate concentrations exceed or approach the US EPA MCL (10 mg N/L). For example, Dubrovsky et al. (2010) found that nitrate concentrations exceeded the drinking water MCL in water samples from 7% of 2388 domestic wells and 3% of water samples from 384 PSWs. They noted that the drinking water standard was exceeded in more than 20% of 406 shallow domestic wells (less than 30 m below the water table) located in rural agricultural areas and 6% for domestic wells in urban areas. The higher exceedances in agricultural areas represent the potential for adverse human health effects for this population (as water typically is not treated), and for future contamination of deeper groundwater pumped for public supplies in these areas. This section focuses on factors affecting nitrate contamination of PSWs and domestic wells.

7.6.1. Drinking Water from Public Supply and Domestic Wells

In recent years, there have been several national studies of groundwater quality from public supply and private domestic wells in the United States, Europe, China, and other areas around the world. In the United States, Toccalino and Hopple (2010) summarized data on the quality of source (untreated) water data from 932 wells located in parts of 41 states. These wells withdraw water from parts of 30 regionally extensive aquifers, which make up about 50% of the principal aquifers in the United States. They also compared source and finished (treated) water for a subset of 94 wells. Water samples were collected once during 1993–2007 and were analyzed following a nationally consistent study design (https://water.usgs.gov/nawqa/, accessed 16 December 2018). Nitrate was detected in about 72% of the source-water samples from public wells and were greater than 10 mg/L (MCL) in 1.9% of the samples. Toccalino et al. (2010) found that nitrate-N concentrations approached the MCL (0.1 < BQ < = 1; BQ is the benchmark quotient, expressed as the ratio of concentration to human-health benchmark) in 42.9% of the samples from all principal aquifer rock types. Nitrate concentrations were positively correlated with dissolved oxygen in samples from most principal aquifer rock types. All samples containing nitrate-N concentrations greater than the MCL were collected from PSWs withdrawing water from unconfined aquifers (mostly from unconsolidated sand and gravel aquifers in the western United States that have high effective porosities and are more vulnerable to nitrate contamination). Toccalino et al. (2010) noted that median nitrate-N concentrations were 1.4 and 0.09 mg N/L in unconfined and confined aquifers (statistically significant difference). The water-quality results from their study were compared with those from a study of domestic wells by DeSimone et al. (2009). Not only was nitrate detected more than twice as frequently at concentrations above the MCL in domestic well samples compared to PSWs, but nitrate concentrations exceed the MCL in 4.4% of domestic well water samples. Even though ammonia was detected in 44% of source-water samples from public wells, the median concentration was 0.01, well below the US EPA taste threshold of 30 mg N/L. Nitrate concentrations tended to be higher when ammonia was not detected, but decreased with increasing ammonia concentrations.

7.6.2. Factors Affecting PSWs to Nitrate Contamination

To better understand the factors that affect the vulnerability of PSWs to contamination, the USGS National Water Quality Assessment Program conducted a comprehensive study from 2001 to 2011, referred to as Transport of Anthropogenic and Natural Contaminants (TANC) to PSWs (Eberts et al., 2013). Ten study areas containing PSWs were chosen in regional aquifer systems that combined accounted for about 50% of the groundwater used for public drinking water supply in the United States in 2000 (Fig. 7.8). In addition, local-scale investigations were conducted in six study areas (the first group of four from 2001 to 2007) and the remaining two from 2006 to 2011) (Fig. 7.8).

The TANC study followed up on the Toccalino and Hopple (2010) study that found contaminant concentrations greater than drinking water standards or human health benchmarks in about 22% of samples from PSWs. PSWs are vulnerable to contamination not only from the previously mentioned key groundwater vulnerability factors (i.e., contaminant input, contaminant mobility and persistence, and intrinsic susceptibility) but also from their location, design, construction, operation, and well-maintenance practices (Eberts et al., 2013). Contaminant movement to a PSW is highly dependent on the location of the well relative to a contaminant source, the pumping rate, and the length and placement of the well screen in the aquifer.

Eberts et al. (2013) also noted several other important hydrogeologic and geochemical controls on the vulnerability of PSWs to contamination. These included sources of recharge in the contributing area to a well, geochemical conditions (particularly oxic vs anoxic conditions) in the aquifer where the water is being withdrawn, the age of groundwater and mixtures of water with different ages, preferential groundwater flow pathways (e.g., short-circuiting of groundwater flow and contaminants from overlying hydrogeologic units or the land surface), and the degree of

136 NITROGEN OVERLOAD

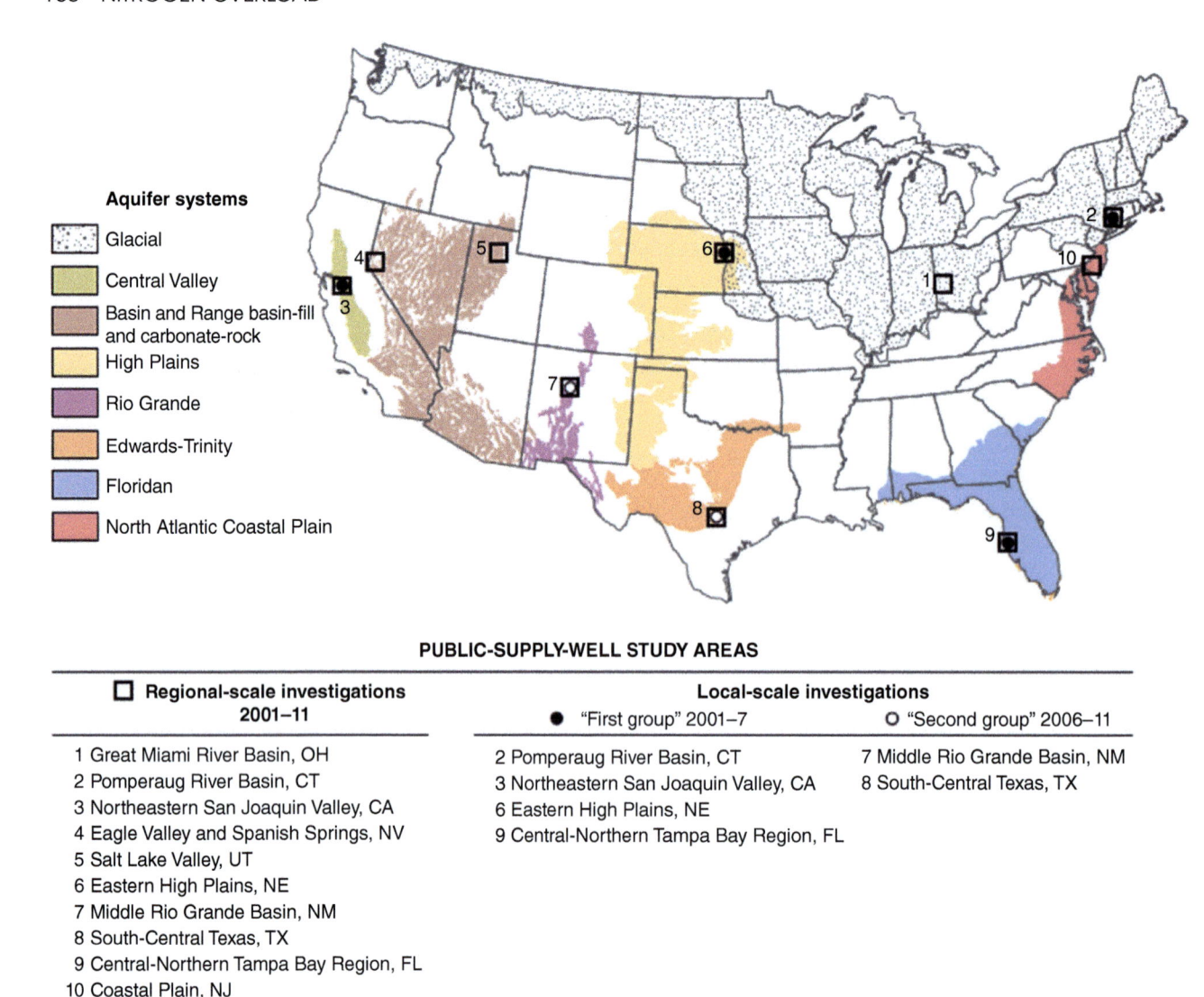

Aquifer systems
- Glacial
- Central Valley
- Basin and Range basin-fill and carbonate-rock
- High Plains
- Rio Grande
- Edwards-Trinity
- Floridan
- North Atlantic Coastal Plain

PUBLIC-SUPPLY-WELL STUDY AREAS

☐ Regional-scale investigations 2001–11	Local-scale investigations	
	● "First group" 2001–7	○ "Second group" 2006–11
1 Great Miami River Basin, OH	2 Pomperaug River Basin, CT	7 Middle Rio Grande Basin, NM
2 Pomperaug River Basin, CT	3 Northeastern San Joaquin Valley, CA	8 South-Central Texas, TX
3 Northeastern San Joaquin Valley, CA	6 Eastern High Plains, NE	
4 Eagle Valley and Spanish Springs, NV	9 Central-Northern Tampa Bay Region, FL	
5 Salt Lake Valley, UT		
6 Eastern High Plains, NE		
7 Middle Rio Grande Basin, NM		
8 South-Central Texas, TX		
9 Central-Northern Tampa Bay Region, FL		
10 Coastal Plain, NJ		

Figure 7.8 Map showing the PSW study areas. *Source*: Eberts et al. (2013)p. 19; Reproduced with permission of United States Geological Survey.

confinement in the aquifer where the PSW is located. The US EPA has established guidelines for source water protection areas for PSWs. The methods can include delineating a fixed radius around a well or computer modeling of contaminant transport to map groundwater protection areas (US EPA, 1994). The NAWQA program delineated recharge areas and groundwater travel times to PSWs using groundwater flow and particle tracking models (MODFLOW and MODPATH).

McMahon et al. (2008) investigated the source, fate, and transport controls on nitrate movement to PSWs in four different hydrogeologic settings, which were part of the larger NAWQA PSW study (Eberts et al., 2013). They analyzed data collected during 2003–2005 from PSWs in four different aquifer systems: Central Valley aquifer system (thick unconsolidated sand and gravel; Modesto, California), High Plains aquifer (thick unconsolidated sand and gravel; Lincoln, Nebraska), glacial aquifer system; (a thin unconsolidated sand and gravel aquifer; Woodbury, Connecticut), and the Floridan aquifer system (karstic carbonate rock aquifer; Tampa, Florida). Nitrogen from urban septic leachate and fertilizer (possibly nonfarm) were the primary sources of elevated nitrate concentrations in PSW capture zones for the glacial aquifer system and the Floridan aquifer system, respectively. In agricultural settings (Central Valley aquifer system and High Plains aquifer), nitrate fluxes to the water table were higher than in urban settings. Based on historical nitrate input from fertilizer usage (Burow and Green, 2008), model simulations showed that nitrate concentrations in a PSW completed in the thin unconsolidated sand and gravel aquifer responded most rapidly, reaching the MCL of 10 mg/L in less than 10 years after nitrate in recharge water reached that level. It took longer (more than 10 years) for nitrate concentrations to reach the MCL in the well in the carbonate-rock aquifer, and more than 60 years to reach the PSWs in the thick unconsolidated sand and gravel aquifers. Other simulations included several urbanization scenarios. One simulation scenario involved reducing the proportion of agricultural land in the recharge area of the well at an annual rate of 2% per year beginning in the 1970s (to represent loss of agricultural land to urban encroachment). Most of the water reaching the wells in the carbonate-rock and thin unconsolidated sand and gravel aquifers was less than 15 years old, thus simulated concentrations peaked and began to decline after a delay of 6–12 years. Changes in nitrate inputs in the thick unconsolidated sand and gravel aquifers were much slower, including lower peak concentrations, due to a significant proportion of old water reaching these wells.

Based on two-component mixing calculations in the agricultural areas, it was found that about 50–85% of the nitrate in water from the PSW likely originated from those modern anthropogenic sources, with the remainder coming from sources in old (>50 years) recharge or sources in young recharge in undisturbed settings such as forests. Denitrification was occurring in the Central Valley aquifer, most likely in a thick reaction zone following a 30-year time lag after recharge. In the High Plains aquifer and the Floridan aquifer system, denitrification occurred more rapidly in thin reaction zones in fine-grained sediments that separated the anoxic PSW producing zones from overlying oxic, high nitrate groundwater. Denitrification did not appear to be a nitrate sink in the glacial aquifer system study area. Based on particle tracking models, nitrate likely reached the PSW in the High Plains aquifer and in the Floridan aquifer system by migration through long well screens that crossed multiple hydrogeologic units (High Plains aquifer) and movement through karst dissolution features in the Floridan aquifer system. In these two systems, preferential flow pathways and short circuiting of water from overlying hydrogeologic units reduced groundwater residence times in the denitrifying zones.

Based on groundwater flow and solute transport models, McMahon et al. (2008) computed water-quality response curves for nitrate for the four PSWs to show differences in how each PSW would respond to 25 years of nitrate contamination of shallow groundwater. Response times varied greatly, ranging from relatively rapid for the glacial aquifer system (Connecticut) and the Floridan aquifer system (Florida) to much longer for the Central Valley aquifer (California) (Fig. 7.9). Differences in response times were related to groundwater-age mixtures. The Connecticut PSW produces mostly young water (average groundwater age of 6 years) and therefore simulated nitrate concentrations increase rapidly after contaminant input (after 10

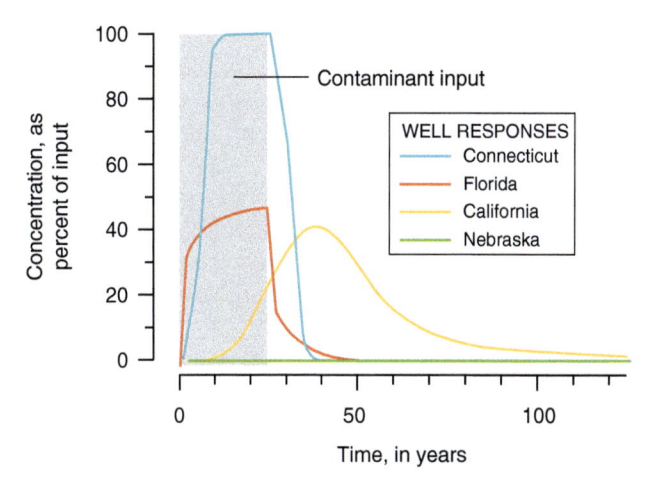

Figure 7.9 Simulated contaminant response curves (to 25 years of nitrate contamination of shallow groundwater) for PSWs in the glacial aquifer system in Connecticut, the karstic Floridan aquifer system in Florida, the Central Valley aquifer system in California, and the High Plains aquifer in Nebraska. *Source*: From Eberts et al., 2013, p.57; public domain; Reproduced with permission of United States Geological Survey.

years) and decrease rapidly after nitrate input stops. Simulated nitrate concentrations in the Floridan aquifer PSW respond more rapidly than the Connecticut PSW because of the fast travel time from the water table to the PSW (more than 20% of the PSW water is less than 1 year old, with an average groundwater age of 13 years). The fast travel time results from high pumping rates (drawing young water from overlying hydrologic units) and short circuiting of water through preferential pathways (mainly dissolution features in the karstic limestone aquifer). However, the Floridan aquifer PSW also produces a large fraction of old water from deeper in the aquifer. Therefore, dilution of the young oxic water with older anoxic water results in lower nitrate concentrations over time. Although, denitrification is limited because of the rapid travel time to the PSW and the mixture of oxic and anoxic water in the PSW. The PSWs in California and Nebraska both produce large fractions of old water, which results in longer time delays between the start of contaminant input and peak concentrations (close to four decades). The simulated contaminant concentrations would increase for about 15 years after contaminant input stops due to a large portion of contaminated groundwater that would continue to travel to the wells at the end of contaminant input. It was concluded that the vulnerability of PSWs to nitrate contamination is affected by complex interactions between contaminant sources, reaction rates, redox conditions, age mixtures, and perturbations of groundwater flow due to PSW pumping.

Differences in nitrate concentrations between PSWs and domestic wells in the same aquifer also are related to the well screen length and pumping rates. Domestic wells typically have short well screens and withdraw water from shallow and localized parts of aquifers where contamination from surface inputs are more likely. In contrast, PSWs usually have long well screens and their high pumping rates capture water from multiple groundwater flow pathways that can originate over large contributing areas. The longer wells screens and higher pumping rates can promote the movement of contaminants from preferential flow pathways.

Preferential flow pathways were observed in all 10 study areas and affected every other factor contributing to the vulnerability of the PSW to contamination (Eberts et al., 2013). These pathways occurred naturally in some areas from fractures in rocks, interconnected high-permeability sediments, or dissolution conduits (Fig. 7.10). Preferential flow pathways were also related to well construction where rapid transport of water from overlying aquifers or from the surface can occur along the well bore. For example, when pumping of a PSW begins, the movement of nitrate and other contaminants in preferential pathways would be substantially increased resulting in less time for denitrification. This would result in less removal or attenuation of nitrate from in situ processes. In addition to less removal of nitrate, rapid movement of water through preferential flow pathways can create favorable conditions for the transport of pathogens to PSWs due to insufficient time for microorganism die-off and removal of microorganisms through filtration or sorption to sediments or the rock matrix.

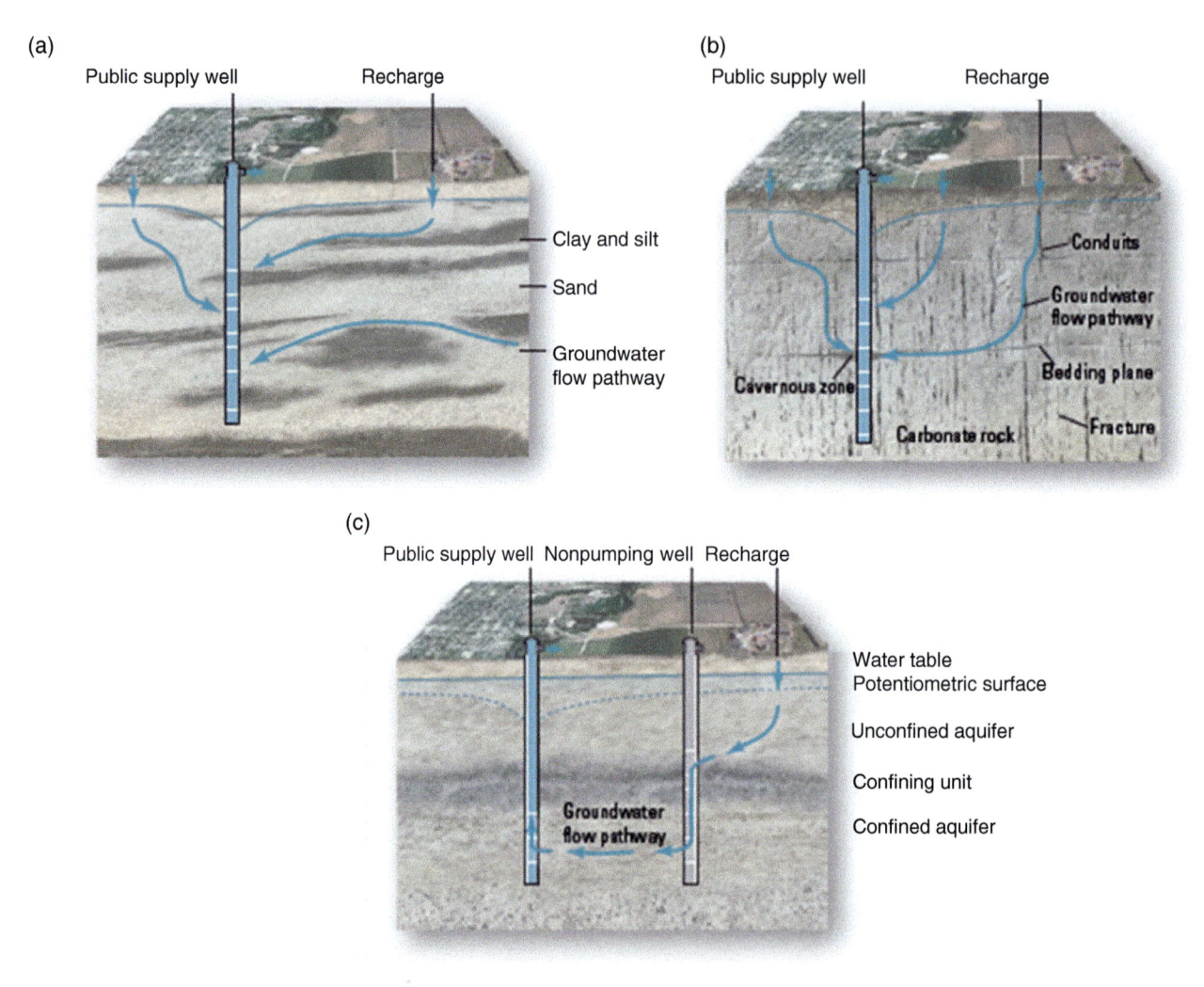

Figure 7.10 Conceptual diagram showing three different ways that preferential flow pathways can occur to PSWs resulting from (a) pumping of the well that draws water and contaminants from the most transmissive material in the aquifer; (b) movement of water through conduits, fractures, bedding planes, or cavernous zones in carbonate rock, and (c) movement of water through well screens that connect one aquifer to another causing contaminants to move rapidly across confining units that normally would prevent flow. *Source*: Modified from colored illustrations on p. 62 of Eberts et al. (2013). Reproduced with permission of United States Geological Survey.

7.7. MODELING NITRATE TRANSPORT AND VULNERABILITY OF GROUNDWATER TO CONTAMINATION

The importance of understanding the factors controlling the fate and transport of nitrate in aquifers cannot be overstated. Humans are increasingly exposed to elevated levels of nitrate in drinking water, particularly in many agricultural areas of the world. Ingested nitrate can lead to the endogenous formation of N-nitroso compounds, which are potent carcinogens, and other health maladies (as discussed in Chapter 5). Also, as we have seen, groundwater flow contributes nitrate to base flow in streams and elevated nitrate concentrations impair stream water quality and ecosystem health. Several approaches have been used to better understand the vulnerability of groundwater, particularly aquifers, to nitrate contamination. Vulnerability of an aquifer to contamination is a function of the following factors: (a) susceptibility, which includes natural features that affect the transport and fate of nitrate, such as soil permeability, dissolved oxygen, and aquifer rock type; (b) reactive nitrogen input, which includes land use, and loading rates from various sources; and (c) mobility and persistence, which includes rate of biochemical transformation and partitioning on soil particles (e.g., ammonium).

A better understanding of the factors governing the vulnerability of groundwater to nitrate contamination is critical for developing effective policy and management strategies to reduce nitrogen input loads, protect groundwater as a safe drinking-water source, and remediate vulnerable groundwater resources. Different approaches for assessing aquifer vulnerability to nitrate contamination have included scoring and index methods (e.g., DRASTIC (**D**epth to water, net **R**echarge, **A**quifer media, **S**oil media, **T**opography, **I**mpact of vadose zone, and hydraulic **C**onductivity), e.g., Antonakos & Lambrakis, 2007; FAVA (Florida Aquifer Vulnerability Assessment), Arthur et al., 2007; ReVA, Green et al., 2005; GWAVA-S, GWAVA-DW, Nolan & Hitt, 2006; see example, Fig. 7.11) for assessing susceptibility to contamination, mass-balance models, statistical methods (e.g., random forest regression (RFR), logistic regression (LR), cluster analysis, weights of evidence, and principal components analysis), process-based computer simulation (numerical) modeling of groundwater flow and solute transport. This chapter focuses on the latter two approaches. Models are simply numerical representations of the complexities of groundwater

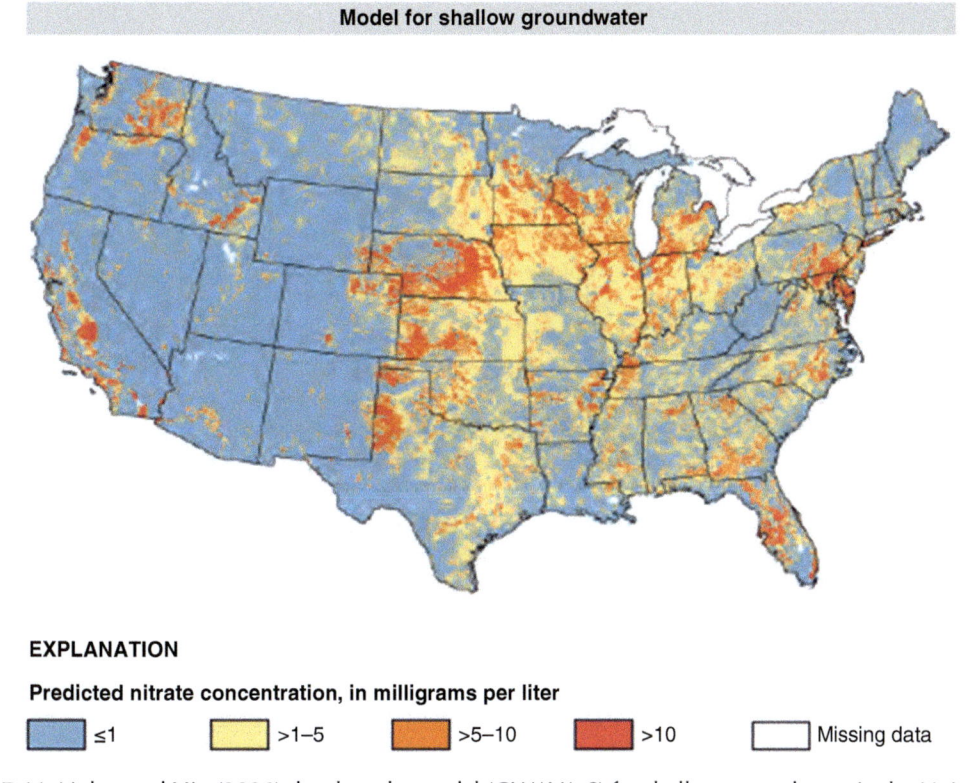

Figure 7.11 Nolan and Hitt (2006) developed a model (GWAVA-S) for shallow groundwater in the United States that predicts ranges of nitrate concentrations (in mg/L) for areas with large nitrogen sources, factors that enhance rapid transport of nitrogen in groundwater, and minimal attenuation processes. Areas where nitrate-N concentrations exceed 10mg/L are shown in red shades. Modified from colored figure 2 in Nolan and Hitt, 2006.

systems. As noted by Anderson et al. (2015), "because the subsurface is hidden from view and analysis is hampered by the lack of field observations, a model is the most defensible description of a groundwater systems for informed and quantitative analyses, as well as forecasting the consequences of proposed actions."

7.7.1. Statistical Methods for Assessing Vulnerability of Groundwater to Nitrate Contamination

Groundwater resources and aquifers are most vulnerable to contamination from recent recharge. Gurdak and Qi (2012) assessed the factors controlling nitrate occurrence related to water recharged less than 60 years ago to 17 principal aquifers of the United States. The principal aquifers studied cover parts of all 48 contiguous states and include the Basin and Range, Biscayne, California Coastal Basins, Cambrian-Ordivician, Central Valley, Coastal Lowlands, Denver Basin, Edwards-Trinity, Floridan, Glacial, High Plains, Mississippi Embayment, New England Crystalline, North Atlantic Coastal Plain, Piedmont and Blue Ridge, Rio Grande, and Valley and Ridge Carbonate aquifer systems. They developed new LR models that tested 87 explanatory variables using data (collected from 1974 to 2010) from the USGS NAWQA program and National Water Information System to identify the most significant source, transport, and attenuation factors that control nonpoint source nitrate concentrations greater than relative background levels in recently recharged groundwater. These models were also used to predict the probability of detecting elevated nitrate in areas beyond the sampling network. Gurdak and Qi (2012) found that the most important factors for predicting elevated nitrate concentrations in the 17 groundwater systems were dissolved oxygen, crops and irrigated cropland, fertilizer application, seasonally high water table, and soil properties that affect infiltration and denitrification. Given some of the limitations in forecasting future groundwater vulnerability conditions due to changing land-use patterns, biogeochemical soil processes, groundwater pumping, and climate, it was concluded that new vulnerability modeling approaches would be necessary for accounting for spatial and temporal variability of the source, transport, and attenuation factors.

Masetti et al. (2008) used a weights-of-evidence statistical modeling technique to analyze anthropogenic and natural factors affecting elevated groundwater nitrate concentrations in the central part of the Po Plain in northern Italy. Their statistical approach analyzed correlations between nitrate data from 69 monitoring wells and several geoenvironmental variables (such as nitrogen fertilizer loading, groundwater recharge, soil drainage, unsaturated zone permeability, groundwater depth, and saturated zone hydraulic conductivity). They found that groundwater recharge and saturated hydraulic conductivity show the best correlation with the presence or absence of nitrate greater than a threshold value of 5 mg/L. The increase in nitrate concentrations with increased groundwater recharge (particularly where recharge was higher than 90 cm/yr) indicated that mass transport of nitrate was dominant compared to dilution in the unsaturated zone. Conversely, it was found that dilution was more important in the saturated zone.

Ransom et al. (2017) developed a hybrid, nonlinear, machine learning model within a statistical learning framework to predict nitrate contamination of groundwater in the Central Valley of California. Machine learning algorithms involve the use of a target function (f) that best maps input variables (X) to an output variable (Y): [e.g., $Y = f(X)$]. Increases in nitrate concentrations in groundwater in this region have impacted the sustainability of this resource. Also, nitrate contamination of drinking water wells has become a serious concern due to increases in nitrogen fertilizer use, manure, and population. Using a database of 145 predictor variables nitrate concentrations was modeled to depths of approximately 500 m below ground surface. The predictor variables (including well characteristics, historical and current field and landscape-scale nitrogen mass balances, historical and current land use, oxidation/reduction conditions, groundwater flow, climate, soil characteristics, depth to groundwater, and groundwater age) were assigned to over 6000 private supply and PSWs measured previously for nitrate and located throughout the study region. They used the boosted regression tree (BRT) method to screen and rank variables to predict nitrate concentrations at various depths (domestic and public well supplies). The novel approach included as predictor variables outputs from existing physically based models of the Central Valley. Ransom et al. (2017) found that the top five most important predictor variables included (a) the probability of manganese concentration to exceed 50 ppb; (b) the probability of dissolved oxygen concentration to be below 0.5 ppm; (c) field-scale adjusted unsaturated zone nitrogen input for the 1975 time period; (d) average difference between precipitation and evapotranspiration during the years 1971–2000; and (e) 1992 total landscape nitrogen input. Their final model included 25 selected variables, which showed that increasing probability of anoxic conditions and increasing precipitation relative to potential evapotranspiration had a corresponding decrease in nitrate concentration predictions. Also, nitrate predictions generally decreased with increasing groundwater age.

Nolan et al. (2014) also evaluated various vulnerability models to map groundwater nitrate concentrations at domestic and PSW depths in the Central Valley, California. They compared three modeling methods regarding their effectiveness in predicting nitrate concentration greater than 4 mg N/L: (a) LR; (b) random forest classification (RFC); and (c) RFR. More information about the RFR method is summarized by Liaw and Wiener (2002). Important processes evaluated by all modeling methods included nitrogen fertilizer input at the land surface, transmission through coarse-textured, well-drained soils, and transport in the aquifer to the well screen. Their modeling approach also included outputs from previous, physically based hydrologic and textural models as predictor variables, which improved the usefulness of the models. The study found that the total percent correct predictions were similar among the three models (69–82%), but the RFR had greater sensitivity (84% for shallow wells and 51% for deep wells). It was concluded that the RFR was more effective in identifying areas with high nitrate concentrations but that the LR and RFC models could be effective in describing bulk conditions in the aquifer. For the RFR models, the two parameters that ranked moderately high-to-high were vertical water fluxes in the aquifer and percent coarse material above the well screen. For the LR models, the average vertical water flux during the irrigation season was highly significant ($p<0.0001$).

Wheeler et al. (2015) developed a RFR model to predict nitrate concentrations in private wells in Iowa. They used 34,084 measurements of nitrate in private wells and systematically evaluated the predictive performance of 179 variables in 36 thematic groups (such as well depth, distance to sinkholes, location, land use, soil characteristics, nitrogen inputs, meteorology, and other factors). Some of the most important variables in their final model (that contained 66 variables in 17 groups) were well depth, slope length within 1 km of the well, year of sample, and distance to nearest animal feeding operation. It was reported that a strong correlation between observed and estimated nitrate concentrations in the training set ($r^2 = 0.77$) and was acceptable for the testing set ($r^2 = 0.38$). They found that the random forest model was considerably better than a traditional LR model or a regression tree in predicting nitrate concentrations. The association between nitrate levels in drinking water and cancer risk in the Iowa participants of the Agricultural Health Study cohort will be assessed using this random forest model. Rodriguez-Galiano et al. (2014) also used a RFR for predictive modeling of nitrate contamination in the Vega de Granada aquifer in an agricultural area of southern Spain. They used a comprehensive GIS database containing 24 parameters related to intrinsic hydrogeologic properties, driving forces, remotely sensed variables, and chemical and physical measurements. They found that electrical conductivity was highly correlated with nitrate concentrations above 50 mg/L, which indicated that irrigation was an important factor. Also, high groundwater vulnerability to nitrate contamination in about 25% of the aquifer's area was related to hydraulic conductivity, hydraulic gradient, depth to groundwater, transmissivity, thickness of unsaturated zone, soil drainage properties, and pH.

In a related study, Zirkle et al. (2016) evaluated the relationship between animal feeding operations (AFOs) and groundwater nitrate concentrations in Iowa. They also evaluated differences in nitrate losses from confined AFOs, open, or mixed types. Their study used LR models and found significant positive associations between the total number of AFOs within 2 km of a well (p trend <0.001), number of open AFOs within 5 km of a well (p trend <0.001), and number of mixed AFOs within 30 km of a well (p trend <0.001), and the log nitrate concentration. They also reported significant increases in log nitrate in the top quartiles for AFO spatial intensity, open AFO spatial intensity, and mixed AFO spatial intensity compared to the bottom quartile (0.171, 0.319, and 0.541 log(mg/L), respectively; all $p<0.001$). An additive model was developed for high-nitrate status that identified statistically significant areas of risk for high levels of nitrate. These results support a relationship between AFOs and groundwater nitrate concentrations and differences in nitrate loss from confined AFOs vs open or mixed types.

Messier et al. (2019) modeled private well groundwater nitrate concentrations in North Carolina by developing multiple "machine-learning" models (https://www.sciencedirect.com/topics/earth-and-planetary-sciences/machine-learning, accessed 22 December 2018) and testing against out-of-sample prediction. Their study incorporated nitrate measurements in approximately 22,000 private wells in North Carolina into various models (including a censored maximum likelihood-based linear model, random forest, gradient boosted machine, support vector machine, neural networks, and kriging). It was found that using a random forest model, their final classification approach predicted three concentrations ranges (<1, 1–5, and ≥ 5 mg/L) with 58 variables. Messier et al. (2019) noted that this final classification model had an overall accuracy of 0.75 and high specificity for the higher two categories and high sensitivity for the lowest category. Their exposure estimates in unmonitored areas were developed for use in the Agricultural Health Study (AHS) cohort to better understand how agricultural exposures and lifestyle factors affect the health of farming populations (https://aghealth.nih.gov/, accessed 5 January 2019).

7.7.2. Numerical Modeling of Nitrogen Fate and Transport in Groundwater

Over the past 50 years, numerical groundwater flow models have helped us to better understand the highly complex patterns of water movement in groundwater systems. More recently, models have been developed to simulate advection, dispersion, and chemical reactions of dissolved constituents (e.g., nitrate) in groundwater systems. These models are powerful tools that can test various future scenarios, such as lag times for water quality restoration efforts, effects of agricultural best management practices, resource protection alternatives and sustainability, and climate change impacts on groundwater nitrate concentrations. It is well beyond the scope of this chapter to discuss the various technical aspects of groundwater flow and solute transport models (e.g., numerical model codes, building a model, calibration, parameter estimation, and sensitivity analysis). Rather, this chapter focuses on some of the more recent applications of different types of numerical models that have been used to investigate the transport and fate of nitrate and other reactive forms of nitrogen in groundwater systems.

Since the 1970s the USGS has been a leader in the development of groundwater modeling software, particularly for simulating the movement of chemical constituents (solute transport) in groundwater. MODFLOW, a three-dimensional finite-difference groundwater flow model was released by USGS in 1983 and has become the most widely used groundwater modeling program in the world (Provost et al., 2009). MODFLOW now includes more than 20 software packages and hydrologic processes that have been added by individuals and groups both within and outside of the USGS. MODPATH uses model-calculated groundwater velocities and flow vectors to delineate particle tracking of groundwater movement and travel times within an aquifer. The USGS website, https://water.usgs.gov/software/lists/groundwater/ (accessed 10 December 2018), provides a list and description of various models used to simulate groundwater flow, transport, geochemical reactions, and groundwater/surface-water interactions. Additional detailed information on the different types of available models can be found elsewhere, for example: https://www.epa.gov/land-research/ground-water-modeling-research (accessed 10 December 2018).

The MODFLOW groundwater flow and transport model was used along with sample analyses and age-dating of groundwater to assess various agricultural practices and nitrate concentrations in groundwater in a glacial outwash aquifer in Minnesota (Puckett & Cowdery, 2002). They calibrated the model against observed water levels at monitoring wells in the study area, and used MODPATH to estimate groundwater travel times and flow paths. Modeling results indicated expected groundwater travel times, and there was good agreement with groundwater ages determined using various transient tracers. Puckett and Cowdery (2002) also used measured nitrate concentrations along with dissolved nitrogen gas to reconstructed historical concentrations. They showed that nitrate concentrations have been increasing in the aquifer since the 1940s and have exceeded the MCL at most sites since the mid-to-late 1960s. The increase in nitrate concentrations was related to a corresponding increase in agricultural fertilizer use. Other studies have combined MODFLOW and MT3DS (transport modeling code that accounts for adsorption and other chemical processes) to simulate different spatial distribution scenarios of nitrate inputs from agricultural practices in a watershed in Brittany, France (Molenat & Gascuel-Odeux, 2002). They found that steady-state simulation results agreed closely with measured nitrate concentrations and showed that reducing nitrate inputs on hillsides was more effective than reducing nitrate in stream water over the entire watershed. Another study combined the models MODFLOW and MT3DMS along with Soil and Water Assessment Tool (SWAT; a watershed modeling code) to simulate nitrogen transport and fate in a watershed with extensive pig-farming operations in Brittany, France, over a 44-month period (Conan et al., 2003). In this area, nitrate concentrations frequently exceeded the European Community drinking water standard (50 mg NO_3/L) in surface and subsurface waters. It was found that the coupled models could assess various nitrogen transformation processes and were effective in simulating stream baseflow and groundwater concentrations. The models also simulated reductions in nitrogen from decreased manure application rates from 210 to 170 kg N/ha, as required by the European Commission Nitrates Directive.

Other studies have combined vadose zone and saturated zone models to investigate nitrate fate and transport in groundwater. For example, Almasri and Kaluarachchi (2007) estimated nitrate leaching to groundwater from both point and nonpoint sources using a soil nitrogen dynamic model. Their work builds on earlier work that used soil transformation models to determine nitrate leaching to groundwater, evaluate the fate and transport of nitrate in groundwater, and develop management options to reduce nitrate and pesticide concentrations in groundwater (e.g., Pesticide Root Zone Model (Carsel et al., 1985), LEACHP (Wagenet & Huston, 1987), GLEAMS (Leonard et al., 1987), and NLEAP (Shaffer et al., 1991)). The modeling framework developed by Almasri and Kaluarachchi (2007) simulates long-term nitrate concentrations related to the existing land-use practices and proposed management scenarios, determines the spatial and temporal nitrate concentrations in groundwater, computes the nitrate mass flux between surface and groundwater at critical stream segments, estimates the spatial and temporal distribution of land surface nitrogen loadings and corresponding nitrate leaching to groundwater, estimates the nitrogen buildup in the soil, and determines the potential for nitrogen attenuation potential in various subareas. The model used MODFLOW (Harbaugh & McDonald, 1996; https://water.usgs.gov/ogw/modflow/, accessed 19 December 2018) for the simulation of the groundwater flow model and on MT3D (Zheng & Wang, 1999) to simulate the nitrate fate and transport processes in groundwater. Their methodology was used for the Sumas–Blaine aquifer of Washington State, US, where extensive dairy industry and berry plantations are concentrated. They found that groundwater denitrification along with manure loadings had the greatest impact on the frequency of MCL exceedances for nitrate (followed by fertilizer loading and atmospheric deposition). For the Sumas–Blaine aquifer, denitrification in groundwater had a higher impact on reducing nitrate mass in groundwater when compared to advection and mechanical dispersion. However, they noted that denitrification in groundwater was highly site-specific. In areas dominated by dairy farms, they noted that reduction of manure loading had a greater impact on reducing nitrate mass buildup in the aquifer compared to reductions in fertilizer loading. They concluded that not all nitrogen management options are efficient in reducing nitrate concentration in groundwater and various management options should be assessed by the model before implementation in the field.

A detailed literature review of groundwater models used to simulate the transport and fate of nitrate in the unsaturated (vadose) zone and the saturated zone was summarized by Hazen and Sawyer (2010). Based on more than 70 reports or journal articles related to modeling of nitrogen fate and transport, they identified six main types of modeling research listed in decreasing order of the number of studies: (a) transport simulations in the vadose or unsaturated zone; (b) transport simulation in the saturated zone; (c) combined simulation of nitrate movement in the vadose zone and saturated zone; (d) denitrification processes; (e) modeling of nitrogen transport at the watershed scale that includes interactions with surface waters; and (f) nitrogen mass-balance models for catchments including groundwater contributing areas to springs. In addition, they found more than 20 modeling codes or solutions that were not specific to any on contaminant but could be used to simulate the movement of nitrogen in the subsurface.

The vadose zone models referred to in Hazen and Sawyer (2010) contain a myriad of numerical or analytical modeling approaches (classified as either deterministic physical models or stochastic, probabilistic models). The physically based deterministic models for nitrogen fate and transport in the vadose zone typically are solutions of the Richards' equation combined with a one-dimensional solution of the advection-dispersion equation for representing vertical flow and transport (assuming horizontal flow vectors in most cases were negligible). Many of the early vadose zone models simulated the movement of nitrogen from fertilizer applications (e.g., Bakhsh et al., 2004; Moreels et al., 2003); although other unsaturated zone models simulated nitrogen movement associated with the land application of treated wastewater (e.g., Reynolds & Iskandar, 1995).

Sanford and Pope (2013) used a nitrogen mass-balance regression model to assess the time needed for the effects of nitrogen-reducing best management practices implemented at the land surface to reach the Chesapeake Bay from groundwater transport to streams. Their model included the distribution of groundwater return times obtained from a regional groundwater flow model for seven watersheds on the Delmarva Peninsula (draining to the eastern side of Chesapeake Bay) in the central mid-Atlantic region. Several different scenarios involving future nitrate load reduction to the water table were evaluated. Based on model results for various nitrate reduction scenarios, they found that it will take several decades to see improvements in water quality from current and future BMPs.

Wang et al. (2012, 2016b) presented an approach to modeling groundwater nitrate at the national scale in the United Kingdom to simulate the impacts of historical nitrate loading from agricultural land on the occurrence of elevated groundwater nitrate concentrations. They developed a process-based component for the saturated zone of significant aquifers in England and Wales. This flow model uses modeled recharge values, along with known aquifer properties and thickness data, and a spatially distributed and temporally variable nitrate input function. Their model was calibrated using national nitrate monitoring data. Model results included time series of annual average nitrate concentrations and spatially distributed nitrate concentration maps from 1925 to 2150 for 28 selected aquifer zones. Their model results indicated that 16 aquifer zones have an increasing trend in nitrate concentration, whereas average nitrate concentrations are declining in the remaining 12 aquifers. The results also showed the trends in fluxes of groundwater nitrate that discharges to rivers through baseflow. An important advantage of their model is the capability to assess the magnitude and timescale of groundwater nitrate response for developing source contribution tools that can be combined with land-management planning strategies for reducing nitrate losses.

Green et al. (2018) developed a regional vertical flux method (VFM) for estimating the reactive transport of nitrate in the vadose zone and groundwater. They applied the regional VFM to 443 well samples in central-eastern Wisconsin, US. The study included chemical measurements of dissolved oxygen, nitrate, nitrogen gas from denitrification, and groundwater age tracers (carbon-14, chlorofluorocarbons, tritium, and tritiogenic helium). The VFM results were consistent with observed chemical data and indicated that (a) travel times through the unsaturated zone constituted a large portion of the transit time to wells and streams, (b) lag times in the unsaturated zone and the depth of the nitrate front were substantially affected by variability in recharge, and (c) since 1945, fractions of manure nitrogen (mainly from injection of liquid manure) leached to groundwater have increased but fertilizer nitrogen leaching decreased. It was estimated that under current conditions, downward migration of nitrate would affect approximately 40% of the shallow aquifer, and about 60% of the shallow aquifer would be protected by denitrification. Based on the results of the study, Green et al. (2018) concluded that large-scale holistic strategies need to be developed to address the diverse and often contradictory nitrogen management decisions (e.g., reducing nitrogen volatilization from manure versus limiting N losses to groundwater).

Fovet et al. (2015) presented an interesting approach for estimating groundwater transit times in two agricultural catchments in southwestern France using lumped parameter models that incorporated long-term time series data for nitrate. Agriculture intensified in these two areas in the 1960s. Their conceptual lumped model represented shallow groundwater flow as two parallel linear stores with double porosity and riparian process by a constant nitrogen removal function. The performance of the model was evaluated based on how well it simulated stream flow, stream nitrate concentrations, and groundwater nitrate concentrations at seasonal and inter-annual time scales. They found that simulated nitrate transit times were highly sensitive to climate variability compared to parameter uncertainty. Their transit times in groundwater (approximately 6–12 years) were consistent numerical modeling results and age-dating analyses.

The computer program PHAST simulates groundwater flow in three-dimensional systems, solute transport, and multicomponent geochemical reactions. It uses the model PhreeqcRM (Parkhurst & Wissmeier, 2015) for the reaction engine. PHAST has the capability to model a wide range of equilibrium and kinetic geochemical reactions. The flow and transport calculations are based on a modified version of the HST3D model (only using constant fluid density and constant temperature). PHAST can be used to model natural and contaminated groundwater systems (e.g., migration of nutrients) at scales ranging from laboratory experiments to local and regional field scales. The following website contains a report documenting the use of the PHAST simulator, including running the simulator, preparing the input files, selecting the output files, and visualizing the results (https://www.usgs.gov/software/phast-a-computer-program-simulating-groundwater-flow-solute-transport-and-multicomponent, accessed 15 December 2018). The website also includes six examples that verify the numerical method and demonstrate the capabilities of the simulator.

There has been an increased emphasis on modeling nitrogen fate and transport from on-site wastewater treatment and disposal systems (OSTDS; also referred to as septic tanks). About 25% of US households use on-site wastewater treatment systems. Several modeling studies have used HYDRUS1D or HYDRUS2D to simulate nitrogen contamination in the vadose zone from various wastewater systems (Geza et al., 2013; Heatwole & McCray, 2007; Huntzinger & McCray, 2003). The Soil Treatment Unit model (STUMOD; Geza et al., 2014) was developed to predict the transport of nitrogen in the vadose zone. STUMOD is linked to a saturated zone model (STUMOD-HPS) that uses the nitrogen results from the vadose zone and simulates the transport and fate of nitrogen in the saturated zone downgradient from the OSTDS. STUMOD contains a steady-state model based on a simplification of the advection-dispersion equation used to simulate nitrogen transport, a Monod function to account for nitrogen concentration effects, ammonium sorption, nitrification, and denitrification. Using default input parameters from extensive literature review, STUMOD calculates nitrogen species concentrations and the fraction of total nitrogen that reaches specified soil depths.

In Florida, about 30% of the population relies on the OSTDS for wastewater treatment, according to the 2010 US Census. In many areas of Florida, nitrate contamination of the surficial aquifer or shallow groundwater has occurred via percolation and subsurface transport from OSTDS. Several studies have used GIS along with other models to assess vulnerability and fate and transport of nitrogen (e.g., Cui et al., 2016). For example, Rios et al. (2013) presented the ArcGIS-based Nitrate Load Estimation Toolkit (ArcNLET). This model is divided into three submodels that include a groundwater flow model with an analytical solution to the advection-dispersion equation, a nitrate fate and transport model, and a load estimation model, all implemented as an extension to ArcGIS: a simplified nitrate transport model to estimate nitrate loads to surface water bodies from OSTDS. Denitrification is modeled in ArcNLET using first-order decay in the analytical solution with the decay constant. It should be noted that these models are highly sensitive to the denitrification rate coefficient and the highly variable nitrogen mass-flux input from septic systems. Denitrification rates also vary considerably temporally, spatially, and with depth based on field and laboratory studies. Both parameters contribute to a high degree of uncertainty in estimating nitrogen loads to groundwater from septic tanks. STUMOD-FL-HPS is a modified version of STUMOD (Geza et al., 2014) that assesses nitrogen transport and fate in the saturated and unsaturated zone under soil conditions (soil treatment units) found in Florida (Hazen & Sawyer, 2015).

Even though the EU Water Framework Directive (Directive 2000/60/EC) was developed to improve water quality, in the United Kingdom, there has been a continuous decline in freshwater quality due to elevated nitrate levels. Nitrate concentrations have shown an increasing trend in many aquifers in the United Kingdom K (Stuart et al., 2007). In response to concerns about nitrate storage and time lag in the unsaturated zone, Wang et al. (2013) developed a national scale model (that they called the "nitrate time bomb" model) to simulate the nitrate transport in the unsaturated zone and predict the loading of nitrate at the water table for the United Kingdom. As it was noted, time lags are rarely considered in nitrate management/restoration strategies and policy development. The nitrate-time bomb model was integrated with two other numerical models (GISGroundwater and the nitrate transport model in the saturated zone N-FM) that were used to provide information on nitrate time lag in the groundwater system at a catchment scale (Wang et al., 2013). They selected the UK Eden Valley as a case study area. The study area has thick but spatially variable Permo-Triassic sandstone unsaturated zones (up to 180 m thick) along with high nitrate concentrations in groundwater. Wang et al. (2013) found that most parts of the study area have been affected by the peak nitrate loading around 1983, and nitrate loads entering the groundwater systems have declined. However, given the large variability in thickness of the unsaturated zone (0–183 m) they estimated considerable variations in lag times (nitrate transport) that range as high as 60 years. In some of the source protection areas, Wang et al. (2013) estimated that nitrate will arrive at the water table within the next 34 years.

Wang et al. (2016a) extended their efforts to model groundwater nitrate at the national scale in the United Kingdom. Their main objective was to simulate the impacts of historical nitrate loading from agricultural land on the evolution of groundwater nitrate concentrations in selected zones of significant aquifer in England and Wales. They developed a simple flow model (an extended "nitrate time bomb" model described above) that uses modeled recharge values and published aquifer properties and thickness information (to estimate nitrate velocity in the unsaturated zone), and a spatially distributed and temporally variable nitrate input function. The model was calibrated using nitrate monitoring data from nationwide studies. Parameter sensitivity analysis was evaluated using Monte Carlo simulations. Nitrate attenuation in the groundwater system was not included in their study, however, attenuation could be included if denitrification or other attenuation information is available for a particular aquifer zone. The model generated time series of yearly average nitrate concentrations and spatially distributed yearly nitrate concentration maps from 1925 to 2150 for 28 aquifer zones. Wang et al. (2016a) found increasing nitrate trends in 16 aquifer zones, whereas the remaining 12 show declining trends. This trend information also was indicative of the changes over time of groundwater nitrate discharging to rivers during baseflow conditions.

A study of groundwater nitrate concentrations on Prince Edward Island (PEI), Canada, used a model (FEFLOW) to assess how forecasted climate change (from global climate models) and its related potential changes in agricultural practices would affect future nitrate concentrations (Paradis et al., 2016). Groundwater is the sole source of potable water on PEI and approximately 6% of domestic wells have nitrate-N concentrations that exceed the 10 mg N/L health threshold for drinking water. Based on several groundwater flow and mass transport simulations to the year 2050, it was found that nitrate-N concentrations would increase by 25–32% over the island system. In addition, the number of domestic wells that would exceed the nitrate drinking water standard would more than double. Paradis et al. (2016) concluded that better agricultural management practices need to be developed to sustain long-term groundwater resources.

Karst aquifers are complex groundwater flow systems that typically have primary, secondary, and tertiary porosity. Groundwater can flow through intergranular porosity (primary or matrix porosity), layers with interconnected pores (secondary porosity), and conduits (tertiary porosity). Kuniansky (2016) compared several models developed to more effectively simulate groundwater flow in karst aquifers. Recent advances in numerical computer codes have been used to better understand dual (primary [matrix] and secondary [fractures and conduits]) porosity groundwater flow processes, as well as characterization and management of karst aquifers. Different types of models have been used to simulate groundwater flow in the Floridan aquifer system, which is composed of a thick sequence of predominantly carbonate rocks in Florida, Georgia, South Carolina, and Alabama. Numerous karst features are present over most of the areas of the aquifer, especially in Florida (numerous sinkholes, large springs, and submerged conduits). These models typically are distributed parameter models that assume water flows through connected pores within the aquifer system and can be simulated using similar mathematical models as for sand and gravel aquifers. Laminar and turbulent flow in submerged conduits have been simulated by hybrid models as a one-dimensional pipe network within the aquifer.

The Upper Floridan aquifer in the Woodville Karst Plain (WKP) in northern Florida was used for comparisons of simulations using a porous-equivalent media model with and without turbulence (MODFLOW-Conduit Flow Process mode 2 and basic MODFLOW, respectively) and a hybrid (MODFLOW-Conduit Flow Process mode 1). Xu et al. (2015) simulated long-term (1966–2018) nitrate-N contamination transport processes using a research version of the Conduit Flow Process (CFPv2) code with the reactive hybrid transport model UMT3D. They simulated groundwater flow in the WKP limestone porous matrix using Darcy's law, and nonlaminar flow within conduits using the Darcy–Weisbach equation. Nitrate-N conduit transport and advective exchanges of groundwater and nitrate-N between conduits and limestone matrix were calculated by CFPv2 and UMT3D, instead of MODFLOW and MT3DMS because Reynolds numbers for flows in conduits exceeded the criteria of laminar flow. They calibrated the numerical model using field observations and then applied the model to simulate nitrate-N transport in the WKP. The results from simulations indicated that the major nitrate-N point sources within the WKP originated from two sprayfields near the City of Tallahassee and septic tanks in the rural area south of the City. Also, the model results indicated that conduit networks control nitrate-N transport and regional contaminant occurrences in the WKP, because nitrate-N is transported rapidly through conduits and travels over large areas.

Based on comparisons of these models, Kuniansky (2016) concluded that the increased effort required to develop a hybrid model and its extensive computational times would not be necessary for simulation of average hydrologic conditions (nonlaminar flow effects on simulated head and spring discharge were minimal). Also, it was noted that simulation of nonlaminar flow (hybrid model) would be needed for large storm events to match daily spring flow hydrographs. A major obstacle in developing hybrid models is the uncertainty in the location of conduit networks and have high-resolution datasets for calibration. Kuniansky (2016) further pointed out that preferential flow through conduits or highly permeable zones would preclude the simulation of contaminant transport.

7.7.3. Mass-Balance Models

Mass-balance models are relatively simple tools that when combined with GIS applications can provide spatial loading estimates of the transport of nitrogen from the land surface, through the soil zone, and into groundwater. These models typically use data on nitrogen inputs from various sources and balance these inputs with observed or measured information on nitrogen outputs in springs, and discharge to streams and other surface waters. Mass-balance models generally do not account for aquifer properties that can affect groundwater flow directions and nitrogen transformations in the aquifer. For example, Otis (2007) combined information on Florida soil characteristics, such as drainage potential, permeability, organic carbon content, and hydraulic conductivity to develop estimates for the percentage of nitrogen reduction through the vadose zone to develop loading rates that can be used as a source input for groundwater models. Also, Katz et al. (2009a) developed a mass-balance based estimate of nitrogen loading to groundwater from a variety of sources (atmospheric deposition, farm and nonfarm fertilizer applications, animal manure, septic tanks, and the land-application of treated municipal wastewater) in karstic spring basins in Florida. These models are particularly valuable for land-use planning and developing action plans for remediating nitrate-impaired waters. The Florida Department of Environmental Protection is using a mass-balance model (NSILT; Eller & Katz, 2017) to identify and quantify the major contributing nitrogen sources in spring basins and other areas that contain water-quality impaired waters due to elevated nitrate levels. NSILT is an ArcGIS- and spreadsheet-based tool that provides spatial estimates of the relative contribution of nitrogen from various sources and takes into consideration the transport pathways and processes affecting the various forms of nitrogen as they move from the land surface to the Upper Floridan Aquifer (the source of water discharging to springs). The information from the NSILT tool is an integral part of basin management action plans to reduce nitrogen inputs to groundwater and restore water quality of springs and surface waters in Florida.

7.7.4. Future Needs for Modeling Reactive Nitrogen in Groundwater Systems

Input parameters used to model the transport and fate of reactive nitrogen in groundwater are subject to considerable uncertainty (e.g., related to variability in nitrogen flux inputs, denitrification rates, preferential flow pathways, changes in land use, storage in the vadose zone, and shallow groundwater, variability in recharge due to climate change), which affect the model output. For example, Stuart et al. (2016) noted that the several modifications to the nitrate time bomb model would make it more applicable to other areas to evaluate future management/restoration scenarios. They suggested incorporating detailed nitrate fate and transport processes in the groundwater system for applying to catchment-scale studies, climate-change scenarios, different land-use/land management options, and complex groundwater flow patterns in karst aquifers. Several research studies have evaluated the process of parameter estimation. McCray et al. (2005) developed an approach for estimating multiple parameters related to the transport and fate of nitrogen from OSTDS. They generated cumulative frequency distributions for certain parameters based on published studies and came up with statistical distributions for effluent concentrations, and rates of nitrification and denitrification.

Sutton et al. (2017) reflected on the achievements and discussed some of the emerging research challenges 6 years after the comprehensive European Nitrogen Assessment was published (Sutton et al., 2017). They concluded that more comprehensive spatial and temporal models need to be developed to be able to assess future scenarios involving reactive nitrogen pollution, nitrogen use, benefits/risks, and more effective management. It was suggested that a more coherent approach would enable more sharing of input datasets (monitoring data) and allow results from one model to feed into other models (e.g., relating reactive nitrogen emissions to transfers and impacts, linking nitrogen economic benefits on regional and global scales, and models to account for impacts on the atmosphere, terrestrial ecosystems, freshwater ecosystems, groundwater, and the coastal zone).

REFERENCES

Alley, W. M., Healy, R. W., LaBaugh, J. W., & Reilly, T. E. (2002). Flow and storage in groundwater systems. *Science, 296*, 1985–1990.

Alley, W.M., Reilly, T.E., & Franke, O.L. (1999). Sustainability of ground-water resources. U.S. Geological Survey Circular 1186. Reston, VA, 86 p.

Almasri, M. N., & Kaluarachchi, J. J. (2007). Modeling nitrate contamination of groundwater in agricultural watersheds. *Journal of Hydrology, 343*, 211–229. https://doi.org/10.1016/j.jhydrol.2007.06.016

Anderson, M. P., Woessner, W. W., & Hunt, R. J. (2015). *Applied groundwater modeling: Simulation of flow and advective transport* (2nd ed.). Amsterdam: Elsevier, Academic Press, 564 p. ISBN: 978-0-12-058103-0

Anderson, T. R., Groffman, P. M., Kaushal, S. S., & Walter, M. T. (2014). Shallow groundwater denitrification in riparian zones of a headwater agricultural landscape. *Journal of Environmental Quality, 43*(2), 732–744. https://doi.org/10.2134/jeq2013.07.0303

Antonakos, A. K., & Lambrakis, N. J. (2007). Development and testing of three hybrid methods for the assessment of aquifer vulnerability to nitrates, based on the drastic model, an example from NE Korinthia, Greece. *Journal of Hydrology, 333*, 288–304. https://doi.org/10.1016/j.jhydrol.2006.08.014

Arthur, J. D., Wood, A. R., Baker, A. E., Cichon, J. R., & Raines, G. L. (2007). Development and implementation of a bayesian-based aquifer vulnerability assessment in Florida. *Natural Resources Research, 16*(2), 93–107. https://doi.org/10.1007/s11053-007-9038-5

Ascott, M. J., Gooddy, D. C., Wang, L., Stuart, M. E., Lewis, M. A., Ward, R. S., & Binley, A. M. (2017). Global patterns of nitrate storage in the vadose zone. *Nature Communications*, 1–7. https://doi.org/10.1038/s41467-017-01321-w

Ascott, M. J., Wang, L., Stuart, M. E., Ward, R. S., & Hart, A. (2016). Quantification of nitrate storage in the vadose (unsaturated) zone: A missing component of terrestrial N budgets. *Hydrological Processes, 30*, 1903–1915.

Bakhsh, A., Hatfield, J. L., Kanwar, R. S., Ma, L., & Ahuja, L. R. (2004). Simulating nitrate drainage losses from a Walnut Creek watershed field. *Journal of Environmental Quality, 33*, 114–123.

Basu, N. B., Jindal, P., Schilling, K. E., Wolter, C. F., & Takle, E. S. (2012). Evaluation of analytical and numerical approaches for the estimation of groundwater travel time distribution. *Journal of Hydrology, 475*, 65–73.

Bechmann, M. (2014). Nitrogen losses from agriculture in the Baltic Sea region. *Agriculture, Ecosystems and Environment, 198*(15), 13–24. https://doi.org/10.1016/j.agee.2014.05.010

Behrendt, H., Bach, M., Kunkel, R., Opitz, D., Pagenkopf, W. G., Scholz, G., & Wendland, F. (2003). *Nutrient emissions into river basins of Germany on the basis of a harmonized procedure*. Federal Environmental Agency (Umweltbundesamt).

Böhlke, J. K., Antweiler, R. C., Harvey, J. W., Laursen, A. E., Smith, L. K., Smith, R. L., & Voytek, M. A. (2009). Multiscale measurements and modeling of denitrification in streams with varying flow and nitrate concentration in the upper Mississippi River basin, USA. *Biogeochemistry, 93*, 117–141. https://doi.org/10.1007/s10533-008-9282-8

Böhlke, J. K., Smith, R. L., & Miller, D. N. (2006). Ammonium transport and reaction in contaminated groundwater: Application of isotope tracers and isotope fractionation studies. *Water Resources Research, 42*, W05411. https://doi.org/10.1029/2005WR004349

Bouraoui, F., Grizzetti, B., & Aloe, A. (2009). *Nutrient discharge from river to seas for year 2000* (EUR Report 24002 EN). Luxembourg: European Commission Joint Research Centre.

Bratton, J. F. (2010). The three scales of submarine groundwater flow and discharge across passive continental margins. *The Journal of Geology, 118*(5), 565–575. https://doi.org/10.1086/655114

Burnett, W. C., Aggarwal, P. K., Sureli, A., Bokuniewicz, H., Cable, J. E., Charette, M. A., et al. (2006). Quantifying submarine groundwater discharge in the coastal zone via multiple methods. *Science of the Total Environment, 367*(2–3), 498–543. https://doi.org/10.1016/j.scitotenv.2006.05.009

Carey, R. O., Hochmuth, G. J., Martinez, C. J., Boyer, T. H., Dukes, M. D., Toor, G. S., & Cisar, J. L. (2013). Evaluating nutrient impacts in urban watersheds: Challenges and research opportunities. *Environmental Pollution, 173*, 138–149. http://dx.doi.org/10.1016/j.envpol.2012.10.004

Carsel, R. F., Mulkey, L. A., Lorber, M. N., & Baskin, L. B. (1985). The pesticide root zone model (PRZM): A procedure for evaluating pesticide leaching threats to ground water. *Ecological Modeling, 30*, 49–69.

Collins, A. R., & Gillies, N. (2013). Constructed wetland treatment of nitrates: Removal effectiveness and cost efficiency. *Journal of the American Water Resources Association, 50*(4), 898–908. https://doi.org/10.1111/jawr.12145

Conan, C., Bouraoui, F., Turpin, N., de Marsily, G., & Bidoglio, G. (2003). Modeling flow and nitrate fate at catchment scale in Brittany (France). *Journal of Environmental Quality, 32*(6), 2026–2032. https://doi.org/10.2134/jeq2003.2026

Corbett, D.R., Burnett, W.C., & Chanton, J.P. (2002). Submarine groundwater discharge: An unseen yet potentially important coastal phenomenon. University of Florida Institute of Food and Agricultural Sciences, Sea Grant Department, SGEB-54, 6 p.

Cui, C., Zhou, W., & Geza, M. (2016). GIS-based nitrogen removal model for assessing Florida's surficial aquifer vulnerability. *Environmental Earth Sciences*, 75(6), 526. https://doi.org/10.1007/s12665-015-5213-x

Davis, H.D., Katz, B.G., & Griffin, D. W. (2010). Nitrate-N movement in groundwater from the land application of treated municipal wastewater and other sources in the Wakulla Springs Springshed, Leon and Wakulla Counties, Florida, 1966–2018. Tallahassee, Florida: U.S. Geological Survey Scientific Investigations Report 2010–5099, 80 p.

DeSimone, L. A., Hamilton, P. A., & Gilliom, R. J. (2009). Quality of groundwater from private domestic wells. *Water Well Journal*. Retrieved from http://info.ngwa.org/GWOL/pdf/091384002.pdf

DeSimone, L.A., McMahon, P.B., & Rosen, M.R. (2014). The quality of our Nation's waters—Water quality in Principal Aquifers of the United States, 1991–2010. U.S. Geological Survey Circular 1360, 151 p., https://dx.doi.org/10.3133/cir1360.

Dieter, C. A., Maupin, M. A., Caldwell, R. R., Harris, M. A., Ivahnenko, T. I., Lovelace, J. K., Barber, N. L., & Linsey, K. S. (2018). Estimated use of water in the United States in 2015. U.S. Geological Survey Circular 1441, 65 p. https://doi.org/10.3133/cir1441.

Dubrovsky, N.M., Burow, K.R., Clark, G.M., Gronberg, J.M., Hamilton P.A., Hitt, K.J., et al. (2010). The quality of our Nation's waters—Nutrients in the Nation's streams and groundwater, 1992–2004. U.S. Geological Survey Circular 1350, 174 p., Retrieved from http://water.usgs.gov/nawqa/nutrients/pubs/circ1350, accessed 20 September 2019.

Durand, P., Breuer, L., Johnes, P. J., Billen, G., Butturini, A., Pinay, G., et al. (2011). Nitrogen processes in aquatic ecosystems. Ch. 7. In M. A. Sutton, C. M. Howard, J. W. Erisman, G. Billen, A. Bleeker, P. Grennfelt, H. van Grinsven, & B. Grizzetti (Eds.), *The European nitrogen assessment: Sources, effects, and policy perspectives*. Cambridge, UK: Cambridge University Press, 664 p.

Eberts, S.M., Thomas, M.A., & Jagucki, M.L. (2013). The quality of our Nation's waters—Factors affecting public-supply-well vulnerability to contamination—Understanding observed water quality and anticipating future water quality. U.S. Geological Survey Circular 1385, 120 p., Retrieved from http://pubs.usgs.gov/circ/1385/, accessed 20 September 2019.

Eller, K., & Katz, B. G. (2017). Nitrogen Source Inventory and Loading Tool (NSILT): An integrated approach toward restoration of water-quality impaired karst springs. *Journal of Environmental Management*, 196, 702–709.

European Commission. (2007). *COM(2007) 120*. Report from the Commission to the Council and the European Parliament on the implementation of the Council Directive 91/676/EEC concerning the protection of the waters against pollution caused by nitrates from agricultural sources for the period 2000–2003.

Fovet, O., Ruiz, L., Faucheux, M., Molenat, J., Sekhar, M., Vertès, F., et al. (2015). Using long time series of agricultural-derived nitrates for estimating catchment transit times. *Journal of Hydrology*, 522, 603–617.

Fu, B., Zhuang, X., Jiang, G., Shi, J., & Lu, Y. (2007). Environmental problems and challenges in China. *Environmental Science and Technology*, 41, 7597–7602.

Garces, E., Basterretxea, G., & Tovar-Sanchez, A. (2011). Changes in microbial communities in response to submarine groundwater input. *Marine Ecology Progress Series*, 438, 47–58. https://doi.org/10.3354/meps09311

Geza, M., Lowe, K. S., Huntzinger, D. N., & McCray, J. E. (2013). New conceptual model for soil treatment units: Formation of multiple hydraulic zones during unsaturated wastewater infiltration. *Journal of Environmental Quality*, 42, 1196–1204. https://doi.org/10.2134/jeq2012.0441

Geza, M., Lowe, K. S., & McCray, J. (2014). STUMOD—a tool for predicting fate and transport of nitrogen in soil treatment units. *Environmental Modeling and Assessment*, 19(3), 243–256. https://doi.org/10.1007/s10666-013-9392-0

Green, C. T., Liao, L., Nolan, B. T., Juckem, P. F., Shope, C. L., Tesoriero, A. J., & Jurgens, B. C. (2018). Regional variability of nitrate fluxes in the unsaturated zone and groundwater, Wisconsin, USA. *Water Resources Research*, 54, 301–322.

Green, C. T., Puckett, L. J., Böhlke, J. K., Bekins, B. A., Phillips, S. P., Kauffman, L. J., et al. (2008). Limited occurrence of denitrification in four shallow aquifers in agricultural areas of the United States. *Journal of Environmental Quality*, 37, 994–1009.

Green, E.A., LaMotte, A.E., Cullinan, K., & Smith, E.R. (2005). Groundwater vulnerability to nitrate contamination in the mid-Atlantic region. U.S. Geological Survey Fact Sheet 2004–3067. Reston, VA, 4 p.

Grizzetti, B., Bouraoui, F., Billen, G., van Grinsven, H., Cardoso, A.C., Thieu, V., et al. (2011). Nitrogen as a threat to European water quality. Ch. 17. In M. A. Sutton, C. M. Howard, J. W. Erisman, G. Billen, A. Bleeker, P. Grennfelt, H. van Grinsven, & B. Grizzetti (Eds.), *The European nitrogen assessment: Sources, effects, and policy perspectives* (pp. 379–404). Cambridge, UK: Cambridge University Press.

Groffman, P. M., Boulware, N. J., Zipperer, W. C., Pouyat, R. V., Band, L. E., & Colosimo, M. F. (2002). Soil nitrogen cycle processes in urban riparian zones. *Environmental Science and Technology*, 36, 4547–4552.

Groundwater Directive. 2006. Directive 2006/118/EC of the European Parliament and of the Council of 12 December 2006 on the protection of groundwater against pollution and deterioration.https://www.eea.europa.eu/policy-documents/groundwater-directive-gwd-2006-118-ec, accessed 29 September 2019.

Gu, B., Ge, Y., Chang, S. X., Luo, W., & Chang, J. (2013). Nitrate in groundwater of China: Sources and driving forces. *Global Environmental Change*, 23, 1112–1121. http://dx.doi.org/10.1016/j.gloenvcha.2013.05.004

Gurdak, J. J., & Qi, S. L. (2012). Vulnerability of recently recharged groundwater in principle aquifers of the United States to nitrate contamination. *Environmental Science and Technology*, 46, 6004–6012.

Harbaugh, A.W., & McDonald, M.G. (1996). User's documentation for MODFLOW-96, an update to the US Geological Survey modular finite-difference ground-water flow model. US Geological Survey Open-File Report 96–485, p. 56.

Harvey, J. W., Bohlke, J. K., Voytek, M. A., Scott, D., & Tobias, C. R. (2013). Hyporheic zone denitrification: Controls on effective reaction depth and contribution to whole-stream balance. *Water Resources Research*, *49*, 6298–6316. https://doi.org/10.1002/wrcr.20492.2013

Hazen and Sawyer, Inc. (2010). Florida onsite sewage nitrogen reduction strategies study: Literature review of nitrogen fate and transport modeling. Association with the Colorado School of Mines, Florida Department of Health, Contract CORCL, Task D.2, pp. 2.1–2.29.

Hazen and Sawyer, Inc. (2015). Florida onsite sewage nitrogen reduction strategies study: Task D Report and STUMOD-FL-HPS user's guide. Florida Department of Health, Contract CORCL, Task D.16, 78 p.

Heatwole, K. K., & McCray, J. E. (2007). Modeling potential vadose-zone transport of nitrogen from onsite wastewater systems at the development scale. *Journal of Contaminant Hydrology*, *91*, 184–201.

Heffernan, J. B., Albertin, A. R., Fork, M. L., Katz, B. G., & Cohen, M. J. (2012). Denitrification and inference of nitrogen sources in the karstic Floridan Aquifer. *Biogeosciences*, *9*, 1671–1690.

Hu, C., Muller-Karger, F. E., & Swarzenski, P. W. (2006). Hurricanes, submarine groundwater discharge, and Florida's red tides. *Geophysical Research Letters*, *33*, L11601. https://doi.org/10.1029/2005GL025449

Huntzinger, D. N., & McCray, J. E. (2003). Numerical modeling of unsaturated flow in wastewater soil absorption systems. *Ground Water Monitoring and Remediation*, *23*(2), 64–72. https://doi.org/10.1111/j.1745-6592.2003.tb00672.x

Jiang, R., Hatano, R., Zhao, Y., Woli, K. P., Kuramochi, K., Shimizu, M., & Hayakawa, A. (2014). Factors controlling nitrogen and dissolved organic carbon exports across timescales in two watersheds with different land uses. *Hydrological Processes*, *28*, 5105–5121. https://doi.org/10.1002/hyp.9996

Katz, B. G., & Griffin, D. W. (2008). Using chemical and microbiological indicators to track the impacts from the land application of treated municipal wastewater and other sources on groundwater quality in a karstic springs basin. *Environmental Geology*, *55*, 801–821. https://doi.org/10.1007/s00254-007-1033-y

Katz, B. G., Griffin, D. W., & Davis, H. D. (2009b). Groundwater quality impacts from the land application of treated municipal wastewater in a large karstic spring basin: Chemical and microbiological indicators. *Science of the Total Environment*, *407*, 2872–2886.

Katz, B. G., Griffin, D. W., McMahon, P. B., Harden, H., Wade, E., Hicks, R. W., & Chanton, J. P. (2010). Fate of effluent-borne contaminants beneath septic tank drainfields overlying a karst aquifer. *Journal of Environmental Quality*, *39*, 1181–1195.

Katz, B. G., Sepulveda, A. A., & Verdi, R. J. (2009a). Estimating nitrogen loading to ground water and assessing vulnerability to nitrate contamination in a large karstic spring basin. *Journal of the American Water Resources Association*, *45*, 607–627.

Kaushal, S. S., Groffman, P. M., Band, L. E., Elliott, E. M., Shields, C. A., & Kendall, C. (2011). Tracking nonpoint source nitrogen pollution in human-impacted watersheds. *Environmental Science and Technology*, *45*, 8225–8232; dx.doi.org. doi:10.1021/es200779e

Kidmose, J., Engesgaard, P., Ommen, D. A., Nilsson, B., Findt, M. R., & Andersen, F. O. (2015). The role of groundwater for lake-water quality and quantification of N seepage. *Ground Water*, *53*(5), 709–721. https://doi.org/10.1111/gwat.12281

Knight, R. L., Payne, V. W. E., Jr., Borer, R. E., Clarke, R. A., Jr., & Pries, J. H. (2000). Constructed wetlands for livestock wastewater management. *Ecological Engineering*, *15*, 41–55.

Kroeger, K. D., & Charette, M. A. (2008). Nitrogen biogeochemistry of submarine groundwater discharge. *Limnology and Oceanography*, *53*(3), 1025–1039. https://doi.org/10.4319/lo.2008.53.3.1025

Kuniansky, E.L. (2016). Simulating groundwater flow in karst aquifers with distributed parameter models—Comparison of porous-equivalent media and hybrid flow approaches. U.S. Geological Survey Scientific Investigations Report 2016-5116, 14 p., http://dx.doi.org/10.3133/sir20165116.

Kyllmar, K., Bechmann, M., Deelstra, J., Iital, A., Blicher-Mathiesen, G., Jansons, V., et al. (2014a). Long-term monitoring of nutrient losses from agricultural catchments in the Nordic-Baltic region: A discussion of methods, uncertainties and future needs. *Agriculture, Ecosystems & Environment*, *198*, 4–12. https://doi.org/10.1016/j.agee.2014.07.005

Kyllmar, K., Stjernman Forsberg, L., Andersson, S., & Mårtensson, K. (2014b). Small agricultural monitoring catchments in Sweden representing environmental impact. *Agriculture, Ecosystems & Environment*, *198*, 25–35. https://doi.org/10.1016/j.agee.2014.05.016

Lawniczak, A. E., Güsewell, S., & Verhoeven, J. T. A. (2009). Effect of N: K supply ratios on the performance of three grass species from herbaceous wetlands. *Basic and Applied Ecology*, *10*(8), 715–725. https://doi.org/10.1016/j.baae.2009.05.004

Lecher, A. L., Mackey, K., Kudela, R., Ryan, J., Fisher, A., Murray, J., & Paytan, A. (2015). Nutrient loading through submarine groundwater discharge and phytoplankton growth in Monterey Bay, CA. *Environmental Science and Technology*, *49*(11), 6665–6673. https://doi.org/10.1021/acs.est.5b00909

Lee, C., Fletcher, T. D., & Sun, G. (2009). Nitrogen removal in constructed wetland systems. *Engineering in Life Sciences*, *9*(1), 11–22.

Leonard, R. A., Knisel, W. G., & Still, D. A. (1987). GLEAMS: Groundwaterloading effects of agricultural management systems. *Transactions of the American Society of Agricultural Engineers*, *30*, 1403–1418.

Liaw, A., & Wiener, M. (2002). Classification and regression by random forest. *R News*, *2*(3), 18–22. ISSN 1609-3631

Lindsey, B.D., Berndt, M.P., Katz, B.G., Ardis, A.F., & Skach, K.A. (2009). Factors affecting water quality in selected carbonate aquifers in the United States, 1993–2005. U.S. Geological Survey Scientific Investigations Report 2008-5240. Reston, VA, 117 p.

Liu, J., Jiang, L. H., Zhang, C. J., Li, P., & Zhao, T. K. (2017). Nitrate-nitrogen contamination in groundwater: Spatiotemporal variation and driving factors under cropland in Shandong Province, China. *3rd International Conference on Water Resource and Environment (WRE 2017) IOP Conference Series: Earth and Environmental Science*, *82*(2017), 012059. https://doi.org/10.1088/1755-1315/82/1/012059

Masetti, M., Poli, S., Sterlacchini, S., Beretta, G. P., & Racchi, A. (2008). Spatial and statistical assessment of factors influencing nitrate contamination of groundwater. *Journal of Environmental Management, 86,* 272–281. https://doi.org/10.1016/j.jenvman.2006.12.023

McCray, J. E., Kirkland, S. L., Siegrist, R. L., & Thyne, G. D. (2005). Model parameters for simulating fate and transport of on-site wastewater nutrients. *Ground Water, 43*(4), 628–639.

McMahon, P. B., Böhlke, J. K., Kauffman, L. J., Kipp, K. L., Landon, M. K., Crandall, C. A., Burow, K. R., & Brown, C. J. (2008). Source and transport controls on the movement of nitrate to public supply wells in selected principal aquifers of the United States. *Water Resources Research, 44,* W04401, 17 p. doi:10.1029/2007WR006252.

McMahon, P. B., Dennehy, K. F., Bruce, B. W., Bohlke, J. K., Michel, R. L., Gurdak, J. J., & Hurlbut, D. B. (2006). Storage and transit time of chemicals in thick unsaturated zones under rangeland and irrigated cropland, High Plains, United States. *Water Resources Research, 42*(3), W03413. https://doi.org/10.1029/2005WR004417

Meals, D. W., Dressing, S. A., & Davenport, T. E. (2010). Lag time in water quality response to best management practices: A review. *Journal of Environmental Quality, 39,* 85–96.

Messier, K. P., Wheeleer, D. C., Flory, A. R., Jones, R. R., Patel, D., Nolan, B. T., & Ward, M. H. (2019). Modeling groundwater nitrate exposure in private wells of North Carolina for the Agricultural Health Study. *Science of the Total Environment, 655,* 512–519.

Ministry of Environmental Protection (MEP), China. (1993). *Quality standard for groundwater*. Beijing: MEP.

Ministry of Environmental Protection (MEP), China. (2010). *China environment yearbook*. Beijing: China Environment Yearbook.

Molenat, J., & Gascuel-Odeux, C. (2002). Modelling flow and nitrate transport in groundwater for the prediction of water travel times and of consequences of land use evolution on water quality. *Hydrological Processes, 16*(2), 479–492. https://doi.org/10.1002/hyp.328

Montiel, D., Dimova, N., Bartolome, A., Prieto, J., Garcia-Orellana, J., & Rodellas, V. (2018). Assessing submarine groundwater discharge (SGD) and nitrate fluxes in highly heterogeneous coastal karst aquifers: Challenges and solutions. *Journal of Hydrology, 557,* 222–242. https://doi.org/10.1016/j.jhydrol.2017.12.036

Moreels, E., De Neve, S., Hofman, G., & Van Meirvenne, M. (2003). Simulating nitrate leaching in bare fallow soils: A model comparison. *Nutrient Cycling in Agroecosystems, 67*(2), 137–144.

Nolan, B. T., Gronberg, J. M., Faunt, C. C., Eberts, S. M., & Belitz, K. (2014). Modeling nitrate at domestic and public-supply well depths in the Central Valley. *Environmental Science and Technology, 48*(10), 5643–5651. https://doi.org/10.1021/es405452q

Nolan, B. T., & Hitt, K. J. (2006). Vulnerability of shallow groundwater and drinking-water wells to nitrate in the. *United States: Environmental Science and Technology, 40*(24), 7834–7840.

Otis, R.J. (2007). Estimates of nitrogen loadings to groundwater from onsite wastewater treatment systems in the Wekiva study area. Wekiva onsite nitrogen contribution study prepared for the Florida Department of Healt, Task 2 Report, 45 p.

Palmeri, L., Bendoricchio, G., & Artioli, Y. (2005). Modelling nutrient emissions from river systems and loads to the coastal zone: Po River case study, Italy. *Ecological Modelling, 184,* 37–53.

Paradis, D., Vigneault, H., Lefebvre, R., Savard, M. M., Ballard, J.-M., & Qian, B. (2016). Groundwater nitrate concentration evolution under climate change and agricultural adaptation scenarios: Prince Edward Island, Canada. *Earth System Dynamics, 7,* 183–202. https://doi.org/10.5194/esd-7-183-2016

Parkhurst, D. L., & Wissmeier, L. (2015). PhreeqcRM: A reaction module for transport simulators based on the geochemical model PHREEQC. *Advances in Water Resources, 83,* 176–189. https://doi.org/10.1016/j.advwatres.2015.06.001

Provost, A.M., Reilly, T.E., Harbaugh, A.W., & Pollock, D.W. (2009). U.S. Geological Survey groundwater modeling software: Making sense of a complex natural resource. U.S. Geological Survey Fact Sheet 2009–3105, 4 p.

Puckett, L. J., & Cowdery, T. K. (2002). Transport and fate of nitrate in a glacial outwash aquifer in relation to groundwater age, land use practices, and redox processes. *Journal of Environmental Quality, 31,* 782–796.

Puckett, L. J., Tesoriero, A. J., & Dubrovsky, N. M. (2011). Nitrogen contamination of surficial aquifers- a growing legacy. *Environmental Science and Technology, 45,* 839–844.

Qiu, J. (2011). China to spend billions cleaning up groundwater. *Science, 334,* 745–745.

Ransom, K. M., Nolan, B. T., Traum, J. A., Faunt, C. C., Bell, A. M., Gronberg, J. A. M., et al. (2017). A hybrid machine learning model to predict and visualize nitrate concentration throughout the Central Valley aquifer, California, USA. *Science of the Total Environment, 601–602,* 1160–1172. http://dx.doi.org/10.1016/j.scitotenv.2017.05.192

Reynolds, C.M., & Iskandar, I.K. (1995). A modeling-based evaluation of the effect of wastewater application practices on groundwater quality. U.S. Army Corps of Engineers Cold Regions Research and Engineering Laboratory Report 95–2, 43 p.

Rios, J. F., Ye, M., Wang, L., Lee, P. Z., Davis, H., & Hicks, R. (2013). ArcNLET, a GIS-based software to simulate groundwater nitrate load from septic systems to surface water bodies. *Computers and Geosciences, 52,* 108–116. https://doi.org/10.1016/j.cageo.2012.10.003

Rivett, M. O., Buss, S. R., Morgan, P., Smith, J. W., & Bemment, C. D. (2008). Nitrate attenuation in groundwater: A review of biogeochemical controlling processes. *Water Research, 42*(16), 4215–4232. https://doi.org/10.1016/j.watres.2008.07.020. Epub 2008 Jul 23

Robertson, W. M., Bohlke, J. K., & Sharp, J. M., Jr. (2017). Response of deep groundwater to land use change in desert basins of the Trans-Pecos region, Texas, USA: Effects on infiltration, recharge, and nitrogen fluxes. *Hydrological Processes, 31*(3), 2349–2364.

Rodriguez-Galiano, V., Mendes, M. P., Garcia-Soldado, M. J., Chica-Olmo, M., & Ribeiro, L. (2014). Predictive modeling of groundwater nitrate pollution using random forest and multisource variables related to intrinsic and specific vulnerability: A case study in an agricultural setting (Southern Spain). *Science of the Total Environment, 476–477,* 189–206. http://dx.doi.org/10.1016/j.scitotenv.2014.01.001

Römheld, V., & Kirkby, E. A. (2010). Research on potassium in agriculture: Needs and prospects. *Plant and Soil, 335*(1–2), 155–180. https://doi.org/10.1007/s11104-010-0520-1

Sanford, W. E., & Pope, J. P. (2013). Quantifying groundwater's role in delaying improvements to Chesapeake Bay water quality. *Environmental Science and Technology, 47*, 13330–13338.

Sawyer, A. H., David, C. H., & Famiglietti, J. S. (2016). Continental patterns of submarine groundwater discharge reveal coastal vulnerabilities. *Science*, aag1058. https://doi.org/10.1126/science.aag1058

Schlesinger, W. H. (2009). On the fate of anthropogenic nitrogen. *Proceedings of the National Academy of Sciences of the USA, 104*, 203–208.

Schreiber, H., Constantinescu, L.T., Cvitanic, I., Drumea, D., Jabucar, D., Juran, S., et al. (2003). Harmonised inventory of point and diffuse emissions of nitrogen and phosphorus for a transboundary river basin (Report 200 22 232). 159 p. Retrieved from https://www.icpdr.org/main/sites/default/files/Danube%20report%20UBA%202003%20color%20si.pdf, accessed 15 December 2018.

Schullehner, J., & Hansen, B. (2014). Nitrate exposure from drinking water in Denmark over the last 35 years. *Environmental Research Letters, s 9*, 095001.

Sebillo, M., Mayer, B., Nicolardot, B., Pinay, G., & Mariotti, A. (2013). Long-term fate of nitrate fertilizer in agricultural soils. *Proceedings of the National Academy of Sciences of the United States, 110*(45), 18185–18189.

Seitzinger, S. P., Mayorga, E., Kroeze, C., Kroeze, C., Beusen, A. H. W., Billen, G., et al. (2010). Global river nutrient export trajectories 1970–2050: A Millennium Ecosystem Assessment scenario analysis. *Global Biogeochemical Cycles*. https://doi.org/10.1029/2009GB003587

Shaffer, M. J., Halvorson, A. D., & Pierce, F. J. (1991). Nitrate leaching and economic analysis package (NLEAP): Model description and application. In R. F. Follet, D. R. Keeney, & R. M. Cruse (Eds.), *Managing N for groundwater quality and farm profitability* (pp. 285–322). Madison, WI: Soil Science Society of America.

Shukla, S., & Saxena, A. (2018). Global status of nitrate contamination in groundwater: Its occurrence, health Impacts, and mitigation measures. In C. M. Hussain (Ed.), *Handbook of environmental materials management*. Springer International Publishing. https://doi.org/10.1007/978-3-319-58538_20-1

Sileika, A. S., Stalnacke, P., Kutra, S., Gaigalis, K., & Berankiene, L. (2006). Temporal and spatial variation of nutrient levels in the Nemunas River (Lithuania and Belarus). *Environmental Monitoring and Assessment, 122*, 335–354.

Spahr, N.E., Dubrovsky, N.M., Gronberg, J.M., Franke, O.L., & Wolock, D.M. (2010). Nitrate loads and concentrations in surface-water base flow and shallow groundwater for selected basins in the United States, water years 1990–2006. U.S. Geological Survey Scientific Investigations Report 2010–5098, 39 p.

Stoliker, D. L., Repert, D. A., Smith, R. L., Song, B., LeBlanc, D. R., McCobb, T. D., et al. (2016). Hydrologic controls on nitrogen cycling processes and functional gene abundance in sediments of a groundwater flow-through lake. *Environmental Science and Technology, 50*(7), 3649–3657. https://doi.org/10.1021/acs.est.5b06155

Stuart, M. E., Chilton, P. J., Kinniburgh, D. G., & Cooper, D. M. (2007). Screening for long-term trends in groundwater nitrate monitoring data. *Quarterly Journal of Engineering Geology and Hydrogeology, 40*, 361–376.

Stuart, M.E., Wang, L., Ascott, M., Ward, R.S., Lewis, M.A., & Hart, A.J. (2016). Modelling the groundwater nitrate legacy. British Geological Survey Groundwater Science Directorate Open Report OR/16/036, 75 p.

Sutton, M. A., Howard, C. M., Brownlie, W. J., Skiba, U., Hicks, W. K., Winiwarter, W., van Grinsven, H., Bleeker, A., Westhoek, H., Oenema, O., de Vries, W., Leip, A. (2017). The European Nitrogen Assessment 6 years after: What was the outcome and what are the future research challenges. In T. Dalgaard, J. E. Oelesen, J. K. Schjoering, et al. (Eds.), *Innovative solutions for sustainable management of nitrogen. Proceedings From the International Conference*, 25–28 June 2017 (pp. 40–49). Aarhus, Denmark: Aarhus University and the Denmark Research Alliance.

Sutton, M. A., Howard, C. M., Erisman, J. W., Billen, G., Bleeker, A., Grennfelt, P., van Grinsven, H., & Grizzetti, B. (2011a). Assessing our nitrogen inheritance. The need to integrate nitrogen science and policies. Ch. 1. In M. A. Sutton, C. M. Howard, J. W. Erisman, G. Billen, A. Bleeker, P. Grennfelt, H. van Grinsven, & B. Grizzetti (Eds.), *The European nitrogen assessment: Sources, effects, and policy perspectives*. Cambridge, UK: Cambridge University Press, 664 p.

Sutton, M. A., Howard, C. M., Erisman, J. W., Billen, G., Bleeker, A., Grennfelt, P., van Grinsven, H., & Grizzetti, B. (Eds.). (2011b). *The European nitrogen assessment: Sources, effects, and policy perspectives*. Cambridge, UK: Cambridge University Press, 664 p.

Szczepaniak, W., Barłóg, P., Łukowiak, R., & Przygocka-Cyna, K. (2013). Effect of balanced nitrogen fertilization in four-year rotation on plant productivity. *Journal of Central European Agriculture, 14*(1), 64–77. https://doi.org/10.5513/JCEA01/14.1.1157

Tesoriero, A. J., Duff, H. H., Saad, D. A., Spahr, N. E., & Wolock, D. M. (2013). Vulnerability of streams to legacy nitrate sources. *Environmental Science and Technology, 47*(8), 3623–3629.

Toccalino, P.L. & Hopple, J.A. (2010). The quality of our Nation's waters—Quality of water from public-supply wells in the United States, 1993–2007. Overview of Major Findings. U.S. Geological Survey Circular 1346. Reston, VA, 58 p.

Toccalino, P.L., Norman, J.E., & Hitt, K.J. (2010). Quality of source water from public-supply wells in the United States, 1993–2007. U.S. Geological Survey Scientific Investigations Report 2010–5024. Reston, VA, 209 p. Retrieved from http://water.usgs.gov/nawqa/studies/public_wells/, accessed 20 September 2019.

U.S. Environmental Protection Agency. (1994). Handbook—Ground water and wellhead protection (EPA/625/R-94-001) [variously paged], accessed 7 May 2012 at http://cfpub.epa.gov/ols/catalog/catalog_records_found.cfm?&FIELD4=CALLNUM&INPUT4=625/R-94-001, accessed 20 September 2019.

Van Drecht, G., Bouwman, A. F., Knoop, J. M., Beusen, A. H. W., & Meinardi, C. R. (2003). Global modeling of the fate of nitrogen from point and nonpoint sources in soils, groundwater, and surface water. *Global Biogeochemical Cycles, 17*(4), 1115. https://doi.org/10.1029/2003GB002060

Van Meter, K. J., & Basu, N. B. (2015). Catchment legacies and time lags: A parsimonious watershed model to predict the effects of legacy storage on nitrogen export. *PLoS One, 10*(5), e0125971. https://doi.org/10.1371/journal.pone.o125971

Van Meter, K. J., & Basu, N. B. (2017). Time lags in watershed-scale nutrient transport: An exploration of dominant controls. *Environmental Research Letters, 12*, 084017. https://doi.org/10.1088/1748-9326/aa7bf4

Van Meter, K. J., Basu, N. B., & Van Cappellen, P. (2017). Two centuries of nitrogen dynamics, legacy sources and sinks in the Mississippi and Susquehanna River basins. *Global Biogeochemical Cycles, 31*(1), 2–23.

Van Meter, K. J., Basu, N. B., Veenstra, J. J., & Burras, C. L. (2016). The nitrogen legacy: Emerging evidence of nitrogen accumulation in anthropogenic landscapes. *Environmental Research Letters, 11*. https://doi.org/10.1088/1748-9326/11/3/035014

Vidon, P., Allan, C., Burns, E., Duval, T. P., Gurwick, N., Inamdar, S., et al. (2010). Hot spots and hot moments inriparian zones: Potential for improved water quality management. *Journal of the American Water Resources Association, 46*, 278–291. https://doi.org/10.1111/j.1752-1688.2010.00420.x

Vidon, P., & Hill, A. R. (2004). Denitrification and patterns of electron donors and acceptors in eight riparian zones with contrasting hydrogeology. *Biogeochemistry, 71*, 259–283. https://doi.org/10.1007/s10533-004-9684-1

Vymazal, J. (2010). Constructed wetlands for wastewater treatment. *Water, 2*, 530–549. https://doi.org/10.3390/w2030530

Wagenet, R. J., & Huston, J. L. (1987). Predicting the fate of nonvolatile pesticides in the unsaturated zone. *Journal of Environmental Quality, 15*, 315–322.

Walvoord, M. A., Phillips, F. M., Stonestrom, D. A., Evans, R. D., Hartsough, P. C., Newman, B. D., & Striegl, R. G. (2003). A reservoir of nitrate beneath desert soils. *Science, 302*(5647), 1021–1024.

Wang, L., Butcher, A., Stuart, M., Gooddy, D., & Bloomfield, J. (2013). The nitrate time bomb: A numerical way to investigate nitrate storage and lag time in the unsaturated zone. *Environmental Geochemistry and Health, 35*, 667–681.

Wang, L., Stuart, M. E., Bloomfield, J. P., Butcher, A. S., Gooddy, D. C., McKenzie, A. A., Lewis, M. A., & Williams, A. T. (2012). Prediction of the arrival of peak nitrate concentrations at the water table at the regional scale in Great Britain. *Hydrological Processes, 26*, 226–239.

Wang, L., Stuart, M. E., Lewis, M. A., Ward, R. S., Skirvin, D., Naden, P. S., Collins, A. L., & Ascott, M. J. (2016a). The changing trend in nitrate concentrations in major aquifers due to historical nitrate loading from agricultural land across England and Wales from 1925 to 2150. *Science of the Total Environment, 542*, 694–705.

Wang, S., Changyuan, T., Xianfang, S., Ruiqiang, Y., Zhiwei, H., & Yun, P. (2016b). Factors contributing to nitrate contamination in a groundwater recharge area of the North China Plain. *Hydrological Processes, 30*(13), 2271–2285.

Ward, M. H., Jones, R. R., Brender, J. D., de Kok, T. M., Weyer, P. J., Nolan, B. T., Villanueva, C. M., & van Breda, S. G. (2018). Drinking water nitrate and human health: An updated review. *International Journal of Environmental Research and Public Health, 15*, 1557. doi:10.3390/ijerph15071557.

Wheeler, D. C., Nolan, B. T., Flory, A. R., DellaValle, C. T., & Ward, M. H. (2015). Modeling groundwater nitrate concentrations in private wells in Iowa. *Science of the Total Environment, 536*, 481–488. https://doi.org/10.1016/j.scitotenv.2015.07.080

Woli, K. P., Hayakawa, A., Kuramochi, K., & Hatano, R. (2008). Assessment of river water quality during snowmelt and base flow periods in two catchment areas with different land use. *Environmental Monitoring and Assessment, 137*(1–3), 251–260. https://doi.org/10.1007/s10661-007-9757-4

Wriedt, G., & Bouraoui, F. (2009). *Towards a large-scale assessment of water availability in Europe*. Luxembourg: European Commission Joint Research Centre.

Xu, Z., Hu, B. X., Davis, H., & Cao, J. (2015). Simulating long term nitrate-N contamination processes in the Woodville Karst Plain using CFPv2 with UMT3D. *Journal of Hydrology, 524*, 72–88. https://doi.org/10.1016/j.jhydrol.2015.02.024

Yoon, G. H. (2005). *Leaching of nitrogen and phosphorus from agricultural soils with different cropping practices and with respect to preferential transport*. Göttingen: Kiel, University, Dissertation. Cuvillier Verlag.

Zarnetske, J. P., Haggerty, R., Wondzell, S. M., Bokil, V. A., & Gonzalez-Pinzon, R. (2012). Coupled transport and reaction kinetics control the nitrate soure-sink function of hyporheic zones. *Water Resources Research, 48*, W11508. https://doi.org/10.0129/2012WR011894

Zheng, C., & Wang, P. P. (1999). *MT3DMS, A modular three-dimensional multi-species transport model for simulation of advection, dispersion and chemical reactions of contaminants in groundwater systems; documentation and user's guide*. Vicksburg, Mississippi: US Army Engineer Research and Development Center Contract Report SERDP-99-1, 169 p.

Zirkle, K. W., Nolan, B. T., Jones, R. R., Weyerd, P. J., Ward, M. H., & Wheeler, D. C. (2016). Assessing the relationship between groundwater nitrate and animal feeding operations in Iowa (USA). Assessing the relationship between groundwater nitrate and animal feeding operations in Iowa (USA). *Science of the Total Environment, 566-567*, 1062–1068. https://doi.org/10.1016/j.scitotenv.2016.05.130

FURTHER READING

Ackerman, J. R., Peterson, E. W., Van der Hoven, S., & Perry, W. I. (2015). Quantifying nutrient removal from groundwater seepage of constructed wetlands receiving treated wastewater effluent. *Environmental Earth Science, 74*, 1633–1645. https://doi.org/10.1007/s12665-015-4167-3

Almasri, M. N., & Kaluarachchi, J. J. (2004a). Implications of on-ground nitrogen loading and soil transformations on groundwater quality management. *Journal of the American Water Resources Association, 40*(1), 165–186.

Almasri, M. N., & Kaluarachchi, J. J. (2004b). Assessment and management of long-term nitrate pollution of ground water in agriculture-dominated watersheds. *Journal of Hydrology, 295*, 225–245.

Billen, G., Silvestre, M., Grizzetti, B., Leip, A., Garnier, J., Voss, M., et al. (2011). Nitrogen flows from European watersheds to coastal marine waters. Ch. 13. In M. A. Sutton, C. M. Howard, J. W. Erisman, G. Billen, A. Bleeker, P. Grennfelt, H. van Grinsven, & B. Grizzetti (Eds.), *The European nitrogen assessment: Sources, effects, and policy perspectives*. Cambridge, UK: Cambridge University Press, 664 p.

Burow, K.R., & Green, C.T. (2008). Spatial and temporal trends in nitrate concentration in the eastern San Joaquin Valley regional aquifer and implications for nitrogen fertilizer management. In *California Plant and Soil Conference: Conservation of Agricultural Resources*, 5–6 February 2008 (pp. 47–52). Visalia, California, accessed 17 June 2010, at http://ucanr.org/sites/calasa/files/320.pdf, accessed 22 December 2018.

Burow, K. R., Nolan, B. T., Rupert, M. G., & Dubrovsky, N. M. (2010). Nitrate in groundwater of the United States, 1991–2003. *Environmental Science and Technology*, *44*(13), 4988–4997.

Capel, P.D., McCarthy, K.A., Coupe, R.H., Grey, K.M., Amenumey, S.E., Baker, N.T., & Johnson, R.L. (2018). *Agriculture—A river runs through it—The connections between agriculture and water quality*. U.S. Geological Survey Circular 1433, 201 p., https://doi.org/10.3133/cir1433.

Conant, R. T., Berdanier, A. A., & Grace, P. R. (2013). Patterns and trends in nitrogen use and nitrogen recovery efficiency in world agriculture. *Global Biogeochemical Cycles*, *27*, 558–566.

Diaz, R. J., & Rosenberg, R. (2008). Spreading dead zones and consequences for marine ecosystems. *Science*, *321*, 926–929.

Erisman, J. W., Galloway, J. N., Dice, N. B., Sutton, M. A., Bleeker, A., Grizzetti, B., et al. (2015). *Nitrogen, too much of a vital resource*. Science Brief. Zeist, The Netherlands: World Wildlife Fund.

Erisman, J. W., van Grinsven , H., Grizzetti, B. Bouraoui, F., Powlson, D., Sutton, M. A., Bleeker, A., & Reis, S. (2011). The European nitrogen problem in a global perspective. Ch. 2. In M.A. Sutton, C.M. Howard, J.W. Erisman Billen, G., Bleeker, A., Grennfelt, P., H. van Grinsven, & B. Grizzetti (Eds.), *The European nitrogen assessment: Sources, effects, and policy perspectives* (pp. 9–31). Cambridge, UK: Cambridge University Press.

Fowler, D., Coyle, M., Skiba, U., Sutton, M. A., Capel, J. N., Reis, S., et al. (2013). The global nitrogen cycle in the twenty-first century. *Philosophical Transactions of the Royal Society B*, *368*, 1621. https://doi.org/10.1098/rtsb.2013.0164

Fowler, D., Steadman, C. E., Stevenson, D., Coyle, M., Rees, R. M., Skiba, U. M., et al. (2015). Effects of global change during the 21st century on the nitrogen cycle. *Atmospheric Chemistry and Physics*, *15*, 13849–13893. https://doi.org/10.5194/acp-15-13849-2015

Galloway, J. N., Dentener, F. J., Capone, D. G., Boyer, E. W., Howarth, R. W., Seitzinger, S. P., et al. (2004). Nitrogen cycles: Past, present and future. *Biogeochemistry*, *70*, 153–226.

Galloway, J. N., Schlesinger, W. H., Levy, H., II, Michaels, A., & Schnoor, J. L. (1995). Nitrogen fixation: Anthropogenic enhancement-environmental response. *Global Biogeochemical Cycles*, *9*(2), 235–252.

Galloway, J. N., Townsend, A. R., Erisman, J. W., Bekunda, M., Cai, Z., Freney, J. R., et al. (2008). Transformation of the nitrogen cycle: Recent trends, questions, and potential solutions. *Science*, *320*, 889–892.

Grizzetti, B., Passy, P., Billen, G., Bouraoui, F., Garnier, J., & Lassaletta, L. (2015). The role of water nitrogen retention in integrated nutrient management: Assessment in a large basin using different modelling approaches. *Environmental Research Letters*, *10*(2015), 065008. http://dx.doi.org/10.1088/1748-9326/10/6/065008

Gu, B., Ju, X., Wu, Y., Erisman, J. W., Bleeker, A., Reis, S., et al. (2017). Cleaning up nitrogen pollution may reduce future carbon sinks. *Global Environmental Change*, *48*, 55–66.

Harrison, J. A., Maranger, R. J., Alexander, R. B., Giblin, A. E., Jacinthe, P. A., Mayorga, E., et al. (2009). The regional and global significance of nitrogen removal in lakes and reservoirs. *Biogeochemistry*, *93*, 143–157.

Henson, W. R., Huang, L., Graham, W. D., & Ogram, A. (2017). Nitrate reduction mechanisms and rates in an unconfined eogenetic karst aquifer in two sites with different redox potential. *Journal of Geophysical Research: Biogeosciences*, *122*, 1062–1077.

Hinkle, S. R., & Tesoriero, A. J. (2014). Nitrogen speciation and trends, and prediction of denitrification extent, in shallow US groundwater. *Journal of Hydrology*, *509*, 343–353.

Howarth, R., Anderson, D., Cloern, J., Elfring, C., Hopkinson, C., Lapointe, B., et al. (2000). Nutrient pollution of coastal rivers, bays, and seas. Issues in Ecology, no. 7, Retrieved from http://www.epa.gov/watertrain/pdf/issue7.pdf.

Howarth, R., Swaney, D., Billen, G., Garnier, J., Hong, B., Humborg, C., et al. (2011). Nitrogen fluxes from the landscape are controlled by net anthropogenic nitrogen inputs and by climate. *Frontiers in Ecology and the Environment*. https://doi.org/10.1890/100178

Jaynes, D. B., Kaspar, T. C., Moorman, T. B., & Parkin, T. B. (2008). In situ bioreactors and deep drain-pipe installation to reduce nitrate losses in artificially drained fields. *Journal of Environmental Quality*, *37*, 429–436.

Katz, B. G. (2019). Nitrate contamination in karst groundwater. In W.B. White & D.C. Culver (Eds.), *Encyclopedia of Caves* (Ch. 91, pp. 756–760), 3rd edition. Boston, MA: Academic Press, Elsevier.

Lassaletta, L., Billen, G., Garnier, J., Bouwman, L., & Valazquez, E. (2016). Nitrogen use in the global food system: Past trends and future trajectories of agronomic performance, pollution, trade, and dietary demand. *Environmental Research Letters*, *11*(9). http://dx.doi.org/10.1088/1748-9326/11/9/095007

Lassaletta, L., Billen, G., Grizzetti, B., Anglade, J., & Garnier, J. (2014). 50-year trends in nitrogen use efficiency of world cropping systems: The relationship between yield and nitrogen input to cropland. *Environmental Research Letters*, *9*(2014), 105011. http://dx.doi.org/10.1088/1748-9326/9/10/105011

Lim, F. Y., Ong, S. L., & Hu, J. (2017). Recent advances in the use of chemical markers for tracing wastewater contamination in aquatic environment: A review. *Water*, *9*, 143. https://doi.org/10.3390/w9020143

Lindsey, B. D., Phillips, S. W., Donnelly, C. A., Speiran, G. K., Plummer, L. N., Böhlke, J. K., Focazio, M. J., & Burton, W. C.

(2003). Residence time and nitrate transport in ground water discharging to streams in the Chesapeake Bay Watershed. U.S. Geological Survey Water-Resources Investigations Report 03-4035, 202 p.

Maxwell, E., Peterson, E. W., & O'Reilly, C. M. (2017). Enhanced nitrate reduction within a constructed wetland system: Nitrate removal within groundwater flow. *Wetlands, 37*(3), 413–422. https://doi.org/10.1007/s13157-017-0877-5

Mee, L. D., Friedrich, J., & Gomoiu, M.-T. (2005). Restoring the Black Sea in times of uncertainty. *Oceanography, 18*, 32–43.

Munn, M.D., Frey, J.W., Tesoriero, A.J., Black, R.W., Duff, J.H., Lee, K., et al. (2018). Understanding the influence of nutrients on stream ecosystems in agricultural landscapes. U.S. Geological Survey Circular 1437, 80 p., https://doi.org/10.3133/cir1437.

Musgrove, M., Opsahl, S. P., Mahler, B. J., Herrington, C., Sample, T. L., & Banta, J. R. (2016). Source, variability, and transformation of nitrate in a regional karst aquifer: Edwards aquifer, central Texas. *Science of the Total Environment, 568*, 457–469.

Phillips, S.W., & Lindsey, B.D. (2003). The influence of ground water on nitrogen delivery to the Chesapeake Bay. U.S. Geological Survey Fact Sheet FS-091-03.

Preston, S.D., Alexander, R.B., Woodside, M.D., & Hamilton, P.A. (2009). SPARROW MODELING enhancing understanding of the Nation's water quality. U.S. Geological Survey Fact Sheet 2009-3019, 6 p.

Pretty, J. N., Mason, C. F., Nedwell, D. B., Hine, R. E., Leaf, S., & Dils, R. (2003). Environmental costs of freshwater eutrophication in England and Wales. *Environmental Science and Technology, 37*(2), 201–208.

Puckett, L. J. (2004). Hydrogeologic controls on the transport and fate of nitrate in ground water beneath riparian buffer zones: Results from thirteen studies across the United States. *Water Science and Technology, 49*(3), 47–53.

Puckett, L. J., Zamora, C. M., Essaid, H. I., Wilson, J. T., Johnson, H. M., Brayton, M. J., & Vogel, J. R. (2008). Transport and fate of nitrate at the ground-water/surface-water interface. *Journal of Environmental Quality, 37*, 1034–1050.

Rupert, M. G. (2008). Decadal-scale changes of nitrate in ground water of the United States, 1988–2004. *Journal of Environmental Quality, 37*, S240–S248.

Scavia, D., & Bricker, S. B. (2006). Coastal eutrophication assessment in the United States. *Biogeochemistry, 79*(1–2), 187–208.

Schullehner, J., Hansen, B., Thygesen, M., Pederson, C. B., & Sigsgaard, T. (2018). Nitrate in drinking water and colorectal cancer risk: A nationwide population-based cohort study. *International Journal of Cancer, 143*, 73–79.

Sutton, M. A., Reis, S., & Butterbach-Bahl, K. (2009). Reactive nitrogen in agro-ecosystems: Integration with greenhouse gas interactions. *Agriculture, Ecosystems and Environment, 133*, 135–138.

Tesoriero, A. J., Liebscher, H., & Cox, S. E. (2000). Mechanism and rate of denitrification in an agricultural watershed—Electron and mass balance along groundwater flow paths. *Water Resources Research, 36*(6), 1545–1559.

Tesoriero, A. J., Saad, D. A., Burow, K. R., Frick, E. A., Puckett, L. J., & Barbash, J. E. (2007). Linking ground-water age and chemistry data along flow paths: Implications for trends and transformations of nitrate and pesticides. *Journal of Contaminant Hydrology, 94*, 139–155.

U.S. Environmental Protection Agency (2004). *Drinking water costs and federal funding; 816-F-04-038*. Washington, DC: U.S. EPA.

U.S. Environmental Protection Agency (2009). *An urgent call to action. Report of the State-EPA Nutrient Innovations Task Group*, 41.

U.S. Environmental Protection Agency (2013). Final aquatic life ambient water quality criteria for ammonia-Freshwater 2013. *Notice in Federal Register, 78*(163), 52192–52194.

U.S. Geological Survey. (1999). *The quality of our Nation's waters—Nutrients and pesticides*. U.S. Geological Survey Circular 1225, 82 p. Retrieved from http://pubs.usgs.gov/circ/circ1225/, accessed 20 September 2019.

Van Meter, K. J., Van Cappellen, P., & Basu, N. B. (2018). Legacy nitrogen may prevent achievement of water quality goals in the Gulf of Mexico. *Science, 22*. https://doi.org/10.1126/science.aar4462

Ward, M. H., de Kok, T. M., Levallois, P., Brender, J., Gulis, G., Nolan, B. T., & VanDerslice, J. (2005). Drinking-water nitrate and health—Recent findings and research needs. *Environmental Health Perspectives, 113*(11), 1607–1614.

Ward, M. H., Mark, S. D., Cantor, K. P., Weisenburger, D. D., Correa-Villasenor, A., & Zahm, S. H. (1996). Drinking water nitrate and the risk of Non-Hodgkin's Lymphoma. *Epidemiology, 7*, 465–471.

Wedin, D. A., & Tilman, D. (1996). Influence of nitrogen loading and species composition on the carbon balance of grasslands. *Science, 274*, 1720–1723.

Weyer, P. J., Cerhan, J., Kross, B. C., Hallberg, G. R., Kantamneni, J., & Breuer, G. (2001). Municipal drinking water nitrate level and cancer risk in older women: The Iowa Women's Health Study. *Epidemiology, 12*, 327–338.

Zhai, Y., Lei, Y., Wu, J., Teng, Y., Wang, J., Zhao, X., & Pan, X. (2017). Does the groundwater nitrate pollution in China pose a risk to human health? A critical review of published data. *Environmental Science and Pollution Research International, 24*(4), 3640–3653. https://doi.org/10.1007/s11356-016-8088-9

Zhu, Y., Ye, M., Roeder, E., Hicks, R. W., Shi, L., & Yang, J. (2016). Estimating ammonium and nitrate load from septic systems to surface water bodies within ArcGIS environments. *Journal of Hydrology, 532*, 177–192. https://doi.org/10.1016/j.jhydrol.2015.11.017

8
Nitrate Contamination in Springs

Springs are unique hydrologic systems that represent the transition from groundwater to surface water at the land surface. Springs occur in a wide variety of geologic settings and support diverse habitats and ecosystems at their source (spring vents, orifices, seeps) and along their spring runs. Spring ecosystems contain rare aquatic and terrestrial plant and animal species that are distinct from other aquatic, wetland, and riparian ecosystems (Stevens et al., 2005). Springs have been important sources of water for human civilization throughout history (Kresic & Stevanovic, 2010). Large springs provide outstanding recreational amenities, as well as historic and cultural values. In their book on *Springs and Bottled Waters of the World*, LaMoreaux and Tanner (2001) compiled a comprehensive summary of the geologic and hydrogeologic setting related to the classification of different types of springs, the development of spring systems throughout history, quantitative analysis of springs, and famous springs around the world and their bottled waters. Because of their unique hydrology, ecology, and biology, springs have been studied by researchers from many different scientific disciplines (such as geochemistry, ecology, biology, cultural anthropology, economics, and sociology). As a result, for nearly 100 years there have been several different classification systems used for quantifying spring discharge and other spring characteristics (such as temperature and other physical properties, geologic origin, biology, geochemistry, and rare and endangered species).

One early classification system that is still widely used presently is based on various physical and chemical variables at the point of discharge of spring water (Meinzer, 1927). Table 8.1 includes average spring flow information grouped by spring magnitude from Meinzer (1927). Springer and Stevens (2008) presented a concise and functional classification system that builds on previous work for 12 different types of springs based on their "spheres of discharge." Their detailed diagrams showing these 12 spring types effectively allows researchers from varied disciplines to share a common language and conceptual visualizations of different spring types. The Springer and Stevens (2008) classification system of spring types also can help water resource managers to develop more effective strategies for springs ecosystem conservation, protection, remediation, and restoration of water-quality impaired springs. Also, in the comprehensive book on *Groundwater Hydrology of Springs*, Kresic and Stevanovic (2010) classified various types of springs (e.g., gravity, artesian, submerged, thermal (warm and hot springs), mineral springs), and noted that numerous other types of classifications have been proposed that focused primarily based on discharge rate and seasonal variability, geologic and geomorphic structure controlling the discharge, water quality, temperature, and hydraulic head. Kresic and Stevanovic (2010) also present strategies for springs protection and sustainability issues, utilization and regulation of springs, and several interesting case studies of major springs around the world.

Springs not only provide exceptional recreational opportunities, natural beauty, and cultural value, but the complex chemical composition can provide important information about groundwater quality in a variety of aquifer types. In many areas, spring water chemistry has been used to evaluate aquifer properties and assess the vulnerability of an aquifer to contamination (e.g., Katz, 2004; Mahon, 2011). Spring water chemistry also provides an early warning system for aquifer water quality degradation that is more effective than wells, which may or may not extract water from zones of contamination or plumes. Several anthropogenic activities have contributed to nitrate contamination of springs and spring runs. Degradation of spring water quality containing elevated nitrate concentrations has been reported in many countries around the world, including China, Croatia, England Ethiopia, France, Ireland, Israel, Jordan, Korea, Slovenia, Spain, Switzerland, Turkey, Uganda, and in many locations in the United States.

Nitrogen Overload: Environmental Degradation, Ramifications, and Economic Costs, Geophysical Monograph 250,
First Edition. Brian G. Katz.
© 2020 American Geophysical Union. Published 2020 by John Wiley & Sons, Inc.

Table 8.1 Classification system for springs (spring magnitude) according to average discharge (in cubic meters per second [m³/s] and million liters per day).

Spring magnitude	Average flow (m³/s)	Flow (million liters/d)
1	>2.8	>246
2	0.28–2.8	>24.6–246
3	0.028–0.28	>2.46–24.6
4	0.0028–0.028	0.246–2.46
5–8	<0.0028	<0.246

The world's largest springs are fed by groundwater in karst areas (Kresic & Stevanovic, 2010). The term "karst" refers to distinctive landforms and hydrology that result from the dissolution of carbonate rocks, such as limestone and dolomite, and other soluble rock types (e.g., marble, gypsum, halite, and some conglomerates). As these rocks dissolve, mainly from reaction with carbonic acid (carbon dioxide dissolved in rainfall and/or soil water), a network of interconnected openings is formed in the rock matrix. The larger openings in the aquifer matrix typically are associated with conduit networks that feed groundwater directly to springs (Fig. 8.1). Karst aquifers occur on all continents and on many oceanic islands, and they are present on over 20% of the land surface (Kresic & Stevanovic, 2010).

Springs in karst environments are highly vulnerable to nitrate contamination from various anthropogenic activities at the land surface. Nitrate can move rapidly from the land surface through dissolution features (e.g., sinkholes, swallets) with little or no attenuation. A thorough understanding of recharge processes and groundwater flowpaths in the contributing area to a spring is essential for assessing sources of nitrate contamination in springs. The following sections in this chapter present examples of springs studies in karst areas of two highly productive aquifers, and a third nonkarst (unconsolidated rock) area. Studies in these areas have combined the use of geochemical, hydrologic, and other tools to evaluate sources of nitrate contamination of springs, changes in spring water quantity over time, and to provide critical information for water-quality restoration and protection efforts.

8.1. FLORIDA'S NITRATE IMPAIRED SPRINGS

Florida has been described as the "land of springs" (Chapelle, 2005; Ringle, 1999; Stamm, 2008) as it contains more than 1000 identified springs. Springs in Florida have been a popular tourist destination for over 130 years and offer outstanding recreational and ecotourism opportunities. They have been described as "bowls of liquid light" and "sanctuaries for the renewal of the human spirit" (University of Florida Water Institute, 2017). Visitors to springs throughout the state provide substantial economic benefits to regional economies. Springs in Florida discharge more than 30 billion liters (8 billion gallons) of groundwater per day, which is greater than the output of any similar size area on earth (Scott et al., 2002). Of the 78 first-magnitude springs in the United States (which discharge more than 2.8 m³/s of water, Table 8.1), 33 of these first magnitude springs are in Florida. Several different spring types have been identified in Florida based on where springs occur and how spring water is discharged to the environment. These include onshore vents, resurgences (river rise), lacustrine vents, sand boils, offshore vents, and offshore seeps (Copeland, 2003). Nutrients in offshore submarine groundwater discharge and offshore springs can be important factors for incidences of harmful algal blooms, or red tides in coastal regions (Hu et al., 2006). As described in more detail later in this chapter, Florida's springs have shown considerable reductions in discharge and water-quality impairment over the past several decades. This impairment has resulted from excessive amounts of nitrate that has fueled the abundance of nuisance algae and algal mats and has caused changes in fish and invertebrate communities.

8.1.1. Hydrogeology of Florida Springs

Most springs occur where the Upper Floridan aquifer (UFA) is unconfined (where limestone occurs close to the surface with a thin mantle of overlying sediments) or where the UFA is semiconfined (where the material of sands and clays overlying the limestone aquifer are less than 30 m thick). Figure 8.2 shows the locations of large springs throughout Florida and southern Georgia that discharge water from the UFA. The UFA is part of the Floridan aquifer system that occurs throughout Florida, and in parts of Georgia, South Carolina, and Alabama, and is one of the most productive aquifers in the world. Scott et al. (2004) present more detailed information about the hydrogeologic units that comprise the Floridan aquifer system. The karstic UFA is composed of limestone and dolomite, which contains fractures and numerous dissolution features including large conduit networks that feed groundwater directly to springs. These dissolution features facilitate nutrient loading to springs. Spring water discharges naturally under artesian pressure from the UFA where the potentiometric surface intersects the ground surface.

The dynamics of groundwater flow and contaminant transport in the UFA are complex, occurring through the aquifer matrix and through large conduits varying as a function of changes in hydrologic conditions. During low flow conditions, water flows to springs predominantly

Figure 8.1 Photo showing variations in porosity and underwater caverns in the UFA near Peacock Springs in northern Florida, USA. *Source*: Photo taken by Mark Long, a world renowned subterranean cave diver and underwater photographer. For scale, note the cave divers with lights in the large openings in the karst limestone. Permission given from Mark Long and Annette Long.

through the matrix, fissures, and small openings, but during high-flow conditions, groundwater flows mainly through major conduits (Martin & Dean, 2001). Water flow through the aquifer is complex due to the highly anisotropic properties of the aquifer matrix (high-fracture porosity with complex intermingling networks of large and small conduits). Cave divers have contributed important information on aquifer porosity and have mapped the location and extent of large conduits in many Florida springs (Kincaid et al., 2004; Ringle, 1999; Werner, 2000). Also, dye trace studies have been useful for measuring rapid groundwater travel times for contaminant transport in conduits feeding springs not only in Florida springs (e.g., Butt & Murphy, 2003; Kincaid et al., 2004, 2005; Kincaid & Werner, 2008) but in many spring systems throughout the world (Benson & Yuhr, 2016; Kresic & Stevanovic, 2010; Trček & Zojer, 2010).

During flood conditions, water from streams and rivers can backflow into some springs. Hensley and Cohen (2017) found that during episodic flow reversals acidic floodwaters containing elevated amounts of dissolved organic carbon (DOC) from adjacent blackwater rivers significantly altered the light environment in spring runs and depleted dissolved oxygen that resulted in a trophic cascade. This cascade created harmful effects on algae consumers, which increased algal proliferation. It was also noted that flow reversal events have become more frequent over time and were related to aquifer declines.

They attributed the increased frequency in reversals to climate variability and groundwater pumping. Although in Florida, springs with low dissolved oxygen levels, algal cover is always high, and reversals have had little effect on algae consumers. In contrast, algal cover typically is low in springs that contain high levels of dissolved oxygen, except where flow reversals occur frequently. These flow reversals and dissolved oxygen levels have important implications regarding spring protection and restoration (Hensley & Cohen, 2017), which is discussed in more detail later in this chapter.

8.1.2. Spring Water Nitrate Contamination

Springs have provided an early warning system for assessing water quality degradation in parts of the UFA. Figure 8.3 shows examples of springs where nitrate-N concentrations have steadily increased over the past 60 years. Various anthropogenic sources (both point and nonpoint [diffuse] sources) have contributed chemical and biological constituents that enter the aquifer through recharge processes. As a result of elevated nitrate-N concentrations, many spring runs, streams, and rivers in Florida have suffered from an increase in nuisance aquatic vegetation, reduced clarity, and algal blooms. Figure 8.4 shows photos documenting the decrease in water clarity in a Florida spring over the last 20 years due to proliferation of algal blooms.

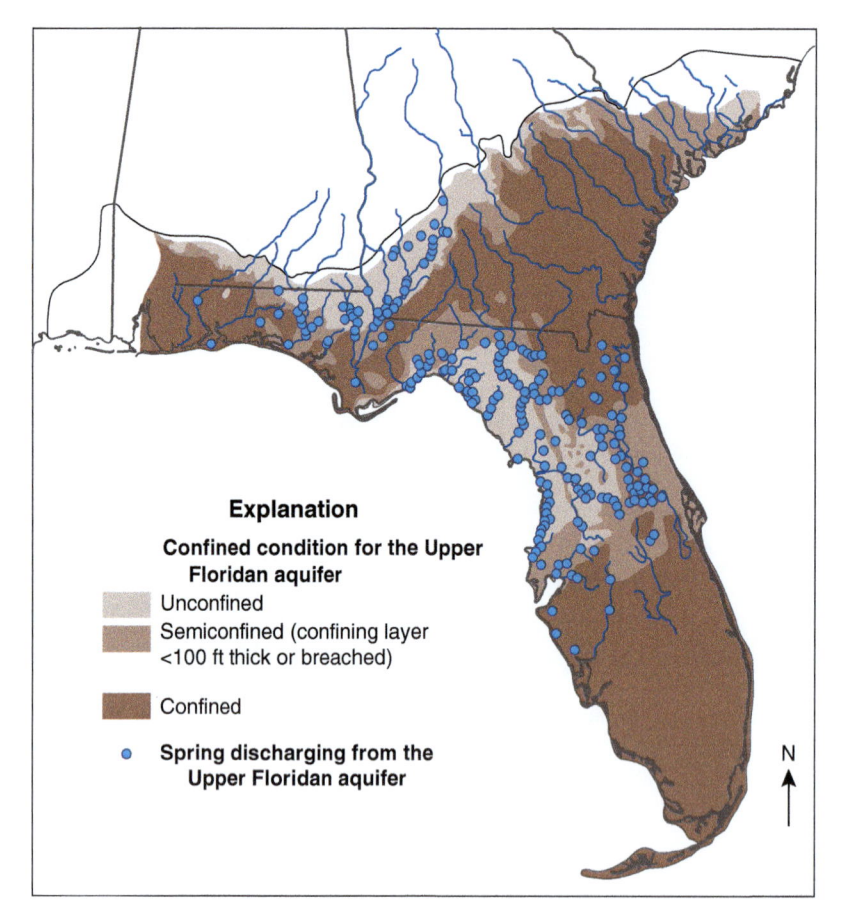

Figure 8.2 Map showing selected springs in the UFA. *Source*: Modified from colored illustration on p. 48 in Berndt et al. 2014. Circular. Reproduced with permission of United States Geological Survey.

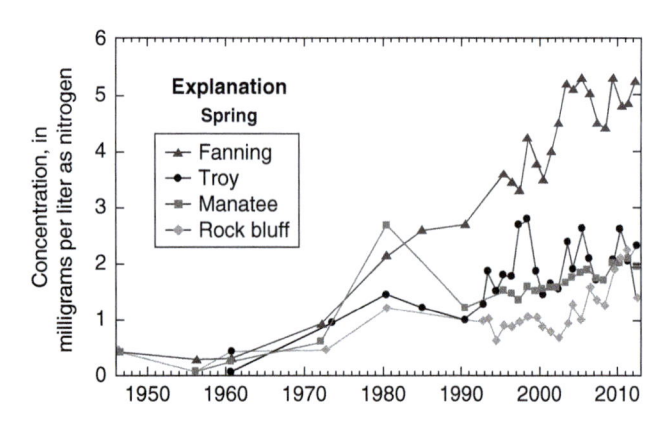

Figure 8.3 Examples of Florida springs in the Suwannee River Basin with nitrate-N concentrations that have increased over the past 60 years. *Source*: From United States Geological Survey Floridan Circular.

Some of these algal blooms produce toxins that are harmful to humans, animals, and ecosystems. The water quality degradation of springs in Florida is directly related to increases in population (increased 10-fold since 1940), agricultural activities, groundwater withdrawals, and changes in land use (Pittman, 2012). In addition, springs have disappeared in several parts of Florida due to heavy pumping of water from the Floridan aquifer. Equally disheartening is the fact that due to droughts and pumping, springs throughout the state have become more saline over time (Copeland et al., 2011), which does not bode well for the future water quality of the Floridan aquifer that supplies drinking water to millions of Florida residents and visitors.

Florida's springs have been studied for nearly 100 years, but during the past 20 years, there have been numerous research studies on Florida's spring systems to determine sources of nitrate contamination. Based on nitrate isotope data and other chemical indicators, the dominant source of nitrate contamination in most Florida springs is synthetic fertilizer applied to cropland (e.g., Heffernan et al., 2012; Katz, 2004) (Fig. 8.5). This is not unexpected as many springs are in areas with varying amounts of agricultural land use. Although, nitrate isotope data for some spring waters have indicated a mix of inorganic and organic sources of nitrogen. These springs are located in areas with animal farming operations (Barnett, 2015; Call & Stephenson, 2003;

Fanning springs

Figure 8.4 Photograph from USGS Circular showing how the clarity of water in Fanning Springs has decreased over the last 20 years mostly due to algae. *Source*: Photo by John Moran. Florida springs are highly sensitive to nutrient inputs. Reproduced with permission of John Moran and United States Geological Survey.

Katz et al., 1999) and where treated municipal wastewater is being applied to the land surface (Davis et al., 2010; Katz et al., 2004, 2009a).

8.1.3. Residence Time of Groundwater in Springsheds and Timescales of Nitrate Contamination

The area where rainfall and recharge travel through the subsurface and flow toward a spring vent is referred to as a springshed (Upchurch et al., 2004). A springshed is analogous to a watershed or surface water basin; however, it encompasses the groundwater contributing area to a spring. Springshed boundaries can change over time depending on rainfall, recharge, and groundwater withdrawals (pumping). In Florida, springsheds typically are delineated by mapping the potentiometric surface of the UFA and designating groundwater flow patterns to a spring. Knowing the springshed area is critical for assessing the loading of nitrate and other forms of nitrogen from various anthropogenic sources.

Figure 8.5 Block diagram showing groundwater flow toward a springs in an agricultural-dominated groundwater contributing area to a spring. *Source*: Reproduced with permission of United States Geological Survey.

The residence time of groundwater within a springshed also is another critical piece of information for assessing timescales and chronology of nitrate contamination, and for estimating response or lag times for water-quality restoration efforts. Groundwater age-dating techniques (for groundwater recharged during the past 50–60 years) have been used to determine the residence time and age distribution of groundwater discharging from springs.

As mentioned briefly in Chapter 4 (methods for assessing sources of nitrate contamination) researchers have used several age-dating techniques and models to assess apparent ages in groundwater. The age-dating techniques involve using various isotopic and other chemical indicators with known time-varying inputs to the atmosphere. These chemical "age" indicators have been released into the atmosphere in low but measurable concentrations (as shown in Fig. 8.6) resulting from various anthropogenic activities, including atmospheric testing of thermonuclear devices (releasing tritium, ^3H, a radioactive isotope of hydrogen), and several industrial processes that have released chlorofluorocarbons (CFCs – trichlorofluoromethane, CCl_3F, CFC-11; dichlorodifluoromethane, CCl_2F_2, CFC-12; and trichlorotrifluoroethane, $C_2Cl_3F_3$, CFC-113), and sulfur hexafluoride (SF_6) (Busenberg & Plummer, 1992, 2000; Katz et al., 2001). Combined measurements of ^3H and its radioactive decay product, ^3He, provide a relatively stable tracer of the initial ^3H input to groundwater that can be used to estimate an apparent $^3H/^3He$ age (Plummer, 2005). Age dating using this method typically is advantageous over CFCs and SF_6 because $^3H/^3He$ ages are not affected by contamination, sorption, or microbial degradation processes that can

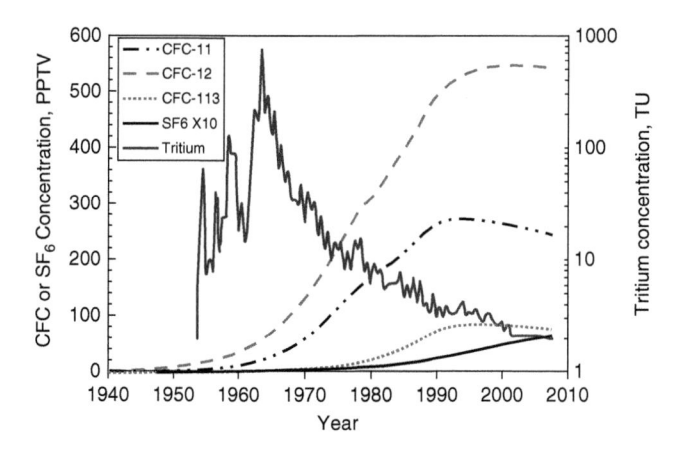

Figure 8.6 Atmospheric input curves for tritium in rainfall (measured in Ocala, Florida), CFCs, and SF_6. *Source*: Modified from Katz (2004). Reproduced with permission of United States Geological Survey.

alter the concentrations of the other tracers. The many intricacies and details associated with these various age-dating techniques are described elsewhere (e.g., Jurgens et al., 2012; Plummer & Busenberg, 1999, 2006; https://water.usgs.gov/lab/3h3he/background/, accessed 22 May 2018).

Rainfall that reaches the land surface transports these chemical "age" tracers into the groundwater system and provides a "time stamp" for the atmospheric conditions present at the time of recharge. As a result, water samples collected from springs and wells can be analyzed for these compounds, and an estimate of the "apparent age" of the water sample can be determined. The apparent age represents an average distribution of residence times for water parcels traveling to the spring along numerous flow paths. Age distributions of spring waters can be evaluated by comparing measurements of multiple tracers for concordance with modeled tracer concentrations. Lumped parameter models (LPMs) have been used to estimate the residence time or mean age of the groundwater age distribution discharging from large springs (Jurgens et al., 2012; Maloszewski & Zuber, 1996; Zuber, 1986). LPMs treat the aquifer system as a homogeneous compartment in which tracer input concentrations are converted to tracer output concentrations according to the system response function used and how groundwater flow to a spring is conceptualized.

Various types of conceptual models of groundwater flow have been used to estimate residence times such as piston-flow, exponential mixing, and binary mixing (Katz et al., 2001, 2005, 2007; Katz & Griffin, 2008; Phelps, 2004; Toth & Katz, 2006). Piston-flow (time-constant flow lines) and exponential mixing (time-varying flow lines) models, or a combination of the two, have been used when groundwater flow and discharge are associated with a single hydrogeologic unit. The binary mixing model typically has been used when the contribution of flow from the spring vent originates from shallow and deep parts of an aquifer with vastly different times of recharge (young and old waters). Lumped-parameter models (Jurgens et al., 2012; Katz et al., 1999, 2001) used in Florida springs studies typically have two important assumptions: (a) a steady-state groundwater flow system, and (b) age tracers are conservative and behave like a water molecule. Although this second assumption is valid for ^3H, which is part of the water molecule, CFCs and SF_6 may or may not be transported in the same way as the water due to sorption and biogeochemical processes that can cause measured concentrations to vary from model results.

Residence times for groundwater discharging to Florida springs are relatively short. They range from several years for third magnitude springs to several decades for first and second magnitude springs based on measurements of CFCs, SF_6, and $^3H/^3He$ and the age distributions modeled using various LPMs (Happell et al., 2006; Katz, 2004; Katz et al., 1999, 2001; Knowles et al., 2010; Phelps, 2004; Toth & Katz, 2006). Several of the springs sampled did not yield valid CFC ages because one or more of the CFCs were contaminated by nonatmospheric sources. In those instances, data were used for $^3H/^3He$ and SF_6, (if not contaminated). Some interesting relationships were found between groundwater residence times, spring magnitude, and groundwater chemistry. For example, nitrate-N concentrations were inversely related to residence time and spring flow. Nitrate and dissolved oxygen concentrations in third-magnitude springs were much higher than in first- or second-magnitude springs (Katz et al., 1999). Spring waters containing high nitrate concentrations have short groundwater residence times, typically less than 10 years and represent flow from shallow parts of the aquifer. In contrast, first- and second-magnitude springs have greater groundwater residence times and lower nitrate concentrations. First- and second-magnitude springs, receive groundwater from large contributing areas with deep flow systems that contain a relatively higher proportion of older water with low concentrations of nitrate. Combined results from age dating and analysis of nitrate isotopes indicate that there was a delayed response of nitrate concentrations in spring waters from historical nitrogen loading data because the mean residence time of groundwater discharging from most springs is on the order of decades (Katz et al., 2001).

It is important to note that residence times estimated using various age tracers are only representative of the hydrologic conditions at the time of sampling (a snapshot in time). Typically, the age tracers are not measured over long periods of time at a single spring. However, Katz (2004) collected water samples annually at Troy Spring

(Fig. 8.7) during 1997–2001, which provided additional insights about nitrate concentrations in spring water and groundwater flow patterns to a first magnitude spring over varying flow conditions. During an extended drought period, October 1998 to August 2000, apparent CFC ages (assuming no contamination from nonatmospheric sources) were greater in samples collected in 1999 and 2000 compared to those in 1997 and 1998. ^{3}H/^{3}He$_{trit}$ apparent ages also increased in samples collected in 2000 (23 years) compared to 1999 (19 years). It is likely that the greater apparent ages in samples from 2000 during the extended drought period (and associated lower spring flow) result from mixtures that contained a higher contribution of older water (presumably from deep flow paths with longer mean transit times). A decrease in nitrate-N and dissolved oxygen concentrations during 1999 and 2000 compared to 1997–1998, also is consistent with a greater contribution of older groundwater to Troy Springs. Similar trends (but with a smaller number of samples) were observed for Manatee Springs located near Troy Springs (Fig. 8.7). Based on age dating of water samples from 31 UFA springs, Happell et al. (2006) noted a large excess of ^{3}He and ^{4}He values that resulted in greater ^{3}H/^{3}He apparent ages than those obtained for CFCs, and they attributed these differences to samples collected at the end of a 4-year drought.

8.1.4. Denitrification in the Upper Floridan Aquifer

In many large springs in Florida, groundwater residence times are long (on the order of decades). Therefore, it is vital to know the fate of nitrate in the UFA and spring waters and the amount of nitrate attenuation mainly resulting from denitrification. Denitrification in the UFA is one possible explanation for the previously described decrease in nitrate-N in Troy Spring during 1999 and 2000. However, during the study, no excess N_2 was measured, dissolved O_2 was present, and organic carbon concentrations were low. More recent studies have investigated the magnitude and rate of denitrification in the UFA. Heffernan et al. (2012) used a combination of nitrate isotope data along with noble gas tracers (Ne, Ar) to estimate

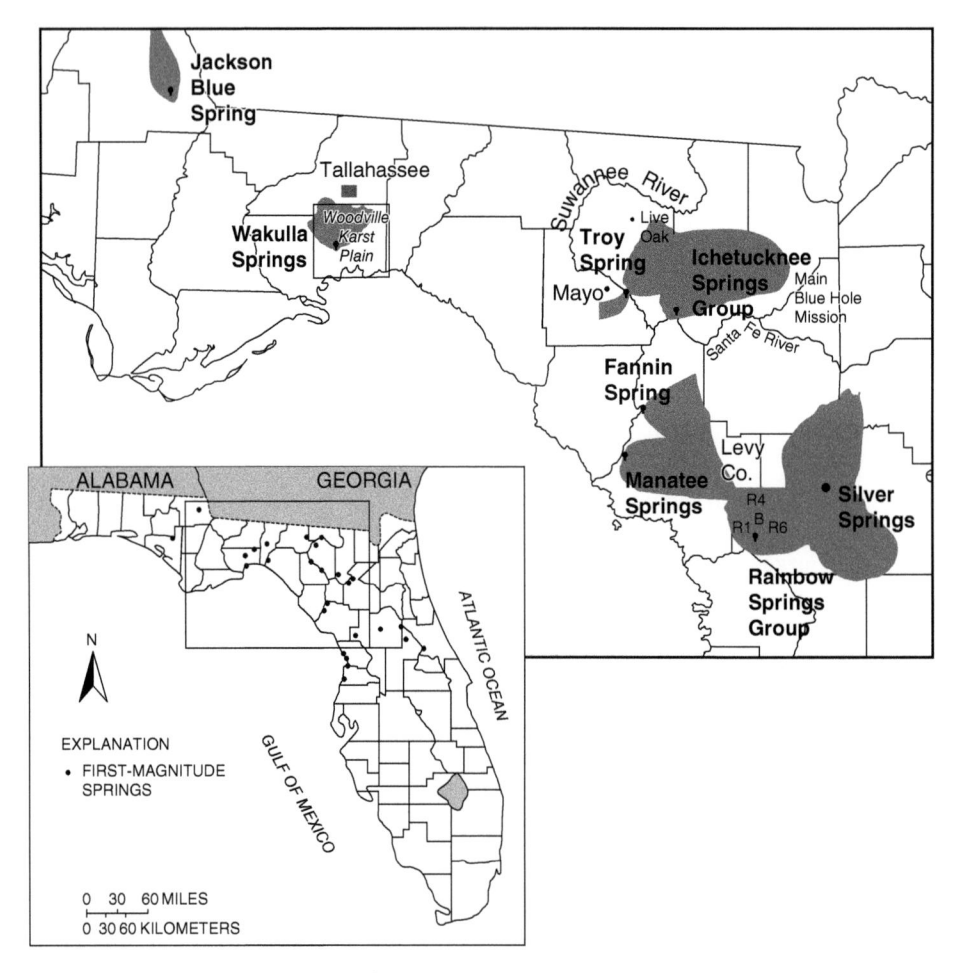

Figure 8.7 Map showing selected springsheds (approximate groundwater contributing areas) to some sampled springs in Florida. *Source*: Modified from Katz (2004). Reproduced with permission of Springer Nature.

excess N_2 gas concentrations 61 UFA springs ($n = 112$ samples). They found that denitrification-derived (excess) N_2 was highly variable and was inversely correlated with dissolved oxygen. Further evidence for denitrification was determined from negative relationships between O_2 and δ^{15}N-NO_3 for a larger dataset of 113 springs, well-constrained isotopic fractionation coefficients, and strong δ^{15}N-NO_3: δ^{18}O-NO_3 covariation further. Even in the organic-matter poor UFA, Heffernan et al. (2012) concluded that denitrification likely is a significant sink for nitrogen leaching to the UFA, removing approximately 32% of the total (flow-weighted) nitrogen discharging from sampled springs. Average volumetric rates of denitrification in the UFA were low compared to direct measurements of N_2 but were within the range reported for agriculturally enriched aquifers with higher nitrate concentrations (Green et al., 2008). Furthermore, Heffernan et al. (2012) noted that aggregate areal rates of denitrification (122 kg/km²/yr) were comparable to the estimated global average for aquifer denitrification (Seitzinger et al., 2006). When adjusting isotope values for denitrification reaction progress, Heffernan et al. (2012) determined that the isotopically enriched nitrate in many springs more likely originated from a fertilizer source rather than urban or animal sources.

A recently completed study by investigators at the University of Florida Water Institute and St. Johns River Water Management District (2017) focused on the sources, transformation, and loss of nitrogen from the land surface to springs in the Silver Springs area in central Florida (Fig. 8.7). These researchers performed laboratory experiments and made field measurements (push-pull tracer tests) in the springshed for Silver Springs (Fig. 8.7) to determine denitrification rates in soil and unsaturated zone profiles and in groundwater. They identified "hotspots" in the springshed, where nitrate applied to the land surface is transported rapidly to springs with little opportunity for dentification to occur. Conversely, other locations showed denitrification "hotspots," where nitrate degradation was measured in some wells. They concluded that rates and timescales for denitrification varied considerably across the springshed but were lowest in areas closest to spring vents. Based on measured concentrations of dissolved gases from two spring vents, they estimated that approximately 17–43% of the nitrate load to the aquifer was lost through denitrification (measured mean rates of 0.22 g N/m²/d).

Multiple measurements of age-dating tracers over time have provided information on how changes in climatic and hydrologic conditions affect residence times and nitrate concentrations. Martin et al. (2016) used time-series measurements of apparent ages (^{3}H/^{3}He, SF$_6$, CFC-11, CFC-12, CFC-113) in six springs over a 16-year period in the Ichetucknee springs complex in northern Florida (Fig. 8.7) to investigate decadal-scale variations in recharge and/or groundwater withdrawals due to pumping. They separated these springs into two groups that included shallow short (Group 1) and deep long (Group 2) flow paths. CFC-113 concentrations indicated a 10–20 year monotonic increase in apparent age from 1997 to 2013, including the flood recession that followed Tropical Storm Debby in mid-2012. Martin et al. (2016) attributed this overall age increase to aquifer recharge that occurred around 1973 for Group 2 springs and around 1980 for Group 1 springs. As described previously by Katz (2004), Martin et al. (2016) noted inverse correlations between apparent age and dissolved oxygen and nitrate concentrations in older water. Also, older waters had higher magnesium and sulfate concentrations, which indicated increased water-rock reactions. Two major El Niño events during the maximum cool phase in the Atlantic Multidecadal Oscillation accounted for excess rainfall (nearly 2 m of rain in excess of the monthly average that fell between 1960 and 2014) and associated groundwater recharge for the decade around 1975. When regional groundwater withdrawals were analyzed, there was a nearly fivefold increase between 1980 and 2005, but groundwater withdrawals (pumpage) represented only 2–5% of decrease in Ichetucknee River flow from springs. Based on these comparisons, Martin et al. (2016) concluded that groundwater withdrawals were less important than decadal-long variations in precipitation.

8.1.5. Water Quality Restoration Efforts for Florida Springs

Due to the severity of the ecosystem damage in Florida springs and spring runs, citizen outrage, and the availability of state funding for springs restoration projects, there has been a considerable amount of springs research in Florida on nitrate sources and loading, transport, transformations, and ecosystem impacts spring systems. Over the last 4 years, the state has provided nearly $268 million for restoration of springs, which has been leveraged with local match funding for a total investment of more than $365 million. In response to water-quality degradation from nutrients, the US EPA approved the State of Florida's numeric nutrient criteria in November 2012 for spring vents, lakes, streams, and south Florida estuaries. The spring vent nitrate-N criterion (an annual geometric mean of 0.35 mg/L for most nitrate-impaired springs) is based on a stressor–response relationship between nitrate and the presence of nuisance algal mats, with the criterion established at a concentration that would prevent nuisance mats from occurring.

In 2016, the Florida Legislature enacted the Florida Springs and Aquifer Protection Act (Part VIII of Chapter 373, Florida Statutes). This Act affords special status and protection to historic first-magnitude springs and to other springs of special significance. For the 24 impaired springs, the Act requires the Florida Department of Environmental Protection (FDEP) to work with local stakeholders to develop and implement a basin management action plan (BMAP) to achieve water quality restoration targets based on an adopted total maximum daily load (TMDL) for each spring. The Act also requires FDEP to delineate priority focus areas where certain activities are prohibited; use the Nitrogen Source Inventory and Loading Tool (NSILT; Eller & Katz, 2017) to identify and quantify the major sources of nitrogen in the groundwater contributing area to a spring and prioritize restoration projects for nutrient load reduction (including cost estimates, schedule). In priority focus areas in springsheds, specialized remediation plans will be developed where nitrogen loads from septic tanks (on-site treatment and disposal systems) account for 20% or more of the total estimated nitrogen inputs. Water-quality restoration targets are developed for 5-year, 10-year, and 15-year milestones, which will ultimately lead to achieving the TMDL in 20 years. The BMAPs contain enforceable sets of projects and management practices designed to reduce current nitrogen loads from a variety of urban and agricultural sources with the goal of improving spring water quality over time. These plans contain strategies designed to reduce nitrate loadings by setting permit limits on wastewater treatment facilities, implementing urban and agricultural best management practices, conservation programs, and other reduction programs. In July 2018, FDEP adopted BMAPs for areas including all Outstanding Florida Springs. These BMAPs will include almost 14 million watershed acres under active basin management, which also will include more than 6.5 million Floridians.

There are several other issues to consider when developing long-term water-quality restoration plans for springs. An important concern for water-quality restoration is the potentially large amount of reactive nitrogen from past land-use practices (legacy nitrogen, mostly in the form of nitrate, but also as organic nitrogen) stored in the vadose (unsaturated zone) (Ascott et al., 2017) and the aquifer (Wang et al., 2013). A recurrent theme in this book is reactive nitrogen that originates from various sources (particularly fertilizer and animal wastes in agricultural areas, and septic tanks and the land application of treated wastewater in suburban and urban areas) can slowly leach below the root zone in soils, move through the unsaturated zone and eventually travel to the aquifer. Several recent studies have shown that groundwater nitrogen applied decades ago can still be found in aquifers (Basu et al., 2012; McMahon et al., 2006; Meals et al., 2010; Puckett et al. 2011; Sanford & Pope, 2013; Sebillo et al., 2013). Given long residence times for groundwater discharging from large springs in Florida, continual discharge of elevated nitrate concentrations may cause ecosystem damage over long periods of time.

Another important consideration is that nitrate may not be the only factor affecting the proliferation of nuisance algae in spring runs (Cohen et al., 2007). A recent comprehensive study of Silver Springs by an interdisciplinary team of researchers at the University of Florida in collaboration with the St. Johns River Water Management District has provided a more in-depth understanding of the controls on the abundance and interactions of nuisance algae and other submerged aquatic vegetation that comprise the primary producer community structure (PPCP) in springs (University of Florida Water Institute, 2017). They concluded that reducing nitrate alone is unlikely to restore PPCP because there are other important drivers, such as the velocity of water movement in spring runs, light and temperature, biota consumers, and thick and mobile benthic sediments that are sources of nonnitrate nutrients (e.g., phosphorus, iron, and sulfide).

8.2. NITRATE IN SPRINGS IN THE EDWARDS AQUIFER, TEXAS

The highly productive Edwards aquifer was the first aquifer to be designated as a sole source aquifer under the Safe Drinking Water Act in the United States. The karstic Edwards aquifer not only supplies water to millions of people in the San Antonio, Texas, area but also supplies artesian water to sustain unique, threatened, and endangered species habitat associated with natural springs. Many water-quality studies have been conducted for the karstic Edwards aquifer in central Texas, where groundwater discharges to large springs (Figs. 8.8 and 8.9). For more detailed information, see http://www.edwardsaquifer.net/geology.html, (accessed 20 May 2018).

8.2.1. Background Information on the Edwards Aquifer Hydrogeology and Ecosystems

Although the Edwards aquifer is a karst system, its geologic framework is vastly different from the UFA. The Edwards Aquifer (Fig. 8.8) crops out within a narrow band along a major fault zone (http://www.edwardsaquifer.net/geology.html; accessed 20 May 2018). Late Cenozoic faulting of Cretaceous-aged carbonates resulted in a series of partially to completely offset blocks of Edwards aquifer rocks and separate the confined and unconfined parts of the aquifer (Musgrove et al., 2014). The aquifer ranges in thickness from 107 to 152 m and is characterized by relatively high transmissivities (Lindgren, 2006). Due to faulting, stratigraphic aquifer units are located progressively deeper in the downdip subsurface. Most

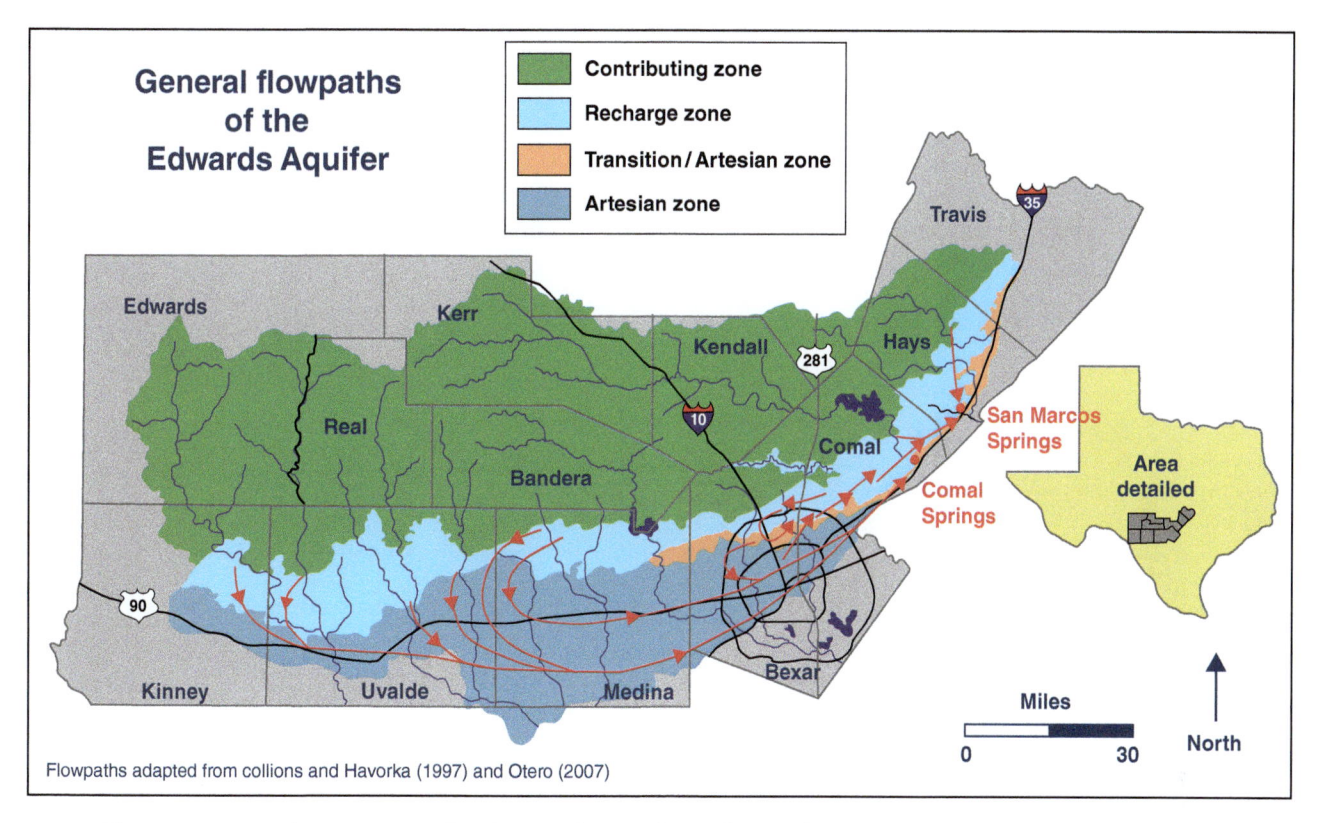

Figure 8.8 Map showing generalized groundwater flowpaths in the Edwards Aquifer, Texas (http://www.edwardsaquifer.net/geology.html, accessed 18 May 2018). *Source*: Reproduced with permission of Gregg Eckhardt.

recharge to the aquifer (60–80%) occurs from losing streams that flow across the unconfined (recharge) zone. Direct infiltration on the recharge zone accounts for most of the remaining recharge to the Edwards Aquifer. Groundwater levels in the aquifer and spring discharge respond rapidly to rainfall and corresponding recharge from losing streams (Musgrove et al., 2016).

Springs, caves, and streams associated with the Edwards aquifer provide habitat for unique flora and fauna. Endangered species in springs, spring runs, and the aquifer include Texas wild rice, fountain darter, Comal Springs dryopid beetle, San Marcos gambusia, Peck's Cave amphipod, Comal Springs riffle beetle, Barton Springs salamander, and the Texas blind salamander. Spring flow and groundwater levels respond rapidly to changes in climate. Global and regional climate models were linked with a hydrologic model and a species vulnerability index to evaluate the vulnerability of karst-related flora and fauna in the Edwards aquifer to climate change during 2011–2050 (Mahler et al., 2015). They projected that spring flow would decrease and the Barton Springs salamander and nine additional species (other salamander species, amphipods, beetles, darter) are vulnerable to climate change. Sixteen species were evaluated for the Edwards aquifer. Increases in nitrate concentrations in Barton Springs and the five streams that provide most of its recharge were likely caused by septic systems and land-applied treated wastewater effluent (Mahler et al., 2011).

8.2.2. Selected Studies of Nitrate Contamination in Edwards Aquifer Springs

Previous studies of spring waters discharging from the Edwards aquifer have looked at temporal variability in geochemistry to investigate the transport of nitrate and other anthropogenic contaminants, and the influence of hydrologic conditions. Mahler and Massei (2007) describe rapid changes in concentrations of anthropogenic contaminants in spring flow in response to storms in the Barton Springs segment of the Edwards aquifer (Fig. 8.8). Mahler et al. (2011) conducted a detailed study of factors affecting nutrients in the Barton Springs segment of the Edwards Aquifer and its contributing zone in the Austin Texas area during 2008–2010. During the dry period in the investigation, Barton Springs was dominated by flow from the aquifer matrix. However, they found some interesting results regarding nitrogen transport and fate during the transition from exceptional dry (drought conditions) to wetter-than-normal conditions. For example, organic nitrogen that was stored in soils during the dry period was nitrified to nitrate when the soils were rewetted, which resulted in elevated concentrations of

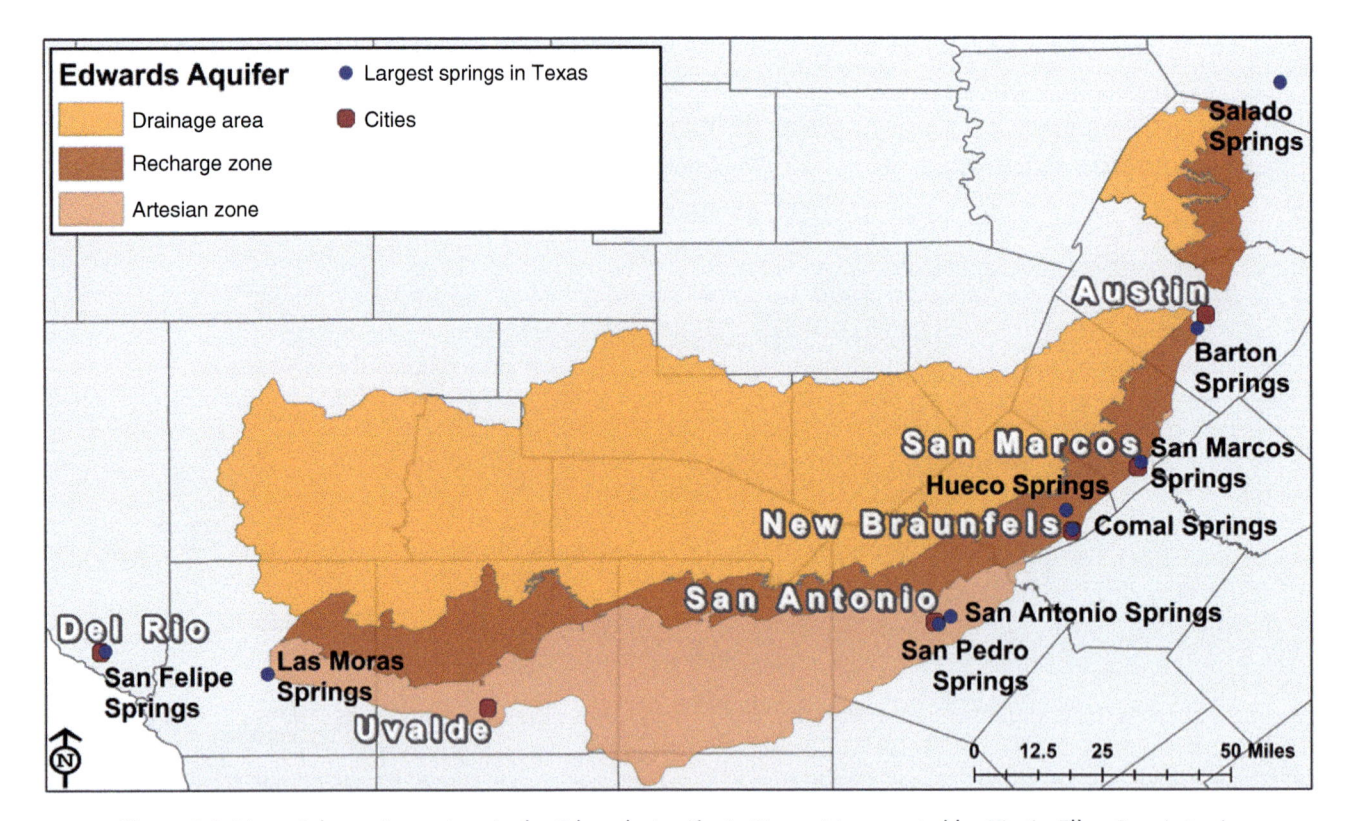

Figure 8.9 Map of the major springs in the Edwards Aquifer in Texas. Map created by Kirstin Eller, San Antonio Water System, San Antonio, Texas. Most of the major Edwards aquifer springs are in the east. San Pedro, San Antonio, Comal, San Marcos, Hueco, and Barton Springs are all located in the eastern third of the Edwards region. In the east there are also many more caves, and the recharge zone is thin at the surface (http://www.edwardsaquifer.net/geology.html, accessed 22 May 2018). *Source*: Map created by Kirstin Eller, San Antonio Water System.

nitrate in streams due to progressive leaching as wet weather continued. Mean monthly loads of organic nitrogen and nitrate in Barton Springs increased during the wet period, in contrast to constant and low loads during the dry period. Average organic nitrogen loads were about six times greater in stream recharge than in Barton Springs discharge, which indicated that organic nitrogen likely was being converted to nitrate within the aquifer. Also, they found that nitrate concentrations measured at Barton Springs during the study period were positively correlated with spring discharge. Based on the findings from this study, the researchers concluded that changes in climatic conditions and streamflow variability have a substantial effect on the fate and transport of nitrogen species from stream recharge to Barton Springs discharge.

Nitrate concentrations have increased over time in both the Barton Springs segment and the San Antonia segment of the Edwards aquifer; however, the rate of change has been different for the two segments (Musgrove et al., 2016). In the Barton Springs segment, nitrate concentrations have increased by about 20% since the 1990s, with the largest increase in the 2000s. In contrast, changes in nitrate concentrations in the San Antonio segment have occurred over a longer timeframe, but have doubled since the 1930s (Musgrove et al., 2016). They found that the isotopic signature of nitrate in both segments are consistent with a human waste (increased generation of wastewaters due to population growth) and or animal waste source. It was noted that these increases in nitrate concentrations, although below drinking water standards, could have serious implications for endangered species in the Edwards aquifer and springs. Rouse et al. (1999) found that nitrate-N concentrations greater than 2.5 mg/L adversely affect some amphibians.

During 2008–2010, Musgrove et al. (2016) also conducted a detailed analysis of nitrogen species loading in recharge (streams) and discharge (springs) for the Barton Springs segment of the Edwards aquifer using nitrate isotopes, other chemical data, and historical data from 1937 to 2007. Short-term variability in nitrate concentrations in the aquifer was related to individual storms and multiannual wet-dry cycles. However, long-term (years to decades) increases in nitrate concentrations and loads were attributed to land-use changes (mainly increases in septic systems and the land application of treated wastewater). Based on an overall mass balance for total nitrogen between recharge and discharge, total nitrogen loads in stream recharge (162 kg/d) were not significantly different than loads in Barton Springs discharge (157 kg/d) for the period of the investigation. However, recharge contained higher concentrations of organic nitrogen and lower concentrations of nitrate than discharge. Musgrove et al. (2010) also concluded that nitrification of organic nitrogen was occurring in the aquifer along with consumption of dissolved oxygen, which likely is an important source for nitrate in the Edwards aquifer and potentially in other oxic karst groundwater systems.

Temporal changes in nitrate concentration have been analyzed in Comal Springs (Fig. 8.8), another large spring system in the Edwards aquifer (Musgrove et al., 2010). During 1938–2006, there was a multidecadal upward trend in Comal Springs for nitrate-N concentration in both filtered and unfiltered water samples (Fig. 8.10). Musgrove et al. (2010) also analyzed nitrate data for filtered samples from Comal Springs during other time periods. They found a statistically significant difference between median nitrate concentration for two time periods: 1.2 mg/L during 1938–1945 ($n = 6$), and 1.9 mg/L ($n = 4$) during 1993–2003. This median nitrate-N value of 1.9 mg/L in the latter period was similar to the median value for regional groundwater samples collected during 1996–2006. Samples for these two time periods (1938–1945 and 1993–2003) were collected during similar aquifer flow conditions based on discharge measurements at Comal Springs; median discharge was 9.49 m^3/s during 1938–1945 and 8.95 m^3/s during 1993–2003. Based on these comparisons of nitrate-N concentrations for the two time periods, they inferred that matrix nitrate-N concentrations in the aquifer have increased over the last 70 years. When the large volume of water in storage in the aquifer is taken into account, this 0.7 mg/L increase in the median nitrate-N concentration at Comal Springs translated to a large increase in the nitrate load in the aquifer. Increases in nitrate concentrations likely are related to leaching of soil nitrate into the aquifer matrix that originates from various anthropogenic activities including spray irrigation of wastewater (in the aquifer contributing zone but not over the recharge zone), application of lawn fertilizers, and leakage from septic and sewer systems. Musgrove et al. (2010) also found statistically significant increases in chloride concentrations at Comal Springs during 1939–2006, which were consistent with anthropogenic sources associated with ongoing urbanization during the 70-year study period.

8.2.3. Protecting Water Quality in the Edwards Aquifer

Protecting water quality in the Edwards Aquifer from nitrate and other contaminants is complicated by several factors. Water quality protection in the recharge zone for the Edwards Aquifer focuses mainly on managing growth, limiting impervious cover, addressing stream water recharge to the aquifer, and reducing groundwater usage, Just recently on 18 May 2018, San Antonio entered Stage 1 water restrictions, which limits outdoor watering with a sprinkler or irrigation system, as required by state law. This was initiated by San Antonio Water Systems (SAWS) because water use has caused aquifer water levels to continue to drop even though there have been cooler temperature and rainfall this year so far (http://www.edwardsaquifer.net/news.html, accessed 25 May 2018). Water waste, such as water running down the street, is prohibited year-round.

Studies also have shown the importance of protecting the quality of water that runs off the Edwards Plateau and ends up as recharge, but controls in the contributing zone have received less attention. Two agencies that monitor and regulate water quality and quantity, the Edwards Aquifer Authority or the Texas Commission on Environmental Quality do not have the power or jurisdiction to make and enforce rules in most of the Edwards aquifer contributing zone. Furthermore, protections for the contributing zone are also complicated by the fact that residents living on the Edwards Plateau are not users of water from the Edwards aquifer. These residents are not likely to implement rules to protect other people's water supply, particularly because they consider Edwards aquifer users not being sufficiently proactive in managing development over their own recharge zone (http://www.edwardsaquifer.net/geology.html; accessed 18 May 2018). Texas also is highly respectful of private property rights, and many residents do not accept the regulation of land use and development. Future protection of water quality of the Edwards aquifer will depend on finding a balance between private property rights and protection of common natural resources.

8.3. NITRATE CONTAMINATION OF SPRINGS IN NORTHEAST SPAIN

Most research on water quality of springs has been focused in karst areas because of their high vulnerability to contamination. However, springs also are vulnerable to nitrate contamination in unconsolidated rock formations in semiarid environments (Mencio et al., 2011). Nitrate pollution from agricultural activities has been studied in the Osona region in northeast Spain since 2004, when a water-quality monitoring program was started to survey nitrate concentrations in springs. Springs in this area have been an important part of the cultural and natural heritage of the region (Mencio et al., 2011).

The hydrogeologic system in the Osona region (about 1250 km^2 in area, and located approximately 60 km north

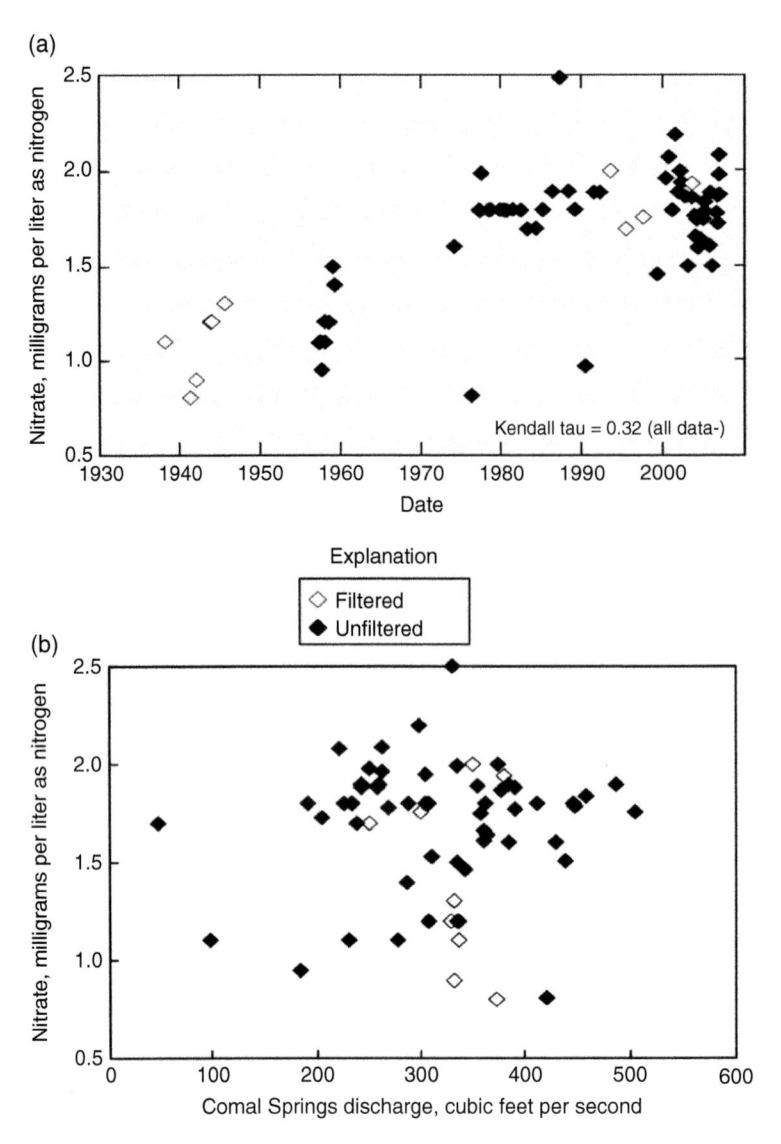

Figure 8.10 (a) Time series of nitrate concentration in Comal Springs discharge, and (b) relation between nitrate concentration at Comal Springs and Comal Springs discharge, San Antonio segment of the Edwards aquifer, south-central Texas, 1938–2006. *Source*: From Musgrove et al. (2010).

of Barcelona) consists of a sequence of Paleogene sedimentary layers (made up of conglomerates, carbonate formations, and alternating calcareous, marl and sandstone units) overlying crystalline (igneous and metamorphic) rocks (Menció et al., 2011). Also, springs in this area occur in the following geological settings: crystalline rocks, pre-quaternary sedimentary rocks, and quaternary sediments (alluvial, colluvial, eluvial, or mass-wasting deposits) that comprise local aquifers. Spring water discharge from a dataset of 130 springs ranged from 0.04 to 1.43 L/second (Boy-Roura et al., 2013). The area is an intensive pig-farming production area (with more than 900 piggeries containing more than 600,000 head of livestock). For decades, manure from the pig farms has been spread as organic fertilizer on row crops (wheat, barley, corn, and sorghum) that are grown on about 25% of the study area. High nitrate-N concentrations were found in springs (average concentrations ranging from 1.8 to 86 mg N/L and in wells (average concentrations ranging from 2.3 to 120 mg N/L) (Boy-Roura et al., 2013). Nitrate isotopes were used to confirm that pig manure was the source of the groundwater nitrate contamination (Vitòria et al., 2008). There were four different types of hydrological responses identified for springs in this area and were based on geology, discharge, conductance, and variations in nitrate concentrations. Highest nitrate-N values were found in springs located in sedimentary formations, while the lowest values were found in springs located in crystalline rocks and forested areas (Boy-Roura et al., 2013). The amount of organic fertilizer (manure) applied over the past several decades was related to the nitrate content in groundwater, although they found the natural variability of nitrate in groundwater to be small during a 1-year intensive monitoring period. Also, during this period, normalized-nitrate mass load plots showed a steady increase in most sampled springs. In an expanded study in Catalonia (northeast Spain), Mencío et al. (2016) studied the relation between nitrate contamination and major ions in groundwater samples from five different aquifer types. They found that despite lithological differences, nitrate contamination homogenizes the overall hydrochemistry; however, fertilization (manure) affects geochemical processes that control the overall major ion composition of groundwater.

8.4. NITRATE IN SPRINGS USED FOR DRINKING WATER

In many parts of the world, particularly in underdeveloped countries, springs can be the main or only source of drinking water. For example, in the suburban areas of Kampala City, Uganda, springs are a major source of water for domestic use due to the availability of springs and the lack of piped water in these areas. High nitrate concentrations are present in many of these springs and mean nitrate-N concentrations ranged from 4.7 to 31 mg N/L in 10 sampled springs (Haruna et al., 2005). High nitrate concentrations in sampled springs were attributed to poor sanitation, most likely from leaching of nitrate from nearby pit latrines. Nitrate contamination of spring water from domestic sewage also has been observed elsewhere in Africa (Haruna et al., 2005).

Roadside springs have been used as a drinking water source in many areas. In Pennsylvania, US, health concerns have been raised about the use of roadside springs for drinking water, which are regularly used by more than 10% of the population (https://extension.psu.edu/roadside-springs, accessed 28 April 2018). During 2013, 35 roadside springs were sampled, and the most common health issues were related to total coliform bacteria, *Escherichia coli* bacteria, and lead. During 2014–2015, additional water quality parameters were analyzed on a subset of 10 of 35 original roadside springs sampled in 2013. Although none of the springs had nitrate-N concentrations that exceeded the maximum contaminant level of 10 mg N/L for drinking water, researchers from Pennsylvania State Extension found that nearly all failed health-based drinking water standards, as many contained *E. coli* bacteria, and some contained the pathogenic parasites *Giardia* and *Cryptosporidium*.

In many areas of Florida, water from the UFA that flows to springs is being withdrawn and sold as bottled "spring" water. The Howard T. Odum Florida Springs Institute (FSI) (http://floridaspringsinstitute.org/, accessed 25 May 2018) has noted that large parts of the UFA in northern Florida have been contaminated with nitrate from various anthropogenic activities. Since the UFA is the source of spring water, they have expressed concerns about people in Florida and elsewhere getting their drinking water from springs or from the aquifer that is the source of bottle water. The FSI compared nitrate-N concentrations in bottled waters from the UFA with water from municipal supplies (Fig. 8.11). Resulting nitrate concentrations in several bottled "spring" waters are well above total maximum daily load target for nitrate-N of 0.35 mg/L.

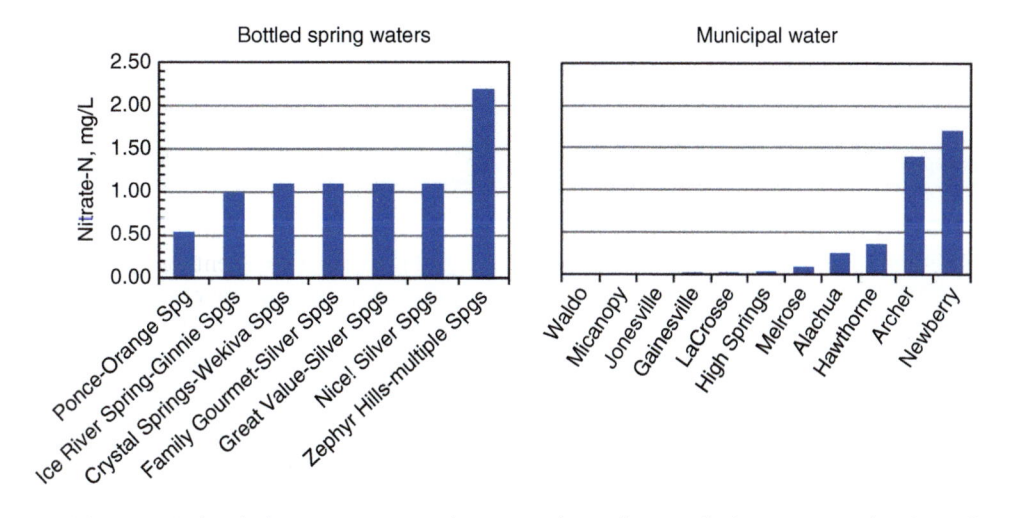

Figure 8.11 Nitrate-N in bottled spring waters and municipal supplies in Alachua County, Florida, collected in March 2016 by the Florida Springs Institute. *Source*: Modified from http://fladefenders.org/wp-content/uploads/2016/06/nitrate-chart.pdf (accessed 25 May 2018). Reproduced with permission of Florida Springs Institute.

8.5. SPRINGS AND WATER QUANTITY ISSUES

Groundwater withdrawals for agricultural, industrial, and drinking water supply in spring basins or springsheds have resulted in significant reductions in spring flow, changes of springs from flowing perennially to ephemerally, or the complete disappearance of springs in many areas. The volume of water withdrawn from an aquifer directly influences water losses to springs and groundwater discharge to surface water bodies; although the effect may develop slowly over time (Kresic & Stevanovic, 2010). For example, springs used to occur across the entire state of Florida; however, in South Florida they were wiped out decades ago by the ditching and draining of the landscape as well as overpumping of the aquifer (Pittman, 2012). Another spring (Kissingen) farther north in the phosphate mining area in central Florida (Kissengen Spring) in Polk County used to discharge more than 110 million liters of water per day in 1930. Twenty years later, it had dried up completely and investigators attributed its disappearance to heavy pumping of groundwater by the phosphate industry. Although water usage was cut back over succeeding decades, Kissengen Spring never flowed again (Pittman, 2012). Several other springs in northern Florida have suffered substantial decreases in flow or have dried up completely due to groundwater withdrawals coupled with a decrease in long-term rainfall. In the Edwards aquifer, San Antonio Springs and San Pedro Springs in San Antonio are dry most of the time due to the large amounts of water withdrawn from the aquifer by users in Bexar county, although these two springs flow when aquifer levels are very high after large amounts of recharge. It is interesting to note that several decades ago Williams (1977) documented the development and alteration of the landscape that caused the disappearance of springs and waterways that had existed in Washington, DC in the 18th century.

Springs all over the world are suffering from several stressors. Increased groundwater withdrawals and increasing nitrate concentrations are having adverse effects on spring water quality and quantity, ecosystems, and human health. Furthermore, these effects are having detrimental impacts on local and regional economies. Effective springs management must include better protection of recharge areas to aquifers by regulating water use and adopting land-use practices that do not adversely affect water quality.

REFERENCES

Ascott, M. J., Gooddy, D. C., Wang, L., Stuart, M. E., Lewis, M. A., Ward, R. S., & Binley, A. M. (2017). Global patterns of nitrate storage in the vadose zone. *Nature Communications*, 1–7. https://doi.org/10.1038/s41467-017-01321-w

Barnett, C. (2015). *Clarity for Florida's Springs: What's behind the steady algal takeover of these aquatic treasures? Discover*, April 2015. Kalmbach Media.

Basu, N. B., Jindal, P., Schilling, K. E., Wolter, C. F., & Takle, E. S. (2012). Evaluation of analytical and numerical approaches for the estimation of groundwater travel time distribution. *Journal of Hydrology, 475*, 65–73.

Benson, R. C., & Yuhr, L. B. (2016). Dye tracing. In *Site characterization in Karst and Pseudokarst Terraines* (pp. 295–306). Dordrecht: Springer. https://doi.org/10.1007/978-94-017-9924-9

Berndt, M.P., Katz, B.G., Kingsbury, J.A., & Crandall, C.A. (2014). *The quality of our Nation's waters—Water quality in the Upper Floridan aquifer and overlying surficial aquifers, southeastern United States, 1993–2010*. U.S. Geological Survey Circular 1355, 72 p., http://dx.doi.org/10.3133/cir1355, accessed 15 May 2018.

Boy-Roura, M., Mencio, A., & Mas-Pla, J. (2013). Temporal analysis of spring water data to assess nitrate inputs to groundwater in an agricultural area (Osona, NE Spain). *Science of the Total Environment, 452–453*, 433–445. https://doi.org/10.1016/j.scitotenv.2013.02.065

Busenberg, E., & Plummer, L. N. (1992). Use of chlorofluoromethanes (CCl3F and CCl2F2) as hydrologic tracers and age dating tools: Example – The alluvium and terrace system of central Oklahoma. *Water Resources Research, 28*, 2257–2283.

Busenberg, E., & Plummer, L. N. (2000). Dating young ground water with sulfur hexafluoride – Natural and anthropogenic sources of SF6. *Water Resources Research, 36*, 3011–3030.

Butt, P. L., & Murphy, G. J. (2003). *Dyal and black sink dye trace studies, Columbia County Florida, May–September 2003*. Tallahassee, FL: Karst Environmental Services for Florida Department of Environmental Protection.

Call, J., & Stephenson, F. (2003). *Spring time in Florida*. Florida State University Research in Review, 18 p. Retrieved from www.research.fsu.edu/researchr/fall2003.

Chapelle, F. H. (2005). *Wellsprings, A natural history of bottle spring waters*. New Brunswick, NJ: Rutgers University Press, 279 p.

Cohen, M.J., Lamsal, S., & Korhnak, L.V. (2007). *Sources, transport and transformations of nitrate-N in the Florida environment*. St. Johns River Water Management District Report SJ2007-SP10, Palatka, FL, 125 p.

Copeland, R. (2003). *Florida spring classification system and spring glossary*. Florida Geological Survey Special Publication No. 52, Tallahassee, FL, 18 p.

Copeland, R., Doran, N.A., White, A.J., & Upchurch, S.B. (2011). *Regional and statewide trends in Florida's spring and well groundwater quality (1991-2003)*. Florida Geological Survey Bulletin No. 69, 393 p.

Davis, H.D., Katz, B.G., & Griffin, D. W. (2010). *Nitrate-N movement in groundwater from the land application of treated municipal wastewater and other sources in the Wakulla Springs Springshed, Leon and Wakulla Counties, Florida, 1966–2018: Tallahassee, Florida*. U.S. Geological Survey Scientific Investigations Report 2010–5099, 80 p.

Eller, K., & Katz, B. G. (2017). Nitrogen source inventory and loading tool: An integrated approach toward restoration of water-quality impaired karst springs. *Journal of Environmental Management, 196*, 702–709.

Green, C. T., Puckett, L. J., Böhlke, J. K., Bekins, B. A., Phillips, S. P., Kauffman, L. J., Denver, J. M., & Johnson, H. M. (2008). *Limited occurrence of denitrification in four shallow aquifers in agricultural areas of the United States. Journal of Environmental Quality*, 37, 994–1009.

Happell, J. D., Opsahl, S., Top, Z., & Chanton, J. P. (2006). Apparent CFC and ^{3}H/^{3}He age differences in water from Floridian Aquifer springs. *Journal of Hydrology, 319*, 410–426.

Haruna, R., Ejobi, F., & Kabagambe, E. K. (2005). The quality of water from protected springs in Katwe and Kisenyi parishes, Kampala city, Uganda. *African Health Sciences, 5*(1), 14–20.

Heffernan, J. B., Albertin, A. R., Fork, M. L., Katz, B. G., & Cohen, M. J. (2012). Denitrification and inference of nitrogen sources in the karstic Floridan Aquifer. *Biogeosciences, 9*, 1671–1690.

Hensley, R. T., & Cohen, M. J. (2017). Flow reversals as a driver of ecosystem transition in Florida's springs. *Freshwater Science, 36*(1), 14–25. https://doi.org/10.1086/690558

Hu, C. M., Muller-Karger, F. E., & Swarzenski, P. W. (2006). Hurricanes, submarine groundwater discharge and Florida's red tides. *Geophysical Research Letters, 11*, 1–5.

Jurgens, B.C., Böhlke, J.K., & Eberts, S.M. (2012). *TracerLPM (Version 1): An Excel® workbook for interpreting groundwater age distributions from environmental tracer data*. U.S. Geological Survey Techniques and Methods Report 4-F3, 60 p. Retrieved from https://pubs.usgs.gov/tm/4-f3/pdf/tm4-F3.pdf, accessed 06 May 2018.

Katz, B. G., Bohlke, J. K., & Hornsby, H. D. (2001). Timescales for nitrate contamination of spring waters, northern Florida. *Chemical Geology, 179*, 167–186.

Katz, B. G., Chelette, A. R., & Pratt, T. R. (2004). Use of chemical and isotopic tracers to assess sources of nitrate and age of ground water, Woodville Karst Plain, USA. *Journal of Hydrology, 289*, 36–61. https://doi.org/10.1016/j.jhydrol.2003.11.001

Katz, B. G., Copeland, R., Greenhalgh, T., Ceryak, R., & Zwanka, W. (2005). Using multiple chemical indicators to assess sources of nitrate and age of groundwater in a karstic spring basin. *Environmental and Engineering Geoscience XI, 4*, 333–346.

Katz, B.G., Crandall, C.A., Metz, P.A., McBride, S., & Berndt, M.P. (2007). *Chemical characteristics, water sources and pathways, and age distribution of ground water in the contributing recharge area of a public-supply well near Tampa, Florida, 2002–05*. U.S. Geological Survey Scientific Investigations Report 2007–5139, 83 p.

Katz, B. G., & Griffin, D. W. (2008). Using chemical and microbiological indicators to track the impacts from the land application of treated municipal wastewater and other sources on groundwater quality in a karstic springs basin. *Environmental Geology, 55*, 801–821. https://doi.org/10.1007/s00254-007-1033-y

Katz, B. G., Griffin, D. W., & Davis, H. D. (2009b). Groundwater quality impacts from the land application of treated municipal wastewater in a large karstic spring basin: Chemical and microbiological indicators. *Science of the Total Environment, 407*, 2872–2886.

Katz, B.G., Hornsby, H.D., Bohlke, J.K., & Mokray, M.F. (1999). *Sources and chronology of nitrate contamination of springwaters, Suwannee River Basin, Florida*. U.S. Geological Survey Water-Resources Investigations Report 99–4252, 54 p.

Katz, B. G., Sepulveda, A. A., & Verdi, R. J. (2009a). Estimating nitrogen loading to ground water and assessing vulnerability to nitrate contamination in a large karstic spring basin. *Journal of the American Water Resources Association, 45*, 607–627. https://doi.org/10.1111/j.1752-1688.2009.00309.x

Kincaid, T. R., Davies, G., DeHan, R., & Hazlett, T. J. (2004). Characterizing rapid point-recharge to the Floridan Aquifer in the Woodville Karst Plain of North Florida – Implications for protecting Wakulla Springs. Abstract Paper No. 31–3. In *53rd Annual Geological Society of America Meeting (Southeastern Section)*, 17–25 March 2004. Washington, DC.

Kincaid, T. R., Hazlett, T. J., & Davies, G. J. (2005). *Quantitative groundwater tracing and effective numerical modeling in karst: An example from the Woodville Karst Plain of North Florida, Sinkholes and the Engineering and Environmental Impacts of Karst (GSP 144)*. In Proceedings of Tenth Multidisciplinary Conference, 24–28 September 2005. San Antonio, TX. doi:https://doi.org/10.1061/40796(177)13.

Kincaid, T.R., & Werner, C. (2008). Conduit flow paths and conduit/matrix interactions defined by quantitative groundwater tracing in the Floridan aquifer. In *Proceedings of the 11th Multidisciplinary Conference on Sinkholes and the Engineering and Environmental Impacts of Karst*, American Society of Civil Engineers. https://doi.org/10.1061/41003(327)28.

Knowles, L., Jr., Katz, B. G., & Toth, D. J. (2010). Using multiple chemical indicators to characterize and determine the age of groundwater from selected vents of the Silver Springs Group, central Florida, USA. *Hydrogeology Journal, 18*, 1825–1838.

Kresic, N., & Stevanovic, Z. (Eds.). (2010). *Groundwater hydrology of springs. Engineering, theory, management, and sustainability*. Jordan Hill, Oxford: Butterworth-Heinemann, Elsevier, 573 p.

LaMoreaux, P. E., & Tanner, J. T. (2001). *Springs and bottled waters of the world*. Berlin, Germany: Springer-Verlag, 315 p.

Lindgren, R.J. (2006). *Diffuse-flow conceptualization and simulation of the Edwards aquifer, San Antonio region, Texas*. U.S. Geological Survey Scientific Investigations Report 2006–5319. Reston, VA: USGS.

Mahler, B. J., & Massei, N. (2007). Anthropogenic contaminants as tracers in an urbanizing Karst aquifer. *Journal of Contaminant Hydrology, 91*(1–2), 81–106.

Mahler, B.J., Musgrove, M., Sample, T.L., & Wong, C.I. (2011). *Recent (2008–10) water quality in the Barton Springs segment of the Edwards aquifer and its contributing zone, central Texas, with emphasis on factors affecting nutrients and bacteria*. U.S. Geological Survey Scientific Investigations Report 2011–5139, 66 p.

Mahler, B.J., Stamm, J.F., Poteet, M.F., Symstad, A.J., Musgrove, M., Long, A.J., & Norton, P.A. (2015). *Effects of projected climate (2011–50) on karst hydrology and species vulnerability—Edwards aquifer, south-central Texas, and Madison aquifer, western South Dakota*. U.S. Geological Survey Fact Sheet 2014–3046.

Mahon, G.L. (2011). *Assessment of Groundwater pathways and contaminant transport in Florida and Georgia using multiple chemical and microbiological indicators*. U.S. Geological Survey Fact Sheet 2011–3070. Retrieved from https://pubs.usgs.gov/fs/2011/3070/, accessed 20 September 2019.

Maloszewski, P., & Zuber, A. (1996). Lumped parameter models for the interpretation of environmental tracer data. In *Manual on mathematical models in isotope hydrology, IAEA-TECDOC 910* (pp. 9–50). Vienna: International Atomic Energy Agency.

Martin, J. B., & Dean, R. A. (2001). Exchange of water between conduits and matrix in the Floridan aquifer. *Chemical Geology, 179*, 145–166.

Martin, J. B., Kurz, M. J., & Khadka, M. B. (2016). Climate control of decadal-scale increases in apparent ages of eogenetic karst spring water. *Journal of Hydrology, 540*, 988–1001. https://doi.org/10.1016/j.jhydrol.2016.07.010

McMahon, P. B., Dennehy, K. F., Bruce, B. W., Bohlke, J. K., Michel, R. L., Gurdak, J. J., & Hurlbut, D. B. (2006). Storage and transit time of chemicals in thick unsaturated zones under rangeland and irrigated cropland, High Plains, United States. *Water Resources Research, 42*(3), W03413. https://doi.org/10.1029/2005WR004417

Meals, D. W., Dressing, S. A., & Davenport, T. E. (2010). Lag time in water quality response to best management practices: A review. *Journal of Environmental Quality, 39*, 85–96.

Meinzer, O.E. (1927). *Large springs in the United States*. U.S. Geological Survey Water Supply Paper 557. Washington, DC: U.S. Government Printing Office, 94 p.

Menció, A., Boy, M., & Mas-Pla, J. (2011). Analysis of vulnerability factors that control nitrate occurrence in natural springs (Osona region, NE Spain). *Science of the Total Environment, 409*, 3049–3058.

Menció, A., Mas-Pla, J., Otero, N., Regas, O., Boy-Roura, M., Puig, R., et al. (2016). Nitrate pollution of groundwater; all right …, but nothing else? *Science of the Total Environment, 539*, 241–251.

Musgrove, M., Fahlquist, L., Houston, N.A., Lindgren, R.J., & Ging, P.B. (2010). *Geochemical evolution processes and water-quality observations based on results of the National Water-Quality Assessment Program in the San Antonio segment of the Edwards aquifer, 1996–2006*. U.S. Geological Survey Scientific Investigations Report 2010–5129, 93 p. Appendixes available online at http://pubs.usgs.gov/sir/2010/5129/, accessed 18 April 2008.

Musgrove, M., Katz, B. G., Fahlquist, L. S., Crandall, C. A., & Lindgren, R. J. (2014). Factors affecting public-supply well vulnerability in two karst aquifers. *Ground Water, 52*(Suppl. 1), 63–75. https://doi.org/10.1111/gwat.12201

Musgrove, M., Opsahl, S. P., Mahler, B. J., Herrington, C., Sample, T. L., & Banta, J. R. (2016). Source, variability, and transformation of nitrate in a regional karst aquifer: Edwards aquifer, central Texas. *Science of the Total Environment, 568*, 457–469.

Phelps, G.G. (2004). *Chemistry of ground water in the Silver Spring Basin, Florida, with an emphasis on nitrate*. U. S. Geological Survey Scientific Investigations Report 2004–5144. Reston, VA.

Pittman, C. (2012). *Florida's vanishing springs. Tampa Bay Times*. Retrieved from https://www.tampabay.com/news/environment/water/floridas-vanishing-springs/1262988

Plummer, L. N. (2005). Dating of young groundwater. In P. K. Aggarwal, J. R. Gat, & K. F. O. Froehlich (Eds.), *Isotopes in the water cycle* (pp. 193–218). Dordrecht, The Netherlands: Springer. https://doi.org/10.1007/1 4020-3023-1_14

Plummer, L. N., & Busenberg, E. (1999). Chlorofluorocarbons. In P. G. Cook & A. L. Herczeg (Eds.), *Environmental tracers in subsurface hydrology*, Ch. 15 (pp. 441–478). Kluwer Academic Press.

Plummer, L. N., & Busenberg, E. (Eds.). (2006). *Use of chlorofluorocarbons in hydrology*. Vienna, Austria: A guidebook. International Atomic Energy Agency, 277 p.

Puckett, L. J., Tesoriero, A. J., & Dubrovsky, N. M. (2011). Nitrogen contamination of surficial aquifers- a growing legacy. *Environmental Science and Technology, 45*, 839–844.

Ringle, K. (1999). Unlocking the labyrinth of north Florida springs. *National Geographic, 195*(3), 40–58.

Rouse, J. D., Bishop, C. A., & Stuger, J. (1999). Nitrogen pollution: An assessment of its threat to amphibian survival. *Environmental Health Perspectives, 107*(10), 799–803.

Sanford, W. E., & Pope, J. P. (2013). Quantifying groundwater's role in delaying improvements to Chesapeake Bay water quality. *Environmental Science and Technology, 47*, 13330–13338.

Scott, T.M., Means, G.H., Means, R.C., & Meegan, R.P. (2002). *First magnitude springs of Florida*. Florida Geological Survey Open-File Report No. 85.

Scott, T.M., Means, G.H., Meegan, R.P., Means, R.C., Upchurch, S.B., Copeland, R.E., et al. (2004). *Springs of Florida*. Florida Geological Survey Bulletin 66. Tallahassee, FL. ISSN 0271-7832.

Sebillo, M., Mayer, B., Nicolardot, B., Pinay, G., & Mariotti, A. (2013). Long-term fate of nitrate fertilizer in agricultural soils. *Proceedings of the National Academy of Sciences of the United States, 110*(45), 18185–18189.

Seitzinger, S., Harrison, J. A., Bohlke, J. K., Bouwman, A. F., Lowrance, R., Peterson, B., et al. (2006). Denitrification across landscapes and waterscapes: A synthesis. *Ecological Applications, 16*(6), 2064–2090. https://doi.org/10.1890/1051-0761(2006)016[2064:DALAWA]2.0.CO;2

Springer, A. E., & Stevens, L. E. (2008). Spheres of discharge of springs. *Hydrogeology Journal*. https://doi.org/10.1007/s10040-008-0341-y

Stamm, D. (2008). *Springs of Florida* (2nd ed.). Sarasota, FL: Pineapple Press, 114 p.

Stevens, L. E., Stacey, P. B., Jones, A., Duff, D., Gourley, C., & Caitlin, J. C. (2005). A protocol for rapid assessment of southwestern stream riparian ecosystems. In C. I. I. van Riper & D. J. Mattson (Eds.), *Fifth conference on research on the Colorado Plateau* (pp. 397–420). Tucson, AZ: University of Arizona Press.

Toth, D. J., & Katz, B. G. (2006). Mixing of shallow and deep groundwater as indicated by the chemistry and age of karstic springs. *Hydrogeology Journal, 14*, 1060–1080.

Trček, B., & Zojer, H. (2010). Recharge of springs. In N. Kresic & Z. Stevanovic (Eds.), *Groundwater hydrology of springs. Engineering, theory, management, and sustainability* (pp. 87–127). Jordan Hill, Oxford: Butterworth-Heinemann, Elsevier.

University of Florida Water Institute. (2017). *Collaborative research initiative on sustainability and protection of springs* (Final Report 2014–2017). Submitted to St. Johns River

Water Management District, July 2017, 9 chapters, variously paged, 1085 p.

Upchurch, S. B., Champion, K. M., Schneider, J. C., Hornsby, D., Ceryak, R., & Zwanka, W. D. (2004). *Defining springshed boundaries and water quality domains near first magnitude springs in North Florida* (pp. 52). Florida Scientist.

Vitòria, L., Soler, A., Canals, A., & Otero, N. (2008). Environmental isotopes (N, S, C, O, D) to determine natural attenuation processes in nitrate contaminated waters: Example of Osona (NE Spain). *Applied Geochemistry, 23*, 3597–3611.

Wang, L., Butcher, A., Stuart, M., Gooddy, D., & Bloomfield, J. (2013). The nitrate time bomb: A numerical way to investigate nitrate storage and lag time in the unsaturated zone. *Environmental Geochemistry and Health, 35*, 667–681.

Werner, C. (2000). *Determination of groundwater flow patterns from cave exploration in the Woodville karst plain, Florida*. Florida Geological Survey Special Publication No. 46, Tallahassee, FL.

Williams, G.P. (1977). *Washington D.C.'s vanishing springs and waterways*. U.S. Geological Survey Circular 752, Washington, DC, 23 p.

Zuber, A. (1986). Mathematical models for the interpretation of environmental radioisotopes in groundwater systems. In P. Fritz & J.-C. Fontes (Eds.), *Handbook of environmental geochemistry: The terrestrial environment* (Vol. 2). New York: Elsevier, 59 p.

FURTHER READING

Albertin, A. R., Sickman, J. O., Pinowska, A., & Stevenson, R. J. (2012). Identification of nitrogen sources and transformations within karst springs using isotope tracers of nitrogen. *Biogeochemistry, 108*, 219–232.

Alley, W. M., Healy, R. W., LaBaugh, J. W., & Reilly, T. E. (2002). Flow and storage in groundwater systems. *Science, 296*, 1985–1990.

Allums, S. E., Opsahl, S. P., Golladay, S. W., Hicks, D. W., & Conner, L. M. (2012). Nitrate concentrations in springs flowing into the lower Flint River Basin, Georgia U.S.A. *Journal of the American Water Resources Association (JAWRA), 48*(3), 423–438. https://doi.org/10.1111/j.1752-1688.2011.00624.x

Boyer, D. G., & Pasquarell, G. C. (1995). Nitrate concentrations in karst springs in an extensively grazed area. *Journal of the American Water Resources Association (JAWRA), 31*(4), 565–573. https://doi.org/10.1111/j.1752-1688.1996.tb04054.x

Cohen, M. J. (2018). Springshed nutrient loading, transport and transformations. In M. T. Brown, K. C. Reiss, M. J. Cohen, J. M. Evans, K. R. Reddy, P. W. Inglett, et al. (Eds.), *Summary and synthesis of the available literature on the effects of nutrients on spring organisms and systems* (pp. 53–134). University of Florida Water Institute.

Dubrovsky, N.M., Burow, K.R., Clark, G.M., Gronberg, J.M., Hamilton P.A., Hitt, K.J., et al. (2010). *The quality of our Nation's waters—Nutrients in the Nation's streams and groundwater, 1992-2004*. U.S. Geological Survey Circular 1350, 174 p. Retrieved from http://water.usgs.gov/nawqa/nutrients/pubs/circ1350, accessed 20 September 2019.

Eckhardt, G. (2010). Case study: Protection of Edwards Aquifer springs, the United States. In N. Kresic & Z. Stevanovic (Eds.), *Groundwater hydrology of springs. Engineering, theory, management, and sustainability* (pp. 526–542). Jordan Hill, Oxford: Butterworth-Heinemann, Elsevier.

Erisman, J. W., Sutton, M. A., Galloway, J. N., Klimont, Z., & Winiwarter, W. (2008). How a century of ammonia synthesis changed the world. *Nature Geoscience, 1*, 636–639.

Florea, L. J., & Vacher, H. L. (2007). Eogenetic karst hydrology: Insights from the 2004 hurricanes, peninsular Florida. *Groundwater, 45*, 439–446.

Florida Springs Task Force. (2000). *Florida's springs—Strategies for protection & restoration*. Florida Department of Environmental Protection. Retrieved from http://archive.floridasprings.org/protection/taskforce/, accessed 18 April 2018.

Focazio, M.J., Plummer, L.N., Bohlke, J.K., Busenberg, E., Bachman, L.J., & Powars, D.S. (1998). *Preliminary estimates of residence times and apparent ages of groundwater in the Chesapeake Bay Watershed, and water-quality data from a survey of springs*. U.S. Geological Survey Water-Resources Investigations Report 97-4225, 75 p.

Galloway, J. N., Townsend, A. R., Erisman, J. W., Bekunda, M., Cai, Z., Freney, J. R., et al. (2008). Transformation of the nitrogen cycle: Recent trends, questions, and potential solutions. *Science, 320*, 889–892.

Galloway, J. N., Winiwarter, W., Leip, A., Leach, A., Bleeker, A., & Erisman, J. W. (2014). Nitrogen footprints: Past, present, and future. *Environmental Research Letters, 9*. https://doi.org/10.1088/1748-9326/9/11/115003

Goldscheider, N. (2010). Delineation of springs protection zones. In N. Kresic & Z. Stevanovic (Eds.), *Groundwater hydrology of springs. Engineering, theory, management, and sustainability* (pp. 305–338). Jordan Hill, Oxford: Butterworth-Heinemann, Elsevier.

Harrington, D., Maddox, G., & Hicks, R. (2010). *Florida springs initiative monitoring network report and recognized sources of nitrate*. Tallahassee, FL: Florida Department of Environmental Protection, 103 p.

Heffernan, J. B., Liebowitz, D. M., Frazer, T. K., Evans, J. M., & Cohen, M. J. (2010). Algal blooms and the nitrogen-enrichment hypothesis in Florida springs—Evidence, alternatives, and adaptive management. *Ecological Applications, 20*(3), 816–829.

Henson, W. R., Huang, L., Graham, W. D., & Ogram, A. (2017). Nitrate reduction mechanisms and rates in an unconfined eogenetic karst aquifer in two sites with different redox potential. *Journal of Geophysical Research: Biogeosciences, 122*, 1062–1077.

Katz, B. G. (2004). Sources of nitrate contamination and age of water in large karstic springs of Florida. *Environmental Geology, 46*, 689–706.

Katz, B. G. (2012). Nitrate contamination in karst groundwater. In W. B. White & D. C. Culver (Eds.), *Encyclopedia of caves* (pp. 564–568). Boston, MA: Academic Press, Elsevier. https://doi.org/10.1016/B978-0-12-383832-2.00146-8

Katz, B. G., Griffin, D. W., McMahon, P. B., Harden, H., Wade, E., Hicks, R. W., & Chanton, J. P. (2010). Fate of effluent-borne contaminants beneath septic tank drainfields overlying a karst aquifer. *Journal of Environmental Quality, 39*, 1181–1195.

Lindsey, B. D., Phillips, S. W., Donnelly, C. A., Speiran, G. K., Plummer, L. N., Böhlke, J. K., Focazio, M. J., & Burton, W. C. (2003). *Residence time and nitrate transport in ground water discharging to streams in the Chesapeake Bay Watershed.* U.S. Geological Survey Water-Resources Investigations Report 03–4035, 202 p.

Lopez, C.B., Jewett, E.B., Dortch, Q., Walton, B.T., & Hudnell, H.K. (2008). *Scientific assessment of freshwater harmful algal blooms.* In *Interagency Working Group on Harmful Algal Blooms, Hypoxia and Human Health of the Joint Subcommittee on Ocean Science and Technology*, Washington, DC, 65 p.

Mahler, B. J., & Garner, B. D. (2009). Using nitrate to quantify quick flow in a karst aquifer. *Ground Water, 47*, 350–360.

Manga, M. (2001). Using springs to study groundwater flow and active geologic processes. *Annual Reviews of Earth and Planetary Sciences, 29*, 201–228.

Panno, S. V., Hackley, K. C., Hwang, H. H., & Kelly, W. R. (2001). Determination of the sources of nitrate contamination in karst springs using isotopic and chemical indicators. *Chemical Geology, 179*, 113–128.

Phillips, S.W., & Lindsey, B.D. (2003). *The influence of ground water on nitrogen delivery to the Chesapeake Bay.* U.S. Geological Survey Fact Sheet FS-091-03.

State-USEPA Nutrient Innovations Task Group. (2009). *An urgent call to action.* State-USEPA Nutrient Innovations Task Group Report to USEPA, August 2009, 170 p.

Sutton, M. A., Howard, C. M., Erisman, J. W., Billen, G., Bleeker, A., Grennfelt, P., van Grinsven, H., & Grizzetti, B. (2011). *The European nitrogen assessment: Sources, effects, and policy perspectives.* Cambridge, UK: Cambridge University Press, 664 p.

U.S. Environmental Protection Agency (2009). *An urgent call to action. Report of the State-EPA Nutrient Innovations Task Group*, 41.

Upchurch, S.B., Chen, J., & Cain, C.R. (2007). *Trends of nitrate concentrations in waters of the Suwannee River Water Management District, 2007.* Prepared by SDII Global Corporation for the Suwannee River Water Management District, 30 May 2007, 36 p.

Wang, S., Changyuan, T., Xianfang, S., Ruiqiang, Y., Zhiwei, H., & Yun, P. (2016). Factors contributing to nitrate contamination in a groundwater recharge area of the North China Plain. *Hydrological Processes, 30*(13), 2271–2285.

9

Co-occurrence of Nitrate with Other Contaminants in the Environment

Many of the sources of reactive nitrogen contamination to the atmosphere, groundwater, and surface waters also contribute other contaminants that can adversely impact human health, ecosystems, and air and water quality. Although many studies focus on identifying sources of nitrate contamination, and its transport and fate in the environment, there have been numerous studies that have identified the co-occurrence of nitrate with other anthropogenic pollutants. Depending on their chemical properties and mode of transport, some contaminants readily dissolve in water and can move with mobile nitrate into groundwater and into surface waters. Other less soluble contaminants may attach to soil particles or other particulate matter and can be transported to surface waters by physical processes such as soil erosion, resuspension of sediments, and stormwater runoff.

Many of these contaminants that co-occur with nitrate or other reactive nitrogen species have been used to identify sources of reactive nitrogen contamination and were discussed in Chapter 4. This chapter presents information on the co-occurrence of nitrogen with other contaminants in the atmosphere, surface waters, groundwater, and associated ecosystems. Selected contaminants include phosphorus (fertilizers, manure, wastewaters), pathogens, trace elements, pesticides, organic compounds including emerging contaminants (pharmaceuticals, hormones, and personal care products), and atmospheric contaminants (i.e., sulfur, particulates, ammonia, and organic nitrogen compounds). Highlights have been selected from some of the extensive research conducted at local, regional, national, and global scales.

9.1. NITROGEN AND PHOSPHORUS

The nitrogen cycle is inextricably linked to the phosphorus cycle. Natural biogeochemical cycles of nitrogen and phosphorus have been so substantially altered by the addition of anthropogenic sources (e.g., fertilizers, manure, and human wastewater) that in many cases, their effects and consequences have been difficult to quantify on a global scale (Sutton et al., 2013). Furthermore, discrepancies in estimating nutrient fluxes have existed among studies due to different approaches and focus. Even with these issues, there have been reliable estimates of global nitrogen and phosphorus fluxes (2000–2010) based on a summary of many reported literature studies (Sutton et al., 2013).

9.1.1. Fertilizers

Fertilizers typically contain nitrogen, phosphorus, and potassium, and manure contains substantial amounts of nitrogen and phosphorus. Potassium in fertilizer and manure typically does not contribute to major pollution concerns. Potassium, unlike nitrogen or phosphorus, generally is not limiting in either freshwater or marine ecosystems, and is only a minor fraction of atmospheric levels of particulate matter. In soils containing clay minerals, potassium can undergo ion exchange with other cations, such as calcium, sodium, or magnesium.

Globally, Sutton et al. (2013) showed that human use of synthetic nitrogen fertilizers has increased by a factor of 9; whereas phosphorus use has tripled from around 1960 to 2010 (Fig. 9.1). They projected a substantial increase of around 40–50% over the next 40 years (Fig. 9.1) based on the needs to feed the growing world population (also accounting for current trends in dietary lifestyles including increasing consumption of animal products). The European Nitrogen Assessment estimated that 85% of harvested reactive nitrogen was used to feed livestock, but only 15% was used to feed people directly (Sutton et al., 2011). This translated to an estimate that a person in the European Union consumed 70% more protein than needed for a healthy diet.

Nitrogen Overload: Environmental Degradation, Ramifications, and Economic Costs, Geophysical Monograph 250,
First Edition. Brian G. Katz.
© 2020 American Geophysical Union. Published 2020 by John Wiley & Sons, Inc.

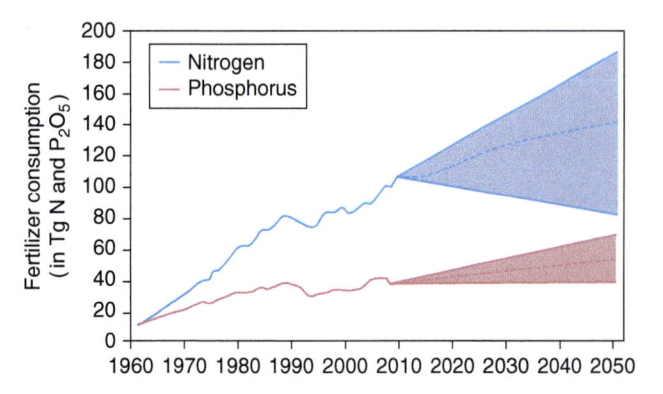

Figure 9.1 Trends in consumption of global mineral fertilizer consumption for nitrogen and phosphorus and future projections based on present-day estimates. *Source*: From Sutton et al. (2013) "Our Nutrient World: The challenge to produce more food and energy with less pollution. Global Overview of Nutrient Management." Centre for Ecology and Hydrology, Edinburgh on behalf of the Global Partnership on Nutrient Management and the International Nitrogen Initiative.

Sutton et al. (2013) also evaluated the overall food chain efficiency of nitrogen and phosphorus and noted that their low efficiency due to over 80% of nitrogen and 25–75% of phosphorus on average (accounting for the amount not stored in agricultural soils) consumed winds up lost to the environment. Nutrient leaching from fertilizer applications is affected by many factors, including fertilizer type, rate, timing, and method of fertilizer application, soil properties (i.e., pH, structure and organic matter content), types of crops, plant fertilizer requirements and uptake of nutrients, agricultural practices; and the amount of animal production. In addition, the quantity of nutrient leaching is dependent on weather conditions and other land use within a watershed. In cropland areas, plant uptake is related to the availability of macro- and microelements in the soil, in particular the mass ratios between elements. Reduction of fertilization level or one of the elements may not reduce leaching of nutrients as a result of the unfavorable ratio of nutrients in soil. A deficiency of phosphorus or potassium limits the uptake of nitrogen by plants, even when the nitrogen level is sufficient (Lawniczak et al., 2016). This suggests that at a low level of fertilization due to shortage of potassium and phosphorus, there may occur loss of nitrogen, which results in water and soil pollution (Lawniczak et al., 2009). Losses of nitrogen and phosphorus to the environment from fertilizers and manure have resulted in various types of pollution including emissions of the greenhouse gas nitrous oxide (N_2O) and ammonia to the atmosphere and excess nitrate and phosphate to surface waters, groundwater, and ecosystems.

Comprehensive nationwide-scale investigations on the sources of nitrogen and phosphorus and their occurrence in streams and groundwater in the United States have been part of the US Geological Survey's National Water Quality Assessment (NAWQA) program (http://water.usgs.gov/nawqa). From 1991 to 2001, the NAWQA program completed interdisciplinary assessments to establish a baseline understanding of water-quality conditions in 51 of the river basins and aquifers in the United States referred to as Study Units by the NAWQA program. As part of the first decade of the NAWQA program, extensive information was collected on major sources of nutrients (nitrogen and phosphorus) from fertilizers, manure, atmospheric deposition, and sewage and tabulated for each county in the United States. Also, these data along with nutrient data for streams and groundwater were summarized by Dubrovsky et al. (2010). During the second phase of the program (2001–2012) 42 of the 51 Study Units were selected to determine the status and trends at surface water and groundwater sites that had been consistently monitored for more than a decade. Also during this second phase of the Program, several priority topics have been addressed, for example, the fate of agricultural chemicals, effects of urbanization on stream ecosystems, and effects of nutrient enrichment on aquatic ecosystems. The NAWQA program has produced several hundred publications that provide a better understanding how human activities affect water quality, links between sources of contaminants and their transport through the hydrologic system, and the potential effects of contaminants on humans and aquatic ecosystems. Many of the results from these studies focused on the co-occurrence of nutrients with other contaminants and are highlighted in this chapter. Additional information on the studies and publications from the NAWQA program is available on the website https://water.usgs.gov/nawqa/bib/ (accessed 2 June 2018).

As is the case in many parts of the world, Dubrovsky et al. (2010) found that commercial fertilizer was the largest single nonpoint source of nutrients in the United States. More than 10 million tons of nitrogen and 2 million tons of phosphorus were applied each year from the 1970s to the late 2000s. Most of the fertilizer applied was for agriculture, but it was estimated that about 2–4% was applied to residential lawns, golf courses, and other recreational areas. In the United States, the use of nitrogen fertilizers had increased by tenfold between about 1950 and the early 1980s; whereas phosphorus fertilizer use increased by fourfold during this period (Fig. 9.2). Maps showing the areal distribution of the input rate of nitrogen from fertilizer and manure across the United States in 1997 are shown in Figure 9.3.

More recent data (US EPA; 2007–2011) indicated similar inputs for nitrogen and phosphorus in fertilizers in the United States as shown in Figure 9.2 for the early 2000s. The increase in use of nitrogen and phosphorus

CO-OCCURRENCE OF NITRATE WITH OTHER CONTAMINANTS IN THE ENVIRONMENT 177

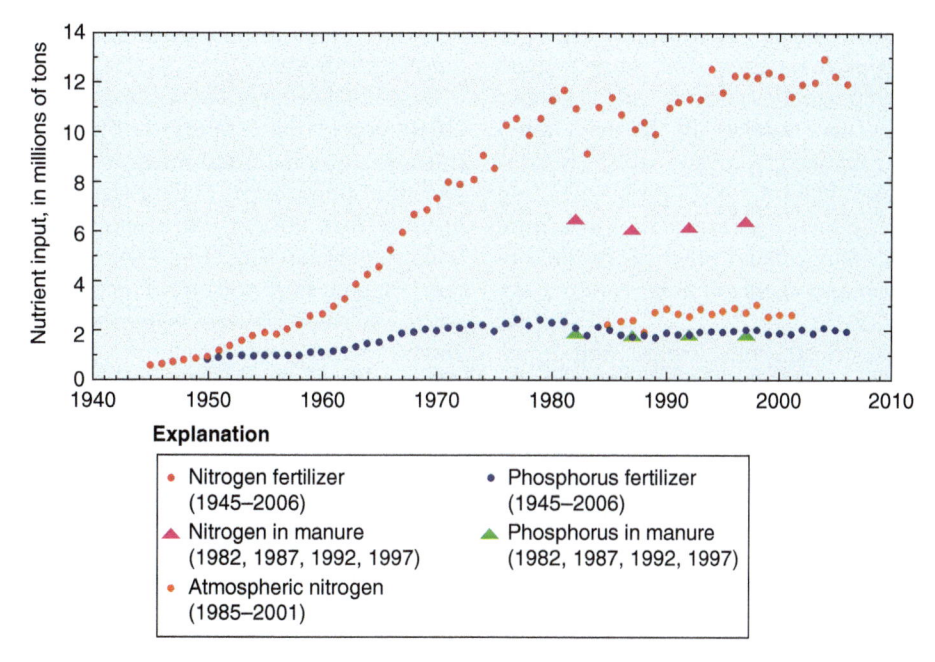

Figure 9.2 The use of nitrogen and phosphorus fertilizers in the United States has increased tenfold and fourfold, respectively, between about 1950 and the early 1980s. Since about 1980, however, applications of nitrogen and phosphorus fertilizers have leveled off and have remained relatively stable. Nitrogen and phosphorus from manure and nitrogen from the atmosphere also have remained relatively stable since the 1980s. *Source*: Modified from Ruddy et al. (2006).

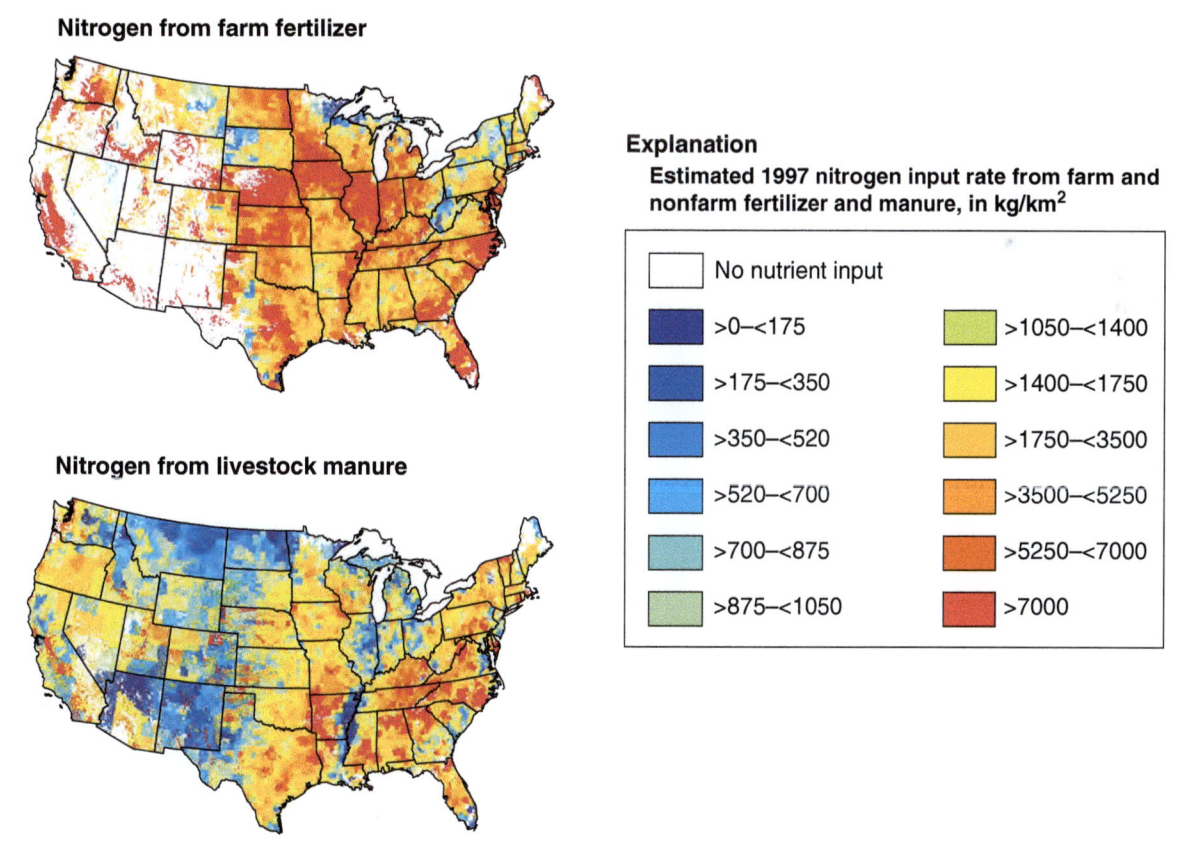

Figure 9.3 Estimated nitrogen input rate from farm and nonfarm fertilizer and manure. *Source*: Modified from colored figure 2–5 in Dubrovsky et al. (2010), public document; I converted pounds/mi^2 to kg/km^2. Reproduced with permission of United States Geological Survey.

fertilizers was related to higher application rates per hectare for major crops (e.g., corn) and more farmed area. Since about 1980, applications of nitrogen and phosphorus fertilizers have leveled off in the United States and have since remained relatively stable. This was partly due to a combination of factors including increased fertilizer costs and growing environmental concerns. Also, better management practices resulted in a decrease or a slower rate of fertilizer usage (Dubrovsky et al., 2010).

Along with usage estimates of nitrogen and phosphorus fertilizers and other nutrient sources, Dubrovsky et al. (2010) summarized nitrate and phosphorus concentrations in stream water and groundwater across the United States during 1991–2002. Sampling sites for streams and groundwater were selected among geographic areas that represented a wide range of physiographic and climatic settings, as well as different land uses associated with a variety of contaminant sources (e.g., agricultural, urban, and natural sources). Samples were collected monthly for two years at 499 stream sites during high-flow and low-flow conditions. More intensive sampling for nutrients (generally weekly or twice monthly) was conducted at a subset of sites during the time of highest runoff and use of agricultural chemicals. Groundwater samples for dissolved nutrients were collected from 5101 wells, including monitoring, domestic, and public-supply wells. Wells were sampled only once because of the comparatively slow rate of change in most groundwater systems, relative to streams.

In US groundwater, Dubrovsky et al. (2010) found that nitrate-N and orthophosphate showed different results for agricultural and urban land-use areas and from major aquifers. Concentrations of nitrate-N in groundwater were higher in agricultural land-use areas than in urban areas. Nitrate-N concentrations were higher in urban areas than in major aquifers. These differences likely are related to well depths. Wells sampled in agricultural and urban areas tend to be shallower than in major aquifers. Concentrations of phosphorus showed little differences among land-use categories and were lower than in streams and were close to background levels. They also noted that dissolved phosphorus concentrations were significantly higher in reduced groundwater than in oxic groundwater, and concentrations also were higher in groundwater with pH values greater than 7.5. Where median phosphorus concentrations were above background levels, dissolved phosphorus concentrations were significantly correlated to pH and to concentrations of ammonia, calcium, magnesium, and iron. In groundwater samples, even though nitrate concentrations exceeded background levels in 64% of 86 shallow aquifer studies sampled in agricultural and urban areas, phosphorus concentrations were very low, as phosphorus tends to be relatively insoluble in water and tends to adsorb onto soils and aquifer materials.

Based on the large NAWQA dataset, Dubrovsky et al. (2010) found that concentrations of all five nutrients (nitrate, ammonia, total nitrogen (TN), orthophosphate, and total phosphorus (TP)) exceeded background levels at more than 90% of 190 streams draining agricultural and urban watersheds (Fig. 9.4). Across the United States nutrient concentrations were higher in agricultural areas compared to urban areas (samples collected during 1992–2004) (Fig. 9.5).

Other factors have influenced nutrient concentrations in streams in the United States. For example, these factors include naturally occurring seasonal fluctuations in climate (which control streamflow conditions), uptake of nutrients by aquatic and riparian vegetation, and human factors (mainly related applications of fertilizer and manure and irrigation) (Dubrovsky et al., 2010). In the eastern half of the United States, TN concentrations at many stream sites were highest in the spring at higher streamflow conditions and at the time when fertilizer is applied. In contrast, TP concentrations were highest in the summer and autumn when streamflow was lowest. In the upper Midwest of the United States, both nitrogen and phosphorus concentrations were highest in the springs during high streamflow. In the western half of the United States, seasonal patterns were less prevalent, which were attributed to highly variable topography and climate and the numerous dams and canals. Phosphorus concentrations were generally highest during the summer (May through July), in western streams (mainly rangeland areas of the interior west United States), most likely related to high streamflows from snow melt and mobilization of sediment-bound phosphorus from erosion. Phosphorus tends to be naturally higher in geologic materials in western rangeland than in other parts of the United States (Fig. 9.6).

The transport of nitrogen and phosphorus in streams to downstream water bodies are controlled by the location of the nutrient sources within a watershed and the characteristics of the stream channel. The USGS has made numerous simulations of nutrient transport in watersheds using the SPARROW (SPAtially Referenced Regressions On Watershed attributes) model (https://water.usgs.gov/nawqa/sparrow/, accessed 10 June 2018), which relates in-stream nutrient loads to the locations of upstream nutrient sources and watershed characteristics that affect the delivery of nitrogen and phosphorus from 62,000 stream reaches to the major rivers and estuaries in the United States. Dubrovsky et al. (2010) noted that the amount of biological processing of nitrogen in streams was dependent on the surface area of the stream bottom (where the organisms live) in relation to the stream's volume, and a greater percentage of the nitrogen delivered

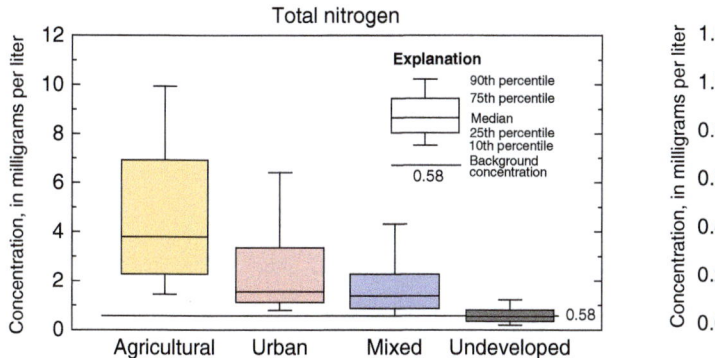

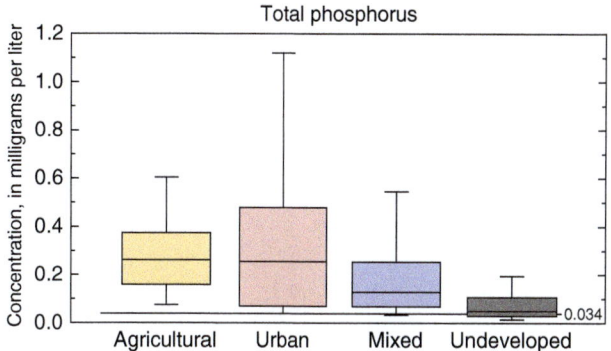

Figure 9.4 Boxplots showing concentrations of total nitrogen and phosphorus in agricultural, urban, mixed, and undeveloped land uses in the United States. Total nitrogen and phosphorus concentrations were highest in agricultural streams (median concentration of about 4 mg/L, which is about 6 times greater than background levels), although concentrations of TP were similar in agricultural and urban streams (median concentration of about 0.25 mg/L, which also is about six times greater than background levels). *Source*: Modified from Dubrovsky et al. (2010), public document. Reproduced with permission of United States Geological Survey.

Estimated 1997 nitrogen inputs from fertilizer and manure, in kg/km²
<515 | 515–2570 | >2,570

Estimated 1997 phosphorus inputs from fertilizer and manure, in kg/km²
<60 | 60–485 | >485

Data on nutrient inputs not available for Alaska and Hawaii

Stream sites By land use
Agricultural Urban

Total nitrogen concentration in streams, in milligrams per liter
Low (<0.66)
Medium (0.66–3.17)
High (>3.17)

Total phosphorus concentration in streams, in milligrams per liter
Low (<0.05)
Medium (0.05–0.28)
High (>0.28)

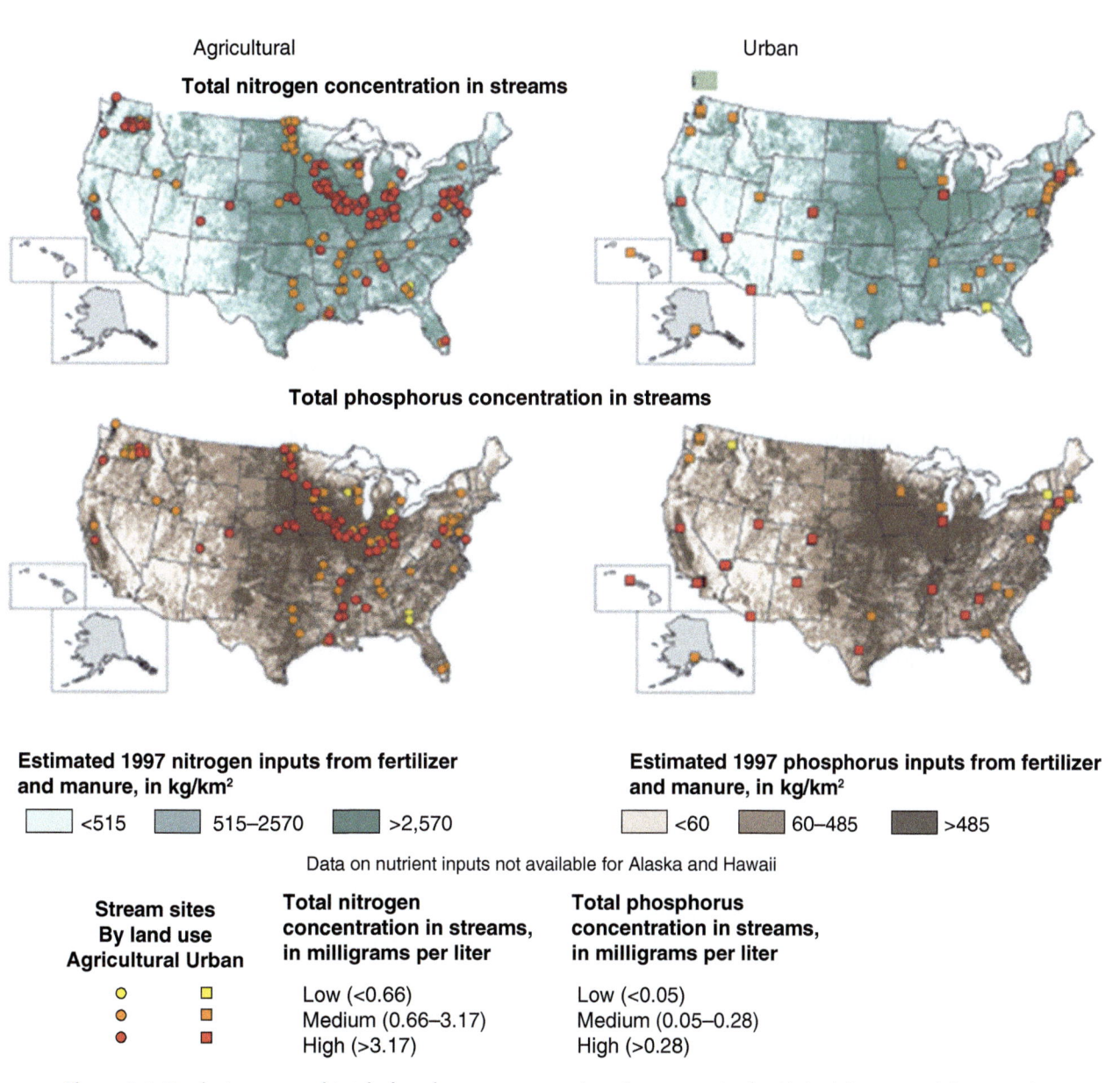

Figure 9.5 Total nitrogen and total phosphorus concentrations in streams in the United States overlain on estimated nitrogen and phosphorus inputs from fertilizer and manure in the United States in 1997. *Source*: Modified from colored figure 1-2 in Dubrovsky et al. (2010), public document; I converted units from pounds/mi² to kg/km². Reproduced with permission of United States Geological Survey.

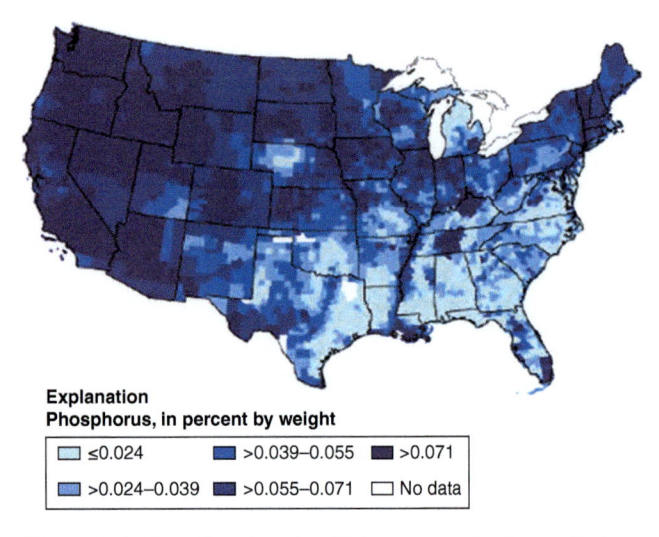

Figure 9.6 Map showing the higher concentrations of phosphorus from natural sources in the western rangelands compared to other parts of the United States. *Source*: Modified from colored figure 4-11 in Dubrovsky et al. (2010), public document. Reproduced with permission of United States Geological Survey.

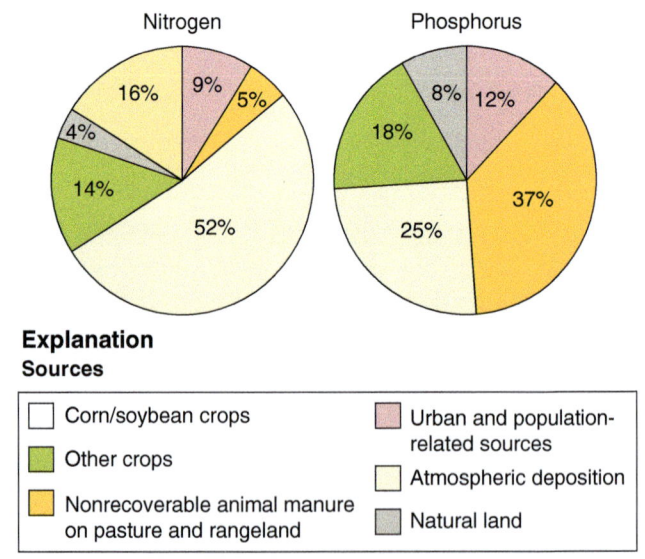

Figure 9.7 Pie charts showing the major sources contributing nitrogen and phosphorus to the Gulf of Mexico. The largest contributor of nitrogen was commercial fertilizer use on corn and soybeans. The largest contributors of phosphorus were corn and soybean cultivation and animal manure on pasture and rangelands. *Source*: From Dubrovsky et al. (2010), based on data from Alexander et al. (2008), public document. Reproduced with permission of United States Geological Survey.

to large deep rivers is typically exported to downstream water bodies than nitrogen delivered to small streams. They also found that corn and soybean cultivation were the largest sources contributing nitrogen annually to the Gulf of Mexico. In contrast, for phosphorus, animal manure applied to pasture and rangelands along with corn and soybean cultivation were the largest contributors to the Gulf of Mexico (Fig. 9.7).

Web-based interactive maps can be accessed for river sites in the United States that show trends for nutrients, including orthophosphate (filtered and unfiltered), TP, ammonia, nitrate, and TN, from 1972 to 2012 (https://nawqatrends.wim.usgs.gov/swtrends/, accessed 10 June 2018); and described in detail by Sprague (2017). An example for TP is shown in Figure 9.8. By clicking on a trend symbol, detailed information can be obtained for the site.

In Europe, Billen et al. (2011) reported modeled nitrogen fluxes from fertilizers and manure (before landscape retention) to five large river basins in Europe. Fluxes of nitrogen (in kg $N/km^2/yr$) were Danube (930), Rhine (2660), Weser (3500), Elbe (2300) and Odra (1430). Nitrogen delivered to coastal systems at the European scale was 4750 kton N/yr, which was an increase of more than fourfold compared to pristine conditions, and approximately an increase of a factor of three with respect to pre-1950 conditions. At the same time, phosphorus delivery increased, but has been decreasing again close to preindustrial levels, as a result of effective phosphorus removal in urban wastewater treatment instituted in many European countries.

Urban stormwater runoff typically contains elevated levels of nitrogen and phosphorus, which are important pollutants of surface waters (Line et al., 2002; Mallin et al., 2009; US EPA, 1983). Stormwater runoff of phosphorus generally occurs from animal wastes, the erosion of particulate phosphorus adsorbed on soils, and surface or subsurface movement of dissolved phosphorus. Nitrogen in stormwater runoff originates from fertilizers (chemical and manure), animal waste disposal (e.g., lagoons at concentrated animal feedlots), leaky sewer lines, and septic tanks. Nutrients (mainly nitrogen and phosphorus) in urban stormwater runoff have contributed to hypoxia in rivers, lakes, and streams (Diaz & Rosenberg, 2008; Mallin et al., 2006) and have caused large algal blooms in coastal waters (Lopez et al., 2008; Scavia & Bricker, 2006) and have created dead zones in Puget Sound, the Chesapeake Bay, Long Island Sound, and many other waterways (Howard, 2014).

9.1.2. Manure N and P

Animal manure is a primary source of nitrogen and phosphorus to surface and groundwater (https://www.epa.gov/nutrient-policy-data/estimated-animal-agriculture-nitrogen-and-phosphorus-manure, accessed 07 May 2019) in agricultural areas. The application of livestock manures also can release emissions of ammonium, methane, or

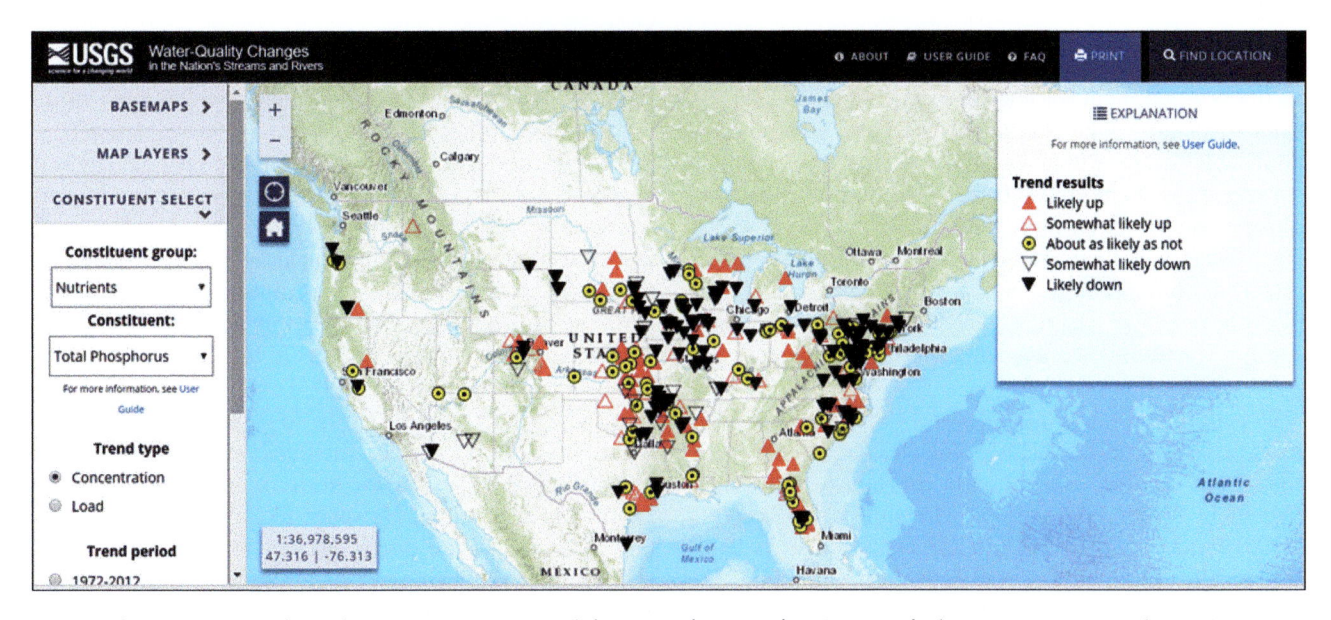

Figure 9.8 Snapshot of an interactive map of the United States showing trends from 1972 to 2012 for TP in selected streams and rivers in the United States (see Sprague (2017) for more information). https://nawqatrends.wim.usgs.gov/swtrends/, accessed 10 June 2018).

sulfide. Manure runoff from cropland and pastures or discharging animal feeding operations and concentrated animal feeding operations (CAFOs) often reaches surface and groundwater systems through surface runoff or infiltration.

Sutton et al. (2013) noted that when livestock is part of the food chain, overall nutrient use efficiency is reduced. This in turn has led to large releases of nitrogen and phosphorus to the environment and requiring more of these nutrients to sustain the human population than would be needed for plant-based foods. Globally, Sutton et al. (2013) estimated 80% of nitrogen and phosphorus in crop and grass harvests that feeds livestock provides only around 20% (15–35%) of the nitrogen and phosphorus in human diets, which further demonstrate the lack of nutrient use efficiency.

Erisman et al. (2011) showed a large increase in manure production in Europe and worldwide since the 1950s. Worldwide manure production for nitrogen in 2005 was approximately 140,000 ktons/yr and about 11,000 ktons N/yr in Europe. Based on N:P ratios in various types of manure, we estimated worldwide production of phosphorus in manure in 2005 was approximately 98,000 ktons/yr, and 7000 ktons/yr in Europe. The worldwide manure amounts were based on the following 2005 animal population estimates for 1000 head of poultry (15,146,608), pigs (917,635), and cattle (1,310,611) (Steinfeld et al., 2006).

In the United States, nitrogen and phosphorus from manure increased from the mid-1940s to 1980, but these amounts have remained relatively stable since the 1980s (Fig. 9.2). In the United States, there were approximately 6 million tons of nitrogen and nearly 2 million tons of phosphorus excreted as manure each year. Dubrovsky et al. (2010) noted that the mass of nitrogen in manure produced by livestock is about one-half as much as that applied in fertilizer and has varied little since data became available in 1982. The mass of phosphorus in manure is about the same as phosphorus applied in fertilizer, and also has varied little since 1982. Heilmann et al. (2014) noted that large-scale farming operations in the United States annually produce amounts of manure that contain the equivalent of 1.2 billion kg of P_2O_5, a measure commonly used to describe phosphorus content in fertilizer (based on 2013 data from the US Department of Agriculture, that included 345 million poultry layers, 41 million swine, and 9.2 lactating cows). The US EPA presents more detailed estimates of nitrogen and phosphorus in manure for different animal types in all states across the United States (https://www.epa.gov/nutrient-policy-data/estimated-animal-agriculture-nitrogen-and-phosphorus-manure).

9.1.3. Wastewater Nitrogen and Phosphorus

Wastewater-effluent discharges from point sources such as municipal wastewater treatment plants in the United States (also referred to as publicly owned treatment works, POTW) can contain large quantities of nitrogen and phosphorus, which degraded water quality in many areas (e.g., https://www.epa.gov/nutrientpollution/sources-and-solutions-wastewater, accessed 07 May 2019). In the United States, wastewater

182 NITROGEN OVERLOAD

treatment for surface water discharges is regulated through federal and state permits, which require secondary treatment (biological oxygen demand [BOD]), total suspended solids (TSS and pH), and adherence to applicable water quality standards. However, only a subset of POTW permits currently contain nitrogen and phosphorus limits. Of more than 16,500 municipal wastewater treatment facilities in the United States, approximately 4% have numeric limits for nitrogen and 9.9% for phosphorus (US EPA, 2009) (https://www.epa.gov/sites/production/files/documents/nitgreport.pdf accessed 2 June 2018).

Depending on the local ecological conditions and their relative contribution, these point discharges can be a significant source of nutrients in many watersheds (Manuel, 2014). For example, Preston and Brakebill (1999) found point-source discharges of nutrients to be statistically significant ($p < 0.005$) for estimating the spatial distribution of TN loading in streams of the Chesapeake Bay watershed in the eastern United States. Wastewater containing laundry detergents was a major source of phosphorus to surface waters from about 1940 to 1970. Contributions of phosphorus in the United States decreased from a peak of 220,000 metric tons in 1967 to less than 10,000 metric tons in 1995 (Dubrovsky et al., 2010). This decrease was attributed to the enactment of State bans on the use of phosphate detergents beginning in the 1970s and the voluntary change by manufacturers in the formulation of detergents. In the Netherlands, wastewater treatment plants contribute about 34% of the annual phosphorus load and about 14% of the annual nitrogen load, while agriculture contributes 62% for phosphorus and 41% for nitrogen.

Maupin and Ivahnenko (2011) developed a national database for the United States that documented point-source nutrient loads to surface waters, based on enhancements to the methods developed by McMahon et al. (2007). They summarized point-source nutrient load data from the United States EPA's Permit Compliance System national database for six regions in the United States for 1992, 1997, and 2002 (Fig. 9.9). Total nitrogen (TN) and total phosphorus (TP) loads to surface waters from municipal and industrial facilities were estimated for approximately 118,250 facilities in 45 states and the District of Columbia. However, in some cases, inconsistent and incomplete discharge locations, effluent flows, and effluent nutrient concentrations limited the use of these data for calculating nutrient loads. They noted that more concentrations were reported for major facilities (discharging more than 3.8 million liters per day), than for minor facilities, and more concentration data were reported for TP than for TN (Table 9.1). Where there were missing concentrations, annual loads were calculated using "typical pollutant concentrations" based on the type and size of facilities. Annual nutrient loads for over 26,600 facilities were calculated for at least one of the three years. Sewage systems represented 74% of all total loads and 58% of all TP loads. This work represents

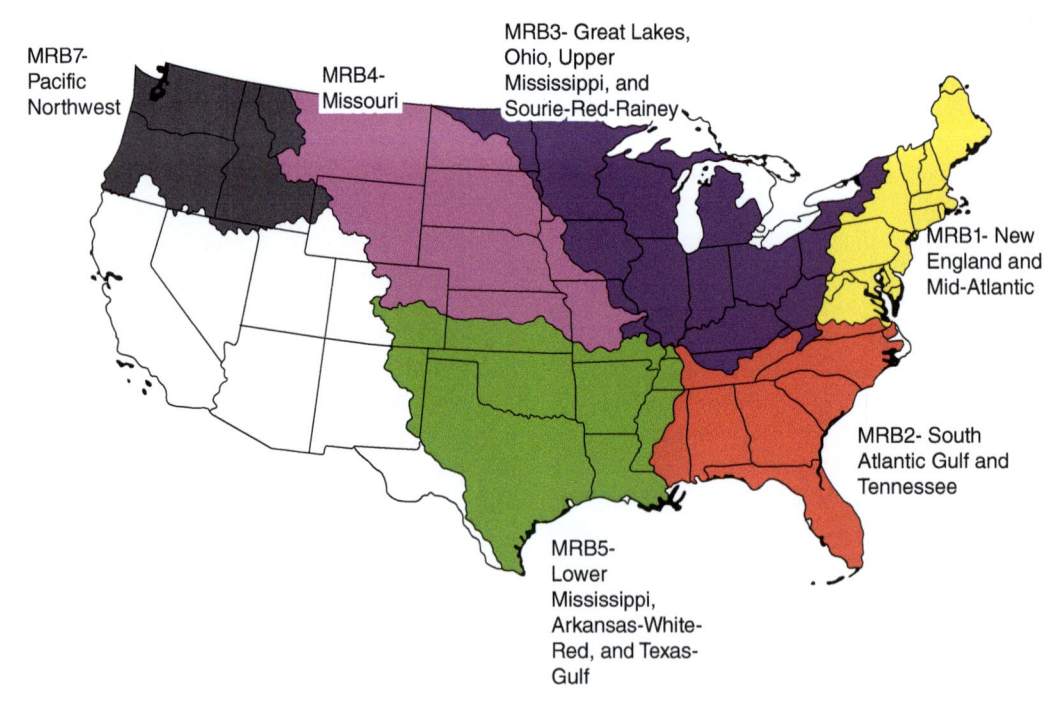

Figure 9.9 Major river basins evaluated for nitrogen and phosphorus load calculations *Source*: Modified from colored figure 1 in Maupin and Ivahnenko (2011). Reproduced with permission of John Wiley & Sons.

Table 9.1 Point-source nutrient loads (2002) for all facilities, and distribution of total nitrogen (A) and total phosphorus (B) loads among major river basins in the United States.

(A) Total nitrogen (TN)

MRB	Total TN load (kg/yr), all dischargers	Total TN areal load (kg/km²/yr), all dischargers	Median facility TN load (kg/yr)	Percentage of total TN load from minor facilities	Percentage of facilities that are sewage systems (SIC 4952)	Percentage of TN loads that were from sewage systems
1	120,799,415	272.2	4985	6.5	70.7	86
2	56,211,267	69.6	1450	8	61.2	77.6
3	252,055,066	183.7	668	13.3	50.5	74.6
4	36,888,359	27.9	906	47.4	63.6	50.7
5	87,153,571	62.9	1470	11.5	64.6	62.6
7	23,074,312	32.1	9992	5.7	59.1	82.4
Total	**576,181,990**	**95.3**	**1291**	**13**	**57.7**	**74.2**

(B) Total phosphorus (TP)

MRB	Total TP load (kg/yr), all dischargers	Total TP areal load (kg/km²/yr), all dischargers	Median facility TP load (kg/yr)	Percentage of total TP load from minor facilities	Percentage of facilities that are sewage systems (SIC 4952)	Percentage of TP loads that were from sewage systems
1	12,845,111	28.9	386	2.4	70.7	85.3
2	8,798,645	10.8	240	12.0	61.2	66.4
3	18,920,674	13.8	109	28.0	50.5	63.6
4	7,915,609	6.0	113	59.6	63.6	27.4
5	11,833,615	8.5	190	22.4	64.6	37.9
7	4,196,267	5.8	1665	14.3	59.1	56.0
Total	**64,509,921**	**10.6**	**161**	**22.7**	**57.7**	**58.7**

Source: From Maupin and Ivahnenko (2011). Reproduced with permission of John Wiley & Sons.

an initial set of data to develop a comprehensive and consistent national database of point-source nutrient loads. These estimated nitrogen and phosphorus loads have been used for water-quality management, watershed modeling, and research efforts at multiple scales in the United States (https://pubs.usgs.gov/sir/2017/5115/sir20175115.pdf, accessed 20 June 2018).

In their article on the need to reevaluate wastewater treatment plant effluent standards for nutrient reduction or nutrient control, Hendriks and Langeveld (2017) point out that eutrophication of freshwater systems is not controlled solely by nitrogen limitation. Eutrophication also could solely be controlled by phosphorus limitation. They state that a low N:P ratio favors the growth of blue-green algae with nitrogen fixing capacities compared to other algae. Also, low nitrate concentrations can contribute to an increase in release of phosphorus from the sediment, which in turn reinforces a low N:P ratio. As the relative abundance of blue-green algae in the algae community increases, this causes a decrease in the grazing pressure of zooplankton because the blue-green algae negatively impact the zooplankton. These results in a dominance of blue-green algae even though there are relatively low nutrient concentrations. The N:P ratio in the freshwater system should stay high enough to prevent a growth advantage for blue-green algae.

9.1.4. Septic Systems

In many areas, septic systems (sometimes referred to as On-Site Treatment and Disposal Systems, OSTDS) are a major nonpoint source of nitrogen and phosphorus. It is estimated that about 25% of the population in the United States relies on a septic system for waste disposal. The average input of nitrogen and phosphorus to a septic tank are 11.2 and 2.7 g per person per day (Lusk et al., 2017). Phosphorus generally is not a concern for groundwater contamination. Most of the organic phosphorus that leaches from a septic tank drainfield is mineralized to orthophosphate (by biological and biochemical processes) or sorbed to soil surfaces. Orthophosphate typically is sorbed by organic matter, clay minerals, and metal oxides and oxyhydroxides in the soil. In anaerobic conditions, phosphorus may precipitate to form minerals with positively charged ions such as iron, calcium, magnesium, and aluminum. Other factors that could result in increased phosphorus concentrations in groundwater include desorption/dissolution, high hydraulic loading rate, or a lack of clogging zone/biomat in the soil beneath the drainfield (Lusk et al., 2017).

Phosphate pollution of surface waters from septic systems can occur if transported in a groundwater plume (especially in shallow groundwater systems), from soil erosion, or short circuiting due to preferential flow paths.

If the surface water system is phosphorus limited, phosphorus levels (as low as 0.03 mg/L) could result in eutrophication. In a literature review, Lusk et al. (2017) noted that 4–25% of all phosphorus loading to surface waters may originate from septic systems. Lakes are particularly susceptible to phosphorus loading from septic systems. Studies in the United States and Canada have indicated that the TP contribution from septic systems to lakes ranged from 4 to 55% (Lusk et al., 2017).

Harman et al. (1996) collected water quality data from 400 groundwater sampling points to delineate a plume from a 44-year old septic system in Ontario, Canada. In parts of the plume, nitrate and phosphate were elevated, and the migration distance of phosphate in groundwater was found to be greater than at other younger septic system sites. Mineral precipitation reactions controlled the attenuation of phosphate in the unsaturated zone.

9.2. NITRATE AND PATHOGENS

In surface water systems, many studies have documented the co-occurrence of nitrate, bacteria, and related pathogens. Urban stormwater runoff has been described as a major pollution source to fresh and brackish receiving waters, especially nutrients and fecal microbial pathogens (Mallin et al., 2009; Smith & Perdek, 2004). One of the most comprehensive studies of urban stormwater runoff was the Nationwide Urban Runoff Program (NURP) conducted during 1978–1983 by the US EPA in cooperation with many federal, state, regional, and local agencies across the United States. NURP considered stormwater runoff as rainfall that falls on impervious surfaces (e.g., roads, driveways, parking lots, rooftops, and other paved surfaces) and flows over the ground surface into retention areas (e.g., ponds, basins, and wetlands) and into receiving water bodies (including streams, rivers, lakes, estuaries, embayments, and aquifers). This program investigated the characteristics of urban runoff in relation to different urban land uses, the effects of urban stormwater runoff on receiving waters, and the performance characteristics and effectiveness of management practices to control pollution loads from urban runoff. There were 28 NURP projects that included 81 specific sites, and water-quality information was collected for more than 2300 separate storm events. Urban stormwater runoff contained significant concentrations of solids, nutrients, organics, and metals. The main sources of pathogens (fecal coliform) and nutrients in stormwater included animal wastes, illegal wastewater connections into storm sewer lines, seepage of groundwater containing pathogens into storm sewer lines, and septic tanks. For extensive information about the results of this program, please visit this site: (https://www3.epa.gov/npdes/pubs/sw_nurp_exec_summary.pdf, accessed 18 June 2018).

Wilkes et al. (2013) collected approximately 690 raw surface water samples during a 6-year period from multiple watersheds in the South Nation River basin, Ontario, Canada. They found that livestock associated *Cryptosporidium* and higher bacterial pathogen occurrence in the fall when stream/river discharge and nitrate concentrations in water were relatively higher. Livestock manure is frequently applied in the spring and/or fall in that region, and field amendments can be introduced into surface water from runoff and artificial subsurface drainage systems, which are common in the area and tend to drain more frequently in fall and spring.

The co-occurrence of nitrate and pathogens (including bacteria, viruses, helminths, and protozoa) in wastewater from septic systems was reviewed by Lusk et al. (2017). They noted that the mobility and fate of pathogens in soils is controlled by physical, biological, and electrostatic/chemical processes. Groundwater contamination from septic system pathogens are related to factors such as septic tank density and the susceptibility of underlying aquifers to contamination. A study by Macler (1996) reported that 60–70% of groundwater in the United States has been contaminated with wastewater-derived viruses and or bacteria at some time. Pathogen removal is greatest for helminth ova, protozoa, and bacteria, but least for viruses. In soils, the fate of pathogens is controlled by sorption to soil surfaces. Lusk et al. (2017) noted that retention of pathogens by soil surfaces is higher than pathogen die-off, and irreversible sorption on soils is a very effective process in protecting groundwater resources from septic system pathogens. Soil beneath most drainfields is generally effective in removing pathogens from septic systems, except in vulnerable environments, such as karst aquifers (Katz et al., 2010), gravel, or fractured rock aquifers (https://www.epa.gov/sites/production/files/2015-06/documents/2006_08_28_sourcewater_pubs_septic.pdf, accessed 07 May 2019). Thus, pathogen retention by soils is much greater than retention of nitrate.

Gosselin et al. (1997) investigated the presence of nitrates, pesticides, and coliform bacteria from septic tanks in domestic wells in an alluvial sediment aquifer in Nebraska, US. They found that nitrate-N in water from domestic wells exceeded 10 mg/L (maximum contaminant level in drinking water) in 39% of samples and coliform bacteria were detected in 26% of the samples.

Verstraeten et al. (2005) studied the occurrence of bacteria and nitrate in domestic wells near septic tank sources. They found that bacteria were not detected in any of the domestic well samples, which indicated there was sufficient residence time to allow for die-off. However, indicator viruses (male-specific coliphages) were detected in 2 of the 19 ground water samples from sandpoint wells drilled within 25 m of a septic field, with depths of 8 and 12 m. They concluded that the detection or abundance of bacteria and nitrate-N alone was not good indicator of septic waste contamination because bacterial transport to groundwater is related to sediment type, travel times, and clogging. Coliphages and other indicator viruses were more indicative of waste contamination of groundwater from human or animal sources because they are smaller, survive longer than bacteria, and typically have a lower tendency to adsorb because of their negative charge.

Based on a study of three septic tank systems in a mantled karst aquifer, Katz et al. (2010) found elevated nitrate concentrations but only sporadic detections of fecal coliforms, enteroviruses, pharmaceuticals, and organic wastewater compounds in soil water from shallow and deep lysimeters in septic tank drainfields. Higher detections of fecal indicators and enteric viruses were found in groundwater at a site where the depth to the limestone aquifer was the shallowest and average daily homeowner water use highest. Katz et al. (2010) concluded that enterovirus data in their study were consistent with the movement of viruses from septic tanks found in other studies, indicating that there is a potential for contamination of drinking water supplies from septic tanks associated with a potential human health risk in karstic aquifer systems.

Tuthill et al. (1998) investigated the relationship of coliform bacteria and nitrate levels to lot size and casing length for wells in unsewered areas ($n = 832$) and different soil groups in Frederick County, Maryland. They found that coliform bacteria and nitrate contamination were negatively correlated with lot size. In addition, coliform bacteria levels were negatively correlated with casing length, and they found an association between nitrate levels and casing length. The study concluded that the casing length required in well construction should be increased in areas where wells may be prone to coliform bacteria contamination.

9.3. TRACE ELEMENT CONTAMINATION ASSOCIATED WITH NITRATE

In two major aquifers of the United States (the High Plains (HP) and Central Valley (CV) (Fig. 9.10) J. Nolan and Weber (2015) collected groundwater geochemical data on nitrate, uranium, and other constituents. The HP and CV aquifers are two of the largest and most productive aquifers in the world and supply drinking water to nearly 6 million people. Water from both aquifers contains naturally high concentrations of dissolved uranium (that exceeded the US Environmental Protection Agency maximum contaminant level (MCL = 30 µg/L). Both aquifers contain unconsolidated sand and gravel composed of silt, sands, poorly sorted clays, and gravel; materials deposited through weathering and alluvial processes.

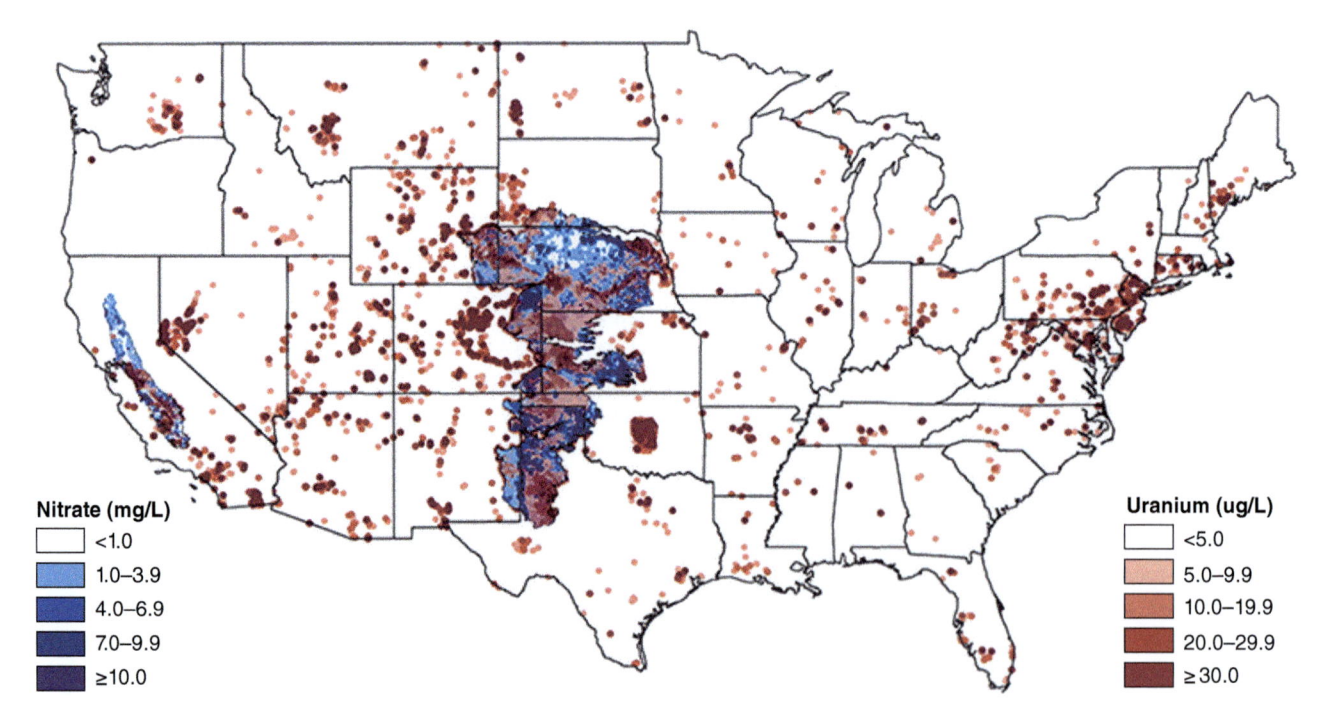

Figure 9.10 Map showing concentrations of nitrate and uranium in the HP and CV aquifers and other locations around the United States. *Source*: From J. Nolan and Weber (2015). Reproduced with permission of American Chemical Society.

They found a moderately strong correlation between uranium and nitrate, as well as between uranium and alkalinity and calcium. J. Nolan and Weber (2015) noted that nitrate can alter uranium solubility by oxidative dissolution of reduced uranium (IV) minerals. They reported that about 78% of areas where U concentrations were interpolated above the MCL were significantly ($p < 0.05$) correlated to the presence of nitrate. Furthermore, shallow groundwater was found to be the most susceptible to co-contamination, and they concluded that nitrate is an important factor that may lead to secondary uranium contamination of groundwater in addition to the known uranium solubility related to alkalinity and calcium.

Banning et al. (2013) also reported enhanced mobility of uranium when nitrate was present in a Quaternary gravel aquifer in Germany. They assembled a large data set containing solid materials (164 samples of sediments and soils) and 114 samples of groundwater to investigate geo- and hydrochemistry, mineralogy, and uranium mobilization potential. Uranium concentrations exceeded the new German drinking water standard (22% of the samples had uranium concentrations greater than 10 μg/L). Even though no anthropogenic uranium input was detected in their study area, they found that agricultural usage of the moor areas in their study area resulted in geogenic release of uranium via nitrate fertilization, surface peat degradation, and erosion.

Guttormsen et al. (1995) show that increased amounts of cadmium (Cd, a potentially hazardous heavy metal) in nitrogen-phosphorus-potassium (NPK) fertilizers applied to Chinese cabbage and carrots in a sandy soil (pH adjusted to 5.5 and 6.5) increased the Cd concentrations in both vegetables, although differences among treatments were not significant. Anthropogenic sources of Cd in terrestrial environments include the application of sewage sludge, and commercial fertilizers, and naturally from weathering of soil minerals. Gray et al. (2002) also reported that different rates of nitrogen fertilizer application affected the Cd concentration in durum wheat grain grown in a study in New Zealand. They report a mean Cd concentration of 66 μg/kg fresh weight, which was 30% below the maximum allowable level.

As mentioned in the previous section, stormwater runoff is an important source of trace elements and nitrate to surface waters and groundwater. In addition to elevated levels of nutrients (nitrogen and phosphorus), all 13 metals on the US EPA's priority pollutant list were found in urban stormwater runoff during the Nationwide Urban Runoff Program in the United States. In particular, copper, lead, and zinc were the most commonly detected metals. Based on the extensive dataset for total copper, lead, and zinc, event mean concentrations (in μg/L) were 34, 144, 160, respectively. High levels of copper and zinc at some sites were attributed to the effect of acid rain on materials used for gutters, culverts, and other metal surfaces (https://www3.epa.gov/npdes/pubs/sw_nurp_exec_summary.pdf, accessed 18 June 2018).

9.4. NITRATE AND PESTICIDES

About 2.7 billion kilograms of pesticides were used worldwide annually in both 2011 and 2012 (Atwood & Paisley-Jones, 2017). Between 2008 and 2012, herbicides accounted for the largest portion of global usage (approximately 50% annually in all years), followed by fumigants, insecticides, and fungicides, respectively. In the United States more than 500 million kilograms of pesticides were used annually in both 2011 and 2012, with herbicides accounting for nearly 50% of total United States pesticide usage in 2011 and nearly 60% of usage in 2012. During 2008–2012, the US pesticide usage accounted for approximately 23% of total global amount of pesticides applied, 25% of the total herbicides applied, 43% of total fumigants applied, 12% of fungicides applied, and 6% of insecticides. Because most of the pesticide usage is for agricultural production, one might expect to find an association between nitrate and pesticide contamination in groundwater. However, their co-occurrence is related to various physical characteristics of soils and aquifers, chemical properties, and aquifer redox conditions. For example, under anoxic conditions, nitrate can undergo denitrification by microbial reduction; whereas pesticides tend to degrade more slowly under these conditions and persist in denitrifying environments. Several studies have detected atrazine residues in the absence of nitrate concentrations (e.g., Koterba et al., 1993).

Barbash and Resek (1996) listed several studies that showed that nitrate concentrations in groundwater generally were not reliable predictors of pesticide concentrations. However, more recent studies have found significant correlations between pesticide and nitrate concentrations in groundwater and have used nitrate concentrations to predict the occurrence of the herbicide in groundwater. A recent joint study by the Minnesota Departments of Health and Agriculture collected water-quality data including nitrate and 135 pesticides and degradates at 108 community public water systems throughout the state (Johnson et al., 2016). They found that 35% of wells sampled had nitrate-N concentrations above background (1.0 mg/L) and 19% of wells had nitrate-N concentrations equal to or greater than 3.0 mg/L. Generally, nitrate concentrations were higher in water from wells with pesticide detections compared to those without any pesticide detections. In fact, the probability of detecting a pesticide was higher where nitrate was detected. However, the relation between the occurrence of nitrate and pesticides was not entirely consistent, as some wells with low or no nitrate concentrations had pesticide detections, and some wells with high nitrate had low pesticide concentrations. Metolachlor ESA (herbicide degradate) was most frequently detected (63% of wells sampled), but other herbicides detected in more than 10 wells included metolachlor, metolachlor OXA, acetachlor ESA, and alachlor ESA. The study concluded that a cumulative assessment of pesticides and other chemicals detected in these community water systems did not pose a health risk of concern (when chemicals were combined that have a common health endpoint and duration period).

Selected examples of the co-occurrence between nitrate and pesticides in groundwater and surface waters are provided below. Burow et al. (1998) investigated the occurrence of nitrate and pesticides in groundwater beneath different agricultural cropland (vineyard, almond, corn, alfalfa, and vegetable land use) in California, United States. They found that the nematicide 1,2-dibromo-3-chloropropane (DBCP) concentrations were positively correlated with nitrate concentrations in the vineyard land use; however, DBCP was not detected frequently enough to assess the co-occurrence with nitrate concentrations in other agricultural settings. Burow et al. (1998) also noted that nitrate and the (DBCP) were correlated with dissolved oxygen concentrations, which indicated that oxic ground water likely is more vulnerable to contamination from both nitrate and pesticides. In a later study, Burow et al. (2008) studied the spatial and temporal trends in concentrations of nitrate and pesticides in groundwater in the eastern San Joaquin Valley, California, to investigate the long-term effects of agricultural and urban development on regional ground-water quality. Regional trends in concentrations of nitrate, DBCP, and the herbicide simazine during the past 20 years decades were generally consistent with known nitrogen fertilizer and pesticide usage. Higher concentrations of nitrate and pesticides were found in the shallow part of the aquifer system where domestic wells typically are screened. Conversely lower concentrations were found in the deep part of the aquifer system where public supply wells have been installed. They also noted that concentrations of nitrate have increased in the shallow and deep parts of the aquifer system since the 1950s. Burow et al. (2008) concluded that since attenuation processes do not appear to affect nitrate concentrations and pesticide detections, contaminant concentrations in the deep part of the aquifer system may increase as higher amounts of contaminated water are withdrawn by these wells over time. Contaminant concentrations in public supply wells likely will show the effects of the past 40–50 year management practices due to the long travel time of water from the shallow aquifer to the deep aquifer.

A comparison of the fate and transport of agricultural chemicals was studied at a diverse group of five small-to intermediate-sized watersheds in representative agricultural settings in the United States with different hydrologic and climatic conditions, and agricultural management practices (Domagalski et al., 2008). The study areas were in California, Washington, Nebraska, Indiana, and Maryland and ranged in size from 5.5 to 950 km^2. The site in Indiana was the only one with tile drains and no irrigation. Crops grown included corn and soybeans (Nebraska, Indiana, and Maryland); row crops, vineyards, orchards, dairies (Washington); and almonds, vineyards, row crops (California). They found that in most settings, overland flow contributed the greatest loads of nitrate and pesticides. However, a substantial amount of the annual load of nitrate and some pesticide degradates to streams was transported during base-flow conditions (mainly groundwater discharge).

In an article about declining water quality in the Mississippi Basin, Howard (2014), reported that about 57% of streams in agricultural areas and 83% of streams in urban areas have pesticide levels that exceed EPA guidelines for aquatic health. About 10% of agricultural streams and 7% of urban streams show pesticide levels above EPA guidelines for human health (https://news.nationalgeographic.com/news/2014/04/140411-water-quality-nutrients-pesticides-dead-zones-science/, accessed 10 June 2018).

As mentioned previously, groundwater and springs in karst areas are highly vulnerable to contamination, especially in agricultural areas. For example, Panno and Kelly (2003) estimated the mass loading of nitrate and herbicides in two large karst springs discharging from groundwater basins in an agricultural area in Illinois, US. Based on data from two years of monitoring, it was estimated that the following loads in kg/yr were discharging from the two springs: nitrate-N, 100,000; atrazine, 39; and alachlor, 2.8. Panno and Kelly (2003) noted that about half of the discharged nitrate originated from fertilizers and the nitrate load represented a 21–31% loss of fertilizers from the groundwater basins. In contrast, pesticide losses were 3.8% of the atrazine applied, and 0.05–0.08% of the alachlor applied.

In another study of spring water quality in a karst area, Kingsbury (2008) investigated the seasonal variability in concentrations of nitrate, pesticides, selected pesticide degradates (metabolites) in two karst springs in the Mississippian Carbonate Aquifer of northern Alabama. One spring with a shallow flow system had a relatively short average groundwater residence time. The larger other spring had a deeper flow system with a longer average groundwater residence time. Seasonal variations of nitrate and pesticide concentrations were similar for the two springs even though other water-quality data indicated differing short-term responses to rainfall at the two springs. Data collected during November to March, when recharge rates increased, indicated that nitrate and residual pesticides in the soil, unsaturated zone, and storage within the aquifer are transported to the spring discharges. Degradates of the herbicides atrazine and fluometuron were detected at concentrations comparable to or greater than the parent pesticides.

Molenat and Gascuel-Odoux (2001) investigated the transport of nitrate and herbicides from shallow groundwater in saprolite to a stream in an agricultural catchment in French Brittany. Based on modeled results, they concluded that shallow groundwater is a major source of stored nitrate as a result of groundwater residence times up to three years. However, some of the herbicides, particularly the triazines, were transported rapidly through the soil to groundwater and groundwater discharge to streams.

In a study of nitrate and pesticides in groundwater in the Mid-Atlantic region during 1985–1996, Ator and Ferrari (1997) found that the occurrence of nitrate and pesticides was related to rock type and land cover. Data from more than 850 sites were analyzed, and the most commonly detected pesticides (in about 50% of samples) included atrazine, metolachlor, prometon, simazine, and desethylatrazine, an atrazine degradate, but rarely at concentrations exceeding established MCLs. Groundwater from carbonate rocks was more vulnerable to contamination and typically had the highest detected nitrate and pesticide concentrations; although groundwater samples from crystalline rocks also contained higher concentrations of nitrate and pesticides than water samples from unconsolidated or siliciclastic aquifers.

Ator and Denver (2012) estimated the regional flux of nitrate and selected herbicides from groundwater to nontidal headwater streams of the Atlantic Coastal Plain from New Jersey through North Carolina. Samples were collected from 174 streams selected randomly during late-winter or spring base-flow conditions. Estimated fluxes were made for 8834 headwater streams in the study area and indicated 21,200 kg/d of nitrate-N, and 5.83, 0.565, and 20.7 kg/d of alachlor, atrazine, and metolachlor (and selected degradates), respectively. Large differences (ranging from less than 2% to about 70%) were found in the percentage of applied nitrate that was transported from groundwater to streams. Ator and Denver (2012) concluded that fluxes of nitrate and herbicides as a percentage of applications during base flow generally was highest in well-drained areas and lowest in areas with abundant poor drainage and anoxic conditions.

Puckett and Hughes (2005) investigated the transport and fate of nitrate and selected pesticides and their degradates in a surficial aquifer and a hydraulically connected stream in a study area in South Carolina, US. Their study area contained a series of low relief, eroded escarpments with low-gradient streams and swamps. Their study demonstrated the important role of hydrologic conditions and riparian zone processes on the fate of nitrate and the persistence of pesticides and their degradates. During low-flow periods, they found that groundwater had the greatest influence on stream water quality with decreased concentrations of nitrate, pesticides, and pesticide degradates (mainly due to long groundwater residence times that favored the reduction of nitrate and pesticides to negligible concentrations). Conversely, during high-flow periods, they found increased concentrations of nitrate, pesticides, and pesticide degradates. The most frequently detected pesticides in low concentrations (medians less than 0.08 µg/L) included metolachlor, atrazine, simazine, alachlor, deethylatrazine, chlorpyrifos, and tebuthiuron.

In a study of pesticides in 23 public supply wells in southern Sweden (Scania), Akesson et al. (2015) found that most pesticide detections were in oxic waters from shallow, unconfined, unconsolidated or fracture-type bedrock aquifers. About 50% of drinking water in Sweden is provided by groundwater. Water from these aquifers tended to be young; however, the authors concluded that further studies would be desirable to assess the extent and effects of time lags on the fate of pesticides in these aquifers.

More than two decades ago, the US Geological Survey was studying the transport and persistence of herbicides, insecticides, and nutrients in the Mississippi River and several major tributaries. Several herbicides were found (Scribner et al., 2006), and atrazine and alachor occasionally exceeded maximum contaminant levels for drinking water. Although nitrate-N and herbicides were found in river samples, they did not have similar seasonal responses. Battaglin and Goolsby (1997) found that elevated concentrations of herbicides occurred for 1–4 months following application in late spring and early summer, but nitrate-N concentrations in unregulated rivers were elevated during fall, winter, and spring months. For more USGS publications on the co-occurrence of nitrate and herbicides in surface and groundwater, please visit the website: https://toxics.usgs.gov/pubs/ (accessed 02 June 2018). In addition to large loads of nitrogen discharged annually from the Mississippi River to the Gulf of Mexico, Clark and Goolsby (2000) reported that hundreds of metric tons of herbicides and herbicide metabolites (mostly atrazine, metolachlor, alachlor and alachlor ESA) were also reaching the Gulf of Mexico. Annual loads of atrazine and cyanazine were about 1–2% of the estimated amount applied annual in the Mississippi River drainage basin; whereas the annual loads of acetochlor, alachlor, and metolachlor generally were less than 1%.

9.5. NITRATE AND ORGANIC CONTAMINANTS IN GROUNDWATER USED FOR DRINKING WATER

In a study of volatile organic compounds (VOCs), pesticides, nitrate, and their mixtures in drinking water in the United States, Squillace et al. (2002) analyzed samples of untreated groundwater from 1255 domestic drinking water wells and 242 public supply wells collected during 1992–1999. Samples were collected from wells distributed geographically across large aquifers, primarily in rural areas. Each sample was analyzed for 60 VOCs, 83 pesticides, and nitrate. Based on previous studies, it was decided that when nitrate-N concentrations were greater than 3 mg/L, it was indicative of anthropogenic contamination. Seventy percent of the samples contained at least one VOC, pesticide, or anthropogenic nitrate; 47% contained at least two compounds; and 33% contained at least three compounds. Concentrations of VOCs and pesticides ranged from about 0.001 to 100 µg/L, with a median of 0.02 µg/L. Squillace et al. (2002) found that water from about 12% of the wells contained one or more compounds that exceeded US Environmental Protection Agency drinking-water standards or human health criteria, in most cases due to nitrate concentrations exceeding the maximum contaminant level in domestic wells. They noted that over 95% of the detections in 402 mixtures resulted from just 14 compounds (seven VOCs, six pesticides, and nitrate); although most samples with these mixtures also contained a variety of other compounds.

In a later study of water quality in domestic wells, DeSimone (2009) collected data for 2167 domestic wells sampled in major aquifers across the United States and showed that as nitrate concentrations increased there was an increased frequency of detecting an organic contaminant (e.g., a pesticide or volatile organic compound) at a concentration that was greater than one-tenth of an MCL (U.S. Environmental Protection Agency, 2006) or USGS Health-Based Screening Level (HBSL) (Toccalino & Norman, 2006; Toccalino et al., 2006). Overall, 84% of samples with concentrations of organic compounds greater than MCLs or HBSLs also had nitrate-N concentrations above background levels (1 mg N/L). When nitrate concentrations exceeded 1 mg/L (the reported background level), there was a five times greater chance of detecting an organic contaminant that exceeded one-tenth of an MCL or HBSL. Likewise, DeSimone (2009) found that the frequency of occurrence of an organic contaminant at a concentration above an MCL or HBSL was about four times greater in groundwater with nitrate concentrations above background levels than in groundwater with nitrate concentrations below background levels. These findings indicate that when concentrations of nitrate are above background levels in a domestic drinking-water supply, additional analysis for organic contaminants should be considered.

9.6. NITRATE AND EMERGING ORGANIC CONTAMINANTS

In Chapter 4 we learned that organic compounds commonly referred to as emerging contaminants have been used to determine or differentiate between various sources of nitrate, particularly animal versus human wastes. Many pharmaceuticals resist degradation, as they are developed to persist until they have acted upon a specific malady and eventually end up in wastewater. Many studies have shown that endocrine-disrupting compounds (such as hormones and antibiotics) from treated wastewater effluent have caused harm to aquatic organisms in surface waters (e.g., Blazer et al., 2014; Vajda et al., 2008).

In a review of trace organic chemicals from septic systems, Lusk et al. (2017) noted that organic compounds that originate from septic tank effluent have contaminated groundwater, surface waters, and finished drinking water. In areas with shallow groundwater systems where domestic wells are used for drinking water supply, septic systems can be an important source of emerging contaminants in groundwater. The persistence of pharmaceuticals and nitrate in ground water from septic systems is highly dependent on oxidation–reduction conditions. For example, Carrara et al. (2008) investigated three high volume septic systems in Ontario, Canada, to assess the release of pharmaceutical compounds from groundwater to drinking water aquifers. They found that the highest concentrations of three pharmaceutical compounds (ibuprofen, gemfibrozil, and naproxen) were detected in anoxic (nitrate reducing) zones of plumes and at greatest distances from the infiltration source area. Y.Y. Yang et al. (2016) investigated the fate and mass balance of 17 emerging contaminants (wastewater markers, hormones, pharmaceuticals, and personal care products) in the drainfield of a septic system. They found low concentrations of 14 detected micropollutants in the leachate (less than 200 ng/L) compared to the septic tank effluent, and most compounds were attenuated in the drainfield with the exception of sucralose (an artificial sweetener that is persistent in the environment).

Seiler et al. (1999) studied the occurrence of human pharmaceuticals in groundwater with elevated nitrate concentrations. They collected water samples from domestic, public supply, and monitoring wells in three communities using septic systems for waste disposal near Reno, Nevada. Median nitrate-N concentrations ranged from 2.1 to 8.8 mg/L in oxic groundwater from the three communities and low concentrations of several pharmaceutical compounds (chlorpropamide, phensuximide, carbamazepine, and caffeine) were detected indicating recharge from domestic waste water. Peeler et al. (2006) found that elevated caffeine and nitrate in surface waters were correlated to population centers and associated wastewater treatment plants in rural areas. They were also able to use multiple markers (nitrate, caffeine, and coliform bacteria) to distinguish between anthropogenic and natural sources of contamination.

Verstraeten et al. (2005) investigated the occurrence of nitrate, pharmaceutical compounds, organic wastewater compounds, and bacteria in shallow sandpoint and cased domestic wells in coarse alluvial sediments in Nebraska. They found that 13 of 26 domestic wells had evidence of septic system seepage based on the co-occurrence of nitrate, antibiotics and other drugs, and dissolved organic carbon. The highest values were found in sandpoint wells within 30 m of a septic system and less than 14 m deep in a shallow thin aquifer.

Based on a study of three septic systems in a mantled karst aquifer, Katz et al. (2010) found elevated nitrate concentrations but only sporadic detections of pharmaceuticals and organic wastewater compounds in soil water collected from shallow and deep lysimeters in septic tank drainfields and in wells beneath the drainfield. Higher detections of pharmaceuticals and organic wastewater compounds were found in groundwater at one site where the depth to the limestone was the shallowest and average daily homeowner water use was highest.

Lim et al. (2017) noted in their review article that even though numerous studies have effectively used PPCPs to detect wastewater contamination in surface water and groundwater, currently, there is no single chemical that could serve as an accurate marker for wastewater contamination for all sites. The most effective way to use markers along with nitrogen compounds would require a better understanding of land-use patterns, types and concentrations of co-occurring contaminants in the waste system, and transport properties.

9.7. AIR POLLUTION: NITROGEN AND OTHER CONSTITUENTS

Emissions of reactive nitrogen species (nitrous oxides, nitric oxide, and ammonia) typically are associated with other air pollutants, such as sulfur dioxide, methane, hydrogen sulfide, and fine particulates. Hertel et al. (2011) noted that both ammonia and gaseous nitrogen oxides lead to formation of aerosol phase compounds that can be transported over long distances (up to more than 1000 km). It was emphasized that there is a considerable lack of information (both quantitatively or qualitatively) about organic nitrogen compounds in the atmosphere, although they can contribute significant amounts of nitrogen in wet deposition, and they are present in gaseous and particulate forms in the atmosphere. Irvine et al. (2017) reported that additional chemical species of nitrogen are emitted to the atmosphere and contribute to TN deposition (e.g., other oxidized forms and reduced and organic nitrogen compounds), but these species are not regulated for the ecological effects caused by TN deposition. Li et al. (2016) found that deposition of reactive nitrogen in the United States has shifted from nitrate-dominated to ammonium-dominated, with ammonia contributing from 19 to 65% in different regions. They concluded that reductions in ammonia emissions are needed for effective lowering of reactive nitrogen deposition in the United States.

Burning fossil fuels produces a significant amount of reactive nitrogen and sulfur. Sutton et al. (2013) estimated that fossil fuel burning constitutes about 20% of human reactive nitrogen production. Emissions of nitrogen oxide (NO_x) to air contribute to particulate matter and ground-level (tropospheric) ozone that adversely affect human health, ecosystems, and food production systems. Doney et al. (2007) describe the impact from atmospheric nitrogen and sulfur deposition on ocean acidification and the inorganic carbon system.

The combination of nitrogen and sulfur deposition and its impact on the health of forests and other ecosystems has been referred to as acid rain (e.g., Bobbink et al., 2010; Greaver et al., 2012; Irvine et al., 2017; Pardo et al., 2011). Some notable effects on forest ecosystems have been declines in tree growth, susceptibility to drought, frost, pest damage, diseases, increased ozone exposure, all of which have led to increased mortality (e.g., Driscoll et al., 2001; Pardo et al., 2011). Over the past 20 years, there has been notable reductions in emissions that have resulted in lower amounts of wet deposition of sulfate and nitrogen oxides ($NO + NO_2$) by an average of 42–44% and 16–27%, respectively (https://www.epa.gov/clean-air-act-overview/progress-cleaning-air-and-improving-peoples-health, accessed 05 June 2018). However, total sulfur deposition over 30 kg/ha/yr and TN deposition over 20 kg/ha/yr have more recently been reported in parts of the USA (Schwede & Lear, 2014; Zhang et al., 2012; https://digitalcommons.unl.edu/cgi/viewcontent.cgi?referer=http://scholar.google.com/&httpsredir=1&article=1219&context=usepapapers, accessed 18 June 2018).

Ammonia emissions to the atmosphere also have increased during the past several decades and can result in the formation of particulate matter through reactions with nitric and sulfuric acids. These compounds in the atmosphere that are harmful to human and animal health can be dispersed over large areas of the world resulting in eutrophication and acidification of terrestrial, freshwater, and marine habitats (Bobbink et al., 2010; Erisman et al., 2015). Also, deposition of ammonia can damage sensitive vegetation, particularly bryophytes and lichens that consume most of their nutrients from the atmosphere (Sheppard et al., 2011). Sala et al. (2000) noted that certain biomes (particularly northern temperate, boreal, arctic, alpine, grassland, savannah, and Mediterranean) are very sensitive to reactive nitrogen deposition because of the limited availability of nitrogen in these systems under natural conditions. Impacts to ecosystems and human health from emissions of ammonia and other reactive nitrogen compounds to the atmosphere and atmospheric deposition of reactive nitrogen are discussed in more detail in Chapters 5 and 6 in this book.

REFERENCES

Akesson, M., Sparrenbom, C. J., Dahlqvist, P., & Fraser, S. J. (2015). On the scope and management of pesticide pollution of Swedish groundwater resources: The Scanian example. *Ambio*, 44(3), 226–238. https://doi.org/10.1007/s13280-014-0548-1

Alexander, R. B., Smith, R. A., Schwarz, G. E., Boyer, E. W., Nolan, J. V., & Brakebill, J. W. (2008). Differences in phosphorus and nitrogen delivery to the Gulf of Mexico from the Mississippi River basin. *Environmental Science and Technology*, 42, 822–830.

Ator, S. W., & Denver, J. M. (2012). Estimating contributions of nitrate and herbicides from groundwater to headwater streams, northern Atlantic Coastal Plain, USA. *Journal of the American Water Resources Association*, 48(6). https://doi.org/10.1111/j.1752-1688.2012.00672.x

Ator, S.W., & Ferrari, M.J. (1997). Nitrate and selected pesticides in Groundwater of the Mid-Atlantic Region. U.S. Geological Survey Water-Resources Investigations Report 97–4139, 8 p.

Atwood, D., & Paisley-Jones, C. (2017). *Pesticides industry sales and usage: 2008–2012 market estimates*. Washington, DC: U.S. Environmental Protection Agency, Office of Pesticide Programs, 32 p. Retrieved from https://www.epa.gov/sites/production/files/2017-01/documents/pesticides-industry-sales-usage-2016_0.pdf, accessed 10 June 2018.

Banning, A., Demmel, T., Rude, R. R., & Wrobel, M. (2013). Groundwater uranium origin and fate control in a river valley aquifer. *Environmental Science and Technology*, 47(24), 13941–13948. https://doi.org/10.1021/es304609e. Epub 2013 Nov 27

Barbash, J., & Resek, E. A. (1996). *Pesticides in ground water: Distribution, trends, and governing factors*. Chelsea, MI: Ann Arbor Press, 588 p.

Battaglin, W. A., & Goolsby, D. A. (1997). Statistical modeling of agricultural chemical occurrence in midwestern rivers. *Journal of Hydrology*, 196, 1–25.

Billen, G., Silvestre, M., Grizzetti, B., Leip, A., Garnier, J., Voss, M., et al. (2011). Nitrogen flows from European watersheds to coastal marine waters. Ch. 13. In M. A. Sutton, C. M. Howard, J. W. Erisman, G. Billen, A. Bleeker, P. Grennfelt, H. van Grinsven, & B. Grizzetti (Eds.), *The European nitrogen assessment: Sources, effects, and policy perspectives*. Cambridge, UK: Cambridge University Press, 664 p.

Blazer, V. S., Iwanowicz, D. D., Walsh, H. L., Sperry, A. J., Iwanowicz, L. R., Alvarez, D. A., et al. (2014). Reproductive health indicators of fishes from Pennsylvania watersheds: Association with chemicals of emerging concern. *Environmental Monitoring and Assessment*, 186, 6471–6491.

Bobbink, R., Hicks, K., Galloway, J., Spranger, T., Alkemade, R., Ashmore, M., et al. (2010). Global assessment of nitrogen deposition effects on terrestrial plant diversity: A synthesis. *Ecological Applications*, 20, 30–59.

Burow, K.R., Shelton, J.L., & Dubrovsky, N.M. (1998). Occurrence of nitrate and pesticides in groundwater beneath three agricultural land-use settings in the eastern San Joaquin Valley, California, 1993–1995. U.S. Geological Survey Water-Resources Investigations Report 97–4284, Washington, DC, 58 p.

Burow, K. R., Shelton, J. L., & Dubrovsky, N. M. (2008). Regional nitrate and pesticide trends in ground water in the eastern San Joaquin Valley, California. *Journal of Environmental Quality, 37*(5), S249–S263.

Carrara, C., Ptacek, C. J., Robertson, W. D., Blowes, D. W., Moncur, M. C., Sverko, E., & Backus, S. (2008). Fate of pharmaceutical and trace organic compounds in three septic system plumes, Ontario, Canada. *Environmental Science and Technology, 42*(8), 2805–2811. https://doi.org/10.1021/es070344q

Clark, G. M., & Goolsby, D. A. (2000). Occurrence and load of selected herbicides and metabolites in the lower Mississippi River. *Science of The Total Environment, 248*(2–3), 101–113. https://doi.org/10.1016/S0048-9697(99)00534-3

DeSimone, L.A. (2009). Quality of water from domestic wells in principal aquifers of the United States, 1991–2004. U.S. Geological Survey Scientific Investigations Report 2008–5227, 127 p., CD, also available at https://pubs.usgs.gov/sir/2008/5227/, accessed 20 September 2019.

Diaz, R. J., & Rosenberg, R. (2008). Spreading dead zones and consequences for marine ecosystems. *Science, 321*, 926–929.

Domagalski, J. L., Ator, S., Coupe, R., McCarthy, K., Lampe, D., Sandstrom, M., & Baker, N. (2008). Comparative study of transport processes of nitrogen, phosphorus, and herbicides to streams in five agricultural basins, USA. *Journal of Environmental Quality, 37*, 1158–1169. https://doi.org/10.2134/jeq2007.0408

Doney, S. C., Mahowald, N., Lima, I., Feely, R. A., Mackenzie, F. T., Lamarque, J.-F., & Rasch, P. J. (2007). Impact of anthropogenic atmospheric nitrogen and sulfur deposition on ocean acidification and the inorganic carbon system. *Proceedings of the National Academy of Sciences USA, 104*(37), 14580–14585. https://doi.org/10.1073/pnas.0702218104

Driscoll, C. T., Lawrence, G. B., Bulger, A. J., Butler, T. J., Cronan, C. S., Eager, C., et al. (2001). Acidic deposition in the northeastern United States: Sources and inputs, ecosystem effects, and management strategies. *BioScience, 51*, 180–198.

Dubrovsky, N.M., Burow, K.R., Clark, G.M., Gronberg, J.M., Hamilton P.A., Hitt, K.J., et al. (2010). *The quality of our Nation's waters—Nutrients in the Nation's streams and groundwater, 1992–2004.* U.S. Geological Survey Circular 1350, 174 p. Retrieved from http://water.usgs.gov/nawqa/nutrients/pubs/circ1350, accessed 20 September 2019.

Erisman, J. W., Galloway, J. N., Dice, N. B., Sutton, M. A., Bleeker, A., Grizzetti, B., Leach, A. M., & de Vries, W. (2015). *Nitrogen, too much of a vital resource.* Science Brief. Zeist, The Netherlands: World Wildlife Fund.

Erisman, J. W., van Grinsven, H., Grizzetti, B. Bouraoui, F., Powlson, D., Sutton, M. A., Bleeker, A., & Reis, S. (2011). The European nitrogen problem in a global perspective. Ch. 2. In M.A. Sutton, C.M. Howard, J.W. Erisman Billen, G., Bleeker, A., Grennfelt, P., H. van Grinsven, & B. Grizzetti (Eds.), *The European nitrogen assessment: Sources, effects, and policy perspectives* (pp. 9–31). Cambridge, UK: Cambridge University Press.

Gosselin, D. C., Headrick, J., Tremblay, R., Chen, X.-H., & Summerside, S. (1997). Domestic well water quality in rural Nebraska: Focus on nitrate-nitrogen, pesticides, and coliform bacteria. *Ground Water Monitoring and Remediation, 17*(2), 77–87.

Gray, C. W., Moot, D. J., McLaren, R. G., & Reddecliffe, T. (2002). Effect of nitrogen fertiliser applications on cadmium concentrations in durum wheat (*Triticum turgidum*) grain. *New Zealand Journal of Crop and Horticultural Science, 30*(4), 291–299. https://doi.org/10.1080/01140671.2002.9514226

Greaver, T. L., Sullivan, T., Herrick, J. D., Lawrence, G., Herlihy, A., Barron, J., et al. (2012). Ecological effects from nitrogen and sulfur air pollution in the United States: What do we know? *Frontiers in Ecology and the Environment.* https://doi.org/10.1890/110049

Guttormsen, G., Singh, B. R., & Jeng, S. A. (1995). Cadmium concentration in vegetable crops grown in a sandy soil as affected by Cd levels in fertilizer and soil pH. *Fertilizer Research, 41*(1), 27–32.

Harman, J., Robertson, W. D., Cherry, J. A., & Zanini, L. (1996). Impacts on a sand aquifer from an old septic system: Nitrate and phosphate. *Ground Water, 34*(6), 1105–1114.

Heilmann, S. M., Molde, J. S., Timler, J. G., Wood, B. M., Mikula, A. L., Vozhdayev, G. V., et al. (2014). Phosphorus reclamation through hydrothermal crbonization of animal manures. *Environmental Science and Technology, 48*, 10323–10329. dx.doi.org/10.1021/es501872k

Hendriks, A. T. W. M., & Langeveld, J. G. (2017). Rethinking wastewater treatment plant effluent standards: Nutrient reduction or nutrient control? *Environmental Science and Technology, 51*, 4735–4737. https://doi.org/10.1021/acs.est.7b01186

Hertel, O., Reis, S., Skjøth, C. A., Bleeker, A., Harrison, R., Cape, J.N., et al. (2011). Nitrogen processes in the atmosphere. In M. A. Sutton, C. M. Howard, J. W. Erisman, G. Billen, A. Bleeker, P. Grennfelt, H. van Grinsven, & B. Grizzetti (Eds.), *The European nitrogen assessment: Sources, effects, and policy perspectives* (pp. 177–207). Cambridge, UK: Cambridge University Press.

Howard, B.C. (2014). Mississippi basin water quality declining despite conservation. *National Geographic,* 12 April 2014. Retrieved from https://news.nationalgeographic.com/news/2014/04/140411-water-quality-nutrients-pesticides-dead-zones-science/, accessed 20 June2018.

Irvine, I. C., Greaver, T., Phelan, J., Sabo, R. D., & Van Houtven, G. (2017). Terrestrial acidification and ecosystem services: Effects of acid rain on bunnies, baseball, and Christmas trees. *Ecosphere, 8*(6), e01857. https://doi.org/10.1002/ecs2.1857

Johnson, H., Schaefer, B., Timm, D., Rindal, D., & Schliep, A. (2016). A 2015 reconnaissance study of pesticide compounds in community public water supply wells. Minnesota Department of Agriculture and Minnesota Department of Health, October 2016, 60 p.

Katz, B. G., Griffin, D. W., McMahon, P. B., Harden, H., Wade, E., Hicks, R. W., & Chanton, J. P. (2010). Fate of effluent-borne contaminants beneath septic tank drainfields overlying a karst aquifer. *Journal of Environmental Quality*, *39*, 1181–1195.

Kingsbury, J. A. (2008). Relation between flow and temporal variations of nitrate and pesticides in two karst springs in northern Alabama. *Journal of the American Water Resources Association*, *44*(2), 478–488. https://doi.org/10.1111/j.1752-1688.2008.00176.x

Koterba, M. T., Banks, W. S. L., & Shedlock, R. J. (1993). Pesticides in shallow groundwater in the Delmarva Peninsula. *Journal of Environmental Quality*, *22*(3), 500–518.

Lawniczak, A. E., Zbierska, J., Nowak, B., Achtenbert, K., Grzeskowiak, A., & Kanas, K. (2016). Impact of agriculture and land use on nitrate contamination in groundwater and running waters in central-west Poland. *Environmental Monitoring and Assessment*, *188*, 1–17.

Lawniczak, A. E., Güsewell, S., & Verhoeven, J. T. A. (2009). Effect of N:K supply ratios on the performance of three grass species from herbaceous wetlands. *Basic and Applied Ecology*, *10*(8), 715–725.

Li, Y., Schichtel, B.A., Walker, J.T., Schwede, D.B., Chen, X., Lehmann, C.M.B., et al. (2016). Increasing importance of deposition of reduced nitrogen in the United States. In *Proceedings of the National Academy of Sciences of the United States*. Retrieved from http://www.pnas.org/content/early/2016/05/04/1525736113.full, accessed 18 June 2018.

Lim, F. Y., Ong, S. L., & Hu, J. (2017). Recent advances in the use of chemical markers for tracing wastewater contamination in aquatic environment: A review. *Water*, *9*, 143. https://doi.org/10.3390/w9020143

Line, D. E., White, N. M., Osmond, D. L., Jennings, G. D., & Mojonnier, C. B. (2002). Pollutant export from various land uses in the upper Neuse River basin. *Water Environment Research*, *74*(1), 100–108. https://doi.org/10.2175/106143002X139794

Lopez, C. B., Jewett, E. B., Dortch, Q., Walton, B. T., & Hudnell, H. K. (2008). Scientific assessment of freshwater harmful algal blooms. In *Interagency working group on harmful algal blooms, hypoxia and human health of the joint subcommittee on ocean science and technology*. Washington, DC, 65 p.

Lusk, M. G., Toor, G. S., Yang, Y., Mechtensimer, S., De, M., & Obreza, T. A. (2017). A review of the fate and transport of nitrogen, phosphorus, pathogens, and trace organic chemicals in septic systems. *Critical Reviews in Environmental Science and Technology*, *47*(7), 455–541.

Macler, B. (1996). Developing the ground water disinfection rule. *Journal of the American Water Works Association*, *88*, 47–55.

Mallin, M. A., Johnson, V. L., & Ensign, S. H. (2009). Comparative impacts of stormwater runoff on water quality of an urban, a suburban, and a rural stream. *Environmental Monitoring and Assessment*, *159*, 475–491. https://doi.org/10.1007/s10661-008-0644-4

Mallin, M. A., Johnson, V. L., Ensign, S. H., & MacPherson, T. A. (2006). Factors contributing to hypoxia in rivers, lakes and streams. *Limnology and Oceanography*, *51*(1), 690–701.

Manuel, J. (2014). Nutrient pollution: A persistent threat to waterways. *Environmental Health Perspectives*, *122*(12), A323. https://doi.org/10.1289/ehp.122-A304

Maupin, M. A., & Ivahnenko, T. (2011). Nutrient loadings to streams of the continental United States from municipal and industrial effluent. *Journal of the American Water Resources Association*, *47*(5), 950–964. https://doi.org/10.1111/j.1752-1688.2011.00576.x

McMahon, G., Tervelt, L., & Donehoo, W. (2007). Methods for estimating annual wastewater nutrient loads in the southeastern United States. U.S. Geological Survey Open-File Report 2007–1040, 81 p. Retrieved from http://pubs.usgs.gov/of/2007/1040/, accessed 20 September 2019.

Molenat, J., & Gascuel-Odoux, C. (2001). Role of shallow groundwater in nitrate and herbicide transport in the Kervidy agricultural catchment (Brittany, France). In *Iimpact of Hitman Activity on Groundwater Dynamics (Proceedings of a symposium held during the Sixth IAHS Scientific Assembly at Maastricht, The Netherlands, July 2001)*. IAHS Publ. no. 269.

Nolan, J., & Weber, K. A. (2015). Natural uranium contamination in major U.S. aquifers linked to nitrate. *Environmental Science and Technology Letters*, *2*(8), 215–220.

Panno, S. V., & Kelly, W. R. (2003). Nitrate and herbicide loading in two groundwater basins of Illinois' sinkhole plain. *Journal of Hydrology*, *290*(3–5), 229–242.

Pardo, L. H., Fenn, M. E., Goodale, C. L., Geiser, L. H., Driscoll, C. T., Allen, E. B., et al. (2011). Effects of nitrogen deposition and empirical nitrogen critical loads for ecoregions of the United States. *Ecological Applications*, *8*, 3049–3082.

Peeler, K. A., Opsahl, S. P., & Chanton, J. P. (2006). Tracking anthropogenic inputs using caffeine, indicator bacteria and nutrients in rural freshwater and urban marine systems. *Environmental Science and Technology*, *40*, 7616–7622.

Preston, S.D., & Brakebill, J.W. (1999). Application of spatially referenced regression modeling for the evaluation of total nitrogen loading in the Chesapeake Bay Watershed. U.S. Geological Survey Water Resources Investigation Report 1999–4054, 12 p. Retrieved from http://md.water.usgs.gov/publications/wrir-99-4054/pdf/SPARROW.pdf, accessed 20 September 2019.

Puckett, L. J., & Hughes, W. B. (2005). Transport and fate of nitrate and pesticides: Hydrogeology and riparian zone processes. *Journal of Environmental Quality*, *34*(6), 2278–2292. https://doi.org/10.2134/jeq2005.0109

Ruddy, B.C., Lorenz, D.L., & Mueller, D.K. (2006). County-level estimates of nutrient inputs to the land surface of the conterminous United States, 1982–2001. U.S. Geological Survey Scientific Investigations Report 2006–5012, 17 p., also available at http://pubs.usgs.gov/sir/2006/5012/, accessed 18 June 2018.

Sala, O. E., Chapin, I. I. I. F. S., Armesto, J. J., Berlow, E., Bloomfield, J., Dirzo, R., et al. (2000). Global biodiversity scenarios for the year 2100. *Science*, *287*, 1770–1774.

Scavia, D., & Bricker, S. B. (2006). Coastal eutrophication assessment in the United States. *Biogeochemistry*, *79*(1–2), 187–208.

Schwede, D. B., & Lear, G. G. (2014). A novel approach for estimating total deposition in the United States. *Atmospheric Environment*, *92*, 207–220.

Scribner, E.A., Goolsby, D.A., Battaglin, W.A., Meyer, M.T., & Thurman, E.M. (2006). Concentrations of selected herbicides, herbicide degradation products, and nutrients in the lower Mississippi river, Louisiana, April 1991 through December 2003. U.S. Geological Survey Data Series 165, 84 p.

Seiler, R. L., Zaugg, S. D., Thomas, J. M., & Howcroft, D. L. (1999). Caffeine and pharmaceuticals as indicators of wastewater contamination in wells. *Ground Water*, *37*(3), 405–410.

Sheppard, L. J., Leith, I. D., Mizunuma, T., Cape, J. N., Crossley, A., Leeson, S., et al. (2011). Dry deposition of ammonia gas drives species change faster than wet deposition of ammonium ions: Evidence from a long-term field manipulation. *Global Change Biology*, *17*(12), 3589–3607.

Smith, J. E., Jr., & Perdek, J. M. (2004). Assessment and management of watershed microbial contaminants. *Critical Reviews in Environmental Science and Technology*, *34*(2), 109–139. https://doi.org/10.1080/10643380490430663

Sprague, L.A. (2017). Water-quality changes in the Nation's streams and rivers. U.S. Geological Survey Web Page. Retrieved from https://nawqatrends.wim.usgs.gov/swtrends/, accessed 20 September 2019.

Squillace, P. J., Scott, J. C., Moran, M. J., Nolan, B. T., & Kolpin, D. W. (2002). VOCs, pesticides, nitrate, and their mixtures in groundwater used for drinking water in the United States. *Environmental Science and Technology*, *36*, 1923–1930.

Steinfeld, H., Gerber, P., Wassenaar, T., Castel, V., Rosales, M., & de Haan, C. (2006). *Livestock's long shadow: Environmental issues and options*. Rome: LEAD/FAO, 26 p.

Sutton, M. A., Howard, C. M., Erisman, J. W., Billen, G., Bleeker, A., Grennfelt, P., van Grinsven, H., & Grizzetti, B. (2011). *The European nitrogen assessment: Sources, effects, and policy perspectives*. Cambridge, UK: Cambridge University Press, 664 p.

Sutton, M. A., Reis, S., Riddick, S. N., Dragosits, U., Nemitz, E., Theobald, M. R., et al. (2013). Towards a climate-dependent paradigm of ammonia emission and deposition. *Philosophical Transactions of the Royal Society B*, *368*. https://doi.org/10.1098/rtsb.2013.0166

Toccalino, P. L., & Norman, J. E. (2006). Health-based screening levels to evaluate U.S. Geological Survey ground-water quality data. *Risk Analysis*, *26*(5), 1339–1348. https://doi.org/10.1111/j.1539-6924.2006.00805.x

Toccalino, P.L., Norman, J.E., Booth, N.L., & Zogorski, J.S. (2006). Health-based screening levels—A tool for evaluating what water-quality data may mean to human health. U.S. Geological Survey, National Water-Quality Assessment Program. Retrieved from https://cida.usgs.gov/hbsl/apex/f?p=104:1, accessed 07 May 2019.

Tuthill, A., Meikle, D. B., & Alavanja, M. C. R. (1998). Coliform bacteria and nitrate contamination of wells in major soils of Frederick, Maryland. *Journal of Environmental Health*, *60*(8), 16–20.

U.S. Environmental Protection Agency. (1983). *Results of the nationwide urban runoff program, executive summary*. Washington, DC: U.S. Environmental Protection Agency.

U.S. Environmental Protection Agency. (2006). 2006 Edition of the drinking water standards and health advisories (EPA822R06013). Washington, DC, 18 p.

U.S. Environmental Protection Agency. (2009). An urgent call to action. Report of the State-EPA Nutrient Innovations Task Group, 41 p.

Vajda, A. M., Barber, L. B., Gray, J. L., Lopez, E. M., Woodling, J. D., & Norris, D. O. (2008). Reproductive disruption in fish downstream from an estrogenic wastewater effluent. *Environmental Science and Technology*, *42*, 3407–3414.

Verstraeten, I. M., Fetterman, G. S., Meyer, M. T., Bullen, T., & Sebree, S. K. (2005). Use of tracers and isotopes to evaluate vulnerability of water in domestic wells to septic waste. *Ground Water monitoring and Remediation*, *25*(2), 107–117.

Wilkes, G., Ruecker, N., Neumann, N. F., Gannon, V. P., Jokinen, C., Sunohara, M., et al. (2013). Spatiotemporal analysis of *Cryptosporidium* species/genotypes and relationships with other zoonotic pathogens in surface water from mixed-use watersheds. *Applied Environmental Microbiology*, *79*, 434–448.

Yang, Y. Y., Toor, G. S., Wilson, P. C., & Williams, C. F. (2016). Septic systems as hot-spots of pollutants in the environment: Fate and mass balance of micropollutants in septic drainfields. *Science of the Total Environment*, *566*, 1535–1544.

Zhang, L., Jacob, D. J., Knipping, E. M., Kumar, N., Munger, J. W., Carouge, C. C., et al. (2012). Nitrogen deposition to the United States: Distribution, sources and processes. *Atmospheric Chemistry and Physics*, *12*, 4539–4554.

FURTHER READING

Alexander, R. B., Smith, R. A., & Schwarz, G. E. (2000). Effect of stream channel size on the delivery of nitrogen to the Gulf of Mexico. *Nature*, *403*, 758–761.

Bradley, J. (2008). New Zealand. In R. J. LeBlanc, P. Matthews, & R. P. Richard (Eds.), *Global atlas of excreta, wastewater sludge, and biosolids management: Moving forward the sustainable and welcome uses of a global resource* (pp. 447–454). Nairobi: United Nations Human Settlements Programme (UN-HABITAT).

Domagalski, J.L., & Johnson, H. (2012). Phosphorus and groundwater: Establishing links between agricultural use and transport to streams. U.S. Geological Survey Fact Sheet 2012–3004, 4 p.

Gallagher, T. L., & Gergel, S. E. (2017). Landscape indicators of groundwater nitrate concentrations: An approach for trans-border aquifer monitoring. *Ecosphere*, *8*(12), 02047. https://doi.org/10.1002/ecs2.2047

Galloway, J. N., Winiwarter, W., Leip, A., Leach, A., Bleeker, A., & Erisman, J. W. (2014). Nitrogen footprints: Past, present, and future. *Environmental Research Letters*, *9*. https://doi.org/10.1088/1748-9326/9/11/115003

Howarth, R., Anderson, D., Cloern, J., Elfring, C., Hopkinson, C., Lapointe, B., et al. (2000). Nutrient pollution of coastal rivers, bays, and seas. Issues in Ecology, no. 7.

Ivahnenko, T. (2017). Evaluation and use of U.S. Environmental Protection Agency Clean Watersheds Needs Survey data to quantify nutrient loads to surface water, 1978–2012. U.S. Geological Survey Scientific Investigations Report 2017–5115, 11 p. https://doi.org/10.3133/sir20175115.

Katz, B. G. (2004). Sources of nitrate contamination and age of water in large karstic springs of Florida. *Environmental Geology*, *46*, 689–706.

Katz, B. G., Chelette, A. R., & Pratt, T. R. (2004). Use of chemical and isotopic tracers to assess sources of nitrate and age of ground water, Woodville Karst Plain, USA. *Journal of Hydrology, 289*, 36–61.

Kiely, T., Donaldson, D., & Grube, A. (2004). Pesticides industry sales and usage—2000 and 2001 market estimates (EPA-733-R-04-001). Washington, DC: USEPA, Office of Prevention, Pesticides, and Toxic Substances.

Nolan, B. T., & Hitt, K. J. (2006). Vulnerability of shallow groundwater and drinking-water wells to nitrate in the United States. *Environmental Science and Technology, 40*(24), 7834–7840.

Preston, S.D., Alexander, R.B., Woodside, M.D., & Hamilton, P.A. (2009). SPARROW MODELING Enhancing understanding of the Nation's water quality. U.S. Geological Survey Fact Sheet 2009–3019, 6 p.

State-USEPA Nutrient Innovations Task Group. (2009). An urgent call to action. State-USEPA Nutrient Innovations Task Group Report to USEPA, August 2009, 170 p.

Sutton, M. A., Erisman, J. W., Dentener, F., & Moeller, D. (2008). Ammonia in the environment: From ancient times to the present. *Environmental Pollution, 156*, 583–604.

Sutton, M. A., Reis, S., & Butterbach-Bahl, K. (2009). Reactive nitrogen in agro-ecosystems: Integration with greenhouse gas interactions. *Agriculture, Ecosystems and Environment, 133*, 135–138.

U.S. Environmental Protection Agency. (2002). Nitrogen—Multiple and regional impacts. U.S. Environmental Protection Agency Clean Air Markets Division Report EPA-430-R-01-006, 38 p.

U.S. Environmental Protection Agency. (2013). Final aquatic life ambient water quality criteria for ammonia-Freshwater 2013. *Notice in Federal Register, 78*(163), 52192–52194.

Wilkes, G., Edge, T. A., Gannon, V. P., Jokinen, C., Lyautey, E., Neumann, N. F., et al. (2011). Associations among pathogenic bacteria, parasites, and environmental and land use factors in multiple mixed-use watersheds. *Water Research, 45*, 5807–5825.

Yang, S., & Gruber, N. (2016). The anthropogenic perturbation of the marine nitrogen cycle by atmospheric deposition; nitrogen cycle feedbacks and the ^{15}N Haber-Bosch effect. *Global Biogeochemical Cycles, 30*(10), 1418–1440.

10

Economic Costs and Consequences of Excess Reactive Nitrogen

As we have learned from information presented in previous chapters, losses of reactive nitrogen from anthropogenic activities to the atmosphere, hydrosphere, and biosphere have been associated with human health problems, degradation of ecosystems, reductions in biodiversity, air pollution, contamination of surface waters and groundwater, acidification of soils and water bodies, algal blooms, fish kills, and disruptions to greenhouse gas balance. These impacts have had enormous economic ramifications that have resulted in huge healthcare costs associated with respiratory/circulatory issues. Other economic impacts include job losses, decreased home values, losses to the fishing and shellfish industries, substantial costs for drinking water treatment, and large costs to restore the quality of impaired surface waters and groundwater. As mentioned in Chapter 5, emissions of ammonia and other reactive nitrogen compounds from agricultural fertilizer use and manure spreading contribute to fine particulate matter ($PM_{2.5}$) in air pollution, which has been recognized as a major cause of premature mortality and health effects in the United States and worldwide. In 2016 a joint study by the World Bank and the Institute for Health Metrics and Evaluation reported that an estimated 3.5 million lives were lost to diseases associated with outdoor and household air pollution. Their study indicated that air pollution (albeit not all related to reactive nitrogen compounds) is the fourth leading risk factor for premature deaths worldwide. The report noted that these deaths cost the global economy an estimated US$225 billion in 2013 (http://www.worldbank.org/en/news/press-release/2016/09/08/air-pollution-deaths-cost-global-economy-225-billion, accessed 19 April 2019). Approximately 90% of people in low- and middle-income countries are exposed to dangerous levels of ambient air pollution.

We certainly cannot discount the significant benefits from nitrogen fertilizer use in developing countries, which has led to an increase in food production along with substantial reductions in malnutrition. This in turn has greatly improved human health in these countries by increasing immune response to parasitic and infectious diseases (Nesheim, 1993). However, one of the major challenges for humanity is to reduce the significant losses of excess reactive nitrogen to the environment from agricultural activities, fossil fuel emissions, and wastewaters while sustainably producing food and maintaining food security throughout the world.

During the past decade numerous studies have published reports and articles that have used various methods to estimate economic losses related to impacts from excess reactive nitrogen in the environment. Many of these studies have focused on the additional costs associated with more effective nitrogen management (particularly in agriculture) and mitigation of adverse effects of reactive nitrogen. This chapter discusses the economic ramifications and consequences associated with impacts from reactive nitrogen on human health, environmental degradation, water treatment for production of potable water, and direct and indirect losses to various economic sectors (e.g., jobs, fishing, housing industry, tourism, recreation, and agriculture). This chapter highlights these economic costs and consequences in three main regions of the world: Europe, the United States, and China. The chapter concludes with a brief discussion of approaches to reduce economic losses associated with reactive nitrogen in the environment.

10.1. ECONOMIC COSTS AND CONSEQUENCES OF REACTIVE NITROGEN IN EUROPE

10.1.1. Cost–Benefit Analyses for Nitrogen Abatement

As part of the European Nitrogen Assessment study (Sutton et al., 2011), Brink et al. (2011) conducted a cost–benefit analysis of reactive nitrogen abatement using data (2000) from four national and regional case studies for 27-member states in the European Union (EU27). The

---

Nitrogen Overload: Environmental Degradation, Ramifications, and Economic Costs, Geophysical Monograph 250, First Edition. Brian G. Katz.
© 2020 American Geophysical Union. Published 2020 by John Wiley & Sons, Inc.

results were compared by expressing the data in euros per kilogram of added reactive nitrogen (from agriculture) or euros per kilogram of reduced reactive nitrogen (using a unit cost approach). They estimated social costs based on "willingness to pay" for human life or health, for ecosystem services and reductions in greenhouse gas emissions. It was found that the total annual damage related to reactive nitrogen in EU27 ranged between 70 and 320 billion euro, which was equivalent to 150–750 euro/capita. Of this estimated total, 75% was related to health damage and air pollution, and this damage cost constituted 1–4% of the average European income. It was also found that social costs of health impacts from NO_x were highest (10–30 euro/kg of reactive nitrogen emission). Other health costs resulted from secondary ammonium particles (2–20 euro/kg N), greenhouse gas balance effects of N_2O (5–15 euro/kg N), from ecosystem impacts via N-runoff (5–20 euro/kg N), and from N-deposition (2–10 euro/kg N). Lower estimated costs were associated with adverse health impacts from nitrate in drinking water (0–4 euro/kg N) and by N_2O via stratospheric ozone depletion (1–3 euro/kg N). The following estimated societal costs to ecosystems and coastal systems from reactive nitrogen emissions (expressed as euro/kg of reactive nitrogen) were from nitrate-N to air (2–10), ammonia-N to air (2–10), and reactive nitrogen to water (5–50), and were related to nitrogen emissions Also, they estimated crop decline losses of 1–3 euro/kg N due to impacts from ozone. Brink et al. (2011) noted there were several uncertainties in these estimates due to dose–response relationships and lack of comparability of "willingness to pay" studies. They also found it difficult in many cases to identify the reactive nitrogen contribution to adverse impacts and abatement measures. For example, it was noted that there was an exceptionally large uncertainty for the cost of reactive nitrogen impacts on surface water and groundwater. Overall, they recommended that NO_x and ammonia abatement needed to be prioritized over the abatement of N_2O emissions; however, it was noted that evidence is somewhat uncertain for health risks associated with airborne ammonium and nitrate particles. Although Sutton et al. (2011) estimated cost–benefits to agriculture ranged from 10 to 100 billion euro/yr, their total estimated environmental costs were 20–150 billion euro/yr, indicating that costs and benefits are about equivalent.

In a follow-up study to the European Nitrogen Assessment (Sutton et al., 2011) Van Grinsven et al. (2013) conducted an extensive cost–benefit analysis of reactive nitrogen (including nitrate, ammonia, N_2O) for Europe (27 member states in the European Union, EU27) and presented their estimated costs and implications with respect to mitigation of nitrogen pollution. Their estimated costs and benefits were based on the methods and concepts described in the European Nitrogen Assessment study (Brink et al., 2011). The economic value of damage from reactive nitrogen accounted for a valuation of health impacts (e.g., treatment, lost productivity, and willingness to pay to reduce risk of premature death or pain or suffering), ecosystem restoration, and reduction in greenhouse gas emissions (N_2O, NH_3). Although they emphasized large uncertainties in their cost estimates due to a combination of various methods, their estimated social costs for adverse effects of reactive nitrogen in 2008 for EU27 ranged from 75 to 485 billion euro/yr (Fig. 10.1). The estimated annual cost of pollution from agricultural nitrogen ranged from 35 to 230 billion euro/yr; whereas the economic benefit from nitrogen use in agricultural production ranged between 20 and 80 billion euro/yr. The estimated cost for ecosystem restoration ranged from 70 to 300 billion euro/yr; however, several important questions were raised regarding restoration costs for terrestrial ecosystems. This large range reflects certain unknown factors such as the willingness of society to pay for restoration costs and the difficulty in estimating the adverse impacts to society. Therefore, the upper limit of the damage cost of reactive nitrogen deposition to terrestrial ecosystems was capped at five times the restoration costs for nitrogen-related ecosystem services.

Van Grinsven et al. (2013) estimated that approximately 60% of the total economic cost was related to emissions to air and 40% to releases to water. They also noted that approximately 60% of the damage costs are related to impacts on ecosystems and 40% to impacts on human health. There also were slight but uncertain estimates of net climate cooling due to nitrogen emissions at the time of their study (2008 data). They emphasized the difficulty in estimating adverse health impacts particularly for ammonium and nitrate in secondary particulate matter. Their low estimate is related to minor health impacts due to inhalation of nitrate and ammonium-containing salt particles; whereas their upper estimate was assumed that secondary airborne particulates are similarly hazardous as human exposure to primary particulates. Their damage cost estimates for 2008 were compared with those in the European Nitrogen Assessment study (2000 data). They found that the new cost estimates were higher than in the previous study (Brink et al., 2011), and these higher estimates were attributed to the inclusion of assessments of damage to aquatic ecosystems from atmospheric deposition, the use of country-specific unit damage cost data in place of using mean values for the EU27, and the use of updated model results for nitrogen loads to rivers and seas (Fig. 10.1). These higher estimated damage costs were found despite a decrease in emissions of nitrate (18%) and ammonia (11%) to the atmosphere compared to the European Nitrogen Assessment (2000) estimates. It was concluded that further reactive nitrogen reductions are needed (especially NH_3 and NO_x emissions from transport

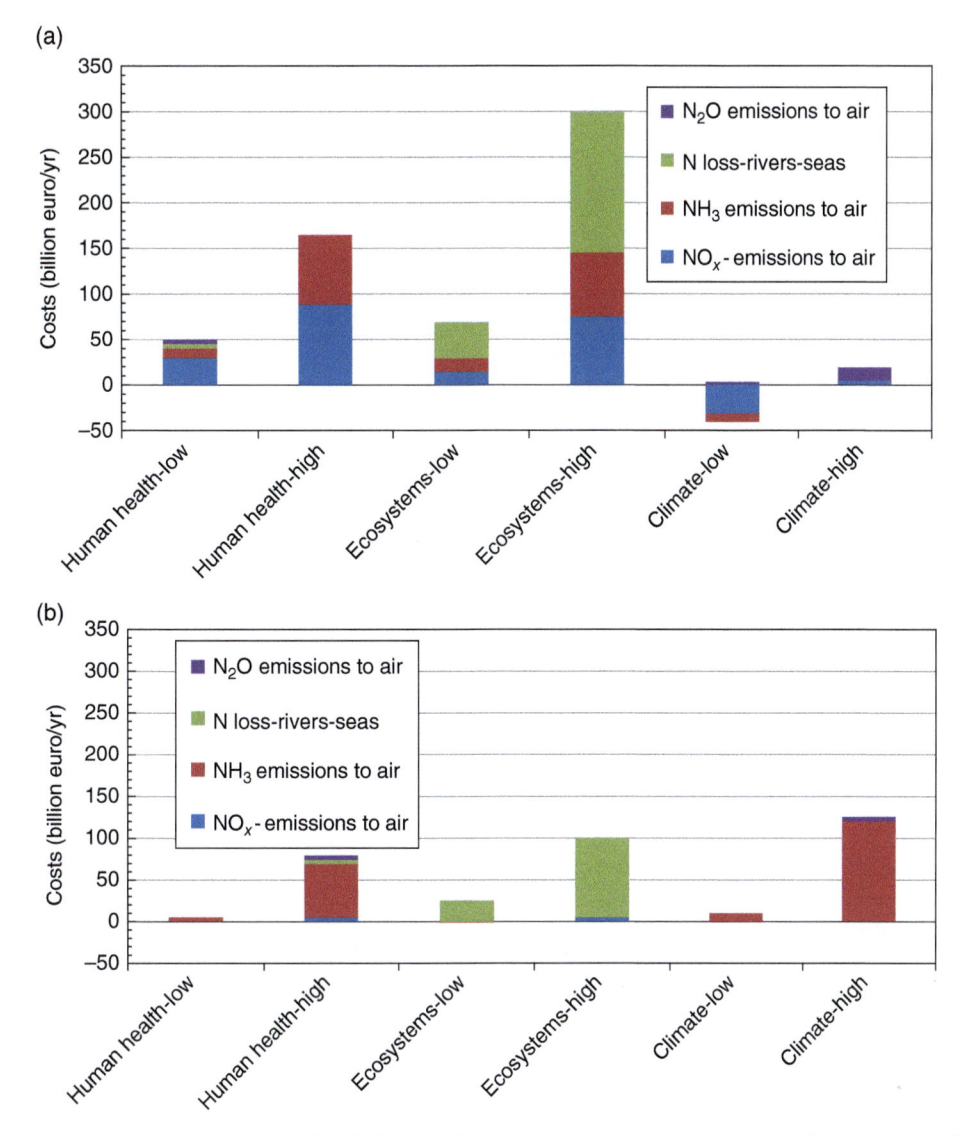

Figure 10.1 Economic costs associated with damage from reactive nitrogen in EU27 member states (a) all sources, (b) agricultural sources. *Source*: Data from Van Grinsven et al. (2013). Reproduced with permission of American Chemical Society.

and agriculture) along with an increase in overall nitrogen use efficiency (NUE) to decrease losses to the environment over the next several decades.

To meet the needs of an increasing world population along with a more affluent diet (particularly increased meat consumption), the Food and Agricultural Organization of the United Nations, the International Fund for Agricultural Development (IFAD), and the World Food Program (WFP) (FAO, 2012) have projected an increased demand in crop production (30%) and livestock production (70%) by 2050. These increases in agricultural practices would likely contribute to increased losses of reactive nitrogen to the environment and global emissions of nitrogen compounds. As many scientists have noted, the main challenges for society are to sustainably produce more nutritional food, guarantee food security, and reduce losses of reactive nitrogen to the environment to prevent further human health maladies and biodiversity loss (Erisman et al., 2015; Sutton et al., 2013). One of the main problems is the lack of agreement among scientists and policymakers on the potential impacts from "intensification" and "extensification" of agriculture and the need for changing diets to a more sustainable food production system (Van Grinsven et al., 2015). Sustainable agricultural "intensification" was defined as "maintaining or increasing food production per hectare without compromising the environment and depleting natural resources." This strategy would involve increasing nitrogen inputs but reducing losses to the environment. They defined sustainable agricultural "extensification" as "decreasing the depletion of natural resources and the environmental impacts while limiting the decrease of food production per hectare." This strategy would involve reducing nitrogen inputs (fertilizers, livestock densities) while minimizing food loss. It was noted that both agricultural "intensification" and "extensification" have important economic consequences.

Van Grinsven et al. (2015) estimated that external costs of losses of nitrogen to the environment from agriculture in the European Union are 0.3–1.9% of the gross domestic product (GDP) in 2008. They examined the plausible benefits (e.g., higher biodiversity, reduced environmental pollution, and lower societal costs) from sustainable "extensification" for agriculture in the EU and the Netherlands by evaluating scenarios that concentrated on reducing nitrogen inputs from fertilizers and livestock densities. They showed that reducing the nitrogen fertilization rate for winter wheat in northwest Europe from 25 to 30% below current nitrogen recommendations would result in a need to compensate producers for a reduction in crop yield by 10–20%. Another scenario involved reducing the pig and poultry sector by 50% in the Netherlands, reducing the dairy sector by 20%, and the synthetic nitrogen fertilizer usage by 40% would result in lowering nitrogen pollution costs from 0.2 to 2.2 billion euro (40%). They noted that this large benefit would make up for the loss of GDP in the primary sector but would not make up for the losses in the supply and processing chain. Van Grinsven et al. (2015) also looked at reducing the consumption and production of animal products by 50% for a 2030 scenario. This dietary change resulted in an estimated 10% reduction of nitrogen pollution with several human health benefits. This course of action would lead to the EU27 exporting more food and reducing land demand outside of Europe in 2030 by more than 100 million hectares (2%). This in turn would balance the increased land demand when agriculture changes to increased organic farming. It was concluded that "extensification" of agriculture in Europe would be sustainable when combined with dietary changes and integration of environmental costs to food prices.

10.1.2. Human Health Costs Related to Nitrate in Drinking Water

As mentioned in Chapter 5, several adverse health effects have been associated with nitrate in drinking water. The European Food Safety Authority (2008) (https://www.efsa.europa.eu/en/corporate/pub/ar08, accessed 28 September 2018) conducted a review of recent epidemiological studies and concluded that there was an overall lack of evidence of an increased risk of colon cancer associated with increased nitrate intake in drinking water. Ward et al. (2018) pointed out the need for future epidemiological studies for evaluating exposures to drinking water nitrate and colon cancer and various other adverse health effects.

To better understand the incidence, risk, and social cost of colon cancer related to nitrate in drinking water, Van Grinsven et al. (2010) evaluated data on colon cancer incidence, nitrogen leaching, and drinking water supply for 11 EU member states. Their cost–benefit analysis study compared health costs with nitrate mitigation costs and social benefits of fertilizer usage and was extrapolated from a case–control. Van Grinsven et al. (2010) found that the risk of colon cancer doubled when humans consume meat in greater than median amounts, and when they are exposed to more than 25 mg/L of nitrate (as NO_3) (5.6 mg/L nitrate as N) in drinking water for more than 10 years. They further estimated that there was a 3% associated increased incidence of colon cancer from nitrate contamination of groundwater used for drinking water in the 11 European states. When averaged over the population for these 11 states, the health loss amounted to 2.9 euro/person, or 0.7 euro/kg of nitrate-N that leaches from fertilizer use. Based on these findings, it was concluded that the current drinking water standards are beneficial to society, although more stringent standards and other measures could be supported. Although they noted that their estimated social costs are somewhat uncertain because only one type of cancer was considered. Their results also pointed out the need for additional improved epidemiological studies.

10.2. ECONOMIC COSTS AND CONSEQUENCES OF REACTIVE NITROGEN IN THE UNITED STATES

10.2.1. Costs Associated with Ecosystem Damages

Dodds et al. (2009) analyzed nutrient data for rivers and lakes throughout the United States and estimated economic losses related to eutrophication. Nutrient data were obtained from the US Environmental Protection Agency (USEPA) STOrage and RETrieval (STORET) data base, the U.S. Geological Survey National Stream Quality Accounting Network (NASQAN), and the National Water Quality Assessment Program (NAWQA). They compared total nitrogen and phosphorus concentrations for the US EPA's 14 nutrient ecoregions (https://www.epa.gov/nutrient-policy-data/ecoregional-nutrient-criteria-rivers-streams, accessed 14 September 2018) with reference conditions (Smith et al., 2003). Total nitrogen data were evaluated for 1587 river stations and for 980 lakes stations. At the time of the study, Dodds et al. (2009) found that total nitrogen and total phosphorus concentrations exceeded median reference values in more than 90% of the rivers and lakes in all nutrient ecoregions. They assessed yearly potential economic costs related to losses in recreational water usage, waterfront real estate, spending on recovery of threatened and endangered species and drinking water treatment. Annual combined economic losses related to eutrophication of U.S. freshwaters were estimated at approximately $2.2 billion (recreational water usage, $1 billion; waterfront property, $0.3–$2.8 billion; recovery of threatened and endangered species, $44 million; and drinking water, $813 million).

Using data from the 2007 U.S. Fish and Wildlife Endangered Species database (www.fws.gov; accessed 14 September 2018) it was found that 156 species were covered in 112 recovery plans that began between 1981 and 2007. Of these, 60 listed species at the time of the study were likely imperiled due to eutrophication (Richter et al., 1997). It was estimated that the annual cost of these 60 recovery plans was $732,800, which translates to $44 million/yr was spent to prevent eutrophication-related losses of aquatic biodiversity. However, they noted that their estimated losses for each category were conservative and likely undervalued because more precise data are needed on algal blooms, fish kills, and associated human health costs to further refine these estimated losses.

Compton et al. (2011) compared the costs of nitrogen-related impacts on ecosystem services using the metric of cost per unit of nitrogen. Adverse effects on human respiratory health from air pollution (e.g., costs of ozone and particulate damages) were estimated at $28/kg NO_x-N. Although it was noted that damages estimated related to losses of productivity, biodiversity, clean water, and recreation were less certain and likely varied between <$2.2–56/kg N. Using the Chesapeake Bay restoration effort as an example, the available damage costs are substantially greater than the projected abatement costs to reduce nitrogen loads to the Bay ($8–15/kg N). It was concluded that by accounting for effects on multiple ecosystem services, this strategy can provide decision-makers with an integrated view of reactive nitrogen sources, damages, and abatement costs with the ultimate goal of reducing N pollution.

Nutrient pollution is a main threat to nearshore ecosystems, such as marshes, mangroves, kelps, and seagrasses. Based on a global assessment of 215 seagrass studies, Waycott et al. (2009) reported seagrasses have been disappearing at a rate of 110 km²/yr since 1980 and the global cover of seagrasses has declined more than 29% during the last century due to nutrient and sediment runoff. Seagrass beds provide important environmental and ecosystem services, such as a sink for global carbon, habitat for fish, birds, and invertebrate species, and filters for water clarity. Waycott et al. (2009) estimated that these services conservatively amounted to $1.9 trillion/yr in terms of nutrient cycling, enhancement of coral reef fish productivity, and important food sources for endangered species. In addition, seagrass meadows support commercial fisheries estimated to be worth about $3500/ha/yr (Waycott et al., 2009).

10.2.2. Costs Associated with the Agricultural Production of Maize in the United States

Maize is a major agricultural crop grown and produced in the United States that is used for animal feed, ethanol biofuel, and human consumption. Geographically, most maize is produced in the "Corn Belt," which includes the top five producing states (Iowa, Illinois, Nebraska, Minnesota, and Indiana). The production of maize generates a substantial amount of air pollution, mainly resulting from emissions of ammonia. Ammonia emissions from fertilizer use and manure spreading contribute to fine particulate matter ($PM_{2.5}$), which is a major cause of premature mortality and health effects in the United States and worldwide. As mentioned previously, a joint study by the World Bank and the Institute for Health Metrics and Evaluation estimated that 3.5 million lives were lost to diseases associated with outdoor and household air pollution. These deaths cost the global economy an estimated US$225 billion in 2013 (http://www.worldbank.org/en/news/press-release/2016/09/08/air-pollution-deaths-cost-global-economy-225-billion, accessed 19 April 2019).

Hill et al. (2019) conducted a detailed study of the health and economic impacts from air emissions of pollutants from farms and the supply chains that produce the chemical and energy inputs used in maize production. They reported that reduced air quality from maize production is associated with 4300 premature deaths annually in the United States, with estimated monetary damages of US$39 billion (range US$14–$64 billion). It was also estimated that average health damages from reduced air quality were equivalent to US$121/ton of harvested maize grain, which was 62% of the US$195/ton decadal average maize grain market price. Hill et al. (2019) also noted that greenhouse gas emissions from maize production resulted in total climate change damages of US$4.9 billion (range US$1.5–$7.5 billion).

10.2.3. Other Economic Losses Associated with Nitrogen Pollution

In 2015, the US EPA released a report that compiled information on cost estimates associated with the impacts and control of nutrient pollution in the United States (US EPA, 2015) (https://www.epa.gov/sites/production/files/2015-04/documents/nutrient-economics-report-2015.pdf, accessed 28 September 2018). The report compiled data from regional and local research studies conducted from 2000 through 2012. There were no estimates of total annual costs for potential losses due to nutrient pollution, eutrophication, and algal blooms for the entire United States (as in the Dodds et al., 2009 study). However, numerous local-scale examples of economic losses and costs from various states were provided for the following categories: tourism and recreation (including decreased restaurant sales, lakeside business closures, decreased tourism spending in local areas, declines in park revenues, and lost labor income); commercial fishing (reduced harvests, fishery closures, and increased processing costs associated with increases in shellfish poisoning risks), property values of waterfront and nearby homes (poor water clarity); human health issues (including respiratory issues, skin rashes, neurotoxic shellfish poisoning from consumption of contaminated shellfish); drinking water treatment costs (e.g., taste and odor, removal of algal toxins); mitigation (e.g., in-lake aeration, alum treatments, biomanipulation, dredging, and herbicide treatments to control nuisance aquatic vegetation); and restoration (developing nutrient criteria, total maximum daily loads, watershed plans, nutrient trading, and offset programs). Costs were also estimated for nutrient pollution control for point and nonpoint sources including municipal wastewater treatment plants, decentralized wastewater treatment systems, industrial wastewater treatment, and urban and residential runoff. The EPA Report concluded with an oft-repeated common-sense message that preventing nutrients from entering the environment is a much more cost-effective strategy for addressing impacts of nutrient pollution than spending enormous amounts of monies for controlling pollution after nutrients are released.

Sobota et al. (2015) spatially assessed potential damage costs related to anthropogenic inputs of reactive nitrogen across the conterminous United States. They estimated the amounts of reactive nitrogen released to the environment (via emissions to the atmosphere, atmospheric deposition, surface freshwater, groundwater, and coastal zones in the early 2000s for large watersheds based on the 8-digit U.S. Geologic Survey Hydrologic Unit Codes; HUC8s; approximately 2200 watersheds with average size of 1813 km^2) (Fig. 10.2). The anthropogenic nitrogen sources included agricultural (synthetic fertilizers, cultivated biological nitrogen fixation, and confined animal feeding operations), wastewater, atmospheric inorganic, and nitrogen deposition (that accounted for NO_x and NH_3 emissions). Sobota et al. (2015) describe the specific details about the methods involving the gridded data and their GIS techniques. Damage costs related to specific nitrogen source inputs were summarized by Sobota et al. (2015) from several previous studies (Birch et al., 2011; Compton et al., 2011; Dodds et al., 2009; Kusiima & Powers, 2010; Van Grinsven et al., 2013). Table 10.1 (from Sobota et al., 2015) lists the nitrogen damage type, system affected, a range of damage costs for each nitrogen source, and associated references.

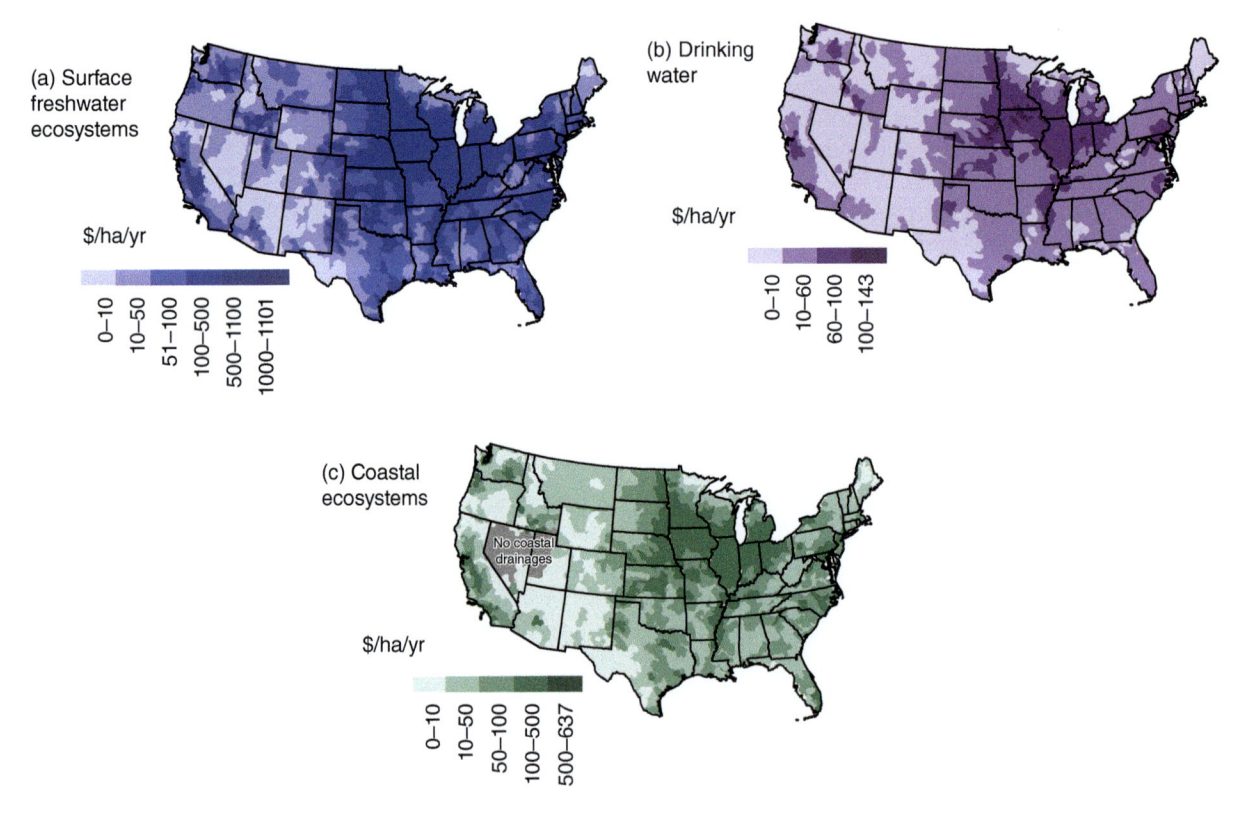

Figure 10.2 Estimated potential damage costs from anthropogenic inputs of reactive nitrogen in the United States to (a) freshwater ecosystems, (b) drinking water, and (c) coastal ecosystems. Light gray area in bottom map denotes "no coastal drainages", darker gray shades indicate high damage costs. *Source*: Modified from colored figure 6 in Sobota et al. (2013). Reproduced with permission of IOP Publishing.

Table 10.1 Anthropogenic nitrogen sources, systems affected, range of potential damage costs from reactive nitrogen (in $/kg; 2008 USD) for each nitrogen damage type, and associated references. Negative values show an economic benefit.

Nitrogen damage type	System affected	Cost ($/kg N) Range	Cost ($/kg N) Median	Reference
Atmospheric NO_x				
Increased incidence of respiratory disease	Air/Climate	12.88–38.63	23.1	Birch et al. (2011), Van Grinsven et al. (2013)
Declining visibility-loss of aesthetics	Air/Climate	0.31–0.31	0.31	Birch et al. (2011)
Increased effects of airborne particulates/ increased carbon sequestration in forests	Air/Climate	−14.17	−4.51	Van Grinsven et al. (2013)
Increased damage to buildings from acid	Land	0.09–0.09	0.09	Birch et al. (2011)
Increased ozone exposure to crops	Land	1.29–2.58	1.51	Birch et al. (2011), Van Grinsven et al. (2013)
Increased ozone exposure to forests	Land	0.89–0.89	0.89	Birch et al. (2011)
Increased loss of plant biodiversity	Land	2.58–12.88	7.73	Van Grinsven et al. (2013)
From atmospheric NH_3				
Increased incidence of respiratory disease	Air/Climate	2.58–25.75	4.93	Birch et al. (2011), Van Grinsven et al. (2013)
Declining visibility-loss of aesthetics	Air/Climate	0.31–0.31	0.31	Birch et al. (2011)
Increased effects of airborne particulates/ increased carbon sequestration in forests	Air/Climate	3.86–1.93	−1.93	Van Grinsven et al. (2013)
Increased damage to buildings from particulates	Land	0.09–0.09	0.09	Birch et al. (2011)
Increased loss of plant biodiversity	Land	2.58–12.88	7.73	Van Grinsven et al. (2013)
From N_2O				
Increased ultravioloet light exposure from ozone on humans	Air/Climate	1.29–3.86	1.33	Compton et al. (2011), Van Grinsven et al. (2013)
Increased emission of a greenhouse gas	Air/Climate	5.15–21.89	13.52	Van Grinsven et al. (2013)
Increased ultraviolet light exposure to crops from ozone	Air/Climate	1.33–1.33	1.33	Birch et al. (2011)
From surface freshwater nitrogen loading				
Declining waterfront property	Freshwater	0.21–0.21	0.21	Dodds et al. (2009)
Loss of recreational use	Freshwater	0.17–0.17	0.17	Dodds et al. (2009)
Loss of endangered species	Freshwater	0.01–0.01	0.01	Dodds et al. (2009)
Increased eutrophication	Freshwater	6.44–25.75	16.1	Compton et al. (2011), Van Grinsven et al. (2013)
Undesirable odor and taste	Drinking water	0.14–0.14	0.14	Kusiima and Powers (2010)
Nitrate contamination	Drinking water	0.54–0.54	0.54	Compton et al. (2011)
Increased colon cancer risk	Drinking water	1.76–5.15	1.76	Van Grinsven et al. (2013)
From groundwater nitrogen loading				
Undesirable odor and taste	Drinking water	0.14–0.14	0.14	Kusiima and Powers (2010)
Nitrate contamination	Drinking water	0.54–0.54	0.54	Compton et al. (2011)
Increased colon cancer risk	Drinking water	1.76–5.15	1.76	Van Grinsven et al. (2013)
From coastal nitrogen loading				
Loss of recreational use	Coastal zone	6.38–6.38	6.38	Birch et al. (2011)
Declines in fisheries and estuarine/marine habitat (range does not include $56/kg N from submerged aquatic vegetation loss in Gulf of Mexico (Compton et al., 2011)	Coastal zone	6.00–26.00	15.84	Compton et al. (2011), Van Grinsven et al. (2013)

Source: Modified from Sobota et al. (2015). Reproduced with permission of IOP Publishing.

For each HUC8, nitrogen inputs were multiplied (referring to as nitrogen leakage estimates) with published coefficients related to nutrient uptake efficiency, leaching losses, and gaseous emissions. Their nitrogen leakage estimates were then scaled with mitigation, remediation, direct damage, and substitution costs related to human health, agriculture, ecosystems, and climate (per kg of N) to estimate annual damage costs (in 2008 $US) attributable to anthropogenic nitrogen sources in each HUC8. Their estimates of reactive nitrogen leakage ranged from <1 to 125 kg N/ha/yr for the HUC8s, with most of the reactive nitrogen was leaked to freshwater ecosystems. Based on median estimates for each watershed, potential damages ranged from $1.94 to $2255/ha/yr, with a median estimate of $252/ha/yr. The most notable damages for all watersheds were related to eutrophication of freshwater ecosystems and respiratory impacts from atmospheric nitrogen pollution. It was cautioned that they were not able to fully evaluate damages from reactive nitrogen due to a substantial amount of missing information related to damage costs from harmful algal blooms (HABs) and drinking water contamination. However, they noted that for the United States, potential health and environmental damages from anthropogenic reactive nitrogen in the early 2000s totaled $210 billion/yr (with a range of $81–$441 billion/yr). Although there are some uncertainties remaining in their calculations, it was felt that these estimates provide stakeholders important information on the adverse effects and tremendous costs associated with pollution and the impetus for making better decisions about reducing nitrogen inputs to the environment.

10.2.4. Harmful Algal Blooms

As of mid-August 2016, the USEPA noted that states across the United States have reported more than 250 health advisories due to HABs that year. The National Aquatic Resource Surveys conducted by the EPA and state and tribal partners in 2012 found that 34% of the lakes surveyed in the United States had high levels of nitrogen associated with harmful ecological impacts. Nonaquatic animals also are affected when consuming waters containing algae. Pets and livestock have died after drinking water containing algal blooms, including 32 cattle on an Oregon ranch in July 2017 (Flesher & Kastanis, 2017).

The economic impacts from HABs in the United States have been reported for more than 20 years. For the period 1987–1992, Anderson et al. (2000) estimated the following annual average economic costs (reported in millions of dollars [$M] in 2000) from HAB impacts in four categories: public health, $22.2 M; commercial fishery, $18.4 M; recreation and tourism, $6.6 M; and monitoring and management, $2.1 M. The estimated average annual total cost for the 6-year period was $49.3 M with a range of $34–$82 M. Public health impacts mainly included human sickness and death from eating seafood tainted with toxins and the dinoflagellate toxin ciguatera. Impacts to commercial fisheries included wild harvest and aquaculture losses of fish and shellfish. Recreation and tourism impacts likely were way underestimated because there was very limited data available. Costs of monitoring and management involve water quality testing for toxins, plankton monitoring, and other management activities. However, it was cautioned that these estimates likely were highly conservative because of the difficulty in obtaining economic impacts. From the 1970s to their 2000 study, it was estimated that cumulative impacts from HABs were likely about $1 billion. It was also noted that effects of economic multipliers (cascading effects) were not included, which could increase the estimates by severalfold. Outbreaks of HABs in some localized areas could easily exceed the estimated annual averages reported. For example, they noted that a 1976 red tide in New Jersey caused estimated economic losses of more than $1 billion (2000 dollars). Hoagland et al. (2002) noted that there have been more than 60,000 incidents of human exposure to algal toxins annually in the United States, which have resulted in approximately 6500 deaths. It was estimated that annual costs approaching US$50 million related to public health and medical treatments, shellfish recalls, fish mortalities, commercial fisheries losses, environmental monitoring and management budgets, decreased tourism, and other related expenses.

10.2.5. Drinking Water Treatment

Numerous studies have documented the enormous costs associated with the treatment of drinking water in the United States (to remove algal toxins and algal decomposition products) and advanced treatment of human wastewaters (municipal wastewater treatment facilities) and management practices to process animal wastes from confined animal feeding operations. Collectively, Federal and State agencies have spent billions of dollars on trying to alleviate water-quality degradation from excess nitrogen; however, recent studies have shown little to no improvement in water quality. In 2004, the Drinking Water Infrastructure Needs Survey estimated that $150.9 billion will need to be spent on drinking water systems to provide safe treatment, storage, and distribution (US EPA, 2004). A 2009 report by the State-USEPA Nutrient Innovations Task Group (2009) noted that current nutrient control strategies are woefully inadequate, and the rate of nutrient pollution will likely increase as the world population continues to increase.

The population of the United States is projected to increase by more than 135 million over the next 40 years. As population grows, more nutrients (mainly reactive nitrogen compounds) will be released to the environment from municipal wastewater discharges and from other sources such as wastes from agricultural livestock, and row-crop runoff are likely to grow. Examples are provided below for two areas in the United States, where significant economic losses have occurred due to nitrogen pollution and eutrophication: The Great Lakes and Florida's fresh and coastal water bodies.

10.2.6. Great Lakes

The Great Lakes in the United States constitute one of the largest freshwater ecosystems in the world. Together they account for about 20% of the Earth's liquid surface fresh water. About 40 million people depend on the Great Lakes for drinking water. Crawford (2017) reported that the eight states and one Canadian province that border the Great Lakes supply 1.5 million jobs and contribute $62 billion in wages to the U.S. economy. Along with water pollution, invasive species have caused significant damage to the ecosystem. For example, the sea lamprey, which spread through all five Great Lakes, killed native fish (e.g., trout, lake sturgeon, and walleye) that drastically reduced commercial fishing. Now attempts are being made to restore a healthier native fishery, but it will likely take decades (Crawford, 2017). The U.S. and Canadian programs to control the lamprey in spawning rivers was costing more than $20 million/yr. In addition, significant losses of wetlands have occurred over the past 50 years. Habitat losses have also hurt migratory bird species.

The Great Lakes Environmental Assessment and Mapping (GLEAM) Project (http://greatlakesmapping.org/great_lake_stressors/7/nitrogen-loading, accessed 20 September 2018) integrates spatial information to help develop restoration, conservation, and management strategies in the Great Lakes. One of their objectives is to spatially show nitrogen loading from various anthropogenic nitrogen sources. Nitrate-nitrogen input to the Great Lakes typically comprises most of the total inorganic nitrogen. Loading to the Great Lakes originates from tributaries and atmospheric deposition. The map below (Fig. 10.3) shows estimated nitrogen inputs from

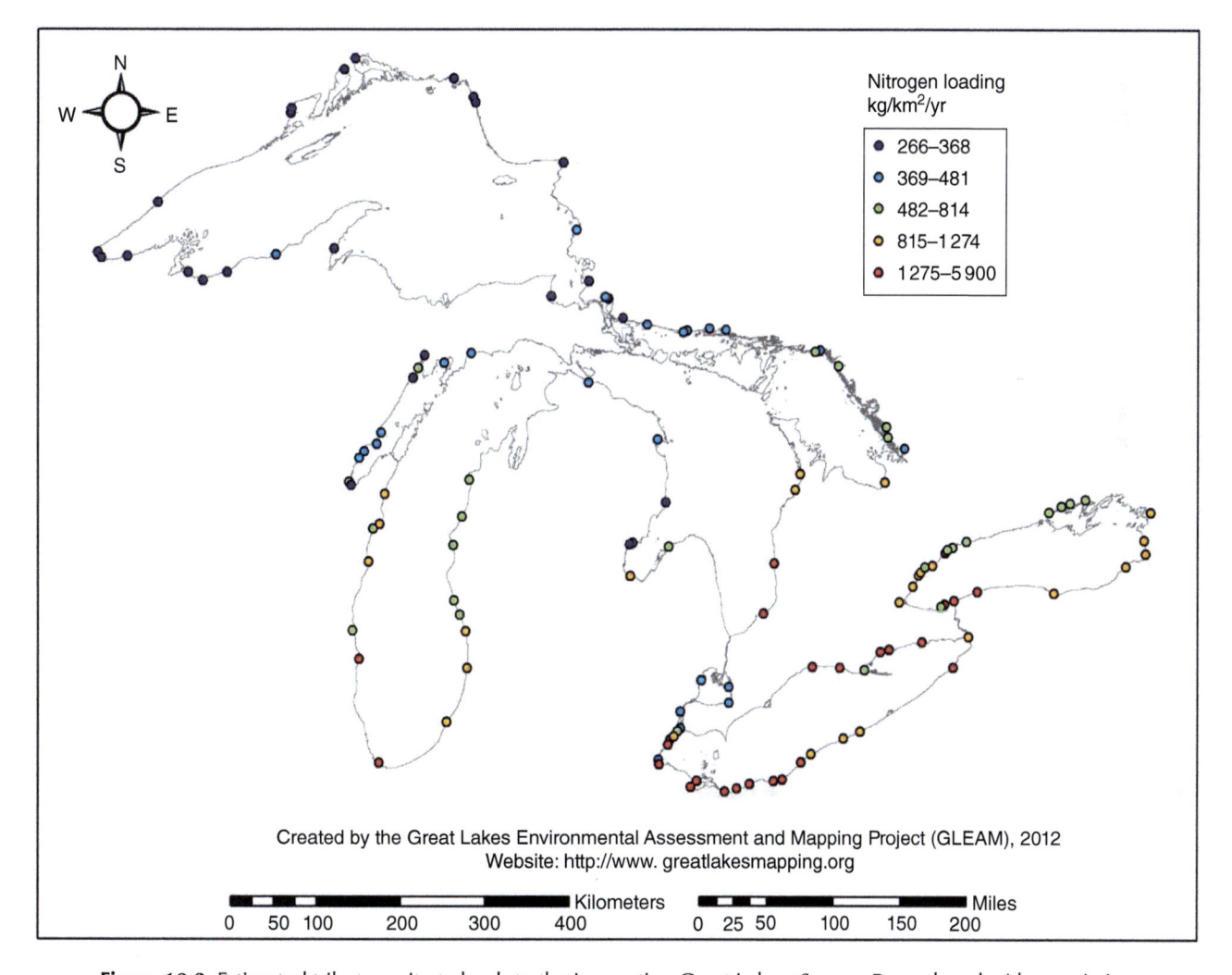

Figure 10.3 Estimated tributary nitrate loads to the Laurentian Great Lakes. *Source*: Reproduced with permission of John Wiley & Sons.

206 NITROGEN OVERLOAD

tributaries (small watersheds were not included due to data unavailability) to the Great Lakes. Tributary nitrate loads were averaged for 1994–2008, except for Lake Erie, where tributary data were available only for 2005. The estimated nitrogen inputs include the most important tributary sources and comprise nearly 60–80% of the total tributary load to each lake. Data used to estimate nitrogen inputs from atmospheric deposition were from the National Atmospheric Deposition Program (NADP, wet nitrate deposition) and the Clean Air Status and Trends Network (CASTNET, dry nitrate deposition).

To estimate nitrogen loading as a stressor to the Laurentian Great Lakes, tributary nitrogen loads were attenuated spatially from river mouths based on the assumption that the nitrogen declines to 10% of its initial value at 15 km and to 1% at 30 km. The spatially attenuated nitrate from tributaries was combined with nitrogen from atmospheric deposition over the lake surface to develop a map of spatial distribution of nitrogen loading (Fig. 10.4). Propagated nitrate was then combined with atmospheric deposition over the lake surface.

Current restoration investments are greater than US$1.5 billion in North America's Laurentian Great Lakes (Allan et al., 2015). Spatial variation was analyzed in five recreational cultural ecosystem services (CES) (including sport fishing, recreational boating, birding, beach use, and park visitation) that generate economic activity in the Great Lakes region. They evaluated the spatial coincidence of these services and identified locations of high total service delivery. They assessed evidence for the economic benefits of service delivery using GDP for tourism and recreation (T&R), two benefits that are the most quantifiable measures of societal benefit from recreational CES. The spatial intersection of new estimates of service delivery was also compared with prior estimates of ecosystem stress to determine if restoration efforts could target stressor alleviation in locations where current service provisioning suggests high potential benefits. They found that recreational CES are widely but unevenly distributed, and that many locations support multiple CES benefits. They noted that a total of US$15.4 billion in GDP was generated within the Great

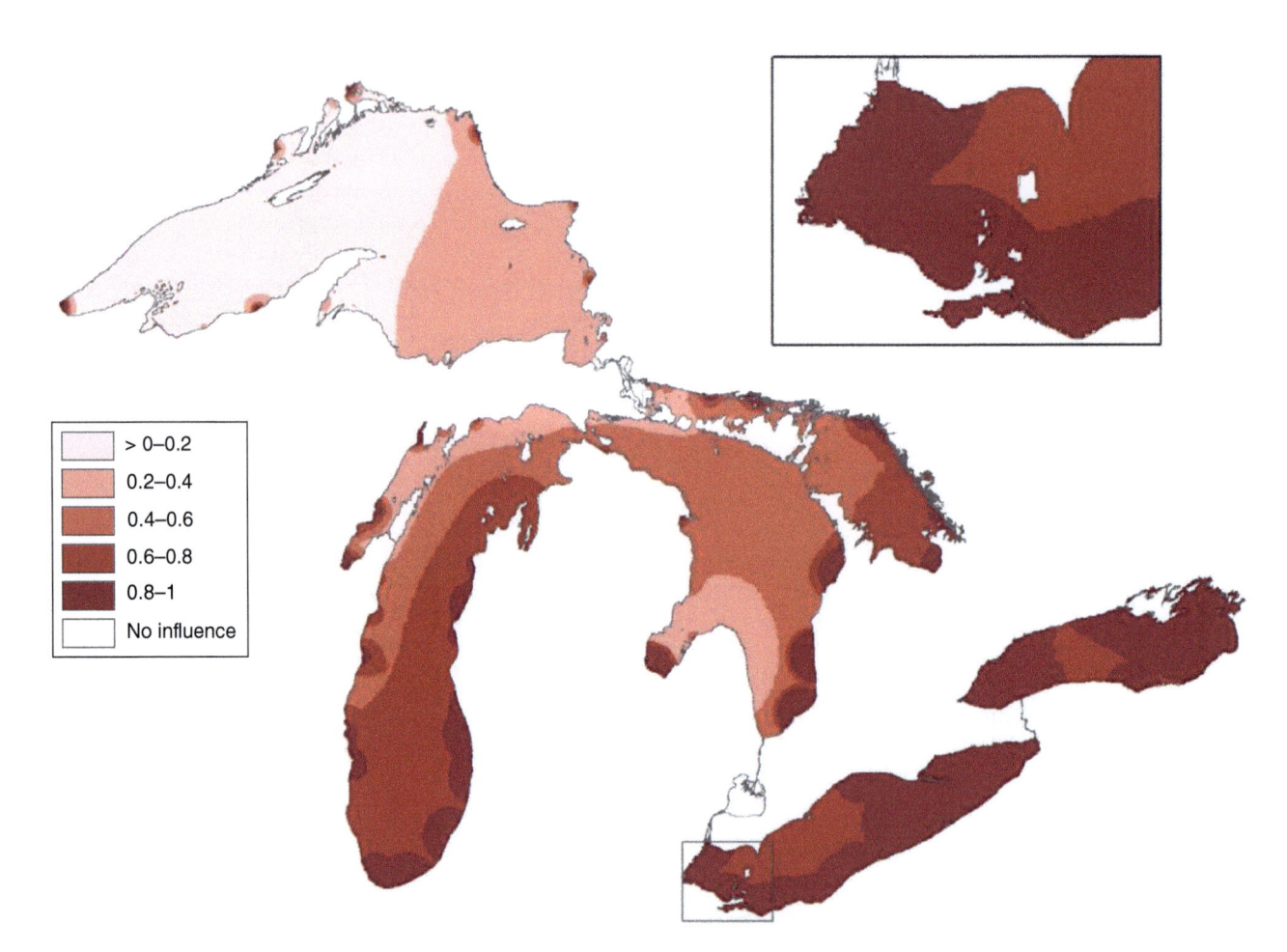

Figure 10.4 Spatial distribution of nitrogen loading as a stressor in the Laurentian Great Lakes (Inset: Western Lake Erie). *Source*: Reproduced with permission of John Wiley & Sons.

Lakes shoreline counties in 2010 in United States, based on lake-associated sectors. Tourism and recreation accounted for US$8.3 billion, or 50.2%, of the total, with the remainder attributable mainly to marine transportation. The combination of the five-service metrics correlated with tourism GDP, which showed that local economies are helped by ecosystem conditions that support CES. Environmental stressors have affected areas with high recreational CES delivery, but their assessment indicated that ecosystem condition or human enjoyment of these recreational CES was resilient even to substantial levels of stress. They concluded that their spatial assessments of recreational CES in combination with ecosystem stress assessments were useful in developing regional-scale restoration efforts.

Toxic algal blooms in Lake Erie have been caused by wastewater, urban stormwater, and agricultural runoff from fertilizers. The Lake Erie region provides drinking water to 3 million Ohio residents. In 2011, a blue-green algal bloom occurred over an area of 4970 km^2 on the surface of Lake Erie. Crawford (2017) noted that the algal blooms have returned every summer since 2011. The city of Toledo, Ohio, uses water from Lake Erie for its public water supply. In August 2014, the city of Toledo warned its 500,000 metro residents not to drink, bathe in, or boil their tap water. Related to this, approximately 60 people were hospitalized with abdominal pain. A state of emergency was declared by the governor and the National Guard distributed thousands of gallons of bottle water to residents (Wolf & Klaiber, 2017). This incident was caused by a massive blue green algae (cyanobacteria) that produced the freshwater toxin microcystin, which is harmful when consumed by humans and animals. Water treatment facilities in Ohio have been spending millions of dollars to upgrade their facilities to treat cyanobacteria blooms. Economic consequences of these HABs have extended loss of revenue from reduced recreational opportunities to capitalization losses associated with homes in lakeshore communities.

Wolf and Klaiber (2017) used a hedonic model approach (used in economics as a way of estimating demand or value related to contributory value from various factors) to link property price impacts to blue-green algae outbreaks for four inland lakes across six counties in Ohio. They included information on extensive time-varying water-quality monitoring data for the four lakes along with detailed housing transaction data from county auditors. The capitalization losses associated with near-lake homes were 11–17% and as high as 22% for lake-adjacent homes for the period 2009–2015. For one of the lakes (Grand Lake Saint Marys), they estimated a loss per house of $17,335, which when multiplied by the total number of single-family residences within 600 m of the lake results in an overall capitalization loss of over $51 million.

The Great Lakes Commission (2014) estimated that Lake Erie provides $10.7 billion in tourism benefits per year along with a $2 billion sports fishery. Also, a recreational boating industry on Lake Erie supports more than 26,000 jobs with an annual economic impact of $3.5 billion. However, adverse effects from HABs threaten these recreational activities and the associated economy. Wolf et al. (2017) investigated the impacts of algal blooms on recreation for eight counties in Ohio that border Lake Erie. They collected data on monthly fishing license counts, algal (cyanobacteria) measurements, rainfall, water temperature, distance to lake from residence, and algal bloom advisories from 2009 to 2014. They found that when mean monthly algae levels exceeded advisory thresholds (20,000 cyanobacteria cells/mL), there was a substantial decrease in monthly permit sales of 10.6%. This would translate to a decrease of 3600 licenses and approximately $2.25–$5.58 million in lost fishing expenditures.

Since 2010, the Great Lakes Restoration Initiative (administered by the US EPA and other federal agencies) has provided more than $2 billion in funding to conservation groups and government entities to clean up and restore the lakes. Programs also are underway to help farmers use fertilizers more efficiently to minimize erosion and runoff of nutrients. The Great Lakes Commission reported a study by the Brookings Institute that concluded that implementation of the Great Lake restoration strategy would result in $80–$100 billion in benefits that would include $50–$125 million in reduced costs to municipalities, $12–$19 billion in increased property values in degraded shoreline areas, and $6.5–$11.8 billion in revenue from tourism, fishing, and recreation (Great Lakes Commission, 2014).

10.2.7. Florida's HABs and Coastal Red Tide

Numerous harmful algal outbreaks in Florida from nutrient pollution after the adoption of the Clean Water Action Plan by the US EPA in 1998 prompted a coalition of environmental organizations represented by EarthJustice to file a lawsuit to force the US EPA to establish numeric nutrient criteria to prevent further degradation of waterbodies. Impacts from these events were variable depending on species and intensity. For example, cyanobacteria and *Karenia brevis* are toxic to humans, pets, and other aquatic species. In humans, they can lead to respiratory problems, neurological harm, or food poisoning. Algal blooms also release wind-blown toxins, noxious odors, and severely degrade the aesthetic value of a waterbody. Massive fish kills have also resulted from algal blooms due to depletion of oxygen levels in waterbodies. Stanton and Taylor (2012) estimated a range of $1.3–$10.5 billion annually as a valuation of Florida's clean water and the price tag on the cost of the state's water pollution. They based their estimate on methodology that used literature information on willingness-to-pay for nonuse values (similar to method used by EPA). However, these values were not based on Florida-specific economic analyses, and may not cover the complete extent and degree of anthropogenic derived threats to Florida's water quality. More studies would be needed at local scales to document the degradation of waterbodies over time to derive a more reliable economic analysis.

In southeast Florida (Martin County), a study by Florida Realtors used information on Secchi disk depth to determine changes in real estate property values. They noted that Secchi disk depth data were more reliable than chlorophyll a, turbidity, and dissolved oxygen in terms of capturing homebuyers' and sellers' perceptions of water quality. They found that changes in the water quality of the southeastern Florida (St. Lucie Estuary, Loxahatchee Estuary, and the portion of the Indian River Lagoon north of the St. Lucie Inlet) (Fig. 10.5) – as measured by changes to 1-year average Secchi disk depth at each monitoring point – resulted in an estimated $488 million reduction in Martin County's aggregate property value between May 2013 and September 2013.

Lake Okeechobee is a very large (1890 km^2) shallow lake that contains elevated nutrient levels and a long history of algal blooms (Fig. 10.5). Anthropogenic activities have contributed large quantities of nutrients to watershed soils, wetlands, and sediments in tributaries and in the lake (Havens, 2013). At times light penetration is greatest and water temperatures are elevated, the lake experiences substantial algal growth, particularly blue-green algae that have adapted to these conditions resulting in large blooms (Paerl & Huisman, 2008). In fact, Havens et al. (2016) noted that certain species of blue-green algae can adjust their vertical position in the water column by regulating their buoyancy thereby finding optimal depths for light and nutrient availability. Cyanobacteria can then bloom in open waters of the lake where light is limited and not favorable for other algae species. Heavy rainfall in May 2018 transported excess nutrients from the local basin into the lake, which in combination with hot summer days and sunlight penetration created optimal conditions for a large algal bloom. Satellite images produced by NOAA show how the cyanobacteria bloom in Lake Okeechobee changed over time, which started in June 2018 and spread to other parts of the lake in July 2018 (Fig. 10.6). The high rainfall amounts may also have contributed to massive algal blooms in the St. Lucie Estuary, as water released from the lake (to prevent damage to the dike) or water from upstream tributaries in the St. Lucie River basin, may have introduced freshwater and nutrients into the estuary causing conditions favorable for algal blooms.

Red tide events have also occurred annually along many parts of the Florida coastline. Tens of miles of beaches on both the east and west coasts of Florida have been closed in 2018 due to airborne irritants from these events that have caused symptoms such as skin irritation, burning and teary eyes, coughing, and sneezing. Officials have warned people with asthma and COPD to avoid these areas. Red tide occurs naturally and is caused by a dinoflagellate, or marine microorganism called *K. brevis*. The severity of red tide events is being fueled by excess nutrients (nitrogen and phosphorus) from various anthropogenic sources. The red tide in Florida is found in bays and estuaries but not in freshwater systems (e.g., lakes and rivers) because *K. brevis* cannot survive in low-salinity waters for very long, so blooms usually remain in salty coastal waters. In 2018, the State of Florida has provided $13 million in grant funding for communities impacted by red tide and blue-green algae. For 2019, the state legislature has allocated $10.8 million to the Florida Department

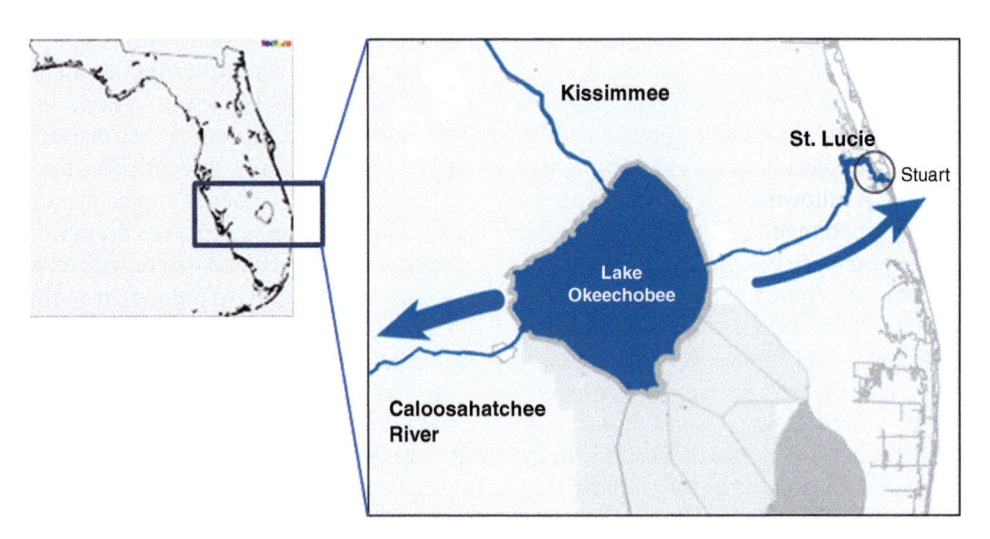

Figure 10.5 Map showing location of Lake Okeechobee, Caloosahatchee River, and St. Lucie River and Estuary in Florida.

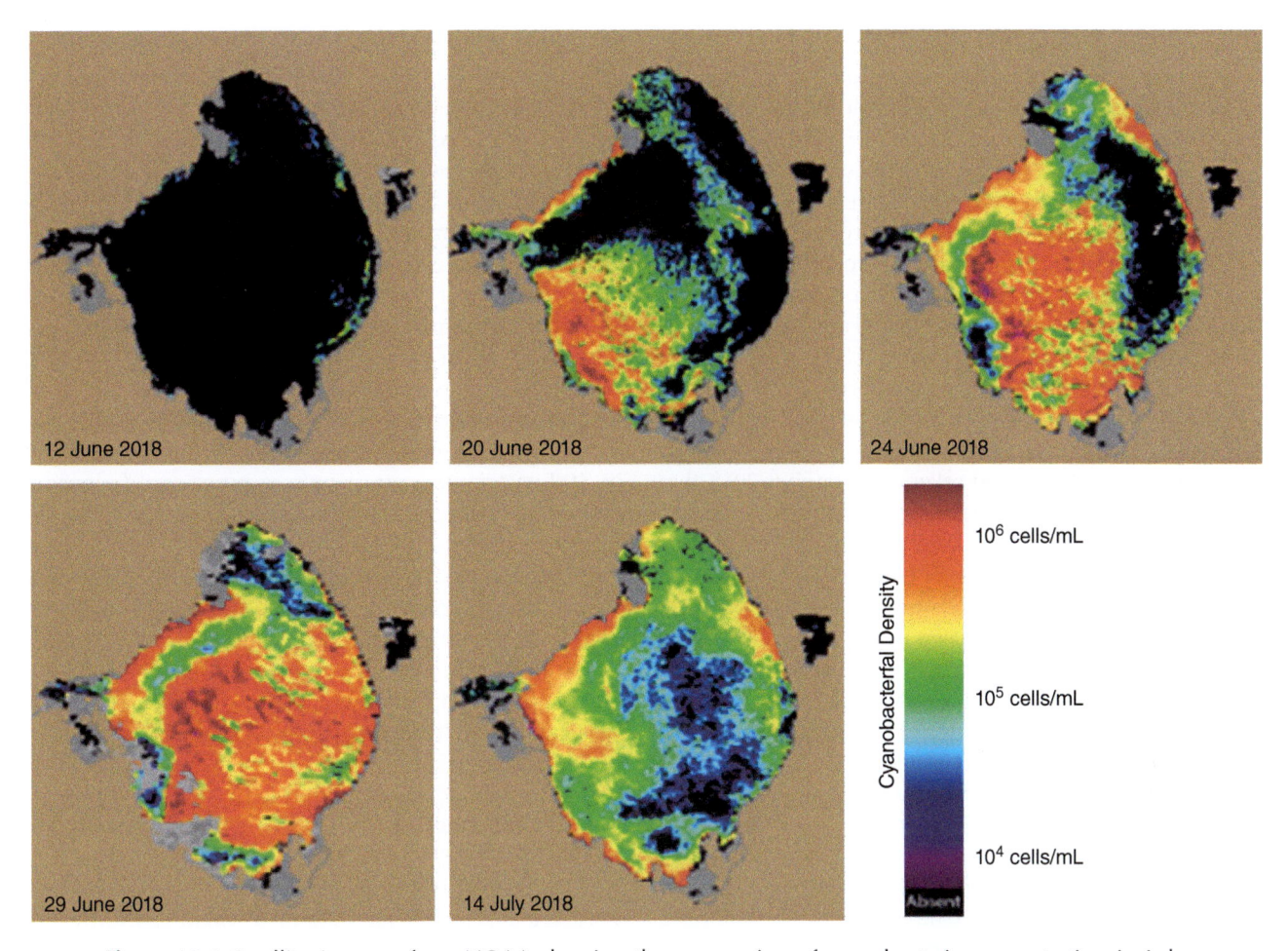

Figure 10.6 Satellite imagery from NOAA showing the progression of cyanobacteria concentration in Lake Okeechobee. 2018. Cyanobacterial density ranged from 10 E04 to 10 E06 (10^4 to 10^6) cells/mL. *Source*: NOAA (2018) and http://blogs.ifas.ufl.edu/extension/2018/07/10/algal-blooms-faq/, accessed 10November2019.

of Environmental Protection (DEP) for a blue-green algae task force that will seek solutions to the toxic blooms that threaten the state marine environment.

In August 2018, the Florida governor announced that the DEP and the U.S. Army Corps of Engineers (Corps) have reached an agreement on a $50 million state investment to fund repairs to the main dike surrounding Lake Okeechobee. This project is designed to reduce the harmful water releases by allowing more water storage in the lake. This funding agreement is on top of a $50 million investment made earlier in 2018, bringing the State of Florida's total investment in this federal project to $100 million. Also announced by the governor in 2018 was that the Florida DEP would be providing $700,000 to Lee County to combat algae and remove it from various tributaries and canals along the Caloosahatchee River. This funding is part of a $3 million grant program by an executive order from the governor. This grant program will provide needed funding to local governments for cleaning up waterways affected by algal blooms.

These are just a small part of the enormous anticipated total costs associated with the effects from the 2018 HABs and red tide events in Florida. However, costs have not been estimated for adverse health effects and treatment of respiratory and other problems; impacts to fisheries; water treatment; and losses related to tourism, jobs, recreation, and businesses. Estimates from previous blooms can provide some idea of the huge economic impacts to come. For example, in southwest Florida (Sarasota County), a red tide resulted in an increase in hospital emergency department diagnoses by 19% for pneumonia, 40% for gastrointestinal illnesses, and 54% for respiratory illnesses compared to a period when red tide was not present (Cheng et al., 2005; Kirkpatrick et al., 2010). The costs of hospital visits for respiratory illness alone ranged between $0.5 and $4 million (Hoagland et al., 2009). No estimates are available for losses to commercial fisheries from HABs; however, Anderson et al. (2000) estimated an annual impact of HABs to commercial fisheries throughout the United

States that ranged from $13 to $25 million with an average annual impact of $18 million (2000 dollars). In terms of economic losses to tourism, a study in the western panhandle of Florida (Okaloosa County, Fort Walton Beach and Destin), estimated economic losses of nearly $6.5 million to local restaurant and hotel sectors from red tides that occurred 1995–2000 (Morgan et al., 2006). Another study in the Fort Walton Beach and Destin area during the same time period (Larkin & Adams, 2007), noted that the impacts from HABs on coastal businesses in (restaurant and lodging sectors) were $2.8 and $3.7 million per month, respectively, which represented a 29–35% decline in average monthly revenues for each sector during months of red tide incidence.

The State of Florida announced in October 2018 that more than $291 million in projects will be spent as part of the State Expenditure Plan (SEP) to restore and protect water quality along Florida's Gulf Coast including environmental restoration projects across 23 counties from Escambia County in the western panhandle to Key West in Monroe County in south Florida. The Florida DEP Secretary Noah Valenstein noted that "this plan that was approved by the Gulf Coast Ecosystem Restoration Council and Gulf Consortium includes a diverse suite of projects and programs that focus on environmental restoration and protection of water quality and coastal resources and promotion of tourism along our Gulf Coast. Florida's counties clearly recognize that healthy ecosystems, fisheries, marine and wildlife habitats, beaches, and wetlands are inextricably linked to our state's economy." For 2019, the newly elected Florida governor has made budget recommendations of $75 million for improving water quality, combating HABs, and restoring water-quality impaired springs with algae fueled by elevated nitrate concentrations.

10.3. ECONOMIC COSTS AND CONSEQUENCES OF REACTIVE NITROGEN IN CHINA

The population in China has increased substantially over the past several decades. To feed the growing population, crop production has increased along with the widespread use of nitrogen and phosphorus fertilizers, which have been supplied at low cost and with subsidies in China (Lu et al., 2015). Liu et al. (2016) estimated average inputs of fertilizer were around 200 kg/ha in 2006, with local values often five times higher. Liu et al. also noted that accompanying the increase in fertilizer nitrogen inputs was a 3.2-fold increase in livestock and a 20.8-fold increase in vehicles since 1980, which has led to substantial increases in nitrogen pollution in China. Ma et al. (2013) analyzed changes in food chain structure, improvements in technology and management, and various combinations of these on food supply and environmental quality using the NUtrient flows in Food chains, Environment and Resources use (NUFER) model. As a result of population growth and dietary changes and current rates of fertilizer use, they projected that nitrogen and phosphorus fertilizer usage in 2030 would both increase by 25%, and nitrogen and phosphorus losses would increase by 44 and 73%, respectively, compared to 2005 (their reference year). Large mineral fertilizer applications in in China also have resulted in substantial soil acidification (Guo et al., 2010). Soil pH declined significantly from the 1980s to the 2000s in main Chinese crop-production areas. In China, the China Geological Survey reported that nitrate pollution of the shallow groundwater is widespread with almost 100% of water samples containing some level of nitrate, and with 30–60% of samples containing concentrations of nitrate-N greater than the national standard (20 mg N/L) (Mateo-Sagasta & Burke, 2008).

The overuse of synthetic fertilizers and leaching of nitrogen from soils has led to adverse impacts to the aquatic environment. In China, eutrophication of lake waters from excess nutrients originating from nonpoint sources has been a serious problem for the past several decades. In 2007, the estimated area of eutrophic lakes in China was 8700 km^2, an increase from 135 km^2 in 1967. Also noted was the increase in the number of coastal "red tides" from 4 in the 1980s to 80 in the 2000s.

Le et al. (2010) conducted an extensive review of the literature and found that 80% of 67 main lakes in China have been polluted to such an extent that they are not healthy for human contact. Each destructive algal bloom that occurred in lakes has caused a direct economic loss of millions to billions of dollars (2010 $US). For example, algal blooms in Lake Tai (area 2250 km^2, 2-m average depth; third largest freshwater lake in China) in the Yangtze Delta plain (eastern China, about 50 km west of Shanghai) have adversely impacted industrial and agricultural production and have forced the shutdown of the entire water supply for a large urban area. In 1998, an algal bloom in the catchment area resulted in an economic loss of around US$6.5 billion, which was 5.9% of the GDP in the catchment for that year.

10.3.1. Air Pollution, Atmospheric Deposition of Nitrogen, and Public Health

Several studies have investigated the increase in nitrogen input to the atmospheric inputs and resulting nitrogen deposition in China. Liu et al. (2013) evaluated N deposition dynamics and their effect on ecosystems across China between 1980 and 2010. They found that the average annual bulk deposition of nitrogen increased by approximately 8 kg N/ha ($P < 0.001$) between the 1980s (13.2 kg N/ha) and the 2000s (21.1 kg N/ha). Nitrogen deposition rates in the industrialized and agriculturally intensified regions of China are as high as the peak levels of deposition in northwestern Europe in the 1980s, before the introduction of mitigation policies. Nitrogen from ammonium (NH_4^+) was the dominant form of N in bulk deposition, but they noted that N deposition from nitrate had the largest rate increase, which corresponded to lower ratios of NH_3 to NO_x emissions since 1980.

Particulate air pollution has had serious effects on human health in China (cardiovascular and respiratory diseases and high mortality and morbidity). Xia et al. (2016) used a model to estimate the monetary value of total output losses resulting from reduced working time caused by diseases related to air pollution across 30 Chinese provinces in 2007. Fine particulate matter ($PM_{2.5}$) pollution was used as an indicator to assess impacts to health caused by air pollution. Their developed model accounted for both direct economic costs and indirect cascading effects throughout an interregional production supply chains. Model results indicated total economic losses of 346.26 billion Yuan (approximately 1.1% of the national GDP) based on the number of affected Chinese employees (72 million out of a total labor population of 712 million) whose work time in years was reduced because of mortality, hospital admissions, and outpatient visits due to diseases resulting from $PM_{2.5}$ air pollution in 2007.

Gu et al. (2012) evaluated reactive nitrogen emissions to the atmosphere in China in 2008, including those from ammonia (NH_3), nitrogen oxides (NO_x), and nitrous oxide (N_2O). They noted that total reactive nitrogen emissions more than doubled over the past three decades in China. However, the increasing trend slowed for NH_3 emissions after 2000, while the trend continued to increase for NO_x and N_2O emissions. In several places in China, hotspots were identified, where reactive nitrogen emissions were about 10 times higher than other areas. Their full-nitrogen cycle analysis quantified and tracked the trajectory of all nitrogen fluxes in an approach referred to as coupled human and natural systems (CHANS). This approach included linking coupled models to represent the interactions between human (e.g., economic and social) and natural (hydrologic, atmospheric, and biological) subsystems. Agricultural sources comprised 95% of total NH_3 emissions; fossil fuel combustion accounted for 96% of total NO_x emissions; and N_2O emissions originated from agricultural (51%) and natural sources (forest and surface water, 39%) in China.

Gu et al. (2012) first used population density for estimating the exposures per unit reactive nitrogen emissions in China. Second, they used the willingness to pay for air pollution mortality and effects reduction. They examined differences in the willingness to pay for health costs between different countries. For example, mortality due to chronic exposure to ozone and secondary particulate matter contributed 67% to total health costs in Europe (Brink et al., 2011). Health damages related to atmospheric reactive nitrogen emissions (total) was estimated at US$19–62 billion (52–60% from ammonia emissions; 39–47% from NO_x emissions). This damage estimate accounted for 0.4–1.4% of China's GDP. These findings provide policymakers with an integrated view of reactive nitrogen sources and their associated health damages for developing strategies to reduce air pollution.

Air deterioration caused by pollution has harmed public health and resulted in enormous economic consequences in China (L. Li et al., 2016). They noted that existing studies on the economic loss caused by a variety of air pollutants in multiple cities were lacking. To understand the effect of different pollutants on public health and to provide the basis of the environmental governance for governments, based on the dose–response relation and the willingness to pay, this study used the latest available data of the inhalable particulate matter (PM_{10}) and sulfur dioxide (SO_2) from January 2015 to June 2015 in 74 cities by establishing the lowest and the highest limit scenarios. The results showed that (a) in the lowest and highest limit scenario, the health-related economic loss caused by PM_{10} and SO_2 represented 1.63 and 2.32% of the GDP, respectively; (b) for a single city, in the lowest and the highest limit scenarios, the highest economic loss of the public health effect caused by PM_{10} and SO_2 was observed in Chongqing; the highest economic loss of the public health effect per capita occurred in Hebei Baoding. The highest proportion of the health-related economic loss accounting for GDP was found in Hebei Xingtai. The main reason is that the terrain conditions are not conducive to the spread of air pollutants in Chongqing, Baoding, and Xingtai, and the three cities are typical heavy industrial cities that are based on coal resources. Therefore, it was necessary to improve the energy structure, use the advanced production process, reasonably control the urban population growth, and adopt the emissions trading system in order to reduce the economic loss caused by the effects of air pollution on public health.

10.3.2. Changes in Food Production in China

There have been major life-style and dietary changes in China during the past several decades. Guo et al. (2017) noted that animal-derived protein consumption has increased from an average of 8% of the total protein consumption in 1961 to 38% in 2011. Associated with this increased consumption of animal-derived protein has been an increase in resource needs (synthetic fertilizer, water, and land) and environmental costs (human health, ecosystems, and water quality) (Tilman & Clark, 2014). To better understand how nitrogen use and impacts have changed over time, Guo et al. (2017) used the N-calculator model along with the NUFER model (Ma et al., 2010). Nitrogen losses related to plant- and animal-emanated food production increased from 2.4 to 19.6 metric tons (MT) N/yr and from 0.6 to 7.5 MT N/yr, respectively. Nitrogen losses originating from food production involving animals increased from 21% in the 1960s to 28% in the 2000s. The following nitrogen losses to the environment were estimated: ammonia (NH_3) emissions and dinitrogen (N_2) emissions through denitrification (each accounting for nearly 40%), nitrogen losses to water systems (20%), and nitrous oxide (N_2O) emissions (1%). The consumption of animal-derived foods along with increased fertilizer usage has resulted in the growing environmental costs of food systems in China.

10.4. REDUCING ECONOMIC COSTS ASSOCIATED WITH MANAGING REACTIVE NITROGEN

10.4.1. Climate Change Ramifications and Economic Costs

Several of the recent studies have presented detailed information on climate change and future impacts from reactive nitrogen on aquatic ecosystems (e.g., Baron et al., 2013; Fowler et al., 2015; Greaver et al., 2012; Irvine et al., 2017; Pinder et al., 2013; Reis et al., 2016; Stuart et al., 2011). These impacts have important economic ramifications associated with increased amounts of reactive nitrogen. For example, the proliferation of blue-green algae would be favored by increased nutrients and higher water temperature (Moss et al., 2011). Warmer water would promote higher densities of fish species in lakes that eat zooplankton, and Havens (2018) noted that *Daphnia* (a zooplankton that is very effective in controlling algae) also is highly sensitive to warming water. With a reduced ability of zooplankton to control algae, blue-green algae would increase and raise the potential for HABs. Higher temperatures would increase the rate of nutrient release from soils that would lead to higher nutrient inputs to lakes and estuaries. Also, the higher water temperatures would enhance bacterial activity leading to depletion of oxygen from the water and stimulate the release of nutrients already present in the bottom sediments (Havens, 2018).

Baron et al. (2013) noted that eutrophication increases with warmer water temperatures, and this would lead to higher costs associated with upgrades to municipal drinking water treatment facilities, bottled water. Also, health costs associated with nitrate in drinking water may lead to disease (Compton et al., 2011). In the United States, more than $200 billion for wastewater management infrastructure needs has been estimated for addressing nutrient control from various sources (USEPA). More eutrophication would also result in economic losses to fisheries, property values, and recreational activities. Freshwater diversity, which has been altered by a combination of habitat loss, homogenization of flow regimes, and eutrophication, will also be diminished by excess reactive nitrogen and changes to thermal properties (Hobbs et al., 2010; Porter et al., 2012).

10.4.2. Using Multiple Metrics to Assess Economic Costs and Management Actions

Several recent studies have shown that a multiple metric approach (including damage costs, health effects, abatement costs, in addition to mass flows) associated with impacts from reactive nitrogen can be used to design more effective management actions than just evaluating mass flows (or fluxes) metrics (e.g., Birch et al., 2011; Compton et al., 2011; US EPA, 2011; Sobota et al., 2013; Horowitz et al., 2016). This approach allows for assessing total economic damage costs (including diverse impacts on ecosystems) and human health impacts, and the cost-effectiveness of mitigation or remediation for each metric. As a result, more effective decision-making and policies could account for the impacts of various forms of reactive nitrogen as they cascade through the atmosphere and biosphere. Two examples showing integrated management approaches using multiple economic metrics are presented below.

There have been considerable expenditures to reduce reactive nitrogen for restoration efforts of the Chesapeake Bay in the United States. However, in recent years, there has been limited progress in reducing nitrogen as indicated using typical metrics, such as nitrogen releases and/or concentrations in water and air. Birch et al. (2011) developed an "Economic Nitrogen Cascade" approach to evaluate multiple metrics including tons of reactive nitrogen, damage costs, human mortality and morbidity, and mitigation costs for the Chesapeake Bay Watershed. The study estimated that about 24% of the reactive nitrogen reaching the bay originated from emissions to the atmosphere, 57% originated from terrestrial inputs, and 19% originated from direct releases to freshwater ecosystems. Even though mass fluxes to the atmosphere were lower than direct releases of reactive nitrogen to terrestrial and aquatic ecosystems, the studies showed that the total damage costs to all ecosystems and human health were higher because of the cascading effects of reactive nitrogen and the resulting human health costs. Unfortunately, as it was noted, their study was limited by the availability of existing valuation studies. For example, no studies were available to evaluate economic damages associated with ozone depletion, greenhouse gas impacts, fertilization benefits, and materials damages from atmospheric nitrogen deposition, freshwater recreational fishing, commercial fishing and ecosystem services throughout the nitrogen cascade. However, they were able to show that abatement costs of reducing reactive nitrogen releases to the atmosphere were lower than other nitrogen abatement options. It was cautioned that nonpoint source pollution from agriculture should still be addressed. However, the overall intent of their study was to provide information to policymakers and the public of the various costs and trade-offs among different approaches so that cost-effective decisions would reflect societal priorities. The multiple metrics approach would result in greater reductions in reactive nitrogen from sources that have contributed to the degraded quality of the Bay's waters and would distribute abatement costs among multiple additional polluters.

A second study evaluated multiple metrics for determining economic impacts from reactive nitrogen, using a two-prong approach for estimating damages from reactive nitrogen: nonhealth costs and health-related costs in the San Joaquin Valley of California (Horowitz et al., 2016). Information from the previous studies (e.g., Birch et al., 2011) was used to estimate the per-kg reactive nitrogen damage costs for nonhealth effects. They estimated health damage costs from air pollution (very small particles, $PM_{2.5}$ concentrations; particulate matter ≤2.5 μm) using the USEPA's BenMAP model (Abt Associates, 2008). Changes in health outcomes and associated economic damage costs were estimated by eliminating each flow using BenMap. The study calculated per-kg economic health-related damages by dividing the total economic damages by the size of the reactive nitrogen flow. They produced maps of different mass and damage flows that provided a geographical distribution of reactive nitrogen flows and economic impacts in the San Joaquin Valley. By comparing the multiple metrics, Horowitz et al. (2016) were able to identify the most effective strategy for intervention and effective policy options. Horowitz et al. (2016) found that reactive nitrogen emissions from mobile sources were smaller than releases of reactive nitrogen from crop agriculture and dairy in the study area. However, as in the Chesapeake Bay Watershed study (Birch et al., 2011), the benefits of abatement of air pollution were found to be greater because of reduced health impacts, and lower abatement costs.

10.4.3. Reducing Economic Costs Associated with Health Effects from Air Pollution

Mentioned previously are the worldwide health impacts and economic losses associated with health effects from air pollution (mainly $PM_{2.5}$) resulting from various reactive nitrogen emissions. Several studies have shown that reductions in air pollution from $PM_{2.5}$ can significantly save lives and have substantial economic benefits. For example, Marshall et al. (2016) noted that thousands of lives have been saved annually in the United States due to policy-driven reductions in $PM_{2.5}$ concentrations. They estimated monetized benefits of $19–$167 billion (as cited in Draft 2012 report to Congress). This is in comparison to annual abatement costs of approximately $7 billion. They concluded that regulations to control $PM_{2.5}$ air pollution have extremely high economic benefits of all federal regulations. Apte et al. (2015) compared linear and nonlinear PM concentration response relationships that describe the risk of PM in relation to several causes of death. They found that relatively small improvements in $PM_{2.5}$ in less-polluted areas, such as North America and Europe, would result in large avoided mortality. In more polluted regions, such as China and India, to keep $PM_{2.5}$-attributable mortality rates constant, $PM_{2.5}$ levels would need to decrease by 20–30% over the next 15 years to balance increases in mortality in aging populations from $PM_{2.5}$ pollution.

10.4.4. Reducing Reactive Nitrogen from Various Institutions

There are hundreds to thousands of various types of institutions in countries around the world that could sustainably reduce their impacts from reactive nitrogen inputs to the environment. A better understanding is needed regarding the connections between nitrogen releases to the environment and associated impacts on social and economic systems. To address this issue, Compton et al. (2017) estimated damage costs related to the release of nitrogen oxides, ammonia, and N_2O to air and nitrogen to water from the consumption of various resources at two universities in the United States. They used a detailed nitrogen footprint approach (discussed in detail in Chapter 11) for the University of Virginia (UVA) and the University of New Hampshire (UNH) that accounted for various types of resource consumption that contributes to nitrogen pollution (such as food purchases, food consumption, sewage, utilities, transportation (university vehicles and commuters), research animals, fertilizer applications, food production and consumption, waste, sewage, and emissions from fuels for utilities and transportation) that contribute to nitrogen pollution. These two institutions were selected because they both had estimated their nitrogen footprints (http://www.n-print.org/, accessed 27 September 2018) and their different use of energy sources (coal at UVA and landfill cogeneration at UNH). Damage estimate values from the aforementioned continental-scale studies from the European Union (Van Grinsven et al., 2013) and the

United States (Sobota et al., 2015) were used. Some of the damages considered were human respiratory health (NO_x, NH_3), air visibility, greenhouse gas emissions, ozone and UV damage to crops (NH_3, N_2O, NO_x), plant diversity declines, closures due to HABs, human health (NO_3), replacement costs associated with bottled water, drinking water treatment, private well water treatment, lake waterfront property values, endangered species protection, eutrophication, and recreational freshwater use.

Compton et al. (2017) estimated the following annual potential damage costs: $11 million at UVA and $3.04 million at UNH. At UVA, the largest component of costs were related to the release of nitrogen oxides to human health, mainly from the use of coal-derived energy. At UNH, the majority of damages was related to food production. Damages from energy production and transportation also were related to distance from power plants and roads resulting. Conversely, food production could have impacts farther away from campus. This type of information would be useful to generate for many other institutions to provide a better understanding of their environmental and health damages due to the release and consumption of various forms of reactive nitrogen. Knowing this information would lead to more informed and effective strategies for reducing nitrogen inputs to the environment.

10.4.5. Dietary Modifications

Changes in the amount of protein consumed in our diets by reducing per capita meat consumption could result in not only health benefits but also lead to more efficient use of nitrogen. This in turn would result in decreased production costs and less pollution of water bodies. Although there are still large areas of the world where increased nitrogen availability is necessary for increasing protein intake (Erisman et al., 2015). T. Li et al. (2017) conducted and extensive analysis of worldwide studies (published from 1980 to 2016) that evaluated different types of enhanced-efficiency fertilizers (EEFs) as a way of better managing nitrogen in agricultural systems and optimizing production costs. Four major types of EEFs (polymer-coated fertilizers PCF, nitrification inhibitors NI, urease inhibitors UI, and double inhibitors DI, i.e., urease and nitrification inhibitors combined) were evaluated in terms of potentially increasing yields and NUE and reducing nitrogen losses. They found that the most effective strategies involved (a) DI use in grassland ($n = 133$), which averaged an 11% yield increase, 33% NUE improvement, and 47% decrease in aggregated nitrogen loss (sum of nitrate, ammonia, and N_2O, totaling 84 kg N/ha); (b) UI use in rice-paddy systems ($n = 100$), with 9% yield increase, 29% NUE improvement, and 41% N-loss reduction (16 kg N/ha). In wheat and maize systems, they found that EEFs were generally less effective. Based on their detailed analysis, they concluded that any potential benefits of EEFs for sustainable agriculture could be most effective when eliminating fertilizer mismanagement along with the implementation of knowledge-based nitrogen management practices.

10.4.6. Agricultural Nitrogen Management

The effectiveness of agricultural best management practices (BMPs) in reducing nutrient loading from nonpoint sources to the environment has been highly variable from one area to another due to differences in the hydrologic setting, climate, biogeochemical cycles, crop fertilizer/animal waste management, and other variables. One example of a partially successful regulated agricultural nitrogen management program is in the Platte River Valley of Nebraska (Ferguson, 2015). Groundwater management districts were formed by the state legislature in the late 1980s in response to widespread nitrate-N contamination of groundwater (reaching concentrations of 30–40 mg N/L with occasional higher concentrations). There was an average fertilizer-N application rate of 154 kg N/ha/yr that was nearly constant from 1967 to 2010, and corn production has increased steadily, which resulted in nearly a factor of two increases in NUE. The increased NUE resulted from a combination of regulatory policies and nonregulatory activities, including (a) credits for nitrogen in legumes, soil, and in irrigation water; (b) practical expectation of crop yields; (c) nitrogen rate recommendations based on economic advantage; (d) more improved timing of fertilizer applications to supply plant-nitrogen needs; and (e) improved corn hybrids (Ferguson, 2015). These measures resulted in a decrease in nitrogen application from 34 kg N/ha in 1988 to 19 kg N/ha in 2012, with an associated decrease in groundwater nitrate-N concentrations on average by 0.15 mg N/L/yr. Also, nitrate residue content in soil decreased by 2.4 kg N/ha/yr. Ferguson (2015) attributed the shift from furrow irrigation to sprinkler irrigation to be responsible for 50% of the change in groundwater nitrate-N concentrations. However, even though there have been 25 years of significant intensive mitigation, the groundwater nitrate-N concentration still remained above 10 mg N/L. The elevated nitrate-N concentrations most likely are related to the prevalence of sandy soils and a shallow aquifer. These conditions can lead to a higher vulnerability to contamination, but possibly a faster recovery when mitigation measures are in place. In areas with less permeable soils and thicker unsaturated zones, results of nitrogen mitigation measures could be much slower. In these other areas, Ferguson (2015) suggested that the next-generation of nutrient management techniques might be required for mitigation, such as fertigation, controlled release fertilizers, and crop canopy sensors for variable rate nitrogen application.

Nonpoint agricultural sources of pollution in the Chaohe River watershed in China have resulted in increasing eutrophication of water in the Miyun reservoir, the source of drinking water for Beijing. A recent study used the Integrated Farming System Model to optimize BMPs at two different farming systems (crop-based mixed farm and dairy-based farms) to improve their economic viability. Geng et al. (2017) reported that total nitrogen losses ranged from 74 to 1390 kg/ha/yr. The optimal BMPs to maximize reductions in nitrogen loss included conservation tillage, improved timing of fertilizer applications, reduced fertilizer and manure applications, buffer strips, and poultry manure storage. The two BMPs that resulted in increased farm profits were conservation tillage (better crop yields resulting from more efficient use of manure) and fertilizer reductions (lower fertilizer consumption).

As has been shown in many studies, NUE (the fraction of nitrogen input harvested in crop products) needs to be improved significantly to address critical issues of food security, environmental degradation, and climate change. By improving NUE, crop productivity increases while reducing economic costs associated with lower fertilizer usage and in turn environmental degradation is reduced. Improving NUE is critical for addressing the 2050 Sustainable Development Goals recently adopted by 193 countries of the U.N. General Assembly (SDSN, 2015). Zhang et al. (2015) analyzed historical global patterns (1961–2011) of agricultural nitrogen use in 113 countries to determine pathways of socioeconomic development and related nitrogen pollution. Their analyses showed that in many countries nitrogen pollution first increased and then decreased with economic growth. Zhang et al. (2015) explored NUE with several factors including fertilizer subsidies in developing countries, new technologies (e.g., improved sed, balanced nutrient amendments, and water management), crop mixes (with N-fixing crops), farm-scale nutrient management practices, fertilizer to crop price ratios, socioeconomic issues affecting farmer decisions, and the fraction of harvested area for fruit and vegetable production. To achieve the 2050 global food demand of 107 Tg N/yr projected by the Food and Agriculture Organization while reducing the nitrogen surplus (as defined by the sum of N inputs (fertilizer, manure, biologically fixed N, and N deposition) minus N outputs) from the current 100 Tg N/yr to a global limit of 50 Tg N/yr would call for substantial increases in NUE. Average global NUE would need to increase from about 0.4 to 0.7 (while the crop yield would need to increase from 74 to 107 Tg N/yr) to meet the 2050 Sustainable Development Goals (SDSN, 2015) of food security and environmental stewardship. Furthermore, continued improvements in NUE will need more cross-disciplinary and cross-sectorial partnerships, which would integrate research and development of innovative agricultural technology and management systems with socioeconomic research. The findings from these cross-collaboration efforts would need to be socially and economically practical and readily usable by farmers. Zhang et al. (2015) also concluded that pollution swapping should be avoided because in many cases, reducing one pollution problem can lead to another (e.g., retaining crop residues can reduce nitrogen runoff but could result in higher N_2O emissions). Progress has been made in reducing nitrogen releases; however, the progress made so far would not be sufficient to meet the projected 2050 goals of both food security and environmental stewardship. Zhang et al. (2015) emphasized the need to train the next generation of interdisciplinary agronomic and environmental scientists that can work together to address food, water, energy, economic, and environmental issues.

10.4.7. Food Production with Less Pollution

Davidson et al. (2015) made several recommendations for producing more food with less pollution and minimizing economic consequences, including (a) creating partnerships at local and regional levels among industry, universities, governments, nongovernmental organizations, crop advisors, and farmers to demonstrate and quantify the most current, economically feasible, best management NUE practices at each area; (b) educating private sector retailers and crop advisors on updated nutrient management practices through professional certification programs by university and government extensions and scientific societies; (c) linking nutrient management to performance-based indicators, including NUE indicators on the farm, and providing useful incentives for farmers for participation and data reporting, and (d) combining information on agronomic, ecological, economic, and social science knowledge of food production with the associated environmental, economic, and social costs to society.

10.5. CONCLUDING REMARKS

Anthropogenic releases of reactive nitrogen to the atmosphere and water resources have resulted in many harmful effects to human health (e.g., cardiopulmonary diseases, respiratory illnesses, cancer risk) and ecosystems (e.g., biodiversity loss, eutrophication, HABs). The annual costs associated with these effects, especially regarding air pollution and human health, are very high and increasingly expensive. Associated costs also need to be considered when assessing other environmental consequences, such as job losses, water treatment, loss of recreational use, decline of commercial fisheries, resource and habitat restoration, and decreases in real estate values. More studies are needed that provide an holistic assessment of the magnitude of total costs and economic consequences to advise decision-makers, environmental resource managers, and other stakeholders of the human health and ecosystem benefits of improved nitrogen management and NUE. The next chapter provides additional information on various approaches and effective strategies that have been developed to reduce reactive nitrogen losses to the environment from anthropogenic sources.

REFERENCES

Abt Associates. (2008). Environmental benefits mapping and analysis program (Version 3.0) Bethesda, MD. In *Prepared for the Environmental Protection Agency, Office of Air Quality Planning and Standards, Air Benefits and Costs Group*. Research Triangle Park, NC.

Allan, J. D., Smith, S. D. P., McIntyre, P. B., Joseph, C. A., Dickinson, C. E., Marino, A. L., et al. (2015). Using cultural ecosystem services to inform restoration priorities in the Laurentian Great Lakes. *Frontiers in Ecology and Environment*, *13*(8), 418–424. https://doi.org/10.1890/140328

Anderson, D.M., Hoagland, P., Kaoru, Y., & White, A.W. Anderson, D.M. (2000). *Estimated annual economic impacts from harmful algal blooms (HABs) in the United States* (Technical Report 2000-09). doi:10.1575/1912/96. Retrieved from https://hdl.handle.net/1912/96.

Apte, J. S., Marshall, J. D., Brauer, M., & Cohen, A. J. (2015). Addressing global mortality from ambient $PM_{2.5}$. *Environmental Science and Technology*, *49*, 8057–8066. http://dx.doi.org/10.1021/acs.est.5b01236

Baron, J. S., Hall, E. K., Nolan, B. T., Finlay, J. C., Bernhardt, E. S., Harrison, J. A., et al. (2013). The interactive effects of excess reactive nitrogen and climate change on aquatic ecosystems and water resources of the United States. *Biogeochemistry*, *114*, 71–92. https://doi.org/10.1007/s10533-012-9788-y

Birch, M. B. L., Gramig, B. M., Moomaw, W. R., Doering, O., & Reeling, C. J. (2011). Why metrics matter: Evaluating policy choices for reactive nitrogen in the Chesapeake Bay watershed. *Environmental Science and Technology*, *45*, 168–174.

Brink, C., Van Grinsven, H., Jacobsen, B. H., Rabl, A., Gren, I.-M., Holland, M., et al. (2011). Costs and benefits of nitrogen in the environment. Ch. 22. In M. A. Sutton, C. M. Howard, J. W. Erisman, G. Billen, A. Bleeker, P. Grennfelt, H. van Grinsven, & B. Grizzetti (Eds.), *The European nitrogen assessment: Sources, effects, and policy perspectives*. Cambridge, UK: Cambridge University Press, 664 p.

Cheng, Y. S., Zhou, Y., Irvin, C. M., Pierce, R. H., Naar, J., Backer, L. C., et al. (2005). Characterization of marine aerosol for assessment of human exposure to brevetoxins. *Environmental Health Perspectives*, *113*, 638–643.

Compton, J. E., Harrison, J. A., Dennis, R. L., Greaver, T., Hill, B. H., Jordon, S. J., et al. (2011). Ecosystem services altered by changes in reactive nitrogen: An approach to inform decision-making. *Ecology Letters*, *14*, 804–815. https://doi.org/10.1111/j.1461-0248.2011.01631.x

Compton, J. E., Leach, A. M., Castner, E. A., & Galloway, J. M. (2017). Assessing the social and environmental costs of institution nitrogen footprints. *Sustainability*, *10*(2). https://doi.org/10.1089/sus.2017.29099.jec

Crawford, A. (2017). Greater Lakes: A $2 billion collaboration between farmers, conservationists, and governments aims to repair the largest flowing freshwater system on Earth. *Nature Conservancy*, *2017*, 47–56.

Davidson, E. A., Suddick, E. C., Rice, C. W., & Prokopy, L. S. (2015). More food, low pollution (Mo Fo Lo Po): A grand challenge. *Journal of Environmental Quality*, *44*, 305–311. https://doi.org/10.2134/jeq2015.02.0078

Dodds, W. K., Bouska, W. W., Eitzmann, J. L., Pilger, T. H., Pitts, K. L., Riley, A. J., et al. (2009). Eutrophication of U.S. freshwaters: Analysis of potential economic damages. *Environmental Science and Technology*, *43*(1), 12–19.

Erisman, J. W., Galloway, J. N., Dice, N. B., Sutton, M. A., Bleeker, A., Grizzetti, B., et al. (2015). *Nitrogen, too much of a vital resource*. Science Brief. Zeist, The Netherlands: World Wildlife Fund, 27 p.

European Food Safety Authority. (2008). Summary of Annual Report 2008. Parma, Italy. Retrieved from https://www.efsa.europa.eu/en/corporate/pub/ar08, accessed 28 September 2018.

Ferguson, R. B. (2015). Groundwater quality and nitrogen use efficiency in Nebraska's central Platte River Valley. *Journal of Environmental Quality*, *44*(2), 449–459. https://doi.org/10.2134/jeq2014.02.0085

Flesher, J., & Kastanis. A. (2017). Toxic algae: Once a nuisance now a severe nationwide threat. *Associated Press*, 16 November 2017.

Food and Agricultural Organization (FAO), International Fund for Agricultural Development (IFAD), World Food Program (WFP). (2012). *The state of food insecurity in the world 2012. Economic growth is necessary but not sufficient to accelerate reduction of hunger and malnutrition*. FAO: Rome.

Fowler, D., Steadman, C. E., Stevenson, D., Coyle, M., Rees, R. M., Skiba, U. M., et al. (2015). Effects of global change during the 21st century on the nitrogen cycle. *Atmospheric Chemistry and Physics*, *15*, 13849–13893. https://doi.org/10.5194/acp-15-13849-2015

Geng, R., Yin, P., Gong, Q., Wang, X., & Sharpley, A. N. (2017). BMP optimization to improve the economic viability of farms in the Upper Watershed of Miyun Reservoir, Beijing, China. *Water*, *9*, 633. https://doi.org/10.3390/w9090633

Great Lakes Commission. (2014). Retrieved from http://lakeerie.ohio.gov/Portals/0/GLRI/Ohio_GLRI%20State%20Factsheet_2014_final_Feb%2028.pdf, accessed 24 September 2018.

Greaver, T. L., Sullivan, T., Herrick, J., Lawrence, G., Herlihy, A., Barron, J., et al. (2012). Ecological effects of air pollution in the U.S.: What do we know? *Frontiers in Ecology and the Environment*, *10*, 365–372.

Gu, B., Ge, Y., Ren, Y., Xu, B., Luo, W., Jiang, H., et al. (2012). Atmospheric reactive nitrogen in China: Sources, recent trends, and damage costs. *Environmental Science and Technology*, *46*(17), 9420–9427. https://doi.org/10.1021/es301446g

Guo, J. H., Liu, X. J., Zhang, Y., Shen, J. L., Han, W. X., Zhang, W. F., et al. (2010). Significant acidification in major Chinese croplands. *Science*, *327*, 1008–1010.

Guo, M., Chen, X., Bai, Z., Jian, R., Galloway, J. N., Leach, A. M., et al. (2017). How China's nitrogen footprint of food has changed from 1961. *Environmental Research Letters*, *12*(2017), 104006.

Havens, K. (2013). *Deep problems in shallow lakes: Why controlling phosphorus inputs may not restore water quality* (SGEF-128). Gainesville, FL: University of Florida Institute of Food and Agricultural Sciences. Retrieved from http://edis.ifas.ufl.edu/sg128

Havens, K. E. (2018). *The future of harmful algal blooms in Florida inland and coastal waters*. Gainesville, FL: Florida Sea Grant College Program and the University of Florida Institute of Food and Agricultural Sciences Extension Report TP-231, 4 p.

Havens, K. E., Hoyer, M. V., & Phlips, E. J. (2016). *Natural climate variability can influence cyanobacteria blooms in Florida Lakes and Reservoirs* (SGEF-234). Gainesville, FL: University of Florida Institute of Food and Agricultural Sciences and Florida Sea Grant College Program. Retrieved from http://edis.ifas.ufl.edu/sg142.

Hill, J., Goodkind, A., Tessum, C., Thakrar, S., Tilman, D., Tilman, D., et al. (2019). Air-quality-related health damages of maize. *Nature Sustainability*, *2*, 397–403. https://doi.org/10.1038/s41893-019-0261-y

Hoagland, P., Anderson, D. M., Kaoru, Y., & White, A. W. (2002). The economic effects of harmful algal blooms in the Unites States: Estimates, assessment issues, and information needs. *Estuaries*, *25*, 819–837.

Hoagland, P., Jin, D., Polansky, L. Y., Kirkpatrick, B., Kirkpatrick, G., Fleming, L. E., et al. (2009). The costs of respiratory illnesses arising from Florida Gulf Coast *Karenia brevis* blooms. *Environmental Health Perspectives*, *117*(8), 1239–1243.

Hobbs, W. O., Telford, R. J., Birks, H. J. B., Saros, J. E., Hazewinkel, R. O., Perren, B. P., et al. (2010). Quantifying recent ecological changes in remote lakes of North America and Greenland using sediment diatom assemblages. *PLoS One*. https://doi.org/10.1371/journal.pone.0010026

Horowitz, A. I., Moomaw, W. R., Liptzin, D., Gramig, B. M., Reeling, C. J., Meyer, J., & Hurley, K. (2016). A multiple metrics approach to prioritizing strategies for measuring and managing reactive nitrogen in the San Joaquin Valley of California. *Environmental Research Letters*, *11*(2016), 064011. https://doi.org/10.1088/1748-9326/11/6/064011

Irvine, I. C., Greaver, T., Phelan, J., Sabo, R. D., & Van Houtven, G. (2017). Terrestrial acidification and ecosystem services: Effects of acid rain on bunnies, baseball, and Christmas trees. *Ecosphere*, *8*(6), e01857. https://doi.org/10.1002/ecs2.1857

Kirkpatrick, B., Bean, J. A., Fleming, L. E., Kirkpatrick, G., Grief, L., Nierenberg, K., et al. (2010). Gastrointestinal ER admissions and Florida red tide blooms. *Harmful Algae*, *9*(1), 82–86.

Krimsky, L., Phlips, E., & Havens, K. (2018). A response to frequently asked questions about the 2018 algae blooms in Lake Okeechobee, the Caloosahatchee, and St. Lucie Estuaries. Florida Sea Grant Program, UF/IFAS Extension, ED2, 7 p.

Kusiima, J. M., & Powers, S. E. (2010). Monetary value of the environmental and health externalities associated with production of ethanol from biomass feedstocks. *Energy Policy*, *38*, 2785–2796.

Larkin, S. L., & Adams, C. M. (2007). Harmful algal blooms and coastal business: Economic consequences in Florida. *Society and Natural Resources*, *20*(9), 849–859. https://doi.org/10.1080/08941920601171683

Le, C., Zha, Y., Li, Y., Sun, D., Lu, H., & Yin, B. (2010). Eutrophication of lake waters in China: Cost, causes, and control. *Environmental Management*, *45*, 662–668. https://doi.org/10.1007/s00267-010-9440-3

Leadership Council of the Sustainable Development Solutions Network (SDSN). (2015). *Indicators and a monitoring framework for sustainable development goals—Revised working draft*. SDSN. Retrieved from http://unsdsn.org/resources, accessed 20 September 2019.

Li, L., Lei, Y., Pan, D., Yu, C., & Si, C. (2016). Economic evaluation of the air pollution effect on public health in China's 74 cities. *Springer Plus*, *5*, 402. https://doi.org/10.1186/s40064-016-2024-9

Li, T., Zhang, W., Yin, J., Chadwick, D., Norse, D., Lu, Y., et al. (2017). Enhanced-efficiency fertilizers are not a panacea for resolving the nitrogen problem. *Global Change Biology*, *2017*, 1–11. https://doi.org/10.1111/gcb.13918

Liu, X., Wang, H., Zhou, J., Hu, F., Zhu, D., Chen, Z., & Liu, Y. (2016). Effect of N fertilization pattern on rice yield, N use efficiency and fertilizer–N fate in the Yangtze River Basin, China. *PLoS One*. https://doi.org/10.1371/journal.pone.0166002

Liu, X., Zhang, Y., Han, W., Tang, A., Shen, J., Cui, Z., et al. (2013). Enhanced nitrogen deposition over China. *Nature*, *494*, 459–462.

Lu, Y., Jenkins, A., Ferrier, R. C., Bailey, M., Gordon, I. J., Song, S., et al. (2015). Addressing China's grand challenge of achieving food security while ensuring environmental sustainability. *AAAS Science Advances*, *1*, e1400039.

Ma, L., Ma, W. Q., Velthof, G. L., Wang, F. H., Qin, W., Zhang, F. S., & Oenema, O. (2010). Modeling nutrient flows in the food chain of China. *Journal of Environmental Quality*, *39*, 1279–1289.

Ma, L., Wang, F., Zhang, W., Ma, W., Velthof, G., Qin, W., Oenema, O., & Zhang, F. (2013). Environmental assessment of management options for nutrient flows in the food chain in China. *Environmental Science and Technology*, *47*, 7260–7268. doi:10.1021/es400456u.

Marshall, J. D., Apte, J. S., Coggins, J. S., & Goodkind, A. L. (2016). Blue skies bluer? *Environmental Science and Technology*. http://dx.doi.org/10.1021/acs.est.5b03154

Mateo-Sagasta, J., & Burke, J. (2008). Agriculture and water quality interactions: A global overview. SOLAW Background Thematic Report – FAO TR08, 46 p.

Morgan, K.L., Larkin, S.L., & Adams, C.M. (2006). Economic impacts of red tide events on restaurant sales. In *Southern Agricultural Economics Association Annual Meeting*, Orlando, FL, 5–8 February 2006.

Moss, B., Kosten, S., Meerhoff, M., Battarbee, R. W., Jeppesen, E., Mazzeo, N., et al. (2011). Allied attack: Climate change and eutrophication. *Inland Waters*, *1*, 101–105.

Nesheim, M. C. (1993). Human health needs and parasitic infections. *Parasitology*, *107*, S7–S18.

Paerl, H. W., & Huisman, J. (2008). Blooms like it hot. *Science*, *320*, 57–58. https://doi.org/10.1126/science.1155398

Pinder, R. W., Bettez, N. D., Bonan, G. B., Greaver, T. L., Schlesinger, W. H., & Davidson, E. A. (2013). Impacts of human alteration of the nitrogen cycle in the US on radiative forcing. *Biogeochemistry*, *114*(1–3), 25–40.

Porter, E. M., Bowman, W. D., Clark, C. M., Compton, J. E., Pardo, L. H., & Soong, J. L. (2012). Interactive effects of anthropogenic nitrogen enrichment and climate change on terrestrial and aquatic biodiversity. *Biogeochemistry*, *114*, 93–120. https://doi.org/10.1007/s10533-012-9803-3

Reis, S., Bekunda, M., Howard, C. M., Karanja, N., Winiwarter, W., Yan, X., Bleeker, A., & Sutton, M. A. (2016). Synthesis and review: Tackling the nitrogen management challenge: From global to local scales. *Environmental Research Letters*, 11(2016), 120205. http://dx.doi.org/10.1088/1748-9326/11/12/120205.

Richter, B. D., Braun, D. P., Mendelson, M. A., & Master, L. L. (1997). Threats to imperiled freshwater fauna. *Conservation Biology*, *11*, 1081–1093.

Smith, R. A., Alexander, R. B., & Schwarz, G. E. (2003). Natural background concentrations of nutrients in streams and rivers of the conterminous United States. *Environmental Science and Technology*, *37*, 3039–3047.

Sobota, D. J., Compton, J. E., & Harrison, J. A. (2013). Reactive nitrogen in US lands and waterways: how certain are we about sources and fluxes? *Frontiers in Ecology and Environment*, *11*, 82–90.

Sobota, D. J., Compton, J. E., McCrackin, M. L., & Singh, S. (2015). Cost of reactive nitrogen release from human activities to the environment in the United States. *Environmental Research Letters*, *10*(2015), 025006. https://doi.org/10.1088/1748-9326/10/2/025006

Stanton, E. A., & Taylor, M. (2012). *Valuing Florida's clean waters*. Stockholm Environmental Institute-U.S. Center, 30 p.

State-USEPA Nutrient Innovations Task Group. (2009). An urgent call to action. State-USEPA Nutrient Innovations Task Group Report to USEPA, August 2009, 170 p.

Stuart, M. E., Gooddy, D. C., Bloomfield, J. P., & Williams, A. T. (2011). A review of the impact of climate change on future nitrate concentrations in groundwater of the UK. *Science of the Total Environment*, *409*, 2859–2873.

Sutton, M. A., Howard, C. M., Erisman, J. W., Billen, G., Bleeker, A., Grennfelt, P., van Grinsven, H., & Grizzetti, B. (2011). *The European nitrogen assessment: Sources, effects, and policy perspectives*. Cambridge, UK: Cambridge University Press, 664 p.

Sutton, M. A., Reis, S., Riddick, S. N., Dragosits, U., Nemitz, E., Theobald, M. R., et al. (2013). Towards a climate-dependent paradigm of ammonia emission and deposition. *Philosophical Transactions of the Royal Society B*, *368*. https://doi.org/10.1098/rstb.2013.0166

Tilman, D., & Clark, M. (2014). Global diets link environmental sustainability and human health. *Nature*, *515*, 518–522.

U.S. Environmental Protection Agency. (2004). *Drinking water costs and federal funding; 816-F-04-038*. Washington, DC: US EPA.

U.S. Environmental Protection Agency (2015). *A compilation of cost data associated with the impacts and control of nutrient pollution*. Washington, DC: US Environmental Protection Agency Report EPA820F-15-096, 110 p.

Van Grinsven, H. J. M., Erisman, J. W., de Vries, W., & Westhoek, H. (2015). Potential of extensification of European agriculture for a more sustainable food system, focusing on nitrogen. *Environmental Research Letters*, *10*, 025002. https://doi.org/10.1088/1748-9326/10/2/025002

Van Grinsven, H. J. M., Holland, M., Jacobsen, B. H., Klimont, Z., Sutton, M. A., & Willems, W. J. (2013). Costs and benefits of nitrogen for Europe and implications for mitigation. *Environmental Science and Technology*. dx.doi.org/10.1021/es303804g

Van Grinsven, H.J.M., Rabl, A., and de Kok, T.M. 2010. Estimation of incidence and social cost of colon cancer due to nitrate in drinking water in the EU: A tentative cost-benefit assessment. *Environmental Health 9*: 58. Retrieved from http://www.ehjournal.net/content/9/1/58, accessed 22 September 2019.

Ward, M. H., Jones, R. R., Brender, J. D., de Kok, T. M., Weyer, P. J., Nolan, B. T., Villanueva, C. M., & van Breda, S. G. (2018). Drinking water nitrate and human health: an updated review. *International Journal of Environmental Research and Public Health,* 15(7), pii: E1557. doi:10.3390/ijerph15071557.

Waycott, M., Duarte, C. M., Carruthers, T. J. B., Orth, R. J., Dennison, W. C., Olyarnik, S., et al. (2009). Accelerating loss of seagrasses across the globe threatens coastal ecosystems. *Proceedings of the National Academy of Sciences*, *106*(30), 12377–12381. https://doi.org/10.1073/pnas.0905620106

Wolf, D., Georgic, W., & Klaiber, H. A. (2017). Reeling in the damages: Harmful algal blooms' impact on Lake Erie's recreational fishing industry. *Journal of Environmental Management, 199*, 148–157.

Wolf, D., & Klaiber, H. A. (2017). Bloom and bust: Toxic algae's impact on nearby property values. *Ecological Economics*, *135*, 209–221.

Xia, Y., Guan, D., Jiang, Z., Peng, L., Schroeder, H., & Zhang, Q. (2016). Assessment of socioeconomic costs to China's air pollution. *Atmospheric Environment*, *139*, 147–156.

Zhang, X., Davidson, E. A., Mauzerall, D. L., Searchinger, T. D., Dumas, P., & Shen, Y. (2015). Managing nitrogen for sustainable development. *Nature*, *528*, 51–59. https://doi.org/10.1038/nature15743

FURTHER READING

Bricker, S., Longstaff, B., Dennison, W., Jones, A., Boicourt, K., Wicks, C., & Woerner, J. (2007). *Effects of nutrient enrichment in the nation's estuaries: A decade of change. NOAA Coastal Ocean Program Decision Analysis Series No. 26*. Silver Spring, MD: National Centers for Coastal Ocean Science.

Cassman, K.G., Doberman, A., & Walters, D.T. (2002). Agroecosystems, nitrogen-use efficiency, and nitrogen management. *Ambio*, *31*(2): 132–140. Retrieved from http://www.jstor.org/stable/4315226, accessed 20 September 2019.

Chatterjee, R. (2009). Economic damages from nutrient pollution create a "toxic debt". *Environmental Science and Technology*. https://doi.org/10.1021/es803044n

Clark, C. M., & Tilman, D. (2008). Loss of plant species after chronic low-level nitrogen deposition to prairie grasslands. *Nature*, *451*, 712–715.

Diaz, R. J., & Rosenberg, R. (2008). Spreading dead zones and consequences for marine ecosystems. *Science*, *321*, 926–929.

Dubrovsky, N.M., Burow, K.R., Clark, G.M., Gronberg, J.M., Hamilton P.A., Hitt, K.J., et al. (2010). *The quality of our Nation's waters—Nutrients in the Nation's streams and groundwater, 1992–2004*. U.S. Geological Survey Circular 1350, 174 p. Retrieved from http://water.usgs.gov/nawqa/nutrients/pubs/circ1350, accessed 20 September 2019.

Galloway, J. N., Winiwarter, W., Leip, A., Leach, A., Bleeker, A., & Erisman, J. W. (2014). Nitrogen footprints: Past, present, and future. *Environmental Research Letters*, *9*. https://doi.org/10.1088/1748-9326/9/11/115003

Gu, B., Ju, X., Wu, Y., Erisman, J. W., Bleeker, A., Reis, S., et al. (2017). Cleaning up nitrogen pollution may reduce future carbon sinks. *Global Environmental Change*, *48*, 55–66.

Havens, K. (2015). Effects of climate change on the eutrophication of lakes and estuaries. UF/IFAS Extension SGEF-189, 3 p.

Henderson, B., Gerber, P., Hilinksi, T., Falcucci, A., Ojima, D. S., Salvatore, M., & Conant, R. T. (2015). Greenhouse gas mitigation potential of the world's grazing lands: Modeling soil carbon and nitrogen fluxes of mitigation practices. *Agriculture, Ecosystems & Environment*, *207*, 91–100.

Hong, B., Swaney, D., & Howarth, R. W. (2011). A toolbox for calculating net anthropogenic nitrogen inputs (NANI). *Environmental Modeling Software*, *26*, 623–633.

Johnson, P. T. J., Townsend, A. R., Cleveland, C. C., Glibert, P. M., Howarth, R. W., McKenzie, V. J., Rejmankova, E., & Ward, M. H. (2010). Linking environmental nutrient enrichment and disease emergence in humans and wildlife. *Ecological Applications*, 20, 16–29.

Joo, Y. J., Li, D. D., & Lerman, A. (2013). Global nitrogen cycle: Pre-anthropocene mass and isotope fluxes and the effects of human perturbations. *Aquatic Geochemistry*, *19*(5–6), 477–500.

Keeler, B. L., Gourevitch, J. S., Polasky, S., Isbell, F., Tessum, C. W., Hill, J. D., & Marshall, J. D. (2016). The social costs of nitrogen. *Science Advances*, *2*, e1600219.

Larkin, S.L., & Adams, C.M. (2013). Economic consequences of harmful algal blooms: Literature summary (Final Report for Gulf of Mexico Alliance Project #00100304). Tallahassee, FL. Retrieved from http://www.fred.ifas.ufl.edu/pdf/Adams-Larkin-LitRev-April2013.pdf, accessed 29 September 2018.

Lassaletta, L., Billen, G., Garnier, J., Bouwman, L., & Valazquez, E. (2016). Nitrogen use in the global food system: Past trends and future trajectories of agronomic performance, pollution, trade, and dietary demand. *Environmental Research Letters*, *11*(9), 095007. http://dx.doi.org/10.1088/1748-9326/11/9/095007

Nielsen, R. (2005). Can we feed the world? Is there a nitrogen limit of food production? Retrieved from http://home.iprimus.com.au/nielsens/nitrogen.html, accessed 21 September 2019.

Nolan, B. T., & Hitt, K. J. (2006). Vulnerability of shallow groundwater and drinking-water wells to nitrate in the United States. *Environmental Science and Technology*, *40*(24), 7834–7840.

Pinder, R.W., Davison, E.A., Goodale, C.L., Greaver, T.L., Herrick, J.D., and Liu, L. 2012. Climate change impacts of U.S. reactive nitrogen. *Proceedings of the National Academy of Science 109* (20): 7671–7675. www.pnas.org/cgi/doi/10.1073/pnas.1114243109, accessed 22 September 2019.

Pretty, J. N., Mason, C. F., Nedwell, D. B., Hine, R. E., Leaf, S., & Dils, R. (2003). Environmental costs of freshwater eutrophication in England and Wales. *Environmental Science and Technology*, *37*(2), 201–208.

Qiu, J. (2011). China to spend billions cleaning up groundwater. *Science*, *334*, 745–745.

Scavia, D., & Bricker, S. B. (2006). Coastal eutrophication assessment in the United States. *Biogeochemistry*, *79*(1–2), 187–208.

U.S. Environmental Protection Agency. (2002). Nitrogen—Multiple and regional impacts. US Environmental Protection Agency Clean Air Markets Division Report EPA-430-R-01-006, 38 p.

U.S. Environmental Protection Agency. (2009). An urgent call to action. Report of the State-EPA Nutrient Innovations Task Group, 41 p.

U.S. Environmental Protection Agency. (2011). Reactive nitrogen in the United States: An analysis of inputs, flows, consequences and management options. A Report of the US EPA Science Advisory Board (EPA-SAB-11-013). Washington, DC, 172 p.

U.S. Environmental Protection Agency. (2013). Final aquatic life ambient water quality criteria for ammonia-Freshwater 2013. *Notice in Federal Register*, *78*(163), 52192–52194.

U.S. Environmental Protection Agency. (2016). *Water quality assessment and TMDL information*. Washington, DC: United States Environmental Protection Agency. Available at https://ofmpub.epa.gov/waters10/attains_index.home, accessed 05 June 2018

Wang, L., Butcher, A., Stuart, M., Gooddy, D., & Bloomfield, J. (2013). The nitrate time bomb: A numerical way to investigate nitrate storage and lag time in the unsaturated zone. *Environmental Geochemistry and Health*, *35*, 667–681.

Wang, L., Stuart, M. E., Lewis, M. A., Ward, R. S., Skirvin, D., Naden, P. S., Collins, A. L., Ascott, M. J. (2016a). The changing trend in nitrate concentrations in major aquifers due to historical nitrate loading from agricultural land across England and Wales from 1925 to 2150. *Science of the Total Environment*, 542, 694–705.

Wang, S., Changyuan, T., Xianfang, S., Ruiqiang, Y., Zhiwei, H., & Yun, P. (2016b). Factors contributing to nitrate contamination in a groundwater recharge area of the North China Plain. *Hydrological Processes*, *30*(13), 2271–2285.

Ward, M. H., de Kok, T. M., Levallois, P., Brender, J., Gulis, G., Nolan, B. T., & VanDerslice, J. (2005). Drinking-water nitrate and health—Recent findings and research needs. *Environmental Health Perspectives*, *113*(11), 1607–1614.

Ward, M. H., Mark, S. D., Cantor, K. P., Weisenburger, D. D., Correa-Villasenor, A., & Zahm, S. H. (1996). Drinking water nitrate and the risk of Non-Hodgkin's Lymphoma. *Epidemiology*, *7*, 465–471.

Zaehle, S., Ciais, P., Friend, A. D., & Prieur, V. (2011). Carbon benefits of anthropogenic reactive nitrogen offset by nitrous oxide emissions. *Nature Geoscience*, *4*, 601–605. https://doi.org/10.1038/ngeo1207

11
Strategies for Reducing Excess Reactive Nitrogen to the Environment

In the previous chapters, we have learned that various forms of reactive nitrogen released into the environment from anthropogenic activities have contributed to human health problems, degradation of ecosystems, reductions in biodiversity, air pollution, contamination of surface waters and groundwater, acidification of soils and water bodies, algal blooms, fish kills, disruptions to greenhouse gas balance, large economic losses (e.g., jobs, home values, fishing and shellfish industries), and substantial economic costs for drinking water treatment. However, one cannot discount the enormous benefits from nitrogen fertilizer use in developing countries, which has led to an increase in food production and substantial reduction in malnutrition. With better nutrition, healthier diets can lead to more efficient immune response to parasitic and infectious diseases (Nesheim, 1993). Although, shortages of food and malnutrition still exist in many large parts of the world.

11.1. MAJOR CHALLENGES

A critical challenge for humanity is to reduce the significant losses of excess reactive nitrogen to the environment from agricultural activities in developed countries, while maintaining sustainable food production and food security in other parts of the world that have limited access to sufficient reactive nitrogen to replenish crop uptake. Globally, more than 90% of the nitrogen used to produce meat and dairy products, along with 80% used to grow plant-based foods, is lost to the environment (Moran, 2016). Based on land-use and nitrogen-budget model simulations used to analyze baseline (reference scenario) and future mitigation strategies for the global agricultural nitrogen cycle, Bodirsky et al. (2014) estimated that reactive nitrogen sources would increase from 185 to 232 Tg/yr in 2050. They indicated that in 2050, reactive nitrogen pollution could rise to 102–156% of the 2010 value. They pointed out that unless rigorous mitigation actions are taken (such as improving nitrogen use efficiency (NUE) in crop and animal production systems, reducing food wastes in households, and lowering consumption of animal products) air, water, and atmospheric pollution of reactive nitrogen would exceed critical environmental thresholds.

Oita et al. (2016) analyzed the flow of reactive nitrogen along international trade routes from centers of food consumption to regions with nitrogen pollution. By combining the estimates of reactive nitrogen creation with an economic information, they found that consumption in China, India, and the United States accounted for 46% of the global nitrogen footprint. Also, based on the amount of reactive nitrogen released during production of commodities that are traded internationally, Oita et al. (2016) found that developed countries (such as Japan, Germany, the United States) have per capita nitrogen footprints that are two times larger than the amount of reactive nitrogen that is released within these countries.

Sutton et al. (2013a) estimated that about 2% of world energy use is for the industrial manufacture of reactive nitrogen, mainly through the Haber–Bosch process (used mainly for fertilizers for agriculture). Also, substantial reductions in reactive nitrogen are needed from other sources, such as emissions from fossil fuel combustion and wastewater disposal. Fig. 11.1 shows the increases in human population and increases in the creation of total anthropogenic reactive nitrogen that have occurred between 1900 and 2012.

In the United States, agriculture accounted for an estimated 65% of nitrogen pollution, followed by emissions from vehicles and power plants (20%) and industry (15%), according to the U.S. Environmental Protection Agency (Moran, 2016). Sutton et al. (2013a) provided a map showing the estimated net anthropogenic nitrogen inputs to the main river basins around the world (Fig. 11.2).

Nitrogen Overload: Environmental Degradation, Ramifications, and Economic Costs, Geophysical Monograph 250, First Edition. Brian G. Katz.
© 2020 American Geophysical Union. Published 2020 by John Wiley & Sons, Inc.

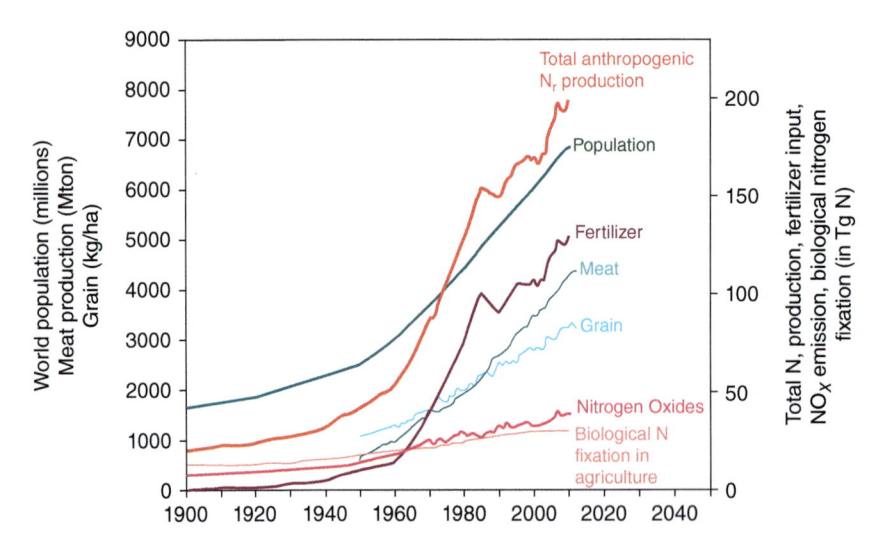

Figure 11.1 Plot showing increases in human population and increases in the creation of total anthropogenic reactive nitrogen between 1900 and 2012. *Source*: With permission from Erisman et al. (2015). Reproduced with permission of World Wildlife Fund.

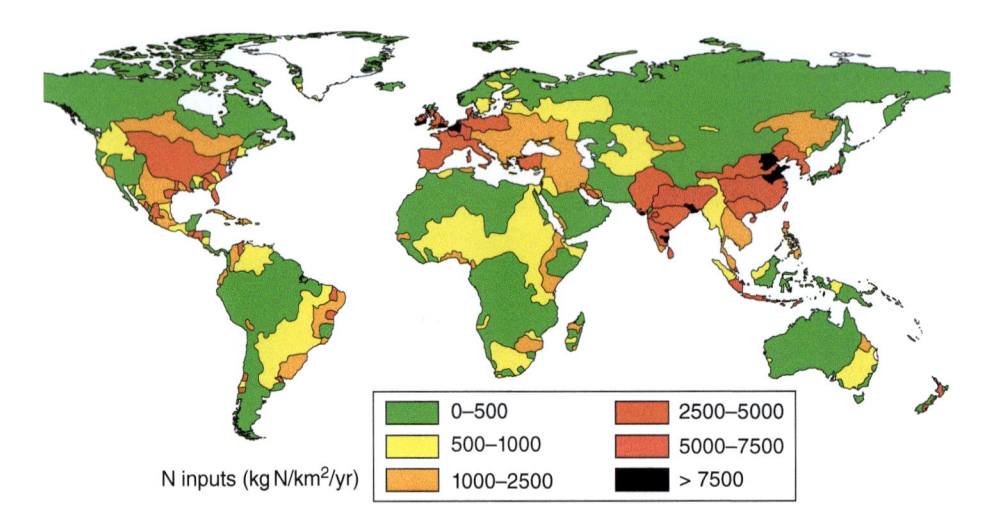

Figure 11.2 Map showing estimated net anthropogenic reactive nitrogen inputs to the main river basins around the world. *Source*: Modified from colored figure ES3 in Sutton et al. (2013a), Our Nutrient World: The challenge to produce more food and energy with less pollution. Global Overview of Nutrient Management. Centre for Ecology and Hydrology, Edinburgh on behalf of the GPNM and the International Nitrogen Initiative.

Reactive nitrogen inputs greater than 5000 kg N/km²/yr were estimated for basins in the United States, Southeast Asia, China, India, and parts of Europe. Research studies indicated that reactive nitrogen inputs from anthropogenic sources would need to be reduced globally by at least 50% to alleviate negative impacts (e.g., de Vries et al., 2015; Steffen et al., 2015) and to "remain a safe operating space for humanity" (Erisman et al., 2015). These negative impacts to human health, the environment, and economies will only get worse if actions are not enhanced over the next decade. Unfortunately, even if current anthropogenic nitrogen inputs are reduced substantially, there could still be a considerable lag time (years to decades, and possibly longer) for the accumulated reactive nitrogen in the atmosphere, hydrosphere, and biosphere to dissipate.

Reactive nitrogen compounds undergo various transformation reactions and cycle between various environmental systems, which has been referred to as the nitrogen cascade (Galloway et al., 2003). Based on the analyses of long-term soil data (1957–2010) from more than 2000 sites throughout the Mississippi River basin, Van Meter et al. (2016, 2017) estimated that the observed accumulation of nitrogen in soils would lead to a biogeochemical lag time of 35 years for removal of 99% of legacy nitrogen, even with a complete termination in fertilizer application. Recent studies show that groundwater nitrogen applied decades ago can still be found in aquifers

(Sanford & Pope, 2013; Sebillo et al., 2013). For example, even though substantial decreases in fertilizer applications have been noted in the Netherlands, Denmark, and Germany (where the nitrogen surplus is back to the level of that of 1970), concentrations of nitrate-N in groundwater have not responded to this decrease of nitrogen surplus (Sutton et al., 2011). Large quantities of nitrate stored in aquifers are released slowly over time, as a function of the groundwater residence time, which can range from weeks to several thousands of years (Alley et al., 2002; Ascott et al., 2017).

The potential adverse effects of climate change also need to be considered when accounting for reactive nitrogen releases to the environment. A recent report by the U.S. Global Change Research Program on climate change impacts in the United States (Melillo et al., 2014) noted that increases in nitrogen loads (along with sediment and other pollutants) to rivers and lakes would likely occur with increasing air and water temperatures, more intense precipitation and runoff, and intensifying droughts.

So why isn't more being done to address the adverse impacts associated with excess reactive nitrogen? Part of this could be due to the role of nitrogen in food security considerations around the world, but as Sutton et al. (2013a) point out, there has been little scientific communication about nitrogen overuse and emissions compared to climate change or biodiversity losses. Other factors hampering this communication include the complex interlinkages among source inputs of various reactive nitrogen forms with effects at local, regional, and global scales. This chapter discusses some of the approaches to reduce nitrogen use, emissions, and deposition to effectively reduce our nitrogen footprint.

During the past two decades, there have been many approaches proposed and implemented to manage anthropogenic nitrogen releases to the environment. For example, since the 1990s, there have been substantial reductions in reactive nitrogen emissions from fossil fuel combustion sources in most developed countries. However, reductions need to be made in several other anthropogenic source categories that contribute significant amounts of reactive nitrogen to the environment. During the past 5 years an increased awareness of the impacts from reactive nitrogen has led to more actions. For example, in November 2013, delegates from several countries participated in the N2013 Conference in Kampala, Uganda, where they crafted and agreed upon the "Kampala Statement for Action on Reactive Nitrogen in Africa and Globally." Four key actions were part of the Kampala Statement that focused on managing reactive nitrogen from anthropogenic sources: (a) reduce nitrogen losses from agriculture, industry, transportation, and energy sectors, (b) improve wastewater treatment methods, (c) encourage more innovation and increased awareness among stakeholders, and (d) identify policy and technology solutions to address global management issues. Reis et al. (2016) summarized information from more than 30 papers published in 2014, 2015, and 2016 as a focus issue in the journal Environmental Research Letters (these articles are available through open access and listed in http://iopscience.iop.org/journal/1748-9326/page/Nitrogen %20management%20challenges, accessed 02 August 2018).

As a way of focusing efforts on decreasing reactive nitrogen releases worldwide, Erisman et al. (2015) identified three main drivers that need to be addressed to achieve global reductions in reactive nitrogen. These drivers include the following:

1. Inefficient and untenable uses of nitrogen fertilizers and manure that contribute to large losses of reactive nitrogen to aquatic ecosystems and the atmosphere.
2. Human population growth, along with a diet shift toward more protein-rich food (meats), has contributed to a large increase in the use of nitrogen fertilizers and their inefficiency.
3. Increased amounts of reactive nitrogen released to the atmosphere from the combustion of fossil fuels due to the higher demand associated with population increases and agricultural activities.

11.2. INCREASING NITROGEN USE EFFICIENCY (NUE) IN AGRICULTURE

Agricultural activities for food production are the largest user of anthropogenic nitrogen in the world. Globally, human use of synthetic nitrogen fertilizers has increased by a factor of nine since the 1960s and increases of around 40–50% have been projected over the next 50 years (Fig. 11.3) (Sutton et al., 2013a). The Food and Agricultural Organization of the United Nations estimated a 50% increase in global fertilizer consumption by 2050 to meet the trends in population, improved diets, and increased consumption of animal products. Accompanying this increase in fertilizer consumption, future scenarios based on population and economic growth indicate nitrogen losses of 70% to the global environment by 2050 (Sutton & Bleeker, 2013).

Recent efforts to reduce reactive nitrogen losses to the environment are focused on improving NUE in agriculture. The Organisation for Economic Co-operation and Development (OECD, http://www.oecd.org/about/, accessed 02 August 2018), which provides opportunities for governments to work together globally to develop solutions to common problems, has defined NUE as the ratio between the amount of fertilizer nitrogen removed from the field by the crop relative to the amount of fertilizer nitrogen applied.

In fact, crops are not all that efficient in taking up nitrogen from fertilizers. Based on information on the full chain of nitrogen summarized from studies on global nitrogen budgets, Sutton et al. (2013a) noted in their publication, "Our Nutrient World: The challenge to produce more food and energy with less pollution," that 80% of nitrogen consumed from agricultural production (e.g., in fertilizers and biological nitrogen fixation) is lost to the environment. Additionally, they noted that livestock further exacerbate the overall agricultural nitrogen inefficiency, as 80% of nitrogen in crop and grass harvests that feed livestock provides only 20% of the nitrogen in human diets. Sutton et al. (2013a) also noted the lack of an intergovernmental framework to address the multiple challenges for nitrogen, phosphorus, and other nutrients. They provided a framework with detailed actions that would maximize nutrient benefits for humanity and also demonstrated the potential for net economic benefits by improving nutrient use efficiency by 20% by the year 2020. Figure 11.4 shows the savings in reactive nitrogen from a 20% relative improvement in NUE for the full food chain.

There have been various approaches for assessing NUE in food production and consumption. For example, a report by the European Union Nitrogen Expert Panel (EUNEP) (2015) presented a user-friendly indicator for NUE for agriculture and food production–consumption systems. Their NUE indicator was based on the mass balance principle (incorporating nitrogen input and output data, N output/N input) (Fig. 11.5). They further noted that NUE values should be interpreted in relation to productivity (nitrogen output) and nitrogen surplus (i.e., the difference between nitrogen input and harvested N output). The calculated NUE values would preferably be between 50% and 90%. Lower values could produce more pollution and higher values could result in mining of soil nitrogen. The EUNEP stated that their proposed NUE indicator would be effective for setting realistic targets and monitoring of progress especially for these sustainable development goals in food and nutrition security, sustainable food consumption and production, marine ecosystems, and terrestrial ecosystems.

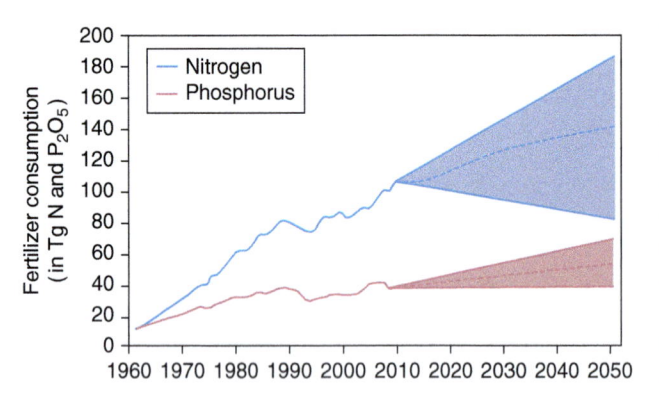

Figure 11.3 From Sutton et al. (2013a). Our nutrient world. Trends in global mineral fertilizer consumption for nitrogen (N) and phosphorus (expressed as P_2O_5) and projected amounts to 2050. Projections are based on present-day decisions explained in this reference. *Source*: Reproduced with permission of Centre for Ecology and Hydrology.

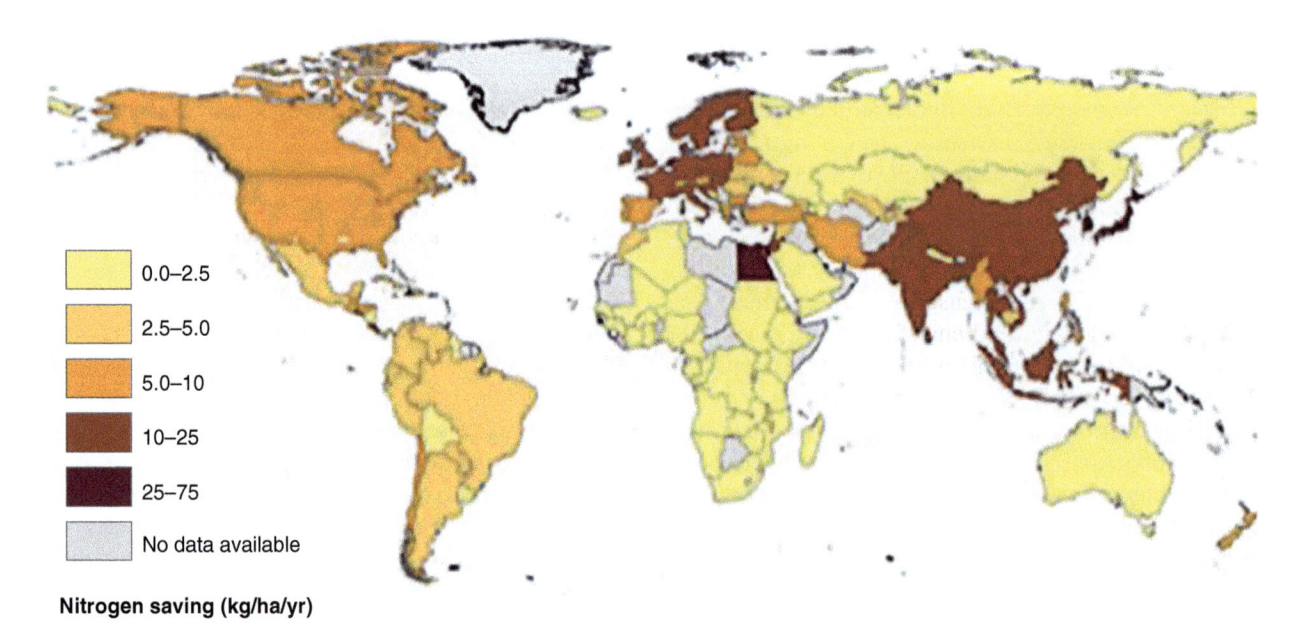

Figure 11.4 Sutton et al. (2013a) showed the savings of reactive nitrogen from a 20% relative improvement in the full-chain NUE. *Source*: Reproduced with permission of Centre for Ecology and Hydrology.

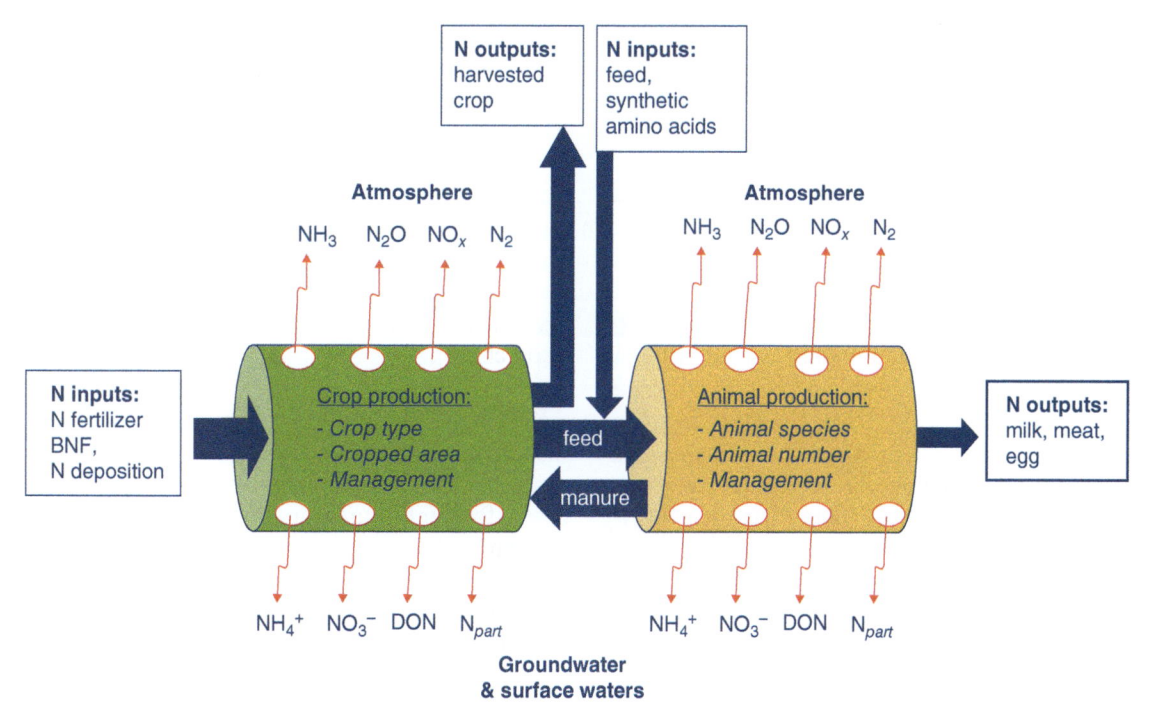

Figure 11.5 A plot from EUNEP (2015) that shows conceptual model for nitrogen input and output mass balance for mixed crop and livestock production systems (total N inputs balance N total outputs after accounting for storage in the system). This model was referred to as "the hole in the pipe" model because it shows the leaks in the nitrogen cycle for crop and animal production. The diagram also shows the interconnection between various components and how changes in one flow can rate affect the others. *Source*: Reproduced with permission of EU Nitrogen Expert Panel.

11.2.1. Enhanced Efficiency Fertilizers

One option for improved management and timing of fertilizer applications and nitrogen uptake is the use of enhanced efficiency fertilizers (EEFs). Li et al. (2017) presented a detailed analysis of worldwide studies published in 1980–2016 that evaluated four major types of EEFs regarding their effectiveness in increasing yield and NUE and reducing N losses. The four types of EEFs evaluated include polymer-coated fertilizers (PCF), nitrification inhibitors (NI), urease inhibitors (UI), and double inhibitors (DI), that is, urease and NI combined. They found that productivity and environmental efficacy was related to both EEF type and cropping systems and were also affected by biophysical conditions. The most effective EEF scenarios were found for grassland and rice paddy systems, as shown in Table 11.1. For wheat and maize systems, it was found that EEFs were generally less effective. They concluded that EEFs could be used for sustainable agricultural production, yet they emphasized the importance of knowledge-based nitrogen management practices and the elimination of fertilizer mismanagement practices. Another study by Xue et al. (2014) in the Taihu Lake region of China found that reduced fertilizer applications to rice–wheat rotation systems (from 510 to 390 kg/ha/yr) along with controlled release fertilizer and the combined use of inorganic and organic fertilizers resulted in less environmental impacts from fertilizer nitrogen.

Although technically not an EEF, another recent approach to reduce nitrogen losses from agriculture (e.g., nitrogen immobilization and mineralization) is the addition of biochar to soils. In their review of biochar and soil nitrogen dynamics, Clough et al. (2013) noted that studies have shown that biochar adsorption of ammonia decreases ammonia and nitrate losses during composting and after manure application. However, reduction in losses varied considerably depending on biochar source material and characteristics. Biochar has a potential as a mitigation agent for reducing nitrogen losses to the environment. However, Clough et al. (2013) pointed out that more research needs to be done to evaluate biochar interactions with nitrogen in studies that cover long-duration time periods.

Reis et al. (2016) pointed out that there needs to be more effective management of nitrogen in soils, particularly in areas with limited nitrogen in soils. For example, in areas of sub-Saharan Africa soil, nitrogen mining

Table 11.1 Results of studies on EEFs.

Crop type	EEF type	n	Yield increase (%)	NUE improvement (%)	N loss decrease (%)	Total N loss (kg N/ha)
Grassland	DI	133	11	33	47	84
Rice paddies	UI	100	9	23	41	16

Source: Data summarized from Li et al. (2017). Reproduced with permission of John Wiley & Sons.

leaves soils with insufficient nitrogen to grow crops, which can lead to food scarcity and nutritional insecurity. Some potential solutions involve more efficient reactive nitrogen cycling in crop-livestock systems, linking fertilizer applications to biophysical environments and socioeconomic status of farmers to increase NUE, and enhancing human-managed biological nitrogen fixation using appropriate legumes (e.g., Woomer et al., 2004).

11.2.2. Reducing Agricultural Emissions to the Atmosphere and Manure Management

Agricultural activities also contribute reactive nitrogen to the atmosphere as N_2O. Several methods have been proposed to reduce N_2O emissions from fertilized agricultural systems. These include the increased use of NI that would slow the conversion of ammonia to nitrate, which would then lead to less accumulation of nitrate in soils (a key substrate for N_2O emissions). Misselbrook et al. (2014) studied the effectiveness of various inhibitors over a range of field conditions in the United Kingdom, and found that dicyandiamide (DCD) additions to fertilizer, cattle urine, and cattle slurries applied to the land surface could reduce agricultural N_2O emissions in the United Kingdom by 20%. They concluded that more cost-effective delivery methods would be needed to entice farmers to use these NI.

Better management of livestock systems would contribute less ammonia and N_2O to the atmosphere, groundwater, and surface-water systems. The United Nations Economic Commission for Europe (UNECE, 2012) proposed several strategies for preventing nutrient losses from manure. These include (a) the use of chemical amendments (e.g., aluminum sulfate) that acidify manures, (b) the cleaning of air streams when using buildings with controlled ventilation to reduce NH_3 emissions. to minimize emissions during animal housing; (c) covering manure storage would decrease NH_3 emissions and also prevent the entry of rainwater and surface runoff of nutrients; (d) during storage treat manures and effluents with chemicals and polymers that would reduce nutrient losses after land application; (e) use of anaerobic digestion methods to generate biogas (CH_4 and CO_2) that can be used for fuel, and these methods could allow recovery of fertilizer products containing nitrogen and phosphorus; and (f) low emission spreading of liquid manures in narrow bands, injection into the soil, and immediate incorporation of manure into the soil would help to considerably reduce NH_3 emissions (40–90% depending on the method, UNECE, 2012), and also prevent surface runoff. Low emission spreading of liquid manures has been a regulatory requirement in the Netherlands and Denmark for the past 10–20 years. Several other management practices have been used to reduce nitrogen losses including composting of manure: improving the fertilizer value of manure by varying animal diets; and wastewater storage, treatment, and spraying directly on crops. Sutton et al. (2013a) emphasized that manures should be applied to conform with the basic principles of "Nutrient Stewardship," which include the determination of available concentrations of N, P, and other nutrients; application at the optimal time at rates that meet crop nutrient requirements, and prevention of nitrogen losses during and after application.

In the United States, large Concentrated Animal Feeding Operations (CAFOs) have to comply with regulatory requirements by the U.S. Environmental Protection Agency (EPA) for permitting of CAFOs. The U.S. EPA and other governmental agencies provide information on voluntary technologies and best management practices to improve the production efficiency of CAFOs and protect the quality of waters in the United States. The U.S. EPA also recommends several other ways that agricultural operations can reduce nutrient pollution including proper fertilizer management, use of cover crops, planting buffers to filter out nutrients, waste composting, conservation tillage (reducing how often fields are tilled), keeping livestock out of surface water bodies to more effectively manage livestock wastes, reduce nutrient loading in drains from agricultural fields, and greater collaboration among stakeholders for protecting water quality in watersheds (https://www.epa.gov/nutrientpollution/sources-and-solutions-agriculture, accessed 28 July 2018). Several states have financial assistance programs to facilitate the voluntary implementation of agricultural management practices to reduce nonpoint source pollution (e.g., from animal feeding operations). Some states provide cost-sharing for installation of best management practices to address issues such as erosion, flooding, poor drainage, stream restoration, and other water quality degradation. Other states offer cost-share and planning assistance for agricultural producers that want to adopt clean water farming practices in vulnerable watersheds.

During the past 20 years, the U.S. EPA has encouraged states and tribes to adopt numeric nutrient criteria for reducing nutrient pollution. EPA's National Numeric Nutrient Criteria Program has developed policies, guidance, provided technical and financial support, and encouraged state and tribal partnerships and accountability by publicly sharing progress in adopting numeric nutrient criteria. Numeric nutrient criteria are a critical tool for protecting and restoring the designated uses of a waterbody from nitrogen and other nutrient pollution. As part of their criteria, the program stipulates monitoring of a waterbody for attaining its designated uses; facilitating development of National Pollutant Discharge Elimination System (NPDES) discharge permits; and helping to develop total maximum daily loads (TMDLs) for restoring nutrient-impaired waters that are not attaining their designated uses (https://www.epa.gov/nutrient-policy-data/state-progress-toward-developing-numeric-nutrient-water-quality-criteria, accessed 24 July 2018).

11.3. NUE FOR THE FOOD PRODUCTION/CONSUMPTION CHAIN

Erisman et al. (2018) summarized results from an extensive review of previous studies that used four different approaches for assessing NUE of food production/consumption, and/or the environmental impact of nitrogen use in food production. These approaches include life cycle analysis (LCA), nitrogen footprint analysis, nitrogen budget analysis, and environmental impact. A brief description of these approaches is included herein (see Erisman et al., 2018, for more details):

1. LCA provides an "assessment of the potential environmental and human health impacts associated with a product, process, or service by compiling an inventory of relevant energy and material inputs and environmental releases and evaluating the potential environmental impacts associated with the identified inputs and releases" (Erisman et al., 2018).

2. A nitrogen footprint approach (discussed later in this chapter) takes into account total nitrogen losses to the environment that results from the production of a defined unit of food.

3. The nitrogen budget analysis accounts for the inputs and outputs of nitrogen across the boundaries of a system, and typically includes information about internal nitrogen fluxes within the system. There are different types of nitrogen budgets; however, Erisman et al. (2018) included these three: (a) the farm-gate nitrogen budget (total inputs/outputs across the farm boundaries); (b) nitrogen balance indicator (difference between available nitrogen to the agricultural system and nitrogen exported and harvested); and (c) the food system waste/loss indicator (estimate of the loss of nitrogen through waste).

4. Environmental impact assessment (EIA), including possible environmental impacts from a proposed project by considering interrelated socioeconomic, cultural, and human health impacts that can be both beneficial and adverse.

The studies reviewed by Erisman et al. (2018) dealt with 17 products for 30 countries and 11 studies used the LCA approach, 6 used the nitrogen footprint approach, 16 used the nitrogen budget approach, and 3 used the EIA approach. Based on results from these studies, Erisman et al. (2018) found a small range (5–15%) for the overall full-chain N efficiency (Nitrogen Use Efficiency for the Food Chain (NUEFC)). On average, only 5–15% of the nitrogen invested in global food production was contained in final consumed food products. The estimated NUEFC was extremely high in some regions (more than 100%; Sutton et al., 2013a), possibly resulting from mining of soil nitrogen.

Taking this a step further, Erisman et al. (2018) assessed the NUE for the whole food production–consumption chain, going beyond the farm level, and accounting for processing and distribution by the food industry. They defined their NUE indicator for the whole food chain, NUE_{FC}, as the available food for consumption of protein nitrogen (protein in food available) for a given human population divided by the newly formed nitrogen that was used to produce the food (see Equation 11.1). More detailed information about the NUE_{FC} calculation methods and rationale for including the various components are described by Erisman et al. (2018). Their NUE_{FC} effectively describes how efficiently the new reactive nitrogen created for food production is converted into the food protein nitrogen that is consumed. Equation used to calculate NUE_{FC}:

$$NUE_{FC} = \frac{\text{Nitrogen food availability}}{\text{Fertilizer} + BNF + atm.\text{deposition} + (\text{import-export}) + \text{changes in stock}}$$

(11.1)

Where BNF is biological nitrogen fixation; atm. dep is atmospheric deposition.

The NUE_{FC} for five different European countries during 1980 to 2011 (Fig. 11.6) were calculated. Ireland had the lowest NUE_{FC}, whereas Italy had the highest. Overall, the estimated NUE_{FC} increased over time, which Erisman et al. (2018) attributed to a reduction in fertilizer use and a steady increase in the availability of protein for consumption, either produced nationally or through import. Factors such as changes in nitrogen stock and import/

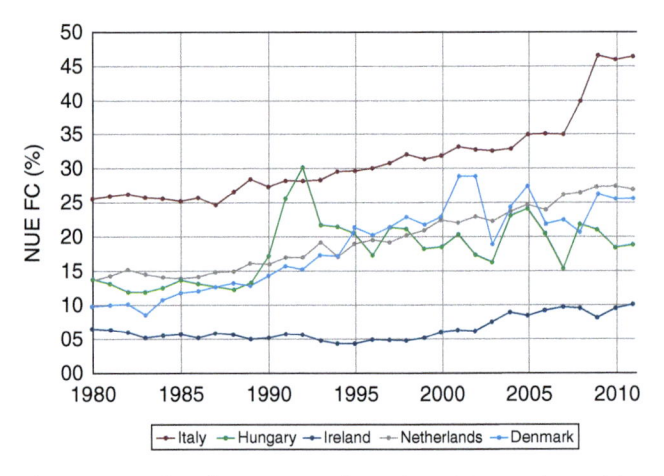

Figure 11.6 Plot showing trends in the NUE$_{FC}$ estimated by Erisman et al. (2018) for five European countries from 1980 to 2011. *Source*: From Erisman et al. (2018). Reproduced with permission of MDPI.

export of nitrogen were not consistent from 1 year to the next. For this reason, using a 3-year moving average for evaluating trends was suggested. Eastern European countries showed an initial increase until 1990, likely related to changes in 1989 after the breakdown of the Berlin Wall and with the reduction in fertilizer subsidies. Generally, it was found that as the total nitrogen input increased, NUE$_{FC}$ decreased. Therefore, the NUE$_{FC}$ is highly dependent on the per capita nitrogen input because the total consumption of food nitrogen increases with the population number. A successful sustainable food production future must include processes that result in a higher NUE$_{FC}$ with lower per capita NUE$_{FC}$. This very useful NUE$_{FC}$ information provides policymakers with strategies for developing a more nitrogen-efficient system for food production by reducing losses and optimizing and increasing NUE. Also consumers will benefit by choosing and consuming products that contain protein (with high nitrogen content) and require very low nitrogen inputs.

11.4. REDUCING REACTIVE NITROGEN EMISSIONS FROM FOSSIL FUEL COMBUSTION

Combustion of fossil fuels results in a substantial amount of reactive nitrogen, approximately 20% of human reactive nitrogen production (Sutton et al., 2013a). This reactive nitrogen contributes emissions of greenhouse gas nitrogen oxide and ammonia to the atmosphere resulting in increased particulate matter and tropospheric ozone that is harmful to human health, ecosystems, and food production systems. For example, high concentrations of reactive nitrogen (NH_3 and NO_2) were found in the air above the Indo-Gangetic plains of India in both rural and urban areas with significant seasonal variability (Singh & Kulshrestha, 2014). These high concentrations document the elevated risk of adverse human health effects from secondary particulate matter concentrations. It was concluded that more stringent air pollution control measures need to be adopted for densely populated regions and urban areas in developing countries. Reis et al. (2016) noted there are other major hotspots for emissions of reactive nitrogen (e.g., the United States, Europe, China). Although tighter emission control policies have been in place and enforced, for example, in the United States, NH_3 emissions have been largely unregulated and this has resulted in increasing concentrations of ammonium in wet deposition (Du et al., 2014).

Davidson and Kanter (2014) evaluated several scenarios extrapolated to the year 2050 to reduce global N_2O emissions. They noted that one scenario involving no changes in current emissions would result in a doubling of N_2O emissions by 2050. Conversely, it was estimated that a 22% reduction in emissions from various sectors (e.g., agriculture, industry, biomass burning, aquaculture, compared to 2005) would be necessary to maintain N_2O concentrations at around 350 ppb by 2050.

Several overarching strategies have been proposed to reduce N_2O emissions. These strategies involve (a) more effective methods for trapping and utilizing wasted emissions of nitrogen oxide (NO_x) to air, (b) development of low-emission and energy efficiency and the subsequent implementation of these systems in the transportation and industrial sectors, and (c) more use of renewable energy systems. These strategies must include economic and technical considerations such as the nitrogen cycle for overall reductions in nitrogen losses. Sutton et al. (2014) elegantly summarized the different viewpoints on ways to reduce N_2O emissions that consider economic performance and a "green economy."

The concept of a green economy was defined by United Nations Environment Programme (UNEP) (2011) and included the "economic opportunities that promote sustainability and improve human well-being and social equity." As Sutton et al. (2014) point out the so-called "green economy" would need to consider issues associated with profitability of production sectors (agricultural industry) and those related to societal welfare. However, a main barrier to change would require bridging or bringing together these two different viewpoints: a "sector view" that considers green actions that are consistent with improved profit, and the "societal view" that incorporates the value of all externalities (e.g., quantification of benefits of N_2O reductions on human health, climate, biodiversity, and ecosystems). Measures to reduce N_2O emissions should focus on improving the NUE, which would benefit the "green economy" by saving fertilizer as a valuable resource. They also pointed out that there would be more economic advantages for industry and combustion sources by improving process efficiency along with developing market share.

Another strategy for improving full chain NUE and reducing N_2O emissions would require less consumption of livestock, which requires the use of fossil fuels for production (Sutton et al., 2014). Furthermore, as an incentive to implement such strategies, the total economic benefits of reduced nitrogen losses would have to significantly exceed the benefits of these actions to the emitting sectors. The additional benefits would include reduced threats to human health and ecosystems through improved air, soil, and water quality. Incentives can help to develop new technologies and to overcome the reluctance to change. Other options noted for increasing NUE were capacity building and training in low N_2O emission approaches, avoidance of environmentally damaging N subsidies, internalizing the price of nitrogen pollution through appropriate levies, subsidies or tradable permits, and improving communication to tout the benefits of clean-nitrogen technologies in the marketplace.

Reductions in NO_x emissions from road transport vehicles have been difficult to quantify. Although the use of catalytic converters and engine management systems have been somewhat effective in reducing nitrogen emissions, Oldenkamp et al. (2016) noted that some manufacturers have changed to less-effective emission control measures. Increased emissions of NO_2 from diesel engines have resulted from less-effective equipment installed in these vehicles (Chen & Borken-Kleefeld, 2014). Countries with emerging economies (e.g., China, India, Russia, and Brazil) likely will have increased emissions due to the increased number of vehicles and energy demands (Liu et al., 2013). For example, China has had a 20.8-fold increase in vehicles since 1980, which have led to significant increases in nitrogen pollution in China (Liu et al., 2013).

More accurate information on vehicle NO_x emissions could be forthcoming. A recent press release from the UNECE (21 June 2018; https://www.unece.org/?id=48872, accessed 13 August 2018) noted that several countries are implementing regulatory requirements for real driving emissions tests, based on new technologies and the work done by the European Union (EU) since 2010. New cars in the EU and Republic of Korea are already approved using this methodology. The EU, Japan, and Korea have been developing regulation that would lead to the establishment of a United Nations Global Technical Regulation on real driving emissions testing (expected to be adopted by 2020). Based on support by the United States, Canada, India, and China, these countries are expected to participate in the development of the regulatory provisions (described as a transparent, data-driven process that is open to inputs from all parties involved).

11.5. REDUCING AND MANAGING REACTIVE NITROGEN IN WASTEWATERS

One of the important global messages in the Kampala Statement addressed managing nitrogen in wastewaters and reducing its impact on the environment by improving treatment of waste: "sewage treatment and solid municipal waste (household wastes) are sources of nitrogen losses that could be reduced by treatment and/or recycling" (Reis et al., 2016). Nitrogen levels in wastewater and sewage tend to be high as proteins contain about 16% nitrogen. Fig. 11.7 shows the variation in daily average protein intake per person in different areas around the world.

Davidson and Kanter (2014) estimated that wastewater and sewage contribute to 3% of the global budget of N_2O. The impacts from wastes and wastewater disposal are highly variable around the world. Bustamente et al. (2015) noted that wastewater was the largest source of total dissolved nitrogen to coastal ecosystems in South America. In Africa, Zhou et al. (2014) were not able to accurately estimate wastewater nitrogen loads to rivers in the Lake Victoria Basin because little information existed on wastewater collection facilities and the paucity of wastewater treatment plants. In India, human wastewater was noted as a source of high levels of ammonia and NO_x concentrations in rural and urban areas (Singh & Kulshrestha, 2014).

11.5.1. Wastewater Treatment Facilities

Wastewater treatment typically involves three phases: primary, where suspended solids and organic matter settle out; secondary, where biodegradable solids and organic matter are removed along with harmful bacteria associated with fecal wastes; and tertiary, where further removal occurs along with additional reduction of reactive nitrogen and phosphorus levels. Advantages of tertiary treatment are denitrification of reactive nitrogen to N_2 and the production of sewage sludge that is enriched in organic nitrogen compounds that could be recycled back to the land as fertilizer. Unfortunately, many systems around the world only use primary wastewater treatment or at most secondary treatment. The denitrification process in tertiary treatment requires a considerable amount of energy. In the Netherlands, improvements in wastewater processing using advanced treatment (e.g., denitrification) has removed up to 78% of nitrogen, which is significantly higher than the overall 5% removal rate in the United States and in many other countries.

Although upgrades to wastewater treatment facilities can be expensive, the U.S. EPA (2015) reported on several low-cost modifications to improve nutrient reduction at nonadvanced wastewater treatment plants based on

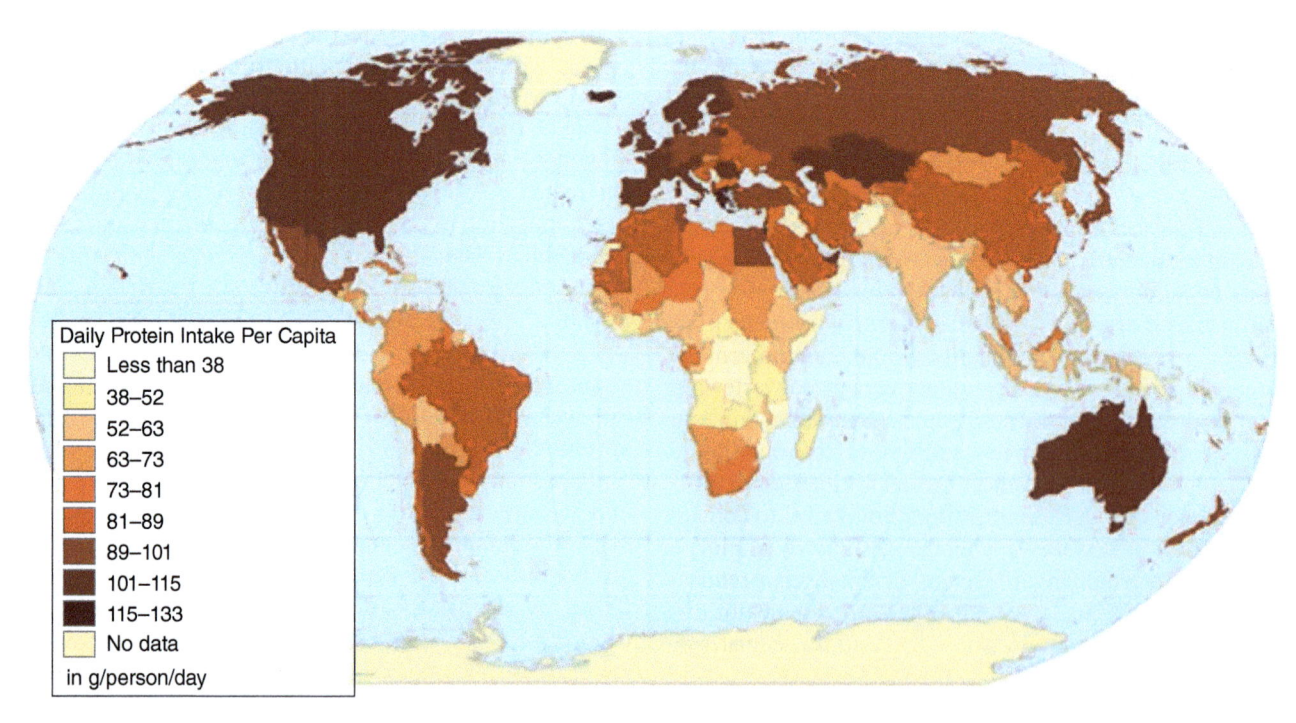

Figure 11.7 Variation in daily average protein intake per person based on regional estimates. *Source*: Modified from colored figure 6.5 in Sutton et al. (2013a) "Our nutrient world." Reproduced with permission of Centre for Ecology and Hydrology.

several case studies. One of their findings was that low-cost nutrient reduction improvements were most feasible for activated sludge plants because excess capacity (volumetric and/or aeration) could be leveraged to enhance nitrification and denitrification without requiring physical infrastructure modifications.

Wastewater treatment upgrades have had substantial improvements in water quality in other areas. For example, in the Chesapeake Bay region in the United States, nitrogen pollution from wastewater treatment plants has decreased by 43%, from 28% of the total nitrogen load to the bay in 1985 to 16% in 2015 (Moran, 2016). Future reduction goals of 39% by 2025 have been set by state and federal agencies from 143,500 metric tons of total nitrogen to 87,000 metric tons. The multistate Chesapeake Bay region had achieved about half of its reduction goal as of 2014 with the nitrogen load to the bay at approximately 112,400 metric tons (Moran, 2016). The Chesapeake Bay Foundation uses a nitrogen footprint calculator developed by Galloway et al. (2014) and Castner et al. (2017) to increase awareness among residents and institutions in the watershed. Shibata et al. (2017) noted that recycling reactive nitrogen from wastewater using new sewage management technologies would also be an effective way to reduce reactive nitrogen releases to the environment.

The use of biochar (as mentioned previously in Section 11.2) may have other benefits for removing nitrogen, as in reducing nitrogen losses from sewage, septage, and manure. For example, De Rozari et al. (2018) noted that soil amended with biochar can be used to improve nutrient removal from wastewater and to increase nitrogen retention and plant growth in agroforestry. Using biochar-amended sands (25% biochar) in wetland mesocosms receiving sewage, they found average removal efficiencies for TN, NO_x-N, and NH_4-N up to 87, 93, and 79%, respectively. The effective nitrogen removal was attributed to mineralization of organic nitrogen and ammonium.

11.5.2. Septic Systems (On-site Wastewater Treatment)

Conventional septic systems are not designed to completely remove nitrogen, which can contribute nitrogen loading to groundwater and waterways. Septic tank systems (including the drainfield) typically remove about 30–50% of the nitrogen that enters the tank, which means that a substantial amount of nitrate can leach to groundwater and be transported to surface water bodies. This issue is especially important in karst aquifers that are highly vulnerable to contamination (Katz, 2019) and in coastal communities, where excess nitrogen releases can lead to toxic algal blooms, beach closures, fish kills, and adverse human health effects. Lusk et al. (2017) indicated that 10–20% of conventional septic systems in the United States fail every year due to improper site selection, design, installation, and lack of maintenance of the system. These failed systems adversely affect public and environmental health. In the United States, EPA estimated that over 2.6 million existing septic systems in nitrogen-sensitive watersheds would be good candidates for advanced septic systems that treat and reduce nitrogen losses to groundwater. Various states in the United States have approved advanced treatment methods (performance-based, innovative systems, advanced treatment units, and alternative treatment technologies) for their decentralized wastewater systems (https://www.epa.gov/septic/advanced-technology-onsite-treatment-wastewater-products-approved-state, accessed 15 August 2018). These different systems can be quite effective in removing nitrogen compared to conventional septic tank systems. For example, a study in Florida showed that a passive nitrogen removal system reduced nitrogen loading by 95% (www.floridahealth.gov/environmental-health/onsite-sewage/research/_documents/rrac/hazensawyervolireportrmall.pdf, accessed 16 August 2018).

In January 2017, as part of Phase I of the Advanced Septic System Nitrogen Sensor Challenge, the U.S. EPA partnered with The Nature Conservancy, U.S. Geological Survey (USGS), and others to request the design of a nitrogen sensor for use in advanced on-site wastewater treatment systems and to monitor their long-term performance. The use of nitrogen sensors in advanced septic systems would allow manufacturers, homeowners, and local and state governments to monitor the performance of these systems in protecting valuable coastal and other water resources. The U.S. EPA has also recommended that homeowners need to properly maintain a septic system (regular pump-outs every 3–5 years, avoid the use of septic system additives, and plant only grass over the drainfield (as trees with extensive root systems could damage the systems pipes or tank).

In summary, upgrading wastewater treatment plants and household septic systems can be expensive, but these modifications can significantly improve the health of water bodies. Moran (2016) included the following quote from professor and research scientist Jan Willem Erisman (Bolk Institute, the Netherlands), "wastewater treatment is clearly part of the solution, especially in limiting outflow of nitrogen to estuaries." He further noted that about 50% of nitrogen that is lost to the environment when humans excrete what they eat can be reduced by appropriate wastewater treatment methods.

11.5.3. Disposal of Food Waste

Reduction in food waste from consumers, retailers, restaurants, and institutional food preparation would result in less-nitrogen release to the environment (Oita et al., 2016; Reis et al., 2016). A potential use of food waste could include anaerobic generation of methane and carbon dioxide biogas that would reduce nitrogen inputs to the environment. Other options for food waste disposal could involve increasing recycling and composting.

11.6. ESTIMATING AND REDUCING NITROGEN FOOTPRINTS

One of the most effective ways for consumers and institutions to reduce use and consumption of reactive nitrogen is to be aware of their nitrogen footprint. A nitrogen footprint has been defined as the total amount of reactive nitrogen that is released to the environment from an entity's consumption patterns (Galloway et al., 2014). Knowledge of the nitrogen footprint provides useful information to make better decisions about resource use and to better understand how offsets and changes in lifestyle would lead to decreases in contributions to nitrogen-related problems from consumer and institution contributions. For example, in an article on how consumers can make better choices using a nitrogen footprint calculator, Stecker (2013) noted that a British vegetarian releases the equivalent of 14 kg N into the environment per year; whereas a daily beef eater in the United Kingdom would contribute three times the nitrogen produced from a vegetarian.

Galloway et al. (2014) summarized and compared five different tools to estimate the nitrogen footprint (N-Calculator, N-Institution, N-Label, N-Neutrality, and N-Indicator) to help manage excess reactive nitrogen in the environment. Attention was mainly focused on consumers and institutions that use the sources that contribute the most amounts of reactive nitrogen. Following a continual increase in the per capita creation of reactive nitrogen from around 1950 to 1980, Galloway et al. (2014) estimated that the global per capita creation of reaction nitrogen has remained relatively constant at about 30 kg/person/yr over the period from approximately 1980–2014. They attributed this relative stability to several processes including decreases in fossil fuel combustion due to beneficial NO_x controls in many developed countries, more efficient agricultural production activities, and a large decrease in fertilizer use especially in Eastern Europe. These decreases in nitrogen emissions were offset by increases in fertilizer use for corn production for biofuels and an increase in meat production and consumption (dietary changes) in large parts of the developing world. Continual increases in reactive nitrogen are likely for the period 2005–2050 (Fig. 11.8).

The studies also compared nitrogen-footprint tools for consumers living in eight different countries (N-Calculator,

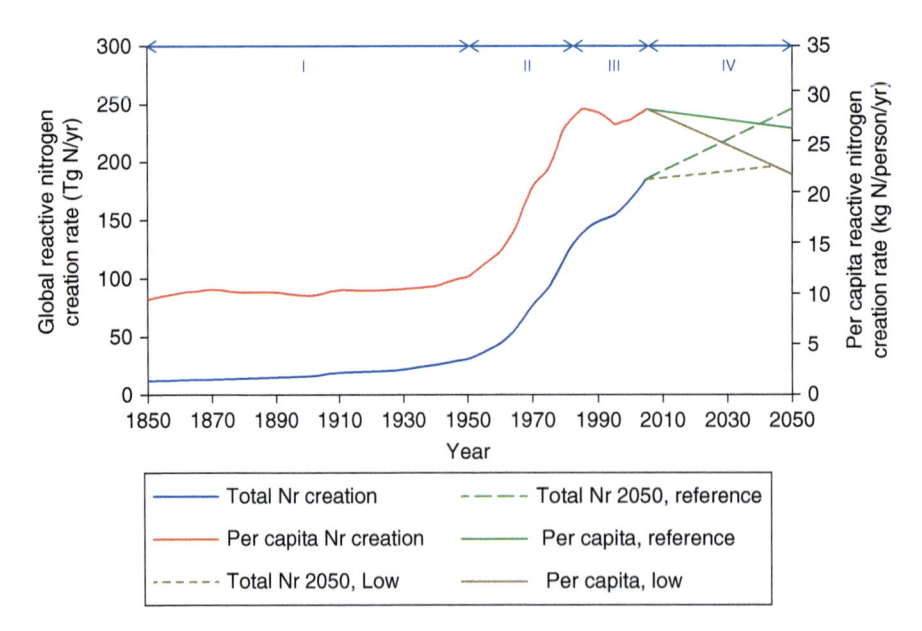

Figure 11.8 Plot from Galloway et al. (2014) showing temporal trends in global anthropogenic reactive nitrogen creation for four time periods (shown and labeled with Roman numerals on top of graph). The left y-axis shows total global creation in Tg N/yr and the right y-axis shows per-capita reactive nitrogen creation rate (kg N/person/yr). Projections to 2050 are based on data from studies presented in Galloway et al. (2014). *Source*: Reproduced with permission of IOP Publishing.

N-Indicator) and the products they buy containing nitrogen (N-Label), for the institutions where people work and receive education (N-Institution), and for events and decision-making regarding offsets (N-Neutrality). Table 11.2 from Galloway et al. (2014) summarizes the annual per capita nitrogen footprints for eight countries using the N-calculator.

Galloway et al. (2014) used the nitrogen footprint tool to calculate the fate of nitrogen consumed in food in different countries. Calculations for the amount of food consumed (total height of column, Fig. 11.9, and subtracting out food waste) were based on the annual per-capita amount of nitrogen consumed in a particular country using Food Supply data from Food and Agricultural Organization of the United Nations Statistical Databases (FAOSTAT) (the Food and Agricultural Organization of the United Nations provides free access to food and agriculture data for over 245 countries and territories from 1961 to the most recent year available; http://www.fao.org/faostat/en/#home, accessed 25 October 2018). Nitrogen consumed in food was further reduced by the percent of reactive nitrogen converted to nitrogen gas (N_2) from denitrification in the wastewater sewage treatment facility (gray portion of column, Fig. 11.9), which yields the estimated net reactive nitrogen loss to the environment (black portion of column, Fig. 11.3).

Findings from the nitrogen footprint calculations showed that developed countries consume about the same amount of nitrogen (5–6 kg N/person/yr), which is greater than the one less-developed country in the study (Tanzania; 2 kg N/person/yr). Advanced wastewater treatment (with conversion of reactive nitrogen to N_2) in some developed countries (e.g., The Netherlands, Germany, and Austria) have substantially decreased nitrogen loads discharged to the environment. Overall, Galloway et al. (2014) found that on average, people in all developed countries overconsumed nitrogen (based on a recommended annual nitrogen intake of 2.5–3.5 kg N/person/yr (WHO, 2007) and U.S. Department of Agriculture (2010) dietary recommendation of 0.8 g protein/person/d and an average healthy body weight of 60–70 kg). Therefore, consumers can decrease their nitrogen footprint from food consumption by reducing the amount of protein to match recommended levels and/or decrease the amount of animal protein in their food intake (e.g., Erisman et al., 2015; Leach et al., 2012; Leip et al., 2014; Westhoek et al., 2014). For an equivalent amount of protein consumption per person globally, Billen et al. (2015) concluded that in the future (by 2050), the fraction of animal protein should not exceed 40% of a total ingestion of 4 kg N/person/yr (or 25% of a total consumption of 5 kg N/person/yr). Moran (2016) also noted that the average person's footprint in the United States is 41 kg N/person/yr, which is much higher than that of many European countries, such as the Netherlands (24 kg) and Austria (20 kg), using estimations from research by Leach et al. (2012). The lower nitrogen foot-

Table 11.2 Nitrogen footprints (kg N/person/yr) calculated by Galloway et al. (2014) for eight countries using the N-calculator. The studies defined "Food consumption" as the nitrogen actually consumed and subsequently excreted; whereas "food consumption, released" is the nitrogen released to the environment after sewage treatment.

	United States	The Netherlands	Germany	United Kingdom	Japan	Austria	Portugal	Tanzania
Food consumption	5.0	1.1	1.6	4.9	3.4	1.1	6.0	2.0
Food production	22	20	18	18	26	16	18	12
Housing	3.0	0.8	1.6	2.0	0.8	0.8	0.7	0.2
Transport	6.0	1.1	1.8	1.1	0.7	1.6	3.5	0.8
Goods and services	2.5	0.5	0.7	1.1	1.0	0.6	0.5	0.2
Total	39	23	24	27	32	20	29	15
Sewage, % denitrification	5	78	67	2	33	79	0	0
Food consumption, released	5.3	5.0	4.9	5.0	5.1	5.2	6.0	2.0

Source: From Galloway et al. (2014). Reproduced with permission of IOP Publishing.

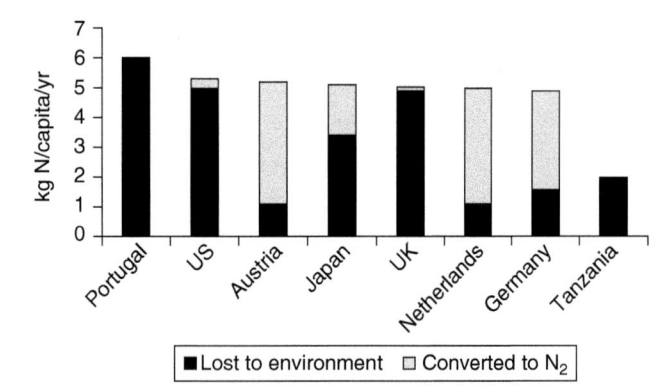

Figure 11.9 Variations in the short-term fate of reactive nitrogen consumed by people and the effect of wastewater treatment in converting reactive nitrogen to nitrogen gas in Portugal, the United States, Japan, the United Kingdom, the Netherlands, Germany, Austria, and Tanzania. *Source*: From Galloway et al. (2014). Reproduced with permission of IOP Publishing.

print of the European countries in this study is likely related to nitrogen removal in advanced sewage treatment processes.

An international N footprint workshop was held in Japan in March 2015 (http://www.n-print.org/japanworkshop, accessed 15 August 2018) where participants exchanged information on results and different methods for calculating nitrogen footprints with the goal of reducing anthropogenic nitrogen loads to the environment locally, regionally, and globally. Based on information from the workshop and to promote global sustainability, Shibata et al. (2017) used the N-Calculator methodology to estimate the average per capita N footprint in ten countries (the United States, Portugal, the Netherlands, the United Kingdom, Germany, Austria, Tanzania, Japan, Taiwan, and Australia). The N footprint varied from 15 kg/person/yr (Tanzania) to 47 kg N/person/yr (Australia) and variation was attributed to differences in the protein consumption rates and food production nitrogen losses. The food sector was dominant in the N footprints for all 10 countries. Shibata et al. (2017) found that the N footprint in countries that import foods and feeds was significantly affected by global connections through trade. Also several N footprint reduction strategies (e.g., for improving N use efficiency, increasing N recycling, reducing food waste, changing dietary choices) were presented. Importantly, the studies identify information gaps, for example, the nitrogen footprint from nonfood goods and soil nitrogen processes.

For further information about nitrogen footprint and ways to calculate it, please visit the website: http://www.n-print.org/, accessed 26 July 2018. The N-PRINT project website is managed by an international group of nitrogen scientists (J.N. Galloway, A. Leach, J.W. Erisman, A. Bleeker, and R. Kohn) and offers assistance with calculating nitrogen footprints for individuals and institutions, posts news articles, and provides links to other nitrogen footprint projects and useful background information on nitrogen.

The Field to Market® organization has created a tool for sustainable agriculture with a goal of minimizing environmental impacts (https://fieldtomarket.org/our-program/, accessed 18 August 2018). The Fieldprint® Platform is a framework that allows brands, retailers, suppliers, and farmers to measure the environmental impacts of commodity crop production and promote continuous improvement. The online tool "Fieldprint Calculator" (https://calculator.fieldtomarket.org/calculator.php, accessed 02 August 2018). The online tool allows for performance calculations on eight "sustainability indicators," including water quality and greenhouse gas emission that address nitrogen, at a specific plot of arable land. Water quality considerations include nitrate runoff into surface water and leaching through the soil; greenhouse gases include nitrous oxide emissions from soil related to fertilizer application rates and farm management practices. In November 2015, Field to Market (along with Together with the Innovation Center for U.S. Dairy) organized a meeting with 350 stakeholders from companies, government agencies, academics, conservation groups, and farmers that are using a comprehensive approach for improving agricultural sustainability. A recent report documented sustainability trends for U.S. agriculture over the past 36 years in eight environmental and five socioeconomic indicators at a national level for ten crops (barley, corn for grain, corn for silage, cotton, peanuts, potatoes, rice, soybeans, sugar beets, and wheat). Eight sustainability metrics were included in the report: biodiversity, energy use, greenhouse gas emissions, irrigation water use, land use, soil carbon, soil conservation, and water quality. (https://fieldtomarket.org/the-alliance/our-history/, accessed 18 August 2018).

11.6.1. China's Nitrogen Footprint and Global Impact

The nitrogen food footprint is much higher in China than in other developed countries and has important global implications to world markets. Fertilizer use in China has increased from 0.5 MT (megatonnes) in the 1960s to 42 MT in 2010 along with a fivefold increase in urea production in the past two decades (Glibert et al., 2014). In addition, emissions of NH_4 to the atmospheric have increased from NH_3 emissions from livestock wastes, fertilizer nitrogen, crop residues, human wastes, and burning. As a result, high rates of atmospheric deposition of reactive nitrogen to coastal seas have occurred (Luo et al., 2014). Based on a vast dataset including an analysis of more than 300 published data sets of nitrogen deposition across China from 1980 to 2010, Liu et al.

(2013) reported a 60% increase in atmospheric nitrogen deposition over the 30-year period. This estimate likely is low given that samples were collected from open-rain collectors and the amount of total nitrogen deposited on plant canopies would be much higher, possibly greater than 80 kg N/ha.

Guo et al. (2017) studied the connection between dietary choices and environmental impacts from reactive nitrogen in China. They combined a nitrogen footprint tool (N-Calculator) with a food-chain model (NUFER, Nutrient flows in Food chains, Environment and Resources use) to estimate the nitrogen footprint of food in China. The NUFER model was used to calculate virtual nitrogen factors (VNFs, in kg N released/kg N in product; nitrogen that was used in the food production process and is not contained in the food product that is consumed) in major food items. The food nitrogen footprint included both the food consumption and food production footprints. The studies report that the average per capita food nitrogen footprint increased from 4.7 kg N/capita in the 1960s to 21 kg N/capita in the 2000s. Also, the national nitrogen food footprint in China increased from 3.4 metric tons (Mt)/yr in the 1960s to 28 Mt/yr in the 2000s, which accounted for 84% of the national food nitrogen footprint. This large increase was attributed to an increase in animal-derived consumption by the growing population. The increasing VNFs indicated that an increasing amount of reactive nitrogen is lost per unit of nitrogen contained in food products consumed by the population during the past five decades. Losses to the environment (23 Mt/yr of reactive nitrogen) from food production in China were attributed to ammonia and nitrogen gas emissions through denitrification (approximately 80%), nitrogen losses to water systems (20%), and nitrous oxide emissions (~1%).

Changes in agricultural and environmental policies in China regarding food security and environmental sustainability have important global implications for reactive nitrogen and world markets. Lu et al. (2015) noted that food security in China is severely affected by limited arable land, deteriorated soil quality, water scarcity and pollution, climate change, and extensive reliance on fertilizers. The overuse of mineral fertilizers has resulted in extensive eutrophication of lakes (with an increase in lake areas undergoing eutrophication increased from 135 km^2 in 1967 to 8700 km^2 in 2007). In coastal areas, excess nutrients have caused a substantial increase in the number of "red tides" (Lu et al., 2015). In response, Chinese leadership has determined that an "ecological red line" must be defined so that a holistic approach can be developed to balance increased food production with environmental sustainability. Also, degraded areas would be identified where environmental restoration and remediation would be appropriate. To reduce the nitrogen food footprint in China, Guo et al. (2017) concluded that reactive nitrogen losses during food production need to be reduced and a shift toward a more plant-based diet was recommended.

The driving forces and impacts of nitrogen flows of food systems in China during 1990–2012 were recently investigated by Gao et al. (2018). Their results showed that a city resident requires 0.5 kg more animal food nitrogen per year and 0.5 kg less plant food nitrogen per year than a rural resident. In 2012, the studies reported that dietary changes associated with income growth, population growth, and rural–urban migration accounted for 52, 31, and 17% of the total increase in animal food nitrogen, respectively. These three factors resulted in 46% of the increased nitrogen use for food production over the past two decades. Another 54% was related to decline in NUE of the food system. Nitrogen loss from food sources was three times higher in urban (502 kg/hm^2) than rural areas (162 kg N/hm^2) in 2012, resulting from inadequate N recovery via solid waste and wastewater treatment in cities. China is facing higher risks of environmental nitrogen pollution with urbanization because of the high demand for animal food nitrogen and higher food-sourced nitrogen losses in cities compared to rural areas.

11.6.2. Increasing Global Awareness of Impacts from Reactive Nitrogen

Several recent international conferences have increased the overall awareness of adverse impacts from reactive nitrogen and the need for multidisciplinary approaches for management strategies that reduce nitrogen releases to the environment (i.e., not just in one sector at the expense of another). A conference was held in 2013 that included about 160 agronomists, scientists, extension agents, crop advisors, economists, social scientists, farmers, representatives of regulatory agencies, and nongovernmental organizations (NGOs), and other agricultural experts. The focus was on the production of more food to meet the nutritional needs of a growing population while minimizing adverse human health and ecological impacts. Davidson et al. (2015) summarized results from 14 papers from conference participants that analyzed the numerous technical, economic, and social impediments to improving NUE in crop and animal production systems with the goals of more food production with low pollution (Mo Fo Lo Po). To accomplish these goals, they identified the following needs: "(a) develop partnerships among private and public sectors to show the most current, economically feasible, best management NUE practices at local and regional scales"; (b) continue to educate private sector retailers and crop advisers; (c) link nutrient management to performance-based indicators on the farm and impacts to the atmosphere, hydrosphere, and biosphere; and (d) increase funding and investments in research, education, extension, and human resources to further develop interdisciplinary knowledge and necessary skills to achieve agricultural sustainability goals.

The UNECE is well aware that most people around the world do not know that nitrogen pollution is one of the most significant problems in our times. They recognize the dire consequences that have resulted from excess reactive nitrogen, including air, soil and water pollution that damages human health, threats to biodiversity of forests and rivers, and coastal and marine pollution, and amplification of the effects of climate change. To improve our understanding of nitrogen flows in the environment and their impacts, UNECE recently joined efforts in a new multistakeholder project led by the International Nitrogen Initiative (INI). At the kick-off meeting held in Lisbon on 27 and 28 April 2015, partners from several countries, international organizations, and academic institutions discussed the next steps for the project, which is to be financed by the Global Environment Facility (GEF). The INI is a scientific partnership that works toward optimizing the sustainable use of nitrogen in areas of the world where food production is currently not supplying people with healthy diets. It is a joint project of the International Geosphere–Biosphere Program and the Scientific Committee on Problems of the Environment (SCOPE). The INI is coordinated by a steering committee and is administered by a chair and six regional center directors that represent Africa, Europe, Latin America, North America, South Asia, and East Asia. The project will provide guidelines to improve nitrogen management at the global and regional levels contributing to the establishment of an International Nitrogen Management System. Demonstration activities will cover transboundary basins around the world, including the Black Sea basin. Here, the work of partners will build on the experience of the Air Convention and on the UNECE Convention on Protection and Use of Transboundary Watercourses and International Lakes (Water Convention).

In some European countries, over 40% of air pollution-related mortality can be attributed to emissions of nitrogen compounds from agriculture. To increase awareness of these issues, nitrogen pollution was discussed during a high-level dialogue on "Nitrogen: Joining up for a Cleaner Environment" during the World Environment Day celebrations in New Delhi, India. World Environment Day, celebrated annually on 5 June, is the principal vehicle of the United Nations for encouraging worldwide awareness and action for the environment. The Task Force on Reactive Nitrogen under the UNECE includes 56 member countries in Europe, North America, and Asia. All interested UN member countries can participate in the work of UNECE. Over 70 international professional organizations and other NGOs participate in UNECE activities.

The UNECE Convention on Long-range Transboundary Air Pollution (Air Convention) discussed the achievements of the Convention in reducing nitrogen emissions in the UNECE region through legally binding emission reduction targets. Also highlighted were remaining challenges, especially when it comes to agricultural emissions. The collective knowledge of the Convention and the lessons learned in the UNECE can help other areas in finding solutions to the problem of nitrogen pollution. (https://www.unece.org/info/media/news/environment/2018/world-environment-day-celebrations-highlight-nitrogen-pollution-including-work-under-the-unece-air-convention/doc.html, accessed 20 July 2018).

The Global Partnership on Nutrient Management (GPNM) is a multistakeholder partnership that includes governments, private sector, scientific community, civil society organizations, and UN agencies that work together to advance effective nutrient management to bring about food security through increased productivity in combination with conservation of natural resources and the environment (https://www.unenvironment.org/explore-topics/oceans-seas/what-we-do/addressing-land-based-pollution/global-partnership-nutrient, accessed 20 July 2018). The UNEP, through the Coordination of Office of the Global Program of Action for the Protection of the Marine Environment from Land-based Activities (GPA), provides the Secretariat of GPNM. The purpose of the Global Overview is to raise awareness of the challenges and further trans-disciplinary dialogue with all stakeholders for future steps. Their report incorporates outcomes from several international meetings, held in 2011 and 2012 (the Global Overview of Nutrient Management Preparatory Workshop (2011); the Global Conference on Land-Ocean Connections (2012); the Third Inter- Governmental Review meeting of the GPA (2012), the United Nations Conference on Sustainable Development (2012); the First Stakeholder Workshop toward Global Nitrogen Assessment (2012); and the 11th Conference of Parties of the Convention on Biological Diversity (2012). (https://www.unenvironment.org/explore-topics/oceans-seas/what-we-do/addressing-land-based-pollution/global-partnership-nutrient-1, accessed 27 July 2018).

Brownlie et al. (2015) described the discussions from the Second BASF Fireside Chat in Germany in March 2015, which addressed solutions for reducing nitrogen pollution from agricultural systems and for increasing productivity. Sixty international stakeholders from various regions provided strategies that included encouraging empowerment of farmers to develop their own solutions and the need for more effective intervention by governments. These challenges are being addressed by the INI and the UNEP with funding from the GEF. These entities are working toward the International Nitrogen

Management System, which will provide coordinated scientific support for international policymaking for addressing reactive nitrogen contamination from agricultural activities. The following shared global solutions for increasing NUE were contributed by delegates from the United States, Brazil, Australia, India, China, sub-Saharan Africa, and EU27: (a) better communication between stakeholders; (b) make better use of manures; (c) use enhanced efficiency fertilizer technologies; and (d) coordinate scientific support for effective international policymaking for managing reactive nitrogen.

The "Towards INMS" (International Nitrogen Monitoring System) program, which was first launched at the 6th INI Conference in Melbourne (December 2016), has moved into its operational phase in October 2017. "*Towards INMS*" is a multimillion USD program that was implemented by the UN Environment with funding through the Global Environment Facility (GEF). The Executing Agency (based at the Centre for Ecology and Hydrology, on behalf of INI) is currently formulating and negotiating subcontracts with more than 70 partners who will be directly engaged in the work. The "Towards INMS" program will be executed through the United Kingdom's Natural Environment Research Council (NERC), and its Centre for Ecology & Hydrology (CEH) (Professor Mark Sutton and his team), on behalf of the International Nitrogen Initiative (INI). The main goals of "*Towards INMS*" are the development of scientific documentation to demonstrate the necessity of more effective practices for global nitrogen management and highlight options to maximize the multiple benefits of better nitrogen use. This project provides an effective way to feature extensive science evidence on excess reactive nitrogen and its adverse impacts on the nitrogen cycle, to develop a process that encourages research, governments, businesses, and civil society working together to build mutual understanding and promote sustainability, and to show how effective management of the global nitrogen cycle will provide quantifiable improvements for the health of oceans, air quality, terrestrial ecosystems, and the overall health of our global society.

11.6.3. Personal Choices for Improving NUE and Reducing Our Nitrogen Footprint

People can make a huge difference in terms of reducing nitrogen releases to the atmosphere, hydrosphere, and biosphere. There are many ways that people around the world can improve their NUE. Nitrogen emissions to the atmosphere can be reduced substantially by conserving energy and transportation (more use of fuel-efficient cars, alternative transit options, shifting to renewable energy sources) and by supporting both air pollution and climate policies. As mentioned previously, many research studies have advocated for reducing the consumption of animal protein, which in turn would lower our nitrogen footprint. For example, Sutton et al. (2013a) emphasized this latter choice, because 80% of nitrogen harvested globally (agricultural biomass harvested, including grazing) is used to feed livestock rather than people, with only 20% directly feeding people. Reay et al. (2011) noted that the average European citizen consumes 70% more protein than is needed for a healthy diet. Emphasizing this point, Sutton et al. (2013a) stated that society's high use of nutrients in many developed countries is not related to "food security," but rather is more tied to "food luxury," and they coined a new word "food luxury," which they defined as "the security of our food luxury." Furthermore, the reduced consumption of animal products in developed countries would lead to a substantial improvement in full-chain NUE and significantly reduce losses of reactive nitrogen to the environment. Also, consumers should reduce food waste by only buying what is needed and eating all purchased food. Any remaining food waste should be composted to return its nutrients and reduce fertilizer use for producing crops (Shibata et al., 2017).

The U.S. EPA had other recommendations for homeowners to improve NUE. For example, to reduce nutrient inputs from lawns and gardens, they recommended that homeowners apply fertilizers at the recommended amount (but not before windy or rainy days), avoid applying fertilizer near waterways, refrain from overwatering lawns and gardens, and properly store unused fertilizers and dispose of empty containers (https://www.epa.gov/nutrientpollution/what-you-can-do-your-yard, accessed 28 July 2018). Also, people can prevent nutrient losses to the environment by picking up pet wastes, avoid walking pets near streams and other waterways, increase water use efficiency, and strive to reduce energy use around the house and while driving. There also are many opportunities to get involved in community efforts to raise awareness about reducing nutrient use and protecting water quality of aquifers and surface water bodies (https://www.epa.gov/nutrientpollution/what-you-can-do-your-community, accessed 28 July 2018).

11.7. CLOSING REMARKS

The material presented in this book has integrated and synthesized recent information on environmental, human health, and economic consequences of excess reactive nitrogen in our environment. In the first chapter, we learned about the various issues and the environmental, health, and economic consequences associated with excess reactive nitrogen. The next chapters provided background information about the nitrogen cycle and how the cycle has been dramatically altered during the past 100 years by anthropogenic activities. Detailed information was presented on the major point and nonpoint sources that contribute reactive nitrogen to the environment at large regional and global scales. Following this background information, information was presented on various factors affecting nitrogen transport and transformation, and innovative tools to identify and quantify sources of nitrate contamination. Several of the next chapters provided detailed information about the ramifications (adverse consequences) from the significant anthropogenic alteration of the nitrogen cycle including human health impacts; degradation of surface waters, groundwater, springs, and ecosystems; and how the interactions between groundwater and surface water compound these problems in lakes, streams, and estuaries; and the co-occurrence of other pollutants with nitrate-nitrogen contamination. An introduction was provided on several software or modeling tools available for assessing the impacts of nitrogen and other co-occurring contaminants on groundwater and surface water systems. Following this information, we learned about the adverse economic consequences associated with environmental pollution, human health maladies, clean-up costs, drinking water treatment costs, losses to commercial fishing, tourism, recreational activities, decreased real estate values, and cascading effects to other socioeconomic sectors.

This chapter presented critical information on effective ways to increase NUE and to reduce our nitrogen footprint (locally and globally) from human activities, as well as practical strategies, and international efforts and recommendations for restoring water quality in impacted surface water bodies, groundwater and springs, and terrestrial and aquatic ecosystems. Three areas were highlighted where ongoing efforts are being focused: increasing NUE and sustainability in agricultural systems, reducing the per capita consumption of animal proteins, and decreasing fossil fuel combustion and replacing with alternative energy sources. In recent years, many countries and regions, particularly the United States, Japan, Europe, have made important strides in identifying and communicating societal threats, and in developing strategies for integrated approaches for tackling excess reactive nitrogen in our environment.

REFERENCES

Alley, W. M., Healy, R. W., LaBaugh, J. W., & Reilly, T. E. (2002). Flow and storage in groundwater systems. *Science*, *296*, 1985–1990.

Ascott, M. J., Gooddy, D. C., Wang, L., Stuart, M. E., Lewis, M. A., Ward, R. S., & Binley, A. M. (2017). Global patterns of nitrate storage in the vadose zone. *Nature Communications*, 1–7. https://doi.org/10.1038/s41467-017-01321-w.

Billen, G., Lassaletta, L., & Garnier, J. (2015). A vast range of opportunities for feeding the world in 2050: Trade-off between diet, nitrogen contamination and international trade. *Environmental Research Letters*, *10*, 025001.

Bodirsky, B. L., Popp, A., Lotze-Campen, H., Dietrich, J. P., Rolinski, S., Weindl, I., et al. (2014). Reactive nitrogen requirements to feed the world in 2050 and potential to mitigate nitrogen pollution. *Nature Communications*, *5*, 3858. https://doi.org/10.1038/ncomms4858.58OI

Brownlie, W. J., Howard, C. M., Pasda, G., Navé, B., Zerulla, W., & Sutton, M. A. (2015). Developing a global perspective on improving agricultural nitrogen use. *Environmental Development*, *15*(1), 145–151.

Bustamente, M. M. C., Martinelli, L. A., Pérez, T., Rasse, R., Ometto, J. P. H. B., Pacheco, F. S., Lins, S. R. M., & Marquina, S. (2015). Nitrogen management challenges in major watersheds of South America. *Environmental Research Letters*, 10, 065007.

Castner, E. A., Leach, A. M., Leary, N., Baron, J., Compton, J. E., Galloway, J. N., et al. (2017). The nitrogen footprint tool network: A multi-institution program to reduce nitrogen pollution. *Sustainability*, *10*(2), 79–88. https://doi.org/10.1089/sus.2017.29098.eac

Chen, Y., & Borken-Kleefeld, J. (2014). Real-driving emissions from cars and light commercial vehicles: Results from 13 years remote sensing at Zurich/CH. *Atmospheric Environment*, *88*, 157–164.

Clough, T. J., Condron, L. M., Kammann, C., & Muller, C. (2013). A review of biochar and soil nitrogen dynamics. *Agronomy*, *3*, 275–293. https://doi.org/10.3390/agronomy3020275

Davidson, E. A., & Kanter, D. (2014). Inventories and scenarios of nitrous oxide emissions. *Environmental Research Letters*, *9*, 105012.

Davidson, E. A., Suddick, E. C., Rice, C. W., & Prokopy, L. S. (2015). More food, low pollution (Mo Fo Lo Po): A grand challenge for the 21st century. *Journal of Environmental Quality*, *44*, 305–311.

De Rozari, P., Greenway, M., & El Hanandeh, A. (2018). Nitrogen removal from sewage and septage in constructed wetland mesocosms using sand media amended with biochar. *Ecological Engineering*, *111*, 1–10. https://doi.org/10.1016/j.ecoleng.2017.11.002

Du, E., de Vries, W., Galloway, J. N., Hu, X., & Fang, J. (2014). Changes in wet nitrogen deposition in the United States between 1985 and 2012. *Environmental Research Letters*, *9*, 095004.

Erisman, J. W., Galloway, J. N., Dice, N. B., Sutton, M. A., Bleeker, A., Grizzetti, B., Leach, A. M., & de Vries, W. (2015). Nitrogen, too much of a vital resource. Science Brief. Zeist, The Netherlands: World Wildlife Fund.

Erisman, J. W., Leach, A., Bleeker, A., Atwell, B., Cattaneo, L., & Galloway, J. N. (2018). An integrated approach to a nitrogen use efficiency (NUE) indicator for the food production-consumption chain. *Sustainability*, *10*, 925. https://doi.org/10.3390/su10040925

EU (European Union) Nitrogen Expert Panel. (2015). *Nitrogen use efficiency (NUE) - An indicator for the utilization of nitrogen in agriculture and food systems*. Wageningen

University, Alterra. Retrieved from http://www.eunep.com/wp-content/uploads/2017/03/Report-NUE-Indicator-Nitrogen-Expert-Panel-18-12-2015.pdf, accessed 03 August 2018.

Galloway, J. N., Aber, J. D., Erisman, J. W., Seitzinger, S. P., Howarth, R. W., Cowling, E. B., & Cosby, B. J. (2003). The nitrogen cascade. *BioScience, 53*, 341–356.

Galloway, J. N., Winiwarter, W., Leip, A., Leach, A., Bleeker, A., & Erisman, J. W. (2014). Nitrogen footprints: Past, present, and future. *Environmental Research Letters, 9*. https://doi.org/10.1088/1748-9326/9/11/115003

Gao, B., Huang, Y., Huang, W., Shi, Y., Bai, X., & Cui, S. (2018). Driving forces and impacts of food system nitrogen flows in China, 1990 to 2012. *Science of the Total Environment, 610-611*, 430–441.

Glibert, P. M., Maranger, R., Sobota, D. J., & Bouwman, L. (2014). The Haber Bosch–harmful algal bloom (HB–HAB) link. *Environmental Research Letters, 9*, 105001.

Guo, M., Chen, X., Bai, Z., Jiang, R., Galloway, J. N., Leach, A. M., et al. (2017). How China's nitrogen footprint has changed from 1961 to 2010. *Environmental Research Letters, 12*(10), 104006.

Katz, B. G. (2019). Nitrate contamination in karst groundwater. In W.B. White & D.C. Culver (Eds.), *Encyclopedia of Caves* (Ch. 91, pp. 756–760), 3rd edition. Boston, MA: Academic Press, Elsevier.

Leach, A. M., Galloway, J. N., Bleeker, A., Erisman, J. W., Kohn, R., & Kitzes, J. (2012). A nitrogen footprint model to help consumers understand their role in nitrogen losses to the environment. *Environmental Development, 1*, 40–66.

Leip, A., Weiss, F., Lesschen, J. P., & Westhoek, H. (2014). The nitrogen footprint of food products in the European Union. *Journal of Agricultural Science, 152*, S20–S33.

Li, T., Zhang, W., Yin, J., Chadwick, D., Norse, D., Lu, Y., et al. (2017). Enhanced-efficiency fertilizers are not a panacea for resolving the nitrogen problem. *Global Change Biology, 24*(2), e511–e521.

Liu, X., Zhang, Y., Han, W., Tang, A., Shen, J., Cui, Z., et al. (2013). Enhanced nitrogen deposition over China. *Nature, 494*, 459–462.

Lu, Y., Jenkins, A., Ferrier, R. C., Bailey, M., Gordon, J. J., et al. (2015). Addressing China's grand challenge of achieving food security while ensuring environmental sustainability. *Science Advances, 1*(1). https://doi.org/10.1126/sciadv.1400039

Luo, X. S., Tang, A. H., Shi, K., Wu, L. H., Li, W. Q., Shi, W. Q., et al. (2014). Chinese coastal seas are facing heavy atmospheric nitrogen deposition. *Environmental Research Letters, 9*, 095007.

Lusk, M. G., Toor, G. S., Yang, Y., Mechtensimer, S., De, M., & Obreza, T. A. (2017). A review of the fate and transport of nitrogen, phosphorus, pathogens, and trace organic chemicals in septic systems. *Critical Reviews in Environmental Science and Technology, 47*(7), 455–541.

Melillo, J.M., Richmond, T.C., & Yohe, G.W., (Eds.). (2014). Climate change impacts in the United States: The third national climate assessment. U.S. Global Change Research Program, 841 p. doi:https://doi.org/10.7930/J0Z31WJ2.

Misselbrook, T. H., Cardenas, L. M., Camp, V., Thorman, R. E., Williams, J. R., Rollett, A. J., & Chambers, B. J. (2014). An assessment of nitrification inhibitors to reduce nitrous oxide emissions from UK agriculture. *Environmental Research Letters, 9*, 115006. https://doi.org/10.1088/1748-9326/9/11/115006

Moran, S.K. (2016). Wastewater is key to reducing nitrogen pollution: Upgrading wastewater treatment plants can dramatically reduce a municipality's nitrogen footprint. *Scientific American*, 2 June 2016. Retrieved from https://www.scientificamerican.com/article/wastewater-is-key-to-reducing-nitrogen-pollution/, accessed 02 August 2018.

Nesheim, M. C. (1993). Human health needs and parasitic infections. *Parasitology, 107*, S7–S18.

Oita, A., Malik, A., Kanemoto, K., Geschke, A., Nishijima, S., & Lenzen, M. (2016). Substantial nitrogen pollution embedded in international trade. *Nature Geoscience, 9*, 111–115.

Oldenkamp, R., van Zelm, R., & Huijbregts, M. A. (2016). Valuing the human health damage caused by the fraud of Volkswagen. *Environmental Pollution, 212*, 121–127.

Reay, D. S., Howard, C. M., Bleeker, A., Higgins, P., Smith, K., Westhoek, H., et al. (2011). Societal choice and communicating the European nitrogen challenge. Ch. 26. In M. A. Sutton, C. M. Howard, J. W. Erisman, G. Billen, A. Bleeker, P. Grennfelt, H. van Grinsven, & B. Grizzetti (Eds.), *The European nitrogen assessment: Sources, effects, and policy perspectives*. Cambridge, UK: Cambridge University Press, 664 p.

Reis, S., Bekunda, M., Howard, C. M., Karanja, N., Winiwarter, W., Yan, X., Bleeker, A., & Sutton, M. A. (2016). Synthesis and review: Tackling the nitrogen management challenge: From global to local scales. *Environmental Research Letters, 11*(2016), 120205. http://dx.doi.org/10.1088/1748-9326/11/12/120205.

Sanford, W. E., & Pope, J. P. (2013). Quantifying groundwater's role in delaying improvements to Chesapeake Bay water quality. *Environmental Science and Technology, 47*, 13330–13338.

Sebillo, M., Mayer, B., Nicolardot, B., Pinay, G., & Mariotti, A. (2013). Long-term fate of nitrate fertilizer in agricultural soils. *Proceedings of the National Academy of Sciences of the United States, 110*(45), 18185–18189.

Shibata, H., Galloway, J. N., Leach, A. M., Cattaneo, L. R., Noll, L. C., Erisman, J. W., et al. (2017). Nitrogen footprints: Regional realities and options to reduce nitrogen loss to the environment. *Ambio, 46*, 129. https://doi.org/10.1007/s13280-016-0815-4

Singh, S., & Kulshrestha, U. C. (2014). Rural versus urban gaseous inorganic reactive nitrogen in the Indo-Gangetic plains (IGP) of India. *Environmental Research Letters, 9*, 125004.

Stecker, T. (2013). Nitrogen footprint calculator points consumers to better choices. *Climatewire*, 15 May 2013.

Steffen, W., Richardson, K., Rockstrom, J., Cornell, S. E., Fetzer, I., Bennett, E. M., et al. (2015). Planetary boundaries: Guiding human development on a changing planet. *Science*, 1259855. https://doi.org/10.1126/science.1259855

Sutton, M. A., & Bleeker, A. (2013). The shape of nitrogen to come. *Nature, 494*, 435–437.

Sutton, M. A., Bleeker, A., Howard, C. M., Bekunda, M., Grizzeetti, B., de Vries, W., et al. (2013a). *Our nutrient world: The challenge to produce more food and energy with less pollution*. Edinburgh, UK: Centre for Ecology and Hydrology, 128 p.

Sutton, M. A., Howard, C. M., Erisman, J. W., Billen, G., Bleeker, A., Grennfelt, P., van Grinsven, H., & Grizzetti, B.

(2011). The European nitrogen assessment: Sources, effects, and policy perspectives. Cambridge, UK: Cambridge University Press, 664 p.

Sutton, M. A., Reis, S., Riddick, S. N., Dragosits, U., Nemitz, E., Theobald, M. R., et al. (2013b). Towards a climate-dependent paradigm of ammonia emission and deposition. *Philosophical Transactions of the Royal Society B, 368.* https://doi.org/10.1098/rtsb.2013.0166

Sutton, M. A., Skiba, U. M., van Grinsven, H. J. M., Oenema, O., Watson, C. J., et al. (2014). Green economy thinking and the control of nitrous oxide emissions. *Environmental Development, 9,* 76–85. http://dx.doi.org/10.1016/j.envdev.2013.10.002

U.S. Environmental Protection Agency. (2015). Case studies on implementing low-cost modifications to improve nutrient reduction at wastewater treatment plants (EPA-841-R-15-004). Washington, DC. Retrieved from http://www2.epa.gov/nutrient-policy-data/reports-andresearch#reports, accessed 15 August 2018.

UNECE. (2012). Draft guidance document for preventing and abating ammonia emissions from agricultural sources. United Nations Economic Commission for Europe, Convention on Long-range Transboundary Air Pollution, Geneva. Retrieved from www.unece.org/fileadmin/DAM/env/documents/2012/EB/N_6_21_Ammonia_Guidance_Document_Version_20_August_2011.pdf, accessed 24 September 2019.

UNEP. (2011). Toward a green economy: Pathways to sustainable development and poverty eradication, a synthesis for policymakers. Retrieved from https://sustainabledevelopment.un.org/content/documents/126GER_synthesis_en.pdf, accessed 24 September 2019.

USDA. (2010). *U.S. Department of Agriculture and U.S. Department of Health and Human Services Dietary Guidelines for Americans, 2010* (7th ed.). Washington, DC: US Government Printing Office.

Van Meter, K. J., Basu, N. B., & Van Cappellen, P. (2017). Two centuries of nitrogen dynamics, legacy sources and sinks in the Mississippi and Susquehanna River basins. *Global Biogeochemical Cycles, 31*(1), 2–23.

Van Meter, K. J., Basu, N. B., Veenstra, J. J., & Burras, C. L. (2016). The nitrogen legacy: Emerging evidence of nitrogen accumulation in anthropogenic landscapes. *Environmental Research Letters, 11.* https://doi.org/10.1088/1748-9326/11/3/035014

de Vries, W., Hettelingh, J.-P., & Posch, M. (Eds.). (2015). *Critical loads and dynamic risk assessments: Nitrogen, acidity and metals in terrestrial and aquatic ecosystems.* Dordrecht: Springer, 662 p. doi: https://doi.org/10.10007/978-94-017-9508-1

Westhoek, H., Lesschen, J. P., Rood, T., Wagner, S., De Marco, A., Murphy-Bokern, D., et al. (2014). Food choices, health and environment: Effects of cutting Europe's meat and dairy intake. *Global Environmental Change, 26,* 196–205.

WHO (World Health Organization). (2007). *Protein and amino acid requirements in human nutrition WHO Technical Report Series Number 935.* World Health Organization.

Woomer, P. L., Man'gat, M., & Tungani, J. O. (2004). Innovative maize-legume intercropping results in above- and below-ground competitive advantages for understorey legumes. *West African Journal of Applied Ecology, 6,* 85–94.

Xue, L., Yu, Y., & Yang, L. (2014). Maintaining yields and reducing nitrogen loss in rice–wheat rotation system in Taihu Lake region with proper fertilizer management. *Environmental Research Letters, 9,* 115010.

Zhou, M., Brandt, P., Pelster, D., Rufino, M. C., Robinson, T., & Butterbach-Bahl, K. (2014). Regional nitrogen budget of the Lake Victoria Basin, East Africa: Syntheses, uncertainties and perspectives. *Environmental Research Letters, 9,* 105009.

FURTHER READING

Alexander, R. B., Smith, R. A., & Schwarz, G. E. (2000). Effect of stream channel size on the delivery of nitrogen to the Gulf of Mexico. *Nature, 403,* 758–761.

Alexandratos, N., & Bruinsma, J. (2012). *World agriculture towards 2030/2050: The 2012 revision.* Food and Agricultural Organization of the United Nations, ESA Working Paper No. 12-03. Rome: FAO, 154 p.

Chatterjee, R. (2009). Economic damages from nutrient pollution create a "toxic debt". *Environmental Science and Technology.* https://doi.org/10.1021/es803044n

Diaz, R. J., & Rosenberg, R. (2008). Spreading dead zones and consequences for marine ecosystems. *Science, 321,* 926–929.

Dubrovsky, N.M., Burow, K.R., Clark, G.M., Gronberg, J.M., Hamilton P.A., Hitt, K.J., et al. (2010). The quality of our Nation's waters—Nutrients in the Nation's streams and groundwater, 1992–2004. U.S. Geological Survey Circular 1350, 174 p. Retrieved from http://water.usgs.gov/nawqa/nutrients/pubs/circ1350, accessed 23 September 2019.

Erisman, J. W., Sutton, M. A., Galloway, J. N., Klimont, Z., & Winiwarter, W. (2008). How a century of ammonia synthesis changed the world. *Nature Geoscience, 1,* 636–639.

Erisman, J. W., van Grinsven, H., Grizzetti, B. Bouraoui, F., Powlson, D., Sutton, M. A., Bleeker, A., & Reis, S. (2011). The European nitrogen problem in a global perspective. Ch. 2. In M.A. Sutton, C.M. Howard, J.W. Erisman Billen, G., Bleeker, A., Grennfelt, P., H. van Grinsven, & B. Grizzetti (Eds.), *The European nitrogen assessment: Sources, effects, and policy perspectives* (pp. 9–31). Cambridge, UK: Cambridge University Press.

Fowler, D., Coyle, M., Skiba, U., Sutton, M. A., Capel, J. N., Reis, S., et al. (2013). The global nitrogen cycle in the twenty-first century. *Philosophical Transactions of the Royal Society B, 368,* 1621. https://doi.org/10.1098/rstb.2013.0164

Fowler, D., Steadman, C. E., Stevenson, D., Coyle, M., Rees, R. M., Skiba, U. M., et al. (2015). Effects of global change during the 21st century on the nitrogen cycle. *Atmospheric Chemistry and Physics, 15,* 13849–13893. https://doi.org/10.5194/acp-15-13849-2015

Galloway, J. N., Dentener, F. J., Capone, D. G., Boyer, E. W., Howarth, R. W., Seitzinger, S. P., et al. (2004). Nitrogen cycles: Past, present and future. *Biogeochemistry, 70,* 153–226.

Galloway, J. N., Townsend, A. R., Erisman, J. W., Bekunda, M., Cai, Z., Freney, J. R., et al. (2008). Transformation of the nitrogen cycle: Recent trends, questions, and potential solutions. *Science, 320,* 889–892.

Grizzetti, B., Bouraoui, F., Billen, G., van Grinsven, H., Cardoso, A.C., Thieu, V., et al. (2011). Nitrogen as a threat to

European water quality. Ch. 17. In M. A. Sutton, C. M. Howard, J. W. Erisman, G. Billen, A. Bleeker, P. Grennfelt, H. van Grinsven, & B. Grizzetti (Eds.), *The European nitrogen assessment: Sources, effects, and policy perspectives* (pp. 379–404). Cambridge, UK: Cambridge University Press.

Houlton, B. Z., Boyer, D., Finzi, A., Galloway, J. N., Leach, A., Liptzin, D., et al. (2013). Intentional versus unintentional nitrogen use in the United States: Trends, efficiency, and implications. *Biogeochemistry*, *114*, 11–23. https://doi.org/10.1007/s10533-012-9801-5

Howarth, R., Swaney, D., Billen, G., Garnier, J., Hong, B., Humborg, C., et al. (2011). Nitrogen fluxes from the landscape are controlled by net anthropogenic nitrogen inputs and by climate. *Frontiers in Ecology and the Environment*. https://doi.org/10.1890/100178

Lassaletta, L., Billen, G., Garnier, J., Bouwman, L., & Valazquez, E. (2016). Nitrogen use in the global food system: Past trends and future trajectories of agronomic performance, pollution, trade, and dietary demand. *Environmental Research Letters*, *11*(9), 095007. http://dx.doi.org/10.1088/1748-9326/11/9/095007

Lassaletta, L., Billen, G., Grizzetti, B., Anglade, J., & Garnier, J. (2014). 50 year trends in nitrogen use efficiency of world cropping systems: The relationship between yield and nitrogen input to cropland. *Environmental Research Letters*, *9*(2014), 105011. http://dx.doi.org/10.1088/1748-9326/9/10/105011

Meals, D. W., Dressing, S. A., & Davenport, T. E. (2010). Lag time in water quality response to best management practices: A review. *Journal of Environmental Quality*, *39*, 85–96.

Musgrove, M., Opsahl, S. P., Mahler, B. J., Herrington, C., Sample, T. L., & Banta, J. R. (2016). Source, variability, and transformation of nitrate in a regional karst aquifer: Edwards aquifer, central Texas. *Science of the Total Environment*, *568*, 457–469.

Pope, C. A., Burnett, R. T., Thun, M. J., Calle, E. E., Krewski, D., Ito, K., & Thurston, G. D. (2002). Lung cancer, cardiopulmonary mortality, and long-term exposure to fine particulate air pollution. *Journal of the American Medical Association*, *287*, 1132–1141.

Puckett, L. J., Tesoriero, A. J., & Dubrovsky, N. M. (2011). Nitrogen contamination of surficial aquifers- a growing legacy. *Environmental Science and Technology*, *45*, 839–844.

Qiu, J. (2011). China to spend billions cleaning up groundwater. *Science*, *334*, 745–745.

Sala, O. E., Chapin, I. I. I. F. S., Armesto, J. J., Berlow, E., Bloomfield, J., Dirzo, R., et al. (2000). Global biodiversity scenarios for the year 2100. *Science*, *287*, 1770–1774.

Scavia, D., & Bricker, S. B. (2006). Coastal eutrophication assessment in the United States. *Biogeochemistry*, *79*(1-2), 187–208.

Sobota, D. J., Compton, J. E., McCrackin, M. L., & Singh, S. (2015). Cost of reactive nitrogen release from human activities to the environment in the United States. *Environmental Research Letters*, *10*(2015), 025006.

State-USEPA Nutrient Innovations Task Group. (2009). An urgent call to action. State-USEPA Nutrient Innovations Task Group Report to USEPA, August 2009, 170 p.

Townsend, A. R., Howarth, R. W., Bazzaz, F. A., Booth, M. S., Cleveland, C. C., Collinge, S. K., et al. (2003). Human health effects of a changing global nitrogen cycle. *Frontiers in Ecology and the Environment*, *1*(5), 240–246.

U.S. Environmental Protection Agency. (2002). Nitrogen—Multiple and regional impacts. U.S. Environmental Protection Agency Clean Air Markets Division Report EPA-430-R-01-006, 38 p.

U.S. Environmental Protection Agency. (2009). An urgent call to action. Report of the State-EPA Nutrient Innovations Task Group, 41 p.

Van Meter, K. J., & Basu, N. B. (2015). Catchment legacies and time lags: A parsimonious watershed model to predict the effects of legacy storage on nitrogen export. *PLoS One*, *10*(5), e0125971. https://doi.org/10.1371/journal.pone.o125971

Van Meter, K. J., & Basu, N. B. (2017). Time lags in watershed-scale nutrient transport: An exploration of dominant controls. *Environmental Research Letters*, *12*, 084017. https://doi.org/10.1088/1748-9326/aa7bf4

Wang, L., Butcher, A., Stuart, M., Gooddy, D., & Bloomfield, J. (2013). The nitrate time bomb: A numerical way to investigate nitrate storage and lag time in the unsaturated zone. *Environmental Geochemistry and Health*, *35*, 667–681.

Wang, L., Stuart, M. E., Lewis, M. A., Ward, R. S., Skirvin, D., Naden, P. S., Collins, A. L., & Ascott, M. J. (2016a). The changing trend in nitrate concentrations in major aquifers due to historical nitrate loading from agricultural land across England and Wales from 1925 to 2150. *Science of the Total Environment*, *542*, 694–705.

Wang, S., Changyuan, T., Xianfang, S., Ruiqiang, Y., Zhiwei, H., & Yun, P. (2016b). Factors contributing to nitrate contamination in a groundwater recharge area of the North China Plain. *Hydrological Processes*, *30*(13), 2271–2285.

Zhang, X., Davidson, E. A., Mauzerall, D. L., Searchinger, T. D., Dumas, P., & Shen, Y. (2015). Managing nitrogen for sustainable development. *Nature*, *528*, 51–59. https://doi.org/10.1038/nature15743

INDEX

Acidification, 5, 71, 96–97, 101
Ackerman, D., 73
Adapting Mosaic, 108
Adenoviruses, 61
AFOs. *See* Animal feeding operations
Aggarwal, P. K., 54
Agricultural Health Study (AHS), 141
Agriculture. *See also* Animal feeding operations; Fertilizers; Manure
　ammonia from, 37–38
　BMPs for, 214–15
　CFCs and, 54
　groundwater for, 120
　　nitrates in, 121
　maize production, 74–75, 201
　NUE in, 223–27, 224f, 225f, 226t
　nutrient management in, 108
　reactive nitrogen from, 42, 97, 199–200
　in U.S., 221–22
　vadose zone and, 132
　wells and, 123f
AHS. *See* Agricultural Health Study
Air pollution
　adverse health effects of, 84
　in China, 72–73, 210–11
　in EU, 236
　reactive nitrogen and, 190–91, 210–11
　adverse health effects of, 71–75
Air Pollution and Health: A European Approach Phase 2 (APHEA-2), 72
Akesson, M., 189
Albertin, A. R., 60
Alexander, R. B., 180f
Algae
　biodiversity loss from, 94
　blue-green, 17, 102, 184
　with dual nitrate isotopes, 54
　fractionation for, 60
　golden, 103
　hypoxia and, 103, 104
　nitrate and, 60, 156
　nitrification from, 129
　reactive nitrogen and, 91
　in springs, 156, 157, 164

Algal blooms, 4, 7. *See also* Harmful algal blooms
　nitrate source identification in, 61
　in springs, 157–58
Allègre, C., 29
Almasri, M. N., 142
ALS. *See* Amyotrophic lateral sclerosis
Alzheimer's disease, 83, 102
American Society of Civil Engineers (ASCE), 30
American Water Works Association (AWWA), 17
Amino acids, 1, 20
Ammonia, 5, 33, 73, 93, 107
　acute criteria for, 99
　adverse human health effects from, 6
　from agriculture, 37–38
　in air pollution, 71–72, 211
　in atmosphere, 59, 94
　biodiversity loss from, 94
　from denitrification, 22
　in drinking water, 75
　nitrification and, 18
　in oceans, 18
　unionized form of, 98–99
Ammonia Monitoring Network (AMoN), 38
Ammonification, 15–16, 50, 57
Ammonium, 17, 18, 93
　in air pollution, 211
　in groundwater, 58–59, 58f, 123–24
　on-site sewage disposal and treatment systems and, 34
　pH and, 34
　in plant and animal decay, 20
　redox and, 123
　from sewer system leaks, 35–36
　in soil, 37
　in STZ, 131
　unionized form of, 98–99
　volatilization and, 20
AMoN. *See* Ammonia Monitoring Network
Amyotrophic lateral sclerosis (ALS), 83, 102
Anderson, D. M., 204, 209
Anderson, M. P., 140

Anderson, T. R., 129
Animal decay, 20
Animal feeding operations (AFOs), 141
　CAFOs, 34, 226
Annamox reactions, 15, 20
APHEA-2. *See* Air Pollution and Health: A European Approach Phase 2
Apte, J. S., 72, 84, 213
Aquifers. *See* Groundwater
ArcGIS-based Nitrate Load Estimation Toolkit (ArcNLET), 144
Artificial sweeteners, in drinking water, 60
ASCE. *See* American Society of Civil Engineers
Ascott, M. J., 6, 62, 83, 132
Atmosphere. *See also* Air pollution
　ammonia in, 59, 94
　reactive nitrogen in, 22, 37–38, 38f
Ator, S. W., 188
Autotrophically oxidize ammonia, 17
AWWA. *See* American Water Works Association

Bacteria, 29. *See also* Cyanobacteria
　anammox reactions and, 20
　as free-living nitrogen fixers, 17
　nitrous oxide from, 19
　in plant and animal decay, 20
　in springs, 169
Bain, D. J., 36
Banning, A., 186
Baron, J. S., 94, 99, 100, 101, 112, 212
Base flow index (BFI), 133
Basin management action plan (BMAP), 164
Basu, N. B., 110, 132, 133
Belitz, K., 60
Best management practices (BMPs), for agriculture, 214–15
BFI. *See* Base flow index
Billen, G., 104, 180, 232
Biodiversity loss, 71
　from algae, 94
　reactive nitrogen and, 91–112, 94f
　in U.S., 201

Nitrogen Overload: Environmental Degradation, Ramifications, and Economic Costs, Geophysical Monograph 250,
First Edition. Brian G. Katz.
© 2020 American Geophysical Union. Published 2020 by John Wiley & Sons, Inc.

Biological nitrogen fixation (BNF), 1, 2, 29, 43
Biological oxygen demand (BOD), 182
Bioreactors, 19, 20
Biosolids, 18, 32–33, 59, 60, 109, 132
Birch, M. B. L., 212
Bladder cancer, 6, 77, 78
Blue-green algae, 17, 102, 184
BMAP. See Basin management action plan
BMMA. See β-N-methylamino-L-alanine
BMPs. See Best management practices
BNF. See Biological nitrogen fixation
Bobbink, R., 92, 111
BOD. See Biological oxygen demand
Bodirsky, B. L., 221
Bohannan, B. J. M., 17
Böhlke, J. K., 37, 58, 123
Boosted regression tree (BRT), 140
Boron, for reactive nitrogen source identification, 54–55
Boron isotopes, for reactive nitrogen source identification, 54–55
Bosch, Carl, 1
Bouraoui, F., 104, 124
Bouwman, A. F., 24
Bowman, W. D., 96–97
Boyer, E. W., 97
Bradley, J., 32
Bratton, J. F., 130
Breitburg, D., 104
Briand, C., 61
Brink, C., 197, 198
Brownlie, W. J., 236
BRT. See Boosted regression tree
Burford, M. A., 111
Burow, K. R., 187
Bustamente, M. M. C., 229

CAFOs. See Concentrated animal feeding operations
Calcium cyanamide, 37
Cancer, 6, 71, 72, 74, 77, 78, 79, 124, 134
Carbon dioxide, 5, 15, 19, 21, 100, 101
Carey, R. O., 36
Carrara, C., 190
Castanas, E., 72
CASTNET. See Clean Air Status and Trends Network
Cation exchange capacity (CEC), 37
CBP. See Chesapeake Bay Partnership
CEAM. See Center for Exposure Assessment Modeling
CEC. See Cation exchange capacity
Center for Exposure Assessment Modeling (CEAM), 106
CES. See Cultural ecosystem services
CFCs. See Chlorofluorocarbons

CFPv2. See Conduit Flow Process
Chang, C. C. Y., 51
CHANS. See Coupled human and natural systems
Charette, M. A., 131
Chesapeake Bay Partnership (CBP), 110
China
  air pollution in, 72–73, 210–11
  food production in, 211–12
  groundwater in, 80
    nitrate in, 134–35
  nitrogen footprint of, 234–35
  on-site sewage disposal and treatment systems in, 35
  ozone in, 74
  reactive nitrogen and economic costs of, 210–12
    in groundwater, 126–27, 126f
    in surface water, 108
  WWTFs in, 32
Chlorofluorocarbons (CFCs), 54, 160, 161–62
Chlorophyl, 1, 131
Clark, C. M., 96
Clark, G. M., 189
Clean Air Status and Trends Network (CASTNET), 38, 206
Climate change, 24, 92, 212, 223
  carbon dioxide and, 15
  denitrification and, 99
  groundwater nitrate and, 142
  maize and, 74
  nitrogen oxides and, 5
  oceans and, 112
CLM. See Community Land Model
Clough, T. J., 225
CMAQ. See Community Multiscale Air Quality
C:N ratio, 16, 25
Cohen, M. J., 157
Collins, A. R., 130
Colon cancer, 6, 124
Colorectal cancer, 77, 78, 134
Combined sewer overflows (CSOs), 31
Community Land Model (CLM), 107
Community Multiscale Air Quality (CMAQ), 38
Compton, J. E., 94, 201, 213, 214
Concentrated animal feeding operations (CAFOs), 34, 226
Conduit Flow Process (CFPv2), 146
Constructed/reclaimed wetlands
  for denitrification, 19
  as reactive nitrogen buffers, 130
  reactive nitrogen in, 100
Coupled human and natural systems (CHANS), 126
Cowdery, T. K., 142
Cowling, E., 21f

Crawford, A., 205, 207
Critical load, of reactive nitrogen, 92, 96
CSOs. See Combined sewer overflows
Cultural ecosystem services (CES), 206–7
CWA. See U.S. Clean Water Act
Cyanobacteria, 29, 102
  as HABs, 82–83, 101
  in lakes, 129

Dahlgren, R. A., 29
Danon, P. N., 76
Davidson, E. A., 91, 215, 228, 229, 235
Dead zones, 5, 109, 109f
Denitrification, 1, 18–19, 129
  climate change and, 99
  dual nitrate isotopes and, 52–53
  fractionation of, 53–54
  of groundwater, 120, 127
  nitrogen isotopes and, 50
  projected changes in, 22–24
  in PWSs, 137
  in rivers, 99
  in springs, 162–63
  stormwater runoff and, 57
Denver, J. M., 188
Deo, R. P., 59
Depth to water, net Recharge, Aquifer media, Soil media, Topography, Impact of vadose zones (DRASTIC), 139
De Rozari, P., 230
Desaulty, A. M., 54
DeSimone, L. A., 80, 120, 121, 122, 124, 135, 189
DeVries, T., 24, 42
Diazotrophs, 17
DIN. See Dissolved inorganic nitrogen
Dinitrogen, 1
Dise, N. B., 93–94
Dissimilatory nitrate reduction to ammonia (DNRA), 15, 19–20
Dissolved inorganic nitrogen (DIN), 97–98, 108, 131
Dissolved organic carbon (DOC), 57, 128, 157
Dissolved organic nitrogen (DON), 97
Divers, M. T., 35–36, 57
DNA, 1, 17
DNRA. See Dissimilatory nitrate reduction to ammonia
DOC. See Dissolved organic carbon
Dodds, W. K., 7, 200–201
DON. See Dissolved organic nitrogen
Doney, S. C., 191
Drainfield failure rates, of septic tanks, 34
DRASTIC. See Depth to water, net Recharge, Aquifer media, Soil media, Topography, Impact of vadose zones

Drinking water, 8
  artificial sweeteners in, 60
  groundwater for, 120
    reactive nitrogen in, 83–84
    VOCs in, 189
  MCL for, 121–22, 123, 123t
  nitrate in
    adverse health effects of, 75–80, 200
    from groundwater, 189
    from PWSs, 134–38, 136f–38f
    from springs, 169
    from wells, 134–38, 136f–38f
  nitrification and, 18
  pharmaceuticals in, 60
  from surface water, 142
  in U.S., 204–5
Dual nitrate isotopes, 51–53
  algae with, 54
  for streamwater, 57
Dubrovsky, N. M., 73, 98–99, 104, 121, 128, 135, 176–78, 180f, 181
Durand, P., 129, 130

Eberts, S. M., 136–37
Ecology and Oceanography of Harmful Algal Blooms (ECOHAB), 103
Ecosystem degradation. *See also specific topics*
  adverse health effects of, 82–83, 83f
  climate change and, 92
EDGAR. *See* Emissions Database for Global Atmospheric Research
EEA. *See* European Environment Agency
EEFs. *See* Enhanced-efficiency fertilizers
Einsiedl, F., 63
ELEMeNT. *See* Exploration of Long-trM Nutrient Trajectories
EMEP. *See* European Monitoring and Evaluation Program
Emerging substances of concern (ESOCs), 59, 189–90
Emissions Database for Global Atmospheric Research (EDGAR), 38, 43
ENA. *See* European Nitrogen Assessment
Enhanced-efficiency fertilizers (EEFs), 214, 225–26
Enteroviruses, 61, 185
Enzymes, 1
EPA. *See* U.S. Environmental Protection Agency
Erisman, J. W., 21, 42–43, 91, 92, 181, 223, 227–28, 228f
ESA. *See* U.S. Endangered Species Act
Eschenbach, W., 62–63
ESOCs. *See* Emerging substances of concern

Estuaries, 20, 35, 101
EU. *See* European Union
Eugster, O., 24
EUNEP. *See* European Union Nitrogen Expert Panel
EUROHARP, 105
European Environment Agency (EEA), 38, 125–26
European Monitoring and Evaluation Program (EMEP), 38
European Nitrogen Assessment (ENA), 8, 9, 41, 91, 175, 197, 198
European Topic Centre on Air and Climate Change Mitigation, 38
European Union (EU)
  air pollution in, 236
  Groundwater Directive in, 125
  ND in, 104, 124–25, 142
  reactive nitrogen economic costs in, 197–200, 199f
  reactive nitrogen in groundwater in, 124–26
  Water Framework Directive of, 145
European Union Nitrogen Expert Panel (EUNEP), 224, 225f
Eutrophication, 3, 25, 101
  from fertilizers, 109
  from reactive nitrogen, 4, 77
    critical load of, 96
  of surface water, 4, 62, 84, 102, 134, 201
Exploration of Long-trM Nutrient Trajectories (ELEMeNT), 110, 133

FAO. *See* Food and Agricultural Organization
FEFLOW, 145
FEGS-CS. *See* Final Ecosystem Goods and Services Classification System
Felix, J. D., 59
Fenech, C., 59
Ferguson, R. B., 214
Fermenters, 29
Fertilizers, 2, 21, 99
  CLM for, 107
  DIN from, 97–98
  dual nitrate isotopes and, 52
  EEFs, 214, 225–26
  groundwater and, 36
  nitrate from, 37
  nitrogen cycle and, 24
  phosphorus in, 175–80, 176f, 177f
  reactive nitrogen from, 36–37, 37t, 40–41, 175–80, 176f, 177f
  in wetlands, 100
Fewtrell, L., 76, 80
Field to Market, 234

Final Ecosystem Goods and Services Classification System (FEGS-CS), 96
Finlay, J. C., 60
Fisheries, 7
Fogg, G. E., 52
Food
  nitrate and nitric acid in, adverse health effects of, 80–82
  production of
    in China, 211–12
    NUE for, 227–28, 228f
    webs, HABs and, 103
Food and Agricultural Organization (FAO), 199, 232
Fossil fuels, 3, 5, 22, 38, 228–29
Fovet, O., 144
Fowler, D., 22, 29, 93, 108
Fractionation, 51, 53–54, 60
Fram, M. S., 60
Free-living nitrogen fixers, 17, 29
Fungi, 20

G6PD deficiency, 77
Galloway, J. N., 1, 2, 3, 8, 21–22, 21f, 29, 230, 231, 232, 232f
Gao, B., 235
Gascuel-Odoux, C., 188
GDP. *See* Gross domestic product
GEF. *See* Global Environment Facility
GEIA. *See* Global Emissions Inventory Activity
GEMS/Water. *See* Global Environment Monitoring System for Water
Geng, R., 215
GGIS. *See* Global Groundwater Information System
GGMN. *See* Global Groundwater Monitoring Network
Gillies, N., 130
GLEAM. *See* Great Lakes Environmental Assessment and Mapping
GLEAMS. *See* Groundwater loading effects of agricultural management systems
Glibert, P. M., 111
Global Burden of Disease Study of Comparative Risk Assessment, 72
Global Conference on Land-Ocean Connections, 236
Global Emissions Inventory Activity (GEIA), 38
Global Environment Facility (GEF), 237
Global Environment Monitoring System for Water (GEMS/Water), 43
Global Groundwater Information System (GGIS), 126

Global Groundwater Monitoring
  Network (GGMN), 126
Global NEWS model, 107
Global Orchestration, 107
Global Partnership on Nutrient
  Management (GPNM), 236
Golden algae, 103
Goolsby, D. A., 189
GPNM. *See* Global Partnership on
  Nutrient Management
Granger, J., 53
Gray, C. W., 186
Great Lakes Commission, 207
Great Lakes Environmental Assessment
  and Mapping (GLEAM), 205–7,
  205f, 206f
Great Lakes Restoration Initiative, 207
GREEN, 107, 107t
Green, C. T., 62, 143–44
Greenhouse gas, 6, 74
Grizzetti, B., 106, 107, 124, 134
Groffmann, P. M., 24, 36
Gross domestic product (GDP), 200
Groundwater (aquifers), 6–7
  age-dating of, 54
    nitrate source identification and,
      61–62
  ammonium in, 58–59, 58f, 123–24
  in China, 80
  Clean Water Act for, 32
  denitrification of, 19, 24, 120, 127
  DOC in, 57
  for drinking water, 120
    VOCs in, 189
  fertilizers and, 36
  flow rate of, 124
  home construction and, 36
  hyporheic zone and, 127–28, 128f
  industry and, 33–34
  landfills and, 33
  legacy reactive nitrogen in, 132–34
    nitrate isotopes for identification of,
      62–63
  mining and, 33
  nitrate in, 20, 50, 80, 119–47, 120f
    in China, 134–35
    climate change and, 142
    drinking water and, 189
    transport and vulnerability
      modeling for, 139–47, 139f
  nitrate isotopes for legacy reactive
    nitrogen source identification in,
    62–63
  nitrate recharge rate in, 37
  on-site sewage disposal and treatment
    systems and, 35
  orthophosphate in, 178
  pharmaceuticals in, 60
  reactive nitrogen in, 39, 119–47, 120f
    in China, 126–27, 126f
    drinking water and, 83–84
    in EU, 124–26
    lakes and, 129–30
    landfills and, 126
    in U.S., 121–24, 122f
    riparian zones and, 128–29
    springs and, 159–62
    surface water and
      reactive nitrogen between, 127–32,
        128f–30f
    transport between, 109–11
Groundwater Directive (EU), 125
Groundwater Hydrology of Springs
  (Kresic and Stevanovic), 155
Groundwater loading effects of
  agricultural management systems
  (GLEAMS), 142
Gruber, N., 24
Gu, B., 126, 127, 211
Gulis, G., 78
Guo, J. H., 211–12
Guo, M., 235
Gurdak, J. J., 140
Guttormsen, G., 186

Haag, D., 100
Haber, Fritz, 1
Haber-Bosch process, 1–2, 17, 21–22
HABs. *See* Harmful algal blooms
Han, D., 108
Hansen, B., 125
Hanukoglu, A., 76
Happell, J. D., 162
Harman, J., 184
Harmful algal blooms (HABs), 49, 63,
  71, 91, 104f, 112
  adverse health effects of, 82–83, 101–4
  chlorophyl and, 131
  in Great Lakes, 207
  nutrient impacts of, 111
  in U.S., 204, 207–10, 208f, 209f
Harvey, J. W., 128
Havens, K. E., 112, 208, 212
HAVs. *See* Hepatitis A viruses
Hazen and Sawyer, Inc., 143
Health-Based Screening Level (HBSL), 189
Heffernan, J. B., 19, 63, 133, 162–63
Heilmann, S. M., 181
Helton, A. M., 112
Hendriks, A. T. W. M., 184
Hensley, R. T., 157
Hepatitis A viruses (HAVs), 61
Herbert, E. R., 100
Hertel, O., 37, 190
Hill, J., 74, 75, 201
Hoagland, P., 204
Holloway, J. M., 29
Home construction, 36

Hopkins, K. G., 36
Hopple, J. A., 135
Horowitz, A. I., 213
Hot moments, denitrification and, 24
Hotspots
  in China, 211
  for denitrification, 24, 129
  for hypoxia, 102f
  for nitrogen fixation, 17
  for sewer leaks, 36
  in springs, 163
Howard, B. C., 188
Howarth, R. W., 97
HST3D, 144
Hu, C., 131
HUC8. *See* Hydrologic Unit Codes
Hughes, W. B., 188
Hurricane Sandy, 30–31
Hydrologic Unit Codes (HUC8), 202–4
HYDRUS1D, 144
Hyporheic zone, 127–28, 128f
Hypoxia, 4–5, 102, 104, 110
  hotspots for, 102f
HYUDRUS2D, 144

IARC. *See* International Agency for
  Research on Cancer
ICWRGC. *See* International Centre for
  Water Resources and Global
  Change
IFAD. *See* International Fund for
  Agricultural Development
IGRAC. *See* International Groundwater
  Resources Assessment Centre
Industry, 33–34
INMS. *See* International Nitrogen
  Monitoring System
International Agency for Research on
  Cancer (IARC), 77
International Centre for Water
  Resources and Global Change
  (ICWRGC), 43
International Fund for Agricultural
  Development (IFAD), 199
International Groundwater Resources
  Assessment Centre (IGRAC),
  125–26
International Nitrogen Monitoring
  System (INMS), 237
Ion indicators, for nitrate contamination
  identification, 55–57
Ion ratios, for nitrate contamination
  identification, 55–57, 55f, 56f
Irrigation, 20, 24, 28, 38, 108, 141, 167
Irvine, I. C., 190
Ivahnenko, T., 31, 182

Jhun, I., 74
Jiang, Z., 73

Johnson, P. T. J., 6, 16, 20, 82, 84
Joo, Y. J., 25
Jordan, S. J., 100, 101

Kaluarachchi, J. J., 142
Kampa, M., 72
Kane, R. P., 6, 74
Kanter, D., 228, 229
Katsouyanni, K., 72
Katz, B. G., 55, 56, 60, 146, 161–62, 161f, 163, 185, 190
Kaupenjohann, M., 100
Kaushal, S. S., 57
Keck, F., 112
Kelly, W. R., 188
Kendall, C., 51, 52, 53, 54, 60, 62
Kerley, C. P., 80
Kingsbury, J. A., 188
Klaiber, H. A., 207
Klotz, M. G., 15
Knobeloch, L., 76, 80
Kresic, N., 155
Kroeger, K. D., 131
Kuniansky, E. L., 145, 146

Lakes, 101, 129–30
LaMoreaux, P. E., 155
Landfills, 33, 56, 59, 126
Land use, 24–25
Langeveld, J. G., 184
LaPointe, B. E., 61
Law, N. L., 36
Le, C., 210
Leach, A. M., 232
Lear, G. G., 38
Lecher, A. L., 131–32
Lee, C., 130
Lefcheck, J. S., 101
Legacy reactive nitrogen
  in groundwater, 132–34
    nitrate isotopes for identification of, 62–63
  in vadose zone, 132–34
Lepori, F., 112
Lerner, D. N., 33–34, 36
L'hinondel, J. L., 76
Li, T., 214, 225
Li, Y., 190–91
Liaw, A., 141
Lightning, 29
Lim, F. Y., 190
Limit of concern (LOC), 99
Lindsey, B. D., 123
Lithium isotopes, for reactive nitrogen source identification, 54
Lithosphere, 29
Liu, C., 55
Liu, J., 127
Liu, X., 210, 234–35

Livestock waste. *See* Manure
LOC. *See* Limit of concern
Lopez, C. B., 103
Lowe, K. S., 34
Lu, C., 36
Lu, Y., 235
Lung cancer, 71
Lusk, M. G., 36, 184, 185, 190, 230–31

Ma, L., 210
Mahler, B. J., 165
Maize, 74–75, 201
Manure (livestock waste), 180–81
  CLM for, 107
  nitrates in groundwater from, 121
  nitrogen isotopes and, 52
  reactive nitrogen from, 34, 42
  reduction of, 226–27
Marshall, J. D., 213
Martin, J. B., 163
Masetti, M., 140
Mass-balance models, 146
Massei, N., 165
Maupin, M. A., 31–32, 182
Maximum contaminant level (MCL)
  for drinking water, 121–22, 123, 123t
  of EPA, 135
  in PWSs, 135, 136–37
  for VOCs, 189
  of WHO, 126
Mayer, B., 63
McCray, J. E., 146–47
McDonnell, J. J., 54
MCL. *See* Maximum contaminant level
McMahon, P. B., 136, 137, 182
McMinn, B. R., 61
Meinzer, O. E., 155
Mencio, A., 169
Mengis, M., 53
MERHAB. *See* Monitoring and Event Response for Harmful Algal Blooms
Messier, K. P., 141
Methanol, 19
Methemoglobinemia, 75–77
Microbial source tracking (MST), 61
Midwest Stream Quality Assessment (MSQA), 106
Milesi, 36
Millot, R., 54
Mineralization, 15–16, 36, 50
Minet, E. P., 57
Mining, 33
Misselbrook, T. H., 226
MITERRA-EU, 105
MODFLOW, 132, 136, 142, 143, 146
MODPATH, 136, 142

Molenat, J., 188
Monitoring and Event Response for Harmful Algal Blooms (MERHAB), 103
Montiel, D., 131
Morales-Suarez-Varela, M. M., 78
Moran, S. K., 231, 232
MSQA. *See* Midwest Stream Quality Assessment
MST. *See* Microbial source tracking
MT3D, 142, 143
MT3DMS, 146
Musgrove, M., 166–67

NADP. *See* National Atmospheric Deposition Program
NADPH. *See* Nicotinamide dinucleotide phosphate
National Aquatic Resource Surveys, 102
National Atmospheric Deposition Program (NADP), 38, 206
National Groundwater Pollution Prevention Plan (NGPPP), 126
National Numeric Nutrient Criteria Program, 227
National Pollutant Discharge Elimination System (NPDES), 32, 227
National Trends Network (NTN), 38
National Water Quality Assessment Program (NAWQA), 40, 55, 106, 120–21, 122f, 124, 135, 136, 140, 176
National Water Quality Inventory (U.S.), 36
Nationwide Urban Runoff Program (NURP), 184
NAWQA. *See* National Water Quality Assessment Program
ND. *See* Nitrates Directive
Neural tube defects, 134
NGPPP. *See* National Groundwater Pollution Prevention Plan
Nicotinamide dinucleotide phosphate (NADPH), 77
Nishikawa, T., 56
Nitrate
  algae and, 60, 156
  contamination identification of, 49–63
  in drinking water
    adverse health effects of, 75–80, 200
    from PWSs, 134–38, 136f–38f
    from wells, 134–38, 136f–38f
  dual nitrate isotopes for, 51–53, 52f
  ESOCs and, 189–90
  from fertilizers, 37
  in foods, adverse health effects of, 80–82

Nitrate (cont'd)
  in groundwater, 20, 50, 80, 119–47, 120f
    in China, 134–35
      climate change and, 142
      drinking water and, 189
      transport and vulnerability modeling for, 139–47, 139f
    home construction and, 36
    from nitrification, 18
    nitrogen isotopes for, 50
    NOC and, 80
    from on-site sewage disposal and treatment systems, 35
    with other contaminants, 175–91, 176f, 177f, 179f–82f, 183t, 186f
    pathogens and, 184–85
    pesticides and, 187–89
    plant uptake by, 20
    in PWSs, 79
    recharge rate, 37
    from sewer system leaks, 35–36
    in springs, 155–70, 156t, 157f–62f, 165f, 166f, 168f, 169f
      drinking water and, 169
    trace elements with, 185–86, 186f
    from urban fertilizers, 36
    in vadose zone, 6, 62, 83
    in wells, 80, 81f
Nitrate isotopes
  for legacy reactive nitrogen identification in groundwater, 62–63
  for nitrate source determination, 57, 58f
Nitrate oxidoreductase (NOR), 17
Nitrates Directive (ND), in EU, 104, 124–25, 142
Nitric acid, 33
Nitric oxide, 19, 22, 80–82
Nitrification, 18
  from algae, 129
  home construction and, 36
  in lakes, 129
  nitrogen isotopes and, 50
  stormwater runoff and, 57
Nitrite, 35–36, 75
Nitrogen. *See also specific types and topics*
  cascade, 222–23
  HABs and, 111
  in manure, 180–81
  in rivers, 107, 107t
  significance of, 1
  in stormwater runoff, 57, 179
  from WWTFs, 181–84, 183t
Nitrogenase, 17
Nitrogen cycle, 1–3, 2f, 16f
  acidification in, 101
  human alteration of, 5, 21–25, 21f, 23f, 23t
  land use and, 24–25
  phosphorus and, 175
  preindustrial, 15–20
  projected changes in, 22–25
Nitrogen dioxide, 42, 51, 72, 74
Nitrogen fixation, 16–18, 17t, 22–25, 23t
  hotspots for, 17
  in oceans, 23f, 101
Nitrogen footprint
  of China, 234–35
  NUE in, 237
  reduction of, 231–37, 233t
Nitrogen isotopes, for reactive nitrogen source identification, 49–53, 50f
Nitrogen oxides, 3
  adverse human health effects from, 6
  in atmosphere, 73
  climate change and, 5
  as greenhouse gas, 6
  ozone and, 74
Nitrogen/oxygen isotopes. *See* Dual nitrate isotopes
Nitrogen Source Inventory and Loading Tool (NSILT), 164
Nitrogen use efficiency (NUE), 199, 214, 221
  in agriculture, 223–27, 224f, 225f, 226t
  for food production, 227–28, 228f
  in nitrogen footprint, 237
Nitrous oxide, 19, 42–43, 94, 176
NLEAP, 142
β-*N*-methylamino-L-alanine (BMAA), 83, 102
N-nitroso compounds (NOC), 79, 80
Nolan, B. T., 80, 140–41
Nolan, J., 186
NOR. *See* Nitrate oxidoreductase
Noroviruses, 61
NPDES. *See* National Pollutant Discharge Elimination System
NSILT. *See* Nitrogen Source Inventory and Loading Tool
NTN. *See* National Trends Network
NUE. *See* Nitrogen use efficiency
NUFER. *See* NUtrient flows in Food chains, Environment and Resources use
NURP. *See* Nationwide Urban Runoff Program
NUtrient flows in Food chains, Environment and Resources use (NUFER), 210, 235
Nutrient Innovations Task Group, 204
Nutrient-related impairment, of surface water, 99–100

Oceans, 108
  acidification of, 101
  ammonia in, 18
  climate change and, 112
  nitrogen fixation in, 23f, 101
  reactive nitrogen in, 3, 101
OECD. *See* Organisation for Economic Co-operation and Development
Oelsner, G. P., 99
Oita, A., 221
Oldenkamp, R., 229
On-site treatment and disposal systems (OSTDS, septic tanks), 144
  dual nitrate isotopes and, 52
  nitrogen in, 184
  pharmaceuticals in, 190
  phosphorus in, 184
  reactive nitrogen from, 34–35
  reduction of, 230–31
Order from Strength, 107
Organic wastewater compounds (OWCs), 55–56, 59–60
Organisation for Economic Co-operation and Development (OECD), 223
Orthophosphate, 178
OSTDS. *See* On-site treatment and disposal systems
Otis, R. J., 146
Ovarian cancer, 6, 77, 79
OWCs. *See* Organic wastewater compounds
Ozone, 6, 72, 74

Paerl, H. W., 112
Pajares, S., 17
Panno, S. V., 55, 188
Paradis, D., 145
Pardo, L. H., 94, 95
Parkinson's disease, 83, 102
Particulate matter (PM), 71–73, 84, 94, 211
Particulate nitrogen (PN), 97
Pasten-Zapata, E., 56
Pathogens, 184–85
Payne, R. J., 92
Peeler, K. A., 190
Pennino, M. J., 79
Peñuelas, J., 5, 97
Pesticides, 187–89
pH, 5, 15
  ammonium and, 34
  biosolids and, 18
  denitrification and, 19
  nitrate uptake and, 20
  nitrification and, 17
  on-site sewage disposal and treatment systems and, 34
  plants and, 93
  soil buffering capacity and, 37
  volatilization and, 20
Pharmaceuticals, 59–60, 190

PHAST, 144
Phosphorus. *See also* Total phosphorus
　in ESOCs, 184
　in fertilizers, 175–80, 176f, 177f
　　HABs and, 111
　in manure, 180–81
　nitrogen cycle and, 175
　in stormwater runoff, 179
　from WWTFs, 181–84, 183t
PhreeqcRM, 144
Plants
　ammonium assimilation by, 20
　decay of, 20
　nitrate uptake of, 20
　nitrogen uptake of, 119
　pH and, 93
　reactive nitrogen and, 93–94
PM. *See* Particulate matter
PN. *See* Particulate nitrogen
Pope, J. P., 110, 143
Potassium, in fertilizers, 175
Powlson, D. S., 76
PPCP. *See* Primary producer community structure
Pretty, J. N., 8
Primary producer community structure (PPCP), 164
Prostate cancer, 78
Proteins, 1
　nitrate uptake and, 20
　nitrogen fixation and, 17
Public water systems (PWSs)
　MCL in, 135, 136–37
　nitrate in, 79
　　drinking water from, 134–38, 136f–38f
Puckett, l. J., 133, 142, 188
PWSs. *See* Public water systems

Qi, S. L., 140
Qiu, J., 8

Random forest classification (RFC), 141
Random forest regression (RFR), 141
Ransom, K. M., 140
RCC. *See* Renal cell carcinoma
Reactive barriers
　adverse human health effects of, 71–84, 72f, 81f, 83f
　for denitrification, 19
Reactive nitrogen, 2, 3. *See also* Legacy reactive nitrogen; Nitrate
　adverse human health impacts from, 5–6, 5f
　from agriculture, 42, 97, 199–200
　　in U.S., 221–22
　in air pollution, 190–91, 210–11
　　adverse health effects of, 71–75
　algae and, 91

in atmosphere, 22, 37–38, 38f
from bacteria, 29
biodiversity loss and, 91–112, 94f
climate change and, 212, 223
in coastal ecosystems, 101
consequences of, 3–5, 4f
contamination identification of, 49–63
creation of, 23f
critical load of, 92, 96
economic costs of, 7–8
　in China, 210–12
　in EU, 197–200, 199f
　reduction of, 212–15
　in U.S., 200–210, 202f, 203t, 205f, 206f, 208f, 209f
in estuaries, 101
eutrophication from, 4, 77
　critical load of, 96
from fertilizers, 36–37, 37t, 40–41, 175–80, 176f, 177f
from fossil fuels, reduction of, 228–29
global awareness of, 235–37
global distribution of, 92, 93f
global scale models for, 107–8
in groundwater, 119–47, 120f
　in China, 126–27, 126f
　drinking water and, 83–84
　in EU, 124–26
　lakes and, 129–30
　landfills and, 126
　surface water and, 127–32, 128f–30f
　in U.S., 121–24, 122f
human alteration of, 21–25
human sources of, 29–38, 31f, 32t, 33f, 35f, 37t, 38f
from irrigation, 28
in lakes, 101
land use and, 24–25
legacy subsurface storage of, 6–7
from manure, reduction of, 226–27
natural sources of, 29, 30f
nonpoint sources of, 34–38, 35f, 37t, 38f
in oceans, 3, 101
from OSTDS, reduction of, 230–31
plants and, 93–94
point sources of, 30–34, 32t, 33f
reduction strategies for, 221–38, 222f, 224f, 225f, 226t, 228f, 230f, 232f, 233t, 234f
in reservoirs, 101
in rivers, 97–99, 98f
　transport models for, 104–6
source information on, 40–43, 41f, 42f
sources of, 29–43
in surface water, 91–112, 98f
　in China, 108
　transport modeling in, 104–11, 106f, 107t, 109f

transport processes of, 38–40, 39f
in watersheds, transport models for, 104–6
in wells, 120–21
in wetlands, 99, 99f
from WWTFs, 123–24, 234
　reduction of, 229–30
Reay, D. S., 237
Redox, 123
Red tide, 207–10, 208f, 209f
Reis, S., 223, 225–26, 228
Ren, L., 73, 74
Renal cell carcinoma (RCC), 78
Reservoirs, 101
RFC. *See* Random forest classification
RFR. *See* Random forest regression
Rhizobium spp., 17
Richard, A. M., 77
Riddick, S., 107
Rios, J. F., 144
Riparian zones
　denitrification in, 24
　groundwater and, 128–29
Rivers
　DIN in, 108
　nitrogen in, 107, 107t
　reactive nitrogen in, 97–99, 98f
　　transport models for, 104–6
　TMDL in, 110
　TP in, 181f
RiverStrahler, 106, 107t
Rivett, M. O., 133
RNA, 1
Rodriguez-Galiano, V., 141
Rueedi, J., 35

Saccon, P., 54
Safe Drinking Water Act (U.S.), 80, 164
Sala, O. E., 5, 59, 94, 191
Salinity transition zone (STZ), 131
Sanford, W. E., 110, 143
SAV. *See* Submerged aquatic vegetation
Saxena, A., 134
Schmidt, T. S., 106
Schueler, T., 36
Schullehner, J., 75, 77–78, 125
Schwede, D. B., 38
Scientific Committee on Problems of the Environment (SCOPE), 9, 236
Scott, T. M., 156
Seal, T., 60
SEARCH. *See* Southeastern Aerosol Research and Characterization
Seiler, R. L., 190
Seitzinger, S. P., 97–98, 101, 107–8, 112
SEM. *See* Structural equation modeling
Septic tanks. *See* On-site treatment and disposal systems
Sewer system leaks, 35–36

SGD. *See* Submarine groundwater discharge
Shibata, H., 230, 234
Shoda, M. E., 99
Shukla, S., 134
Silva, S. R., 51
Silver chloride, 51
Simkin, S. M., 96
Skin cancer, 6, 74
Skinner, K. D., 31–32
Sobota, D. J., 202, 202f
Soil
　acidification of, 96–97
　buffering capacity of, pH and, 37
Soil and Water Assessment Tool (SWAT), 107, 107t, 142
Soil organic nitrogen (SON), 109
Soil Treatment Unit model (STUMOD), 144–45
SON. *See* Soil organic nitrogen
Southeastern Aerosol Research and Characterization (SEARCH), 38
Spahr, N. E., 127
SPAtially Referenced Regressions on Watershed (SPARROW), 94, 106, 178
Spoelstra, J., 60
Sprague, L. A., 180
Springer, A. E., 155
Springs, 7–8
　age-dating for, 163
　algae in, 156, 157, 164
　algal blooms in, 157–58
　denitrification in, 162–63
　groundwater and, 159–62
　hotspots in, 163
　hydrogeology of, 156–57
　nitrate in, 155–70, 156t, 157f–62f, 165f, 166f, 168f, 169f
　drinking water and, 169
Springs and Bottled Waters of the World (LaMoreaux and Tanner), 155
Spruill, T. B., 56
Squillace, P. J., 189
SRB. *See* Susquehanna River Basin
Stable isotopes, for nitrate source identification, 60–61
Stanek, L. W., 72
Stanton, E. A., 207
State-U.S. EPA Nutrient Innovations Task Group, 8
Stecker, T., 231
Stein, L. Y., 15
Stets, E. G., 99
Stevens, L. E., 155
Stevnovic, Z., 155
Stoliker, D. L., 129
Stomach cancer, 78
STOrage and RETrieval (STORET), 200

Stormwater runoff, 36, 57, 179
Streamwater
　dual nitrate isotopes for, 57
　groundwater nitrates and, 84, 127
　reactive nitrogen in, 29, 39
　TN in, 179f
　TP in, 179f, 181f
Strous, M., 20
Structural equation modeling (SEM), 106
Stuart, M. E., 146
STUMOD. *See* Soil Treatment Unit model
STZ. *See* Salinity transition zone
Submarine groundwater discharge (SGD), 130–32, 130f
Submerged aquatic vegetation (SAV), 111
Sucralose, 60
Sui, Q., 60
Sulfur dioxide, 211
Sun, S., 73
Surface water. *See also specific types*
　acidification of, 101
　in China, reactive nitrogen in, 108
　drinking water from, 142
　eutrophication of, 4, 62, 84, 102, 134, 201
　groundwater and
　　reactive nitrogen between, 127–32, 128f–30f
　　transport between, 109–11
　nutrient-related impairment of, 99–100
　pharmaceuticals in, 59–60
　reactive nitrogen in, 91–112, 98f
　transport modeling in, 104–11, 106f, 107t, 109f
Susquehanna River Basin (SRB), 110
Sutton, M. A., 22, 80, 91, 94, 134, 147, 175–76, 181, 191, 198, 221–22, 223, 224, 224f, 226, 228, 237
SWAT. *See* Soil and Water Assessment Tool

Takaya, N., 19
TANC. *See* Transport of Anthropogenic and Natural Contaminants
Tanner, J. T., 155
Taylor, M., 207
TDEP. *See* Total deposition
Terrestrial Ocean aTomosphere Ecosystem Model (TOTEM), 25
Tesoriero, A. J., 133
Tian, L., 73
TMDL. *See* Total maximum daily loads
TN. *See* Total nitrogen
TNF-β. *See* Tumor necrosis factor-beta
Toccalino, P. L., 135

Toor, G. S., 57
Total deposition (TDEP), 38
Total maximum daily loads (TMDL), 7, 99, 106, 110, 164
Total nitrogen (TN), 99
　in groundwater, 178
　in streamwater, 179f
　in WWTFs, 31–32, 182
Total phosphorus (TP), 99
　in groundwater, 178
　in OSTDS, 184
　in rivers, 181f
　in streamwater, 179f, 181f
　in WWTFs, 31–32, 182
TOTEM. *See* Terrestrial Ocean aTomosphere Ecosystem Model
Townhill, B. L., 111
Townsend, A. R., 5, 6, 82
TP. *See* Total phosphorus
TPCs. *See* Typical pollutant concentrations
Trace elements, 185–86, 186f
Transport of Anthropogenic and Natural Contaminants (TANC), 135
Tritium, 121
Tumor necrosis factor-beta (TNF-β), 78
Typical pollutant concentrations (TPCs), 31

Ultraviolet-B radiation, 6
UN. *See* United Nations
UNEP. *See* United Nations Environment Programme
UNEP-WCMC. *See* United Nations Environment Programme, World Conservation Monitoring Center
UNESCO. *See* United Nations Educational, Scientific and Cultural Organization
Unionized form, of ammonia and ammonium, 98–99
United Nations (UN), FAO in, 199, 232
United Nations Educational, Scientific and Cultural Organization (UNESCO), 43
United Nations Environment Programme (UNEP), 43, 228
United Nations Environment Programme, World Conservation Monitoring Center (UNEP-WCMC), 92
United States (U.S.). *See also specific agencies and laws*
　biodiversity loss in, 201
　drinking water in, 204–5
　HABs in, 204, 207–10, 208f, 209f
　maize air quality-related damages in, 201

reactive nitrogen in
    from agriculture, 221–22
    economic costs of, 200–210, 202f, 203t, 205f, 206f, 208f, 209f
    groundwater and, 121–24, 122f
    red tide in, 207–10, 208f, 209f
Urea, 33, 37
Urothelial cancer, 78
Urquhart, E. A., 103
U.S. Clean Water Act (CWA), 32, 105
U.S. Department of Agriculture, 232
U.S. Endangered Species Act (ESA), 96
U.S. Environmental Protection Agency (EPA), 6
    on air pollution, 71
    on biosolids, 32
    CEAM of, 106
    CWA and, 105
    on drinking water contaminants, 75
    Great Lakes Restoration Initiative of, 207
    on HABs, 101–2, 204
    on manure, 181
    MCL of, 135
    National Aquatic Resource Surveys of, 102
    National Pollutant Discharge Elimination System of, 32
    NPDES of, 32
    on NUE, 237
    NURP of, 184
    Nutrient Innovations Task Group of, 204
    on nutrient-related impairment, 99–100
    on PWSs, 136
    on springs, 163
    STORET of, 200
    TDEP for, 38
    on VOCs, 74
    WWTFs and, 30
U.S. Geological Survey (USGS)
    HBSL of, 189
    HUC8 of, 202–4
    MODFLOW of, 132, 136, 142, 143, 146
    MODPATH of, 136, 142
    NAWQA of, 40, 55, 106, 120–21, 122f, 124, 135, 136, 140, 176
    on nitrate in groundwater, 119–20
    SPARROW of, 94, 105–6, 178

Vadose zone
    legacy reactive nitrogen in, 132–34
    nitrate in, 6, 62, 83
Van Drecht, G., 127
Van Grinsven, H. J. M., 198, 199f, 200
Van Meter, K. J., 110, 132, 133, 134, 222
Van Metre, P. C., 109
Velthof, G. L., 105
Verstraeten, I. M., 190
Vertical flux method (VFM), 143–44
Virtual nitrogen factors (VNFs), 235
Vitamins, 1
Vitoria Minana, I., 79
VNFs. See Virtual nitrogen factors
Volatile organic compounds (VOCs), 74, 189
Volatilization, 20, 50
Volcanic eruption ammonia, 29
Volkmer, B. G., 78
Vrzel, J., 61
Vymazal, J., 130

Wakida, F. T., 33–34, 36
Wang, L., 145
Wang, Q., 108
Wankel, S. D., 53
Ward, B. B., 29
Ward, M. H., 75, 76, 77, 78–79, 80–81, 84, 134, 200
Wastewater treatment facilities (WWTFs)
    anammox ractions in, 20
    for denitrification, 19
    nitrates in groundwater from, 121
    nitrogen from, 181–84, 183t
    phosphorus from, 181–84, 183t
    reactive nitrogen from, 30–32, 40–41, 123–24, 234f
    reduction of, 229–30
Water Framework Directive (EU), 145
Water Pollution Prevention and Control Action Plan (China), 108
Waycott, M., 111, 201
Weber, K. A., 185, 186
Wells
    agriculture and, 123f
    nitrate in, 80, 81f

drinking water from, 134–38, 136f–38f
    reactive nitrogen in, 120–21
Wetlands. See also Constructed/reclaimed wetlands
    DNRA in, 19–20
    reactive nitrogen and, 99, 99f
    buffers of, 129–30
WFP. See World Food Program
Wheeler, D. C., 141
WHO. See World Health Organization
Wiener, M., 141
Wilkes, G., 185
Williams, G. P., 170
WMO. See World Meteorological Organization
Wolf, D., 207
World Food Program (WFP), 199
World Health Organization (WHO), 6
    on air pollution, 71–72
    on groundwater nitrates, 134
    MCL of, 126
    on methemoglobinemia, 75
    on on-site sewage disposal and treatment systems, 35
    on PM, 73
World Meteorological Organization (WMO), 43, 126
Wu, Y., 73
WWTFs. See Wastewater treatment facilities

Xia, Y., 211
Xu, Z., 146
Xue, D., 51
Xue, L., 225

Yang, D., 72
Yang, Y., 57
Yang, Y. Y., 190
Young, M., 55

Zeman, C., 76, 78
Zhai, Y., 78
Zhang, N., 72–73
Zhang, Q. H., 32
Zhang, X., 215
Zhou, M., 229
Zirkle, K. W., 141
Zooplankton, 184